Springer Series in Optical Sciences   Volume 21

Edited by Arthur L. Schawlow

Springer Series in Optical Sciences

Editorial Board: J. M. Enoch   D. L. MacAdam   A. L. Schawlow   T. Tamir

Laser
Spectroscopy IV

Proceedings of the Fourth International Conference
Rottach-Egern, Fed. Rep. of Germany,
June 11–15, 1979

Editors
H. Walther and K. W. Rothe

With 411 Figures

Springer-Verlag Berlin Heidelberg New York 1979

Professor Dr. HERBERT WALTHER
Dr. KARL WERNER ROTHE

Sektion Physik der Universität München,
D-8046 Garching bei München, Fed. Rep. of Germany

ISBN 3-540-09766-X Springer-Verlag Berlin Heidelberg New York
ISBN 0-387-09766-X Springer-Verlag New York Heidelberg Berlin

Offset printing: Beltz Offsetdruck, Hemsbach/Bergstr. Bookbinding: J. Schäffer oHG, Grünstadt.
2153/3130-543210

Preface

The Fourth International Conference on Laser Spectroscopy (FICOLS) was held in the Hotel Überfahrt in Rottach-Egern, Tegernsee, June 11-15, 1979. Rottach-Egern is a well-known health resort situated on the southern end of Lake Tegernsee. As with the previous laser spectroscopy conferences in Vail, Megève, and Jackson, the purpose of FICOLS was to provide an informal setting where an international group of scientists active in laser spectroscopy could discuss current problems and developments in the field. The program consisted essentially of invited lectures with appropriate time provided for the latest postdeadline results.

The conference was attended by 340 scientists representing 25 countries: Austria, Australia, Brazil, Canada, Peoples Republic of China, Denmark, Finland, France, Germany (FRG), Germany (GDR), Great Britain, India, Italy, Japan, Netherlands, New Zealand, Norway, Pakistan, Poland, Spain, Sweden, Switzerland, U.S.A., U.S.S.R., and Yugoslavia.

Unfortunately five of our colleagues from Japan who planned to attend the meeting could not come due to an interruption of airline schedules. Their absence was a distinct loss to the conference. However, their papers will be published in one of the forthcoming issues of the journal Applied Physics.

Numerous people have contributed to making the conference a success. Especially we would like to thank the members of the steering committee for their advice concerning the program. In addition, we appreciate the help of the staff of the Sektion Physik, Universität München and the Projektgruppe für Laserforschung der Max-Planck-Gesellschaft zur Förderung der Wissenschaften e.V. Financial support of the meeting was provided by the Deutsche Forschungsgemeinschaft, the Universität München, the Max-Planck-Gesellschaft zur Förderung der Wissenschaften e.V., as well as from the companies Spectra-Physics, Coherent, Lambda Physik, Heyden & Sons, Kristall-optik, Balzers, Cryophysics, Oriel, Carl Baasel, Rottenkolber, and Nucletron.

August 1979   Herbert Walther
                                                      Karl Werner Rothe

Obituary for **Sergio Pereira Da Silva Porto**

Aram Mooradian

Sergio Porto died of a heart attack at age 53, shortly after attending the Fourth Laser Spectroscopy Conference in Rottach-Egern. Porto was one of the outstanding contributors to the field of Laser Spectroscopy and was best known for pioneering the field of Laser Raman Spectroscopy. He together with his coworkers at Bell Telephone Laboratories was the first to effectively utilize lasers for Raman Spectroscopy and made the first observations of Raman scattering from F-centers and magnons as well as resonant Raman scattering from phonons in solids. Porto provided the dynamic synergism between the laser community and industry that made Raman spectroscopy a major area of scientific research. He also carried out some of the early pioneering work in solid state lasers. He has over 140 professional publications and 60 invited papers and was a fellow of the Optical Society of America, the Brazilian Academy of Science, and the American Physical Society. Porto was perhaps the most famous Brazilian scientist in history. He provided not only technical but inspirational leadership wherever he was. He was particularly active in promoting research and educational opportunities in the United States for talented South American students in physics and engineering. His death is an enormous loss for Brazil as well as the international scientific community. He breathed vitality and enthusiasm into his science and radiated warmth to his friends and colleagues. He will be missed.

He was born in Niteroi, Brazil and received the B.Sc. and Licenciado in chemistry in 1946 and 1947, respectively, from Fac. Filosofia University, Brazil. In 1954 he received the Ph.D. in physics from The Johns Hopkins University where he studied under G. H. Dieke. From 1954-60 Porto taught physics at the Instituto Technologica da Aeronautica in Sao Paulo, Brazil. He was a member of the technical staff and supervisor of the Quantum Electronics Department of Bell Telephone Laboratories, Murray Hill, New Jersey from 1960-67. Porto was professor of Physics and Electrical Engineering at the University of Southern California, Los Angeles, California from 1967-73. In 1974, Porto returned to Brazil to establish a quantum electronics group at the University of Campinas and rose to become Vice-President of the University.

Sergio Porto is survived by his wife Hilta, and four children: Sergio Bruno, Marcia, Ivan Pedro, and Paulo Rogerio.

Contents

Introduction

Historical Perspectives in Laser Spectroscopy

G.W. Series

J.J. Thomson Physical Laboratory, The University of Reading, Whiteknights
Reading RG6 2AF, U.K.

1. Introduction

To be invited to give the Introductory talk at a Conference of this nature
is, indeed, a great honour, and in accepting such an invitation one is very
conscious of the challenge it presents. For, to survey the field from the
published literature is to condemn oneself to be at least a year out of date,
and to attempt to summarise what we shall hear in the next few days is, for
obvious reasons, impossible.

What I have decided to do, therefore, is to look back at some of the historical antecedents of laser spectroscopy, dwelling particularly on studies
of radiative interactions, and then to mention briefly, with particular examples, some fields of laser spectroscopy which are, apparently, not represented
in the conference programme. This latter part of my talk will serve the purpose of making the Proceedings of the Conference a little more completely
representative of laser spectroscopy at the present time. That part of my
talk which deals with historical antecedents may be of value to some of our
younger colleagues, who will not themselves be able to recall the days when
a source of intense radiation was a high pressure mercury lamp.

2. Radiative Interactions

In the historical context I see the studies of interactions between radiation
and matter as stemming very directly from the great variety of radio-frequency
resonance studies which proliferated after World War II: magnetic and electric
resonances of atoms and molecules in beams; electron spin resonance (although
the main lines of progress here were in the context of solid state physics);
nuclear magnetic resonance, with its extraordinary variety of highly ingenious
pulse techniques for the study of transients and the extraction of information
about the nuclear environments; and above all, from studies associated with
the resonance fluorescence of atoms: optical pumping, optical radio-frequency
double resonance, and level-crossing experiments.

Those who recall that 'lasers' were first called 'optical masers' may quar-
rel with me here, and point out that optical masers sprang more directly from
the family of microwave devices which included principally the ruby maser -
solid state magnetic resonance, and the ammonia maser - molecular beam, electric resonance. But in these microwave devices spontaneous emission played
no part, whereas in resonance fluorescence spontaneous emission is the dominant interaction, and it is precisely the role of spontaneous emission in relation to the stimulated interactions which, in my view, gives a special interest
to the study of laser radiation interacting with atoms.

3. Comparison Between Radio-Frequency and Laser Spectroscopy

Let me enumerate what I see as the main points of contrast between radio-frequency spectroscopy and laser spectroscopy, points where the quantitative difference – a factor of a million or so in wavelength – leads to qualitative differences:

(i) in laser spectroscopy spontaneous emission needs to be reckoned with;

(ii) in laser spectroscopy the wavelength of the radiation is small in relation to the size of the sample, and propagation effects need to be considered, whereas in radio-frequency spectroscopy the wavelength is large, and the phase of the field is uniform over the sample;

(iii) in laser spectroscopy the atoms or molecules under study have a very much more complicated energy level structure than the Zeeman or hyperfine multiplets which are the typical structures examined by radio-frequency methods.

Though I am perhaps riding a hobby-horse, I make no apology for making a special point of spontaneous emission. In my view, spontaneous emission is the most fascinating of radiative interactions. The subject finds expression in our conference in the papers on super-radiance, or super-fluorescence. My experience of trying to understand why it is that excited atoms, or assemblies of atoms, spontaneously emit light is that one cannot avoid taking into account the universe at large. That is part of the attraction of the optical radio-frequency double resonance technique I mentioned a few minutes ago. It combines the simplicity of the radio-frequency interaction work with the profundity of the spontaneous emission problem, and it yields a rich harvest in purely spectroscopic information, to boot.

4. Optical Radio-Frequency Double Resonance

Return with me, if you will, to the early nineteen-fifties, to the publication of that famous paper by BROSSEL and BITTER [1]: an optical radio-frequency double-resonance experiment in mercury, a paper which marked the birth of Doppler-free spectroscopy applied to excited state structures. (I am distorting history by conveniently ignoring absorption- and emission-spectroscopy from atomic beams, but this is not because I do not attach a high value to that work.) Figs. 1 and 2 recall the experimental arrangement and the term scheme, and Fig. 3 some experimental resonance curves. Notice three things about these curves:

(i) they are not Doppler-broadened, so that a g-value could be obtained from them of far greater accuracy than could be achieved by a conventional spectroscopic study of the Zeeman effect in the 254 nm line, whose Doppler width is about 500 G;

(ii) at the higher r.f.fields they show broadening (power broadening) and shifts (Bloch-Siegert shifts), both of which are well-known in laser spectroscopy;

(iii) their shape is not that of a simple, bell-shaped Lorentzian resonance curve.

I draw your attention to item (iii), the shape of the resonance curves.

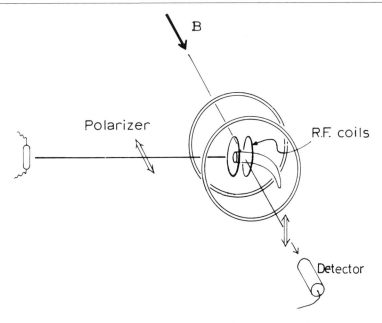

Fig.1 Schematic of apparatus used in the experiment of BROSSEL and BITTER.

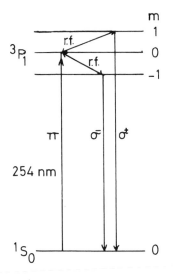

Fig.2 Term diagram and transitions (mercury: even isotopes).

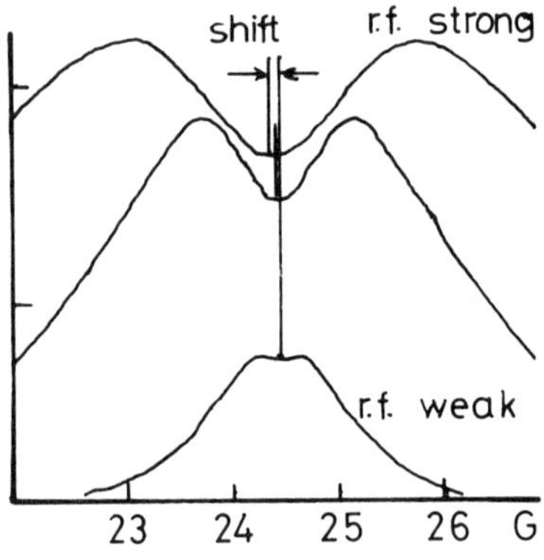

Fig.3 Experimental resonance curves showing broadening and shifts.

This can be understood at various levels of comprehension. At one level it represents simply the distribution of population over the space-quantized levels of a spin-1 system: at another it is analogous to the resonance response curve of two highly-coupled circuits: but, more profoundly, it is one of the five components of the axially-symmetric second-rank tensor which exhibits the geometrical, rather than the dynamical aspects of the experiment.

I think this last is the most mature way of looking at the double-peaked curve. It brings out the point that the essence of the matter is not the spin-1 nature of the excited state in mercury, but that the geometrical structure of the experiment is such as to impose an alignment (as opposed to an orientation) on the mercury vapour, and to monitor this alignment. Of course, alignment cannot be induced in anything less than spin-1, but had the experiment been performed on a system of higher spin, this particular geometrical arrangement would not have shown up anything more elaborate than an alignment.

5. Multi-Level Structures in Laser Spectroscopy

At this point I want to return to laser spectroscopy. Every participant in this conference will be aware of the analogy between the magnetic resonance of a spin-$\frac{1}{2}$ system and the electric dipole resonance of the fictitious 2-level atom, an analogy spelt out in precise terms in 1957 by FEYNMAN, VERNON and HELLWARTH [2]. The solution of the magnetic problem by the use of rotating co-ordinate frames in 3-dimensional physical space finds its analogue in the electric problem through the use of rotations in an abstract space. I want to ask whether the analogy between the magnetic and the electric resonance phenomena might not profitably be pressed beyond the spin-$\frac{1}{2}$, 2-level case to the case of more complicated structures.

The magnetic resonance of a spin of arbitrary magnitude has been studied in various ways. Among recent authors, PANCHARATNAM [3] used the method of irreducible spherical tensor analysis, and evolved a model of an ellipsoid, rotating in physical space, to represent the motion of the tensor of alignment under a driving field. The ellipsoid changes its shape as the magnetic fields are varied through resonance. The question then, is this: might there be some profit in developing a model of a multi-level system undergoing electric resonance based on the rotation of an ellipsoid in the abstract space described by FEYNMAN, VERNON and HELLWARTH? Would a study of such a model give additional insights into the interactions of laser light with multi-level systems? I can, at least, draw your attention to a particular aspect of the analogy which I believe is not well known among laser spectroscopists.

You will recall the Autler-Townes effect: the development of a structure in a 2-level atom interacting with a sinusoidal field, a structure which may be probed by some additional radiation field weakly interacting with the atom, so 'dressed'. Fig. 4 shows you the structure, Fig. 5 shows how it was demonstrated in an experiment by DELSART and KELLER [4]. A two-level structure is irradiated by a strong, resonant laser field. The induced doublet in the upper of the two original levels is probed by monitoring the absorption of light from a weak laser, tuned through a transition to a higher atomic level. The structure of upper and lower levels together is shown up in the well-known triplet of the dynamical Stark effect, shown schematically at the bottom of Fig. 4. Here, the probe is the relatively weak perturbation responsible for spontaneous emission.

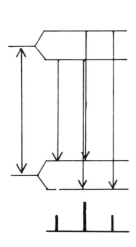

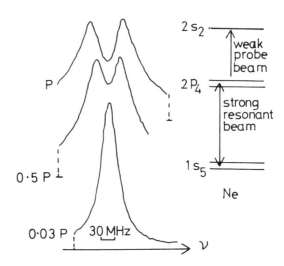

Fig.4 Structure in a
2-level system interacting
with a sinusoidal field.
3-lined spectrum generated
by spontaneous emission
(dynamical Stark effect).

Fig.5 The structure in
the upper level (2p₄) probed
by weak, tunable laser in
absorption (Delsart and
Keller [5]).

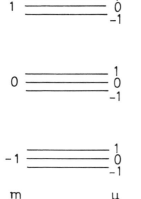

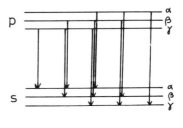

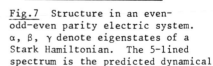

Fig.6 Structure in a
spin-1 magnetic system
interacting with a sinusoidal
field. m, μ are good quantum
numbers in different frames.

Fig.7 Structure in an even-
odd-even parity electric system.
α, β, γ denote eigenstates of a
Stark Hamiltonian. The 5-lined
spectrum is the predicted dynamical
Stark effect.

The Autler-Townes effect for a spin-1 system in magnetic resonance was much studied in the years following BROSSEL and BITTER'S experiment. You see, in Fig. 6, that the three original levels are each split into three by interaction with a strong, sinusoidal field, making nine levels in all. By analogy, one would predict a similar structure, Fig. 7, in a three-level system of alternating parity interacting with a sinusoidal electric field. The dynamical Stark effect one would predict for this system of levels is shown at the bottom of Fig. 7. This is oversimplified in relation to real atoms because equally spaced levels with equal transition probabilities between them have been assumed. One can readily move to a more realistic situation. These structures have been predicted by WHITLEY and STROUD [5], by SOBOLEWSKA [6] and by COHEN-TANNOUDJI and REYNAUD [7]. I am not aware of any experimental demonstration.

6. The Propagation of Light and the Polarization Tensor

So much for what we can learn about laser interactions by recalling the BROSSEL-BITTER experiment. But I cannot leave this field of radiative interactions without recalling studies on the interaction of the atomic system with optical, as distinct from radio-frequency radiation. The propagation of laser light through a vapour has a great deal in common with the forward-scattering of resonance radiation in optical pumping and resonance fluorescence experiments: optical phase relations are of primary importance here: they are fully taken into account by attributing to the medium a macroscopic polarization, formed by summing the electric dipole moments of individual atoms. One develops a theory in terms of the time-dependent, complex, refractive indices of birefringent macroscopic media. The dispersion functions encountered in the theory describing single-beam experiments meet the Doppler barrier in relation to excited state structures, but the two-beam velocity-selective techniques that are now so widely used in laser spectroscopy lead to an active electric dipole, coherent over a macroscopic volume, whose oscillations are not dephased through the Doppler shifts of its atomic constituents. The classification of bulk polarization according to the dominant symmetries, whether geometrical (dipolar, quadrupolar, etc.) or atomic (Zeeman, hyperfine, etc.) which was worked out for the optical pumping situation (see, for example, the review by HAPPER [8]) might well prove useful in the realm of laser spectroscopy.

7. Selected Topics in Laser Spectroscopy

Turning away now from radiative interactions, I conclude by pointing to some areas of laser spectroscopy where, through the use of specific properties of laser radiation, outstanding progress has been made. The list will, of necessity, reflect my own interests and the limitations of my contacts; and, further by way of self-defence, I have to add that, although I must ignore many of the topics we shall be hearing about in the next few days, I do not under-rate their importance.

I would draw your attention, first of all, to the potentiality that tunable lasers have given us for the systematic study of sequences of states. One may cite the studies of Rydberg states of atoms and molecules, which find a place in our conference programme. But I do not find reference in the programme to the systematic studies of ro-vibronic states of molecules. Outstanding in this field is the work of LEHMANN and his co-workers on radiative and pre-dissociation lifetimes in I_2 [9] and on g-values in Se_2 [10]. It is significant, I think, that in these studies LEHMANN has coupled the techniques of level-crossing and quantum beats with those of laser excitation. Again, exemplifying the marriage of different techniques in systematic and wide-ranging studies I cite the work of SVANBERG and his co-workers in studying hyperfine structures, g-values and lifetimes and electric polarizabilities in excited states of the alkalis [11]. His experimental results have provided the material for noteworthy developments in the theory of atomic structure by LINDGREN and ROSEN [12]. An example of the application of lasers to beam-foil spectroscopy is the work of SILVER and his colleagues on important intervals in one- and two-electron ions of high Z [13]. In this context also I would mention the use of lasers as state-selectors in atomic beam magnetic resonance experiments - a topic that does appear on our programme.

The use of short-pulsed lasers to study lifetimes and collisional damping in atoms and fast relaxation processes in molecules is a subject that is represented in our conference, but one that merits particular mention.

The word 'resonance' has occurred many times in my talk. As spectroscopists we are accustomed to probing matter with fine, sharp-edged tools. Let me conclude by showing you that blunt instruments also have their uses in spectroscopy.   7 Joules from a Nd:YAG laser focussed down on to a micro-ballon of glass, 70µ in diameter. Within a tenth of a nanosecond, a compressed plasma. There, in Fig. 8, you have the spectrum of one- and two-electron ions of silicon in a spectral region more remote than anything that appears on our program. This spectrogram was taken by FAWCETT and colleagues at the Central Laser Facility, Rutherford Laboratory, near Reading [14].   I show it with the permission of the Director. I am grateful to the authors, to the American Institute of Physics and to the Institute of Physics for permission to copy Figs. 3 and 6 respectively.

WAVELENGTH
Å

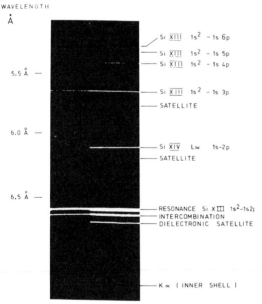

Fig.8 Far X-ray spectrum of one- and two-electron ions of Si, generated by a laser-produced plasma [13].

References

1. J. Brossel and F. Bitter, Phys. Rev. **86**, 308 (1952).

2. R.P. Feynman, F.L. Vernon and R.W. Hellwarth, J. Appl. Phys. **28**, 49 (1957).

3. S. Pancharatnam, Proc. Roy. Soc. **A330**, 265 and 271 (1972).

4. C. Delsart and J.-C. Keller, J. Phys. B **9**, 2769 (1976).

5. R.M. Whitley and C.R. Stroud Jr., Phys. Rev. **A14**, 1498 (1976).

6. B. Sobolewska, Opt. Comm. **20**, 378 (1977).

7. C. Cohen-Tannoudji and S. Reynaud, J. Phys. B **10**, 2311 (1977).

8. W. Happer, Rev. Mod. Phys. **44**, 169 (1972).

9. J.-C. Lehmann, Contemp. Phys. **19**, 449 (1978).

10. G. Gouedard and J.-C. Lehmann, J. Phys. (Paris) **40**, L119 (1979).

11. S. Svanberg, Physica Scripta **14**, 39 (1976), and references there given.

12. I. Lindgren and A. Rosén, Case Studies in Atomic Physics, **4**, 93-398 (1974).

13. E. Träbert, A. Armour, S. Bashkin, N.A. Jelley, R. O'Brien and J.D. Silver, J. Phys B (in press) (1979).

14. Report LD/78/04, Section 2.2, of the Rutherford Laboratory (1978).

Part I

Fundamental Physical Applications
of Laser Spectroscopy

An Improved Test of the Isotropy of Space Using Laser Techniques

A. Brillet[1] and J.L. Hall[2]

Joint Institute for Laboratory Astrophysics,
National Bureau of Standards and University of Colorado
Boulder, CO 80309 USA

This paper describes a new laser version of the Michelson-Morley [1] experiment, in which we achieved a sensitivity $\Delta c/c = (1.5 \pm 2.5) \times 10^{-15}$ [2]. This appears to constitute the most precise test of special relativity yet realized.

Experimental tests of special relativity have been numerous, and often difficult to compare as concerns their meaning and accuracy. Basically, to make such comparisons clear, one needs to have in mind a model of a possible type of failure of the conventional theory. What is needed is a "theory of all possible theories" so that, for example, the Lorentz-Einstein special theory of relativity represents just one point in a suitable hyperspace. Experiments would then serve the role of localizing acceptable theories to the vicinity of this point. However such a more general theory is difficult to invent if it is to include electromagnetic phenomena in a self-consistent way and still be a generalization of the Lorentz transform. Thus up to now most progress has been made by restricting attention to kinematical phenomena. Useful conceptual advances were made in the postwar period by ROBERTSON [3] along these lines. More recently, theoretical papers by MANSOURI and SEXL [4] have appeared proposing a "Test Theory of Special Relativity" which expands Robertson's ideas in a concrete and useful way. This new theory considers (small) additional velocity dependences beyond the Lorentz transformation and is therefore able to make connection and comparisons between previously unrelated experimental "tests of special relativity." Its particular strength is that it enables one to assess the power of the numerous experiments to restrict the acceptable domain in the parameter hyperspace mentioned before. As this theory of Mansouri and Sexl restricts itself to kinematic aspects only and becomes cumbersome for very high velocities, it cannot be regarded as a permanent, complete test theory. But it does show clearly the significance of the several classical relativity experiments and helps one to see what kind of new measurements are needed. Thus it is useful for us to present its basic outline here in a simplified form.

---

[1] Permanent Address: Lab. de l'Horloge Atomique, Orsay, France. JILA Visitor, 1978-79 with partial support by a NATO fellowship.

[2] Staff Member, Quantum Physics Division, National Bureau of Standards.

Consider the following assumptions,

1) Existence of a preferred reference frame Σ in which there is no preferred direction.

2) In Σ the speed of electromagnetic waves is c, independent of direction, wavelength, source motion ... .

3) Existence of a second frame S moving with the velocity v along the x axis of Σ.

4) The coordinate transformation between Σ and S is linear.

After a few simplifications relating to orientation of the axes, the general form of the coordinate transformation can be shown to be [4]

$$t = a(v)T + \varepsilon x + \varepsilon'y + \varepsilon'z$$
$$x = b(v)(X-vT)$$
$$y = d(v)Y \tag{1}$$
$$z = d(v)Z$$

where the capital letters refer to Σ. The parameters ε, ε' are determined by clock synchronization procedures, whereas a(v), b(v) and d(v), which contain the real physics of the problem, are to be determined by experiments or by a more fundamental theory.

The velocity of a light ray propagating in S at an angle θ relative to the x axis is given by [4]

$$C(\theta) = C * \{\cos \theta \; \varepsilon b(1-v^2) + va \cos \theta + \varepsilon'(1-v^2)b \sin \theta$$

$$- a[\cos^2 \theta + b^2d^2(1-v^2)\sin^2\theta]^{1/2}\} *$$

$$* \{\cos^2\theta(\varepsilon^2b - \varepsilon^2bv^2 - a^2/b + 2\varepsilon va) + \sin^2\theta(\varepsilon'^2b - v^2\varepsilon'^2b -$$

$$- d^2ba^2) + 2 \sin \theta \cos \theta * \varepsilon'(\varepsilon b - v^2\varepsilon b + va)\}^{-1} . \tag{2}$$

In this equation only we use the notation $v/c \to v$. Let us use the expansions [5] $a(v) = 1+\alpha(v/c)^2+...$, $b(v) = 1+\beta(v/c)^2+...$, $d(v) = 1+\delta(v/c)^2+...$ (Special relativity corresponds to $\alpha=-1/2$, $\beta=1/2$, $\delta=0$.) As an example of the power and utility of this approach, MANSOURI and SEXL [4] show that the time dilation factor, α, can be determined by first order experiments, using frequency standards synchronized by transport. α can also be deduced from high resolution Mössbauer rotor experiments [6] as well as from direct measurement [7]. β and δ can only be studied through second order experiments. In the case of Einstein's synchronization, one gets to second order:

$$c/c(\theta) = 1 + \left(\beta + \delta - \frac{1}{2}\right)\left(\frac{v}{c}\right)^2 \sin^2\theta + (\alpha - \beta + 1)\left(\frac{v}{c}\right)^2 . \tag{3}$$

The term $[\beta+\delta-(1/2)]$ is measured by "Michelson-Morley" experiments [1,2] and $(\alpha-\beta+1)$ by "Kennedy-Thorndike" [8] experiments. Assuming the velocity of earth relative to "the preferred frame" to be v = 300 km/s [9], the most precise experiments up to now lead to the following limits:

a) $\alpha = -\frac{1}{2} \pm 10^{-7}$ [6] (Mössbauer rotor)

b) $\beta+\delta = \frac{1}{2} \pm 10^{-5}$ [10] (Michelson Morley)

c) $\alpha-\beta = 1.02 \pm 2 \times 10^{-2}$ [8] (Kennedy-Thorndike)   .

Although the determination of $(\alpha-\beta)$ was by far the least precise, we felt more able to improve by a large factor the precision of the measurement of $(\beta+\delta)$, mainly because the $\sin^2\theta$ term allows modulation of the effect by simply rotating the experiment in the laboratory. By contrast, the "Kennedy-Thorndike" term $(\alpha-\beta-1)$ will show its principal variations only with a time scale of a year, and the precision of the measurements is then strongly degraded by low frequency noise (drifts).

Our "spatial isotropy" experiment to determine $[\beta+\delta-(1/2)]$ has been designed to be clear in its interpretation and free of spurious effects. Its principle may be understood by reference to Fig. 1. A He-Ne laser (λ = 3.39 µm) wavelength is servostabilized so that its radiation satisfies optical standing-wave boundary conditions in a highly stable, isolated Fabry-Perot interferometer. Because of the servo, length variations of this cavity -- whether accidental or cosmic -- appear as variations of the laser wavelength. They can be read out with extreme sensitivity as a frequency shift by optically heterodyning a portion of the laser power with another highly stable laser, provided in our case by a CH_4-stabilized [11] laser. To separate a potential cosmic cavity-length variation from simple drift, we arranged to rotate the direction of the cavity length by mounting the length etalon, its laser and optical accessories, onto a 95-cm × 40-cm × 12-cm granite slab which, along with servo and power-supply electronics, may be continuously rotated. (The frequency readout beam comes from a beam splitter up along the rotation axis and is directed over to the CH_4-stabilized laser. Electrical power comes to the rotating table through Hg-filled channels and a contactor pin assembly below the table.) The table rotation angle is sensed via 25 holes pierced in a metal band under the table. A single, separate hole provides absolute resynchronization each turn. The laser beat frequency is counted for 0.2 sec under minicomputer control after each synchronizing pulse, scaled and transfered to storage and display. A genuine spatial anisotropy would be manifest as a beat-frequency variation $\propto P_2(\cos\theta)$. The associated laser-frequency shift may be conveniently expressed as a vector amplitude at twice the table rotation frequency, f, of 1 per ~10 sec. Furthermore, its component in the plane perpendicular to Earth's spin axis should precess 360° in 12 h.

Our fundamental etalon of length is an interferometer which employs fused-silica mirrors "optically contacted" onto a low-expansion glass-ceramic [12] tube of 6-cm o.d. × 1-cm wall × 30.5-cm length. The choice of 50-cm mirror radii provides a well-isolated TEM_{∞} mode. Dielectric coatings at the mirrors' centers provide an interferometric efficiency of 25% and a fringe width ≈4.5 MHz. The interferometer mounts inside a massive, thermally isolated Aℓ vacuum envelope. The environmental temperature is stable to 0.2°C, but no further thermal servo was employed.

Fringe distortion due to optical feedback is prevented by a cascade of three yttrium-iron-garnet Faraday isolators, each having a return loss ≳26 dB. The laser is frequency modulated ≈2.5 MHz peak to peak at 45 kHz. Both first-harmonic and third-harmonic locking were tried, the unused one being a useful diagnostic for adjustment of the Faraday isolators. Based on the 200-μW available fringe signal, the frequency noise of the cavity-stabilized laser is expected (and observed) to be about 20 Hz for a 1-sec measurement, using a first-harmonic lock.

Our CH_4-stabilized "telescope-laser" frequency reference system achieves a comparable stability [11]. The random noise of the beat signal in a typical 20-min data block is observed to be ~3 Hz, compared with the laser frequency of almost 10^{14} Hz. To ensure absolute isolation of the cavity-stabilized and CH_4-stabilized lasers, the latter actually is used to phase lock a "local-oscillator" laser offset by 120 MHz. The ~35-MHz beat of this isolation laser with the cavity-stabilized laser is the measured quantity. See Fig. 1. After each measurement the computer checks that the beat frequency is near its optimum value and readjusts the frequency synthesizer if necessary.

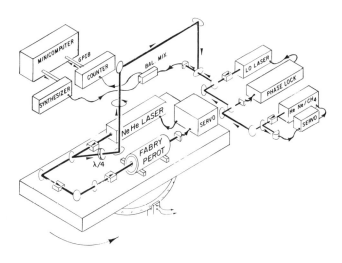

Fig. 1. Schematic of isotropy-of-space experiment. A He-Ne laser (3.39 μm) is servostabilized to a transmission fringe of an isolated and highly stable Fabry-Perot resonator, with provision being made to rotate this whole system. A small portion of the laser beam is diverted up along the table rotation axis to read out the cavity length via optical heterodyne with an "isolation laser" which is stabilized relative to a CH_4-stabilized reference laser. The beat frequency is shifted and counted under minicomputer control, these frequency measurements being synchronized and stored relative to the table's angular position. After 30 minutes of signal averaging the data are Fourier transformed and printed out, and the experiment is reinitialized.

Representative time series data are shown in Fig. 2. In the upper curve one can see a long-term downward trend, which corresponds to a drift rate of ∼-50 Hz/sec. The superimposed sinewave has an amplitude of about 200 Hz and a period of 1 cycle per table revolution. It arises from gravitational stretching of the interferometer because the mounting of the interferometer is inevitably somewhat unsymmetrical about its center and the rotation axis is not absolute vertical. As this spurious signal occurs at 1/2 the frequency of the "anisotropy signal," we can reasonably project it out -- along with the drift -- by fitting these terms and subtracting them from the data. The resulting data are plotted in the lower part of Fig. 2 with 20-fold increased sensitivity. Note that another spurious signal is also clearly present, this one at the interesting frequency of 2 cycles per revolution. It is possible that this term arises via nonlinear elastic response to the 1 cycle per revolution perturbation. Centrifugal stretching of the interferometer due to rotation gives -10 kHz at f = (1 turn)/13 sec and implies a compliance ∼10 times that of the bulk material. (Thus small synchronous speed variations could also give rise to synchronous acceleration forces.) As will be seen below, the spurious signal at 2 cycles per revolution basically controls the useful sensitivity of our experiment.

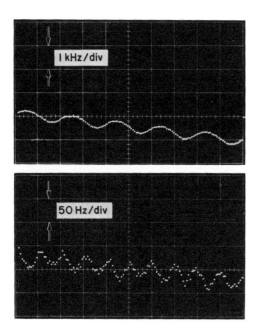

Fig. 2. Upper curve shows time series data for a typical 600 sec run (5 turns per sweep, averaged for 10 sweeps). The general downward trend of the beat frequency represents a drift rate of ∼-50 Hz/sec. Superimposed on this trend is a sinusoidal variation of some 200 Hz amplitude with a period of 1 cycle per table revolution. This "sine-wave" signal is caused by a varying gravitational stretching of the interferometer. See text. Lower curve shows data at 20-fold higher sensitivity after this spurious signal and drift have been removed. Note that a signal at the interesting frequency of 2 cycles per revolution is now visible.

We find that taking data in blocks of N table rotations (N ≃ 8-50) is helpful in minimizing the cross coupling of the first two noise sources into the interesting Fourier bin at 2 cycles per table revolution (actually at 2N cycles per N table revolutions). Typically 10-20 blocks of N revolutions were averaged together in the minicomputer before calculating the amplitude and phase of the signal at the second harmonic of the table rotation frequency. The average result is an amplitude of cos 2θ of ≃17 Hz (2×10^{-13}) with an approximately constant phase in the laboratory frame. A number of such 1/2-h averages spanning a 24-h period are illustrated in Fig. 3 as radius vectors from the origin to the open circles. The noise level of each such average was estimated by computing the noise at the nearby Fourier bins of 2N ± 1 cycles per N table revolutions. For a 1/2-h average (N = 10, averaged 10 times) the typical noise amplitude was 2 Hz with a random phase.

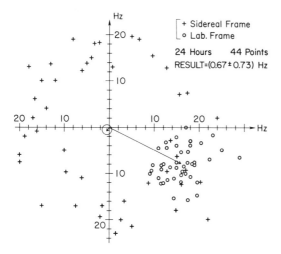

Fig. 3. Second Fourier amplitude from one day's data. The vector Fourier component at twice the table rotation rate is plotted as the radius vector from the origin to the open circles. After precessing these vectors by their appropriate sidereal angles they are plotted as the (+). For the 24-h block of data the average "ether drift" term is 0.67±0.73 Hz, corresponding to $\Delta v/v = (0.76 \pm 0.83) \times 10^{-14}$.

To discriminate between this persistent spurious signal (17-Hz amplitude at 2f) and any genuine "ether" effect, we made measurements for 12 or 24 sidereal hours. We must rotate each vector to obtain its phase relative to a fixed sidereal axis prior to further averaging. Averaging after this rotation leads, as shown in Fig. 3, to a typical 1-day result below 1±1 Hz. We believe Fig. 3 makes a convincing case that the origin of our spurious cos 2θ "signal" is to be sought in the domain of laboratory physics. Further experiments are underway to test the nonlinear elasticity conjecture, as well as to measure the paramagnetic and gravitational gradient contributions. A number of 12- and/or 24-h averages are shown in Fig. 4. We felt that averages for 24 h were sometimes quieter than 12-h

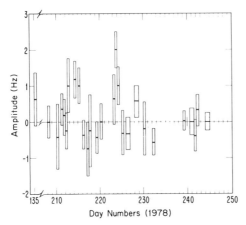

Fig. 4. Averaged data of isotropy-of-space experiment. Data such as those in Fig. 2 were averaged in blocks of 12 h (thinner bars) or 24 h (thicker bars). For completeness this figure includes data from diagnostic experiments before day 225. The data after day 238 represent near-ideal automatic operation of the present apparatus. A 1-Hz amplitude represents ~1.1 × 10^{-14} fractional frequency shift. The reference axis for the projection is the direction identified by Smoot et al.[11.0-h R.A. (right ascension, 6° dec. (declination)], Ref. 9.

averages, an effect which may be related to the observed 24-h period of the floor tilt (≈μrad). The data in Fig. 4 include most of the points taken during various diagnostic experiments. The data taken after day 238 correspond to approximately "ideal" automated operation of the present apparatus. The lack of any significant signal or day dependence allows us to perform an overall average. This final result of our experiment is a null "ether drift" of 0.13±0.22 Hz, which represents a fractional frequency shift of (1.5±2.5) × 10^{-15}.

From Eq. (3), we get $[\beta+\delta-(1/2)](v/c)^2 = (1.5\pm2.5) \times 10^{-15}$. This limit represents a 4000-fold improvement over the most sensitive previous experiment [10]. This advance is due to smaller spurious signals in our experiment (2×10^{-13} instead of 10^{-9}), to superior data-processing techniques and to superior long-term stability of the length etalon and reference laser.

To evaluate the precision with which this result verifies special relativity, one has to make assumptions concerning the direction and the amplitude of $\vec{v}$, the velocity of earth relative to "the preferred frame." We may conservatively use the Earth's velocity around the sun, which gives a null result ($\beta + \delta - 1/2) = \pm5 \times 10^{-7}$, some 10^6 times smaller than the classical prediction. This result includes the the sensitivity reduction factor 0.43 × associated with our 40° latitude and the data processing technique we used. (This factor had always been overlooked by previous authors.)

The recent discovery of a pure $P_1(\cos\theta)$ anisotropy in the cosmic blackbody radiation [9] was interpreted as a Doppler shift produced by motion of the earth (400 km/s) relative to the "preferred inertial frame" in which the blackbody radiation is isotropic. If this velocity is considered to be the relevant one, our sensitivity is $\beta + \delta - 1/2 = \pm2.5 \times 10^{-9}$ and constitutes the most precise test yet of the Lorentz transformation.

The present sensitivity limit arises from two sources: the finite averaging time and some mechanical problems. To improve our result another decade by simple averaging would require 15 years. The same improvement should be possible in a few months averaging with improved mechanical design (rotation speed stabilized to 10^{-4}, better design and mounting of the cavity, taking into account nonlinear elastic mechanical response, and active computer-driven piezo stabilization of the rotation axis to 1 arc sec). plus better vacuum inside the interferometer, and more effective stabilization of its housing temperature. Given success in reducing the systematic effects, higher laser power and better detectors could be used to reduce the random noise level which is presently far above the shot noise limit.

Further extensions of this experiment may diverge in two different ways:

- An increase in sensitivity by two orders of magnitude from the present level, which would require all the above improvements plus somewhat better laser stability, would certainly be a useful advance in experimental tests of special relativity. It also would bring us to the domain where general relativistic effects may be observable, at least in principle. For example the gradient of the earth's gravity over the size of our apparatus produces a vertical spatial anisotropy of the type we could measure. The expected frequency shift would be $\Delta\nu/\nu \simeq \frac{1}{2}(GM/R)(1/c^2)(L/R) \simeq 3 \times 10^{-17}$, if the reference etalon were rotated from a horizontal to a vertical position [13]. However it is evident that the gravity force itself would produce a large and probably disastrous background signal. Thus we presume that improved laboratory experiments of the present type will be basically interesting as useful advances in testing the assumptions of special relativity.

However the high resolution length readout techniques themselves may be of interest for gravity wave detectors and other highly sensitive applications as being perhaps more suitable than simple optical interferometry to read out with extreme resolution and dynamic range the positions of two free test masses.

In the other direction, it is attractive to consider applying these high sensitivity readout techniques to a modern Kennedy-Thorndike type of experiment. Improvement by a large factor, say 1000-fold, would certainly be interesting to relativity theorists. However, careful study of Ref. 8 suggests that this level of improvement will be very difficult to attain, although more modest improvements may surely be expected. Extreme care would be necessary to avoid drifts since, as we noted earlier, the only modulation arises from the earth's motion. For example, temperature stability of 50 μK for 1 month and a variation of the creep rate below 10^{-11} per day^2 would allow a ~20-fold improvement on the results of Ref. 8.

The authors are grateful to R. Mansouri for making us aware of his theoretical work and for stimulating conversations with us (separately). We also are grateful to our colleagues, especially P. L. Bender and J. E. Faller for numerous useful discussions about precision measurement techniques. The work has been jointly sponsored by the National Bureau of Standards, the National Science Foundation, the Office of Naval Research and by NATO through a grant to one of us (A.B.).

References

1. A. A. Michelson and E. W. Morley, Am. J. Sci. 34, 333 (1887).
2. A. Brillet and J. L. Hall, Phys. Rev. Lett. 42, 549 (1979).
3. H. P. Robertson, Rev. Mod. Phys. 21, 378 (1949); H. P. Robertson and T. W. Noonan, Relativity and Cosmology (Saunders, Philadelphia, 1968).
4. R. Mansouri and R. U. Sexl, General Relativity and Gravitation, Vol. 8, 1977, pp. 497-513; 515-524; 809-814.
5. This kind of idea was already discussed in Ref. 3 from a somewhat different aspect, but was only recently crystallized by the work of Ref. 4.
6. G. R. Isaak, Phys. Bull., p. 255 (1970); and D. C. Champeney, G. R. Isaak and A. M. Khan, Phys. Lett. 7, 241 (1963).
7. H. E. Ives and G. R. Stilwel, J. Opt. Soc. Am. 28, 215 (1938); 31, 369 (1941).
8. R. J. Kennedy and E. M. Thorndike, Phys. Rev. 42, 400 (1932).
9. G. F. Smoot, M. V. Gorenstein and R. A. Muller, Phys. Rev. Lett. 39, 898 (1977); and references therein.
10. T. S. Jaseja, A. Javan, J. Murray and C. H. Townes, Phys. Rev. 133, A1221 (1964).
11. J. L. Hall, in Fundamental and Applied Laser Physics: Proceedings of the 1971 Esfahen Symposium, ed. by M. S. Feld, A. Javan and N. Kurnit (Wiley, New York, 1973), p. 463.
12. CER-VIT is a registered trademark of Owens Illinois Inc., Toledo, Ohio.
13. One can understand the expected anisotropy as a manifestation of the gravitational redshift with height or, equivalently, in terms of a time delay of light signals propagating through a gravitational gradient. Direct measurements of the latter type, based on tracking data from the Mars-Viking lander, have recently measured this term to 0.2% [14].
14. R. Reasenberg et al., submitted to Ap. J. Lett.; private communication.

Suggestion and Analysis for a New Optical Test of General Relativity

M.O. Scully

Institute for Theoretical Physics and Optical Sciences Center,
University of Arizona
Tucson, AZ 85721, USA

Experiments in general relativity are notoriously difficult. As Misner, Thorne and Wheeler put it in their classic book [1], "For the first half century of its life, general relativity was a theorist's paradise but an experimentalist's hell". Recently, however, the tools of modern technology in the hands of heroic researchers have led to new tests of general relativity. We here suggest and analyze possible new tests [2] of general relativity using the techniques of precision ring laser interferometry. A ring laser mounted on a rotating platform is analyzed and two experiments suggested. This apparatus would measure the curvature parameter γ and the drag parameters Δ_1 and Δ_2. It is our hope that the present article will stimulate the critical discussions necessary in order to decide the ultimate feasibility of the proposed experiments. Possible applications of such studies to the fields of inertial guidance and geophysics are apparent.

As is well known, light injected into a rotating ring interferometer experiences a differential phase shift between the wave propagating in the same sense (+) or against (-) the direction of rotation [3]. The magnitude of the differential phase, $\Delta\Phi$, between the + and - running waves is given by

$$\Delta\Phi = \int_0^{2\pi} k_\phi^+ d\phi - \int_{2\pi}^0 k_\phi^- d\phi \qquad (1)$$

where $k_\phi^\pm$ is the wave vector for the $\pm$ running wave and ϕ is the angular coordinate. For the present purposes it is sufficient to assume that the light travels in a circular path, and is polarized in the $\hat{z}$ direction, perpendicular to the plane of the ring. In such a case the only non-vanishing component of the vector potential is

$$A_3^{\pm}(\phi,t) = A_0 \exp\left(-i\int k_{\mu}^{\pm}dx^{\mu}\right).$$ (2)

Hence, our fields depend only on the temporal ($x^0 = t$) and angular ($x^1 = \phi$) variables. Since we are driving our interferometer by an external signal of fixed frequency $k_0^{\pm} = \omega$, we need a dispersion relation to determine $k_1^{\pm}$ in the presence of an arbitrary gravitational field. In order to obtain such a relation we write Maxwell's equations in proper covariant form for the vector potential $A_{\sigma}(\phi,t)$ in the presence of an arbitrary metric field $g^{\mu\nu}$ as

$$\frac{\partial}{\partial x^{\nu}}\sqrt{|g|}\left(g^{\mu\rho}g^{\nu\sigma} - g^{\mu\sigma}g^{\nu\rho}\right)\frac{\partial A_{\sigma}}{\partial x^{\rho}} = 0, \quad \mu = 0,1,2,3$$ (3)

where $g = \det g^{\mu\nu}$. The contravariant metric tensor $g^{\mu\nu}$ is, of course, the matrix inverse of the covariant metric tensor $g_{\mu\nu}$ which may be read from the line element appropriate to this problem, namely

$$ds^2 = g_{00}\,dx^0dx^0 + 2g_{01}\,dx^0dx^1 + g_{11}\,dx^1dx^1.$$ (4)

Inserting (2) into (3) we obtain the required relation relating $k_{\phi}^{\pm}$ to ω:

$$k_{\phi}^{\pm} = \left\{-\frac{g_{01}}{g_{00}} \pm \left[\left(\frac{g_{01}}{g_{00}}\right)^2 - \frac{g_{11}}{g_{00}}\right]^{\frac{1}{2}}\right\}\omega.$$ (5)

Then from Eqs. (1) and (5) we obtain the differential phase shift expression

$$\Delta\phi = -2\omega\int_0^{2\pi}\left(\frac{g_{01}}{g_{00}}\right)d\phi.$$ (6)

Likewise, we may operate the ring interferometer by "fixing" the cavity modes $k_1^{\pm} = k$ and measuring the differences between the frequencies $k_0^{\pm} = \omega^{\pm}$. Following a logical development similar to that leading to (6) we find the frequency difference between the two running waves to be given by

22

$$\Delta\omega = -2kc \int_0^{2\pi} \left(\frac{g_{01}}{g_{00}} \right) d\phi \bigg/ \int_0^{2\pi} \left(\frac{-g_{11}}{g_{00}} \right)^{\frac{1}{2}} d\phi \ . \tag{7}$$

The usual Sagnac result may be obtained by transforming the Minkowski line element, appropriate to an inertial frame, to one rotating at a rate Ω. In this case the Sagnac frequency difference is found to be

$$\Delta\omega = \left(\frac{4A}{\lambdabar P} \right) \Omega, \tag{8}$$

where A is the area enclosed by the ring interferometer, $\lambdabar$ is the reduced wavelength of the injected laser radiation, and P is the perimeter of the ring.

As is clear from Eq. (6), a kind of generalized "Sagnac" effect might be expected whenever we have an off-diagonal metric tensor [4]. In particular recall that the line element for the case of a spherical earth of mass $M_\oplus$, radius $r_\oplus$ and rotating at a rate $\Omega_\oplus$ in "parametrized post-Newtonian (PPN) formalism [5]", is

$$ds^2 = \left(1 - \frac{r_s}{r} \right) c^2 dt^2 - \left(1 + \gamma \frac{r_s}{r} \right) \left(dr^2 + r^2 d\theta^2 + r^2 \sin^2\theta \ d\phi^2 \right)$$

$$+ 2ac \frac{r_s}{r} \left(\frac{7\Delta_1}{4} + \frac{\Delta_2}{4} \right) d\phi dt, \tag{9}$$

where the Schwarzschild radius $r_s = 2M_\oplus G/c^2$ and $a = \frac{2}{5} r_\oplus^2 \Omega_\oplus \sin^2\theta/c$. The values of the curvature parameter γ and frame dragging parameters Δ_1 and Δ_2, as predicted by various theories, are summarized in the paper by Ni [6].

Let us now turn to the analysis of an experiment based on the line element (9) which, although idealized, contains the essentials of real, and hopefully realizable, tests of general relativity. Consider our ring laser interferometer to be rotating at a rate Ω, a co-latitude θ, and at a distance $\tilde{r} = r \sin\theta$ all relative to the earth's axis. Ultimately we will consider an earth-bound experiment so that $\Omega \to \Omega_{\oplus}$ and $\tilde{r} \to r_{\oplus} \sin\theta$, but added insight is afforded by considering this more general case. Finally the ring laser is allowed to "spin" about its own axis [7] at a rate Ω'. The second rotation, at Ω', is included in the analysis since it will assist us in making precision measurements, as discussed later. Thus we must transform our metric (9) first to a frame rotating about the earth's axis at a rate Ω and then to our platform, which is rotating at a rate Ω'. As the light is constrained to a circular path with radial coordinate ρ in a frame fixed in our rotating platform, we need the metric coefficients $g_{\mu\nu}$ $\mu = 0,1$, where dx^0 and dx^1 now refer to times and angles measured on the platform. The relevant components of the metric tensor in the doubly rotating frame are then given by

$$g_{00} = c^2 \left(1 - \gamma_0^2 \frac{r_s}{r}\right) + \gamma(\tilde{r}\Omega)^2 \frac{r_s}{r} +$$

$$+ 2ac \frac{r_s}{r}\left(\frac{7\Delta}{4} + \frac{\Delta}{4}\right)\left[\gamma_0^2 \Omega + \frac{\gamma_0 \Omega'}{\tilde{r}^2}(\rho^2 + R\rho \cos\beta)\right]$$

$$- \left(1 + \gamma \frac{r_s}{r}\right)\left[\rho^2(\Omega'^2 + 2\gamma_0 \Omega'\Omega) + 2\gamma_0 \Omega\Omega'R\rho \cos\beta\right] \tag{10a}$$

$$g_{00} = -\left(1 + \gamma \frac{r_s}{r}\right)\left[\rho^2(\Omega' + \gamma_0 \Omega) + \gamma_0 \Omega R\rho \cos\beta\right]$$

$$+ ac\gamma_0 \frac{r_s}{r}\left(\frac{7\Delta}{4} + \frac{\Delta}{4}\right)\left[\rho^2 + R\rho \cos\beta\right]\frac{1}{\tilde{r}^2} \tag{10b}$$

$$g_{11} = -\left(1 + \gamma \frac{r_s}{r}\right)\rho^2 \tag{10c}$$

where $\tilde{r}_s = r_s \sin\theta$, R is the distance from the earth's axis to the center of the ring and $\beta = \phi' + \Omega't$, where ϕ' denotes an angle measured in the rotating platform. The special relativistic factor γ_0 equals $(1 - (R\Omega/c)^2)^{-\frac{1}{2}}$.

Inserting Eqs. (10,a,b,c) into (7), assuming R>>ρ and neglecting terms smaller than $\frac{r_s}{r}\Omega_{\oplus}$ we obtain the frequency differential for the present

24

problem

$$\Delta\omega = \kappa \left[\Omega + \Omega' - \frac{1}{2}(\gamma+1) \frac{r_s}{r_o} \Omega \sin^2\theta_o \right.$$

$$\left. - \frac{2}{5} \frac{r_s r_\oplus^2}{r_o^3} \left(1 - \frac{3}{2}\sin^2\theta_o\right)\left(\frac{7\Delta_1}{4} + \frac{\Delta_2}{4}\right)\Omega_\oplus \right] \qquad (11)$$

where κ is a constant scale factor and r_o and θ_o are radial and angular coordinates to the center of the ring. The various terms in this expression arise from and reflect the following physical processes: The first two terms, $\Omega + \Omega'$ obviously result from the double rotation of our ring laser. The next two terms arise from the fact that these rotations take place in space curved by a source whose strength is proportional to r_s. Finally the term proportional to $\Omega_\oplus$ is a result of gravitational drag due to the earth's rotation. This dipolar "drag term" falls off as $1/r^3$ which is to be compared with the $1/r$ dependence of the other two general relativistic corrections.

In what follows we will restrict our attention to earthbound experiments so that $r_o = r_\oplus$ and $\Omega = \Omega_\oplus$ in (11). The result (11) has been derived with explicate experiments in mind to which we now turn.

Consider first an experiment in which we rotate our turntable at the modest rate of $\Omega' \sim 1$ Hz, then keeping leading terms in (11) we have the frequency difference

$$\Delta\omega = \frac{\Delta\omega_o}{\Omega_\oplus} \left[\Omega_\oplus + \Omega' + \frac{1}{2}(\gamma+1) \frac{r_s}{r_o} \Omega' \sin^2\theta_o \right], \qquad (12)$$

where $\Delta\omega_o$ is the frequency splitting when $\Omega' = 0$. The third term in square brackets is of order $10^{-5} \Omega_\oplus$. As discussed later, it should be possible to measure [8] and separate the various contributions to $\Delta\omega$ in Eq. (12).

The second experiment is "designed" to observe the effects of gravitational drag. In order to avoid worrying about the gyroscopic scale factor, κ, consider the situation wherein we choose Ω' to produce a null $\Delta\omega$. From

Eq. (11) we see that this occurs when

$$\Omega' = -\Omega_\oplus + \left[\tfrac{1}{2}(\gamma+1)\sin^2\theta_o + \frac{2}{5}\left(1 - \frac{3}{2}\sin^2\theta_o\right)\left(\frac{7\Delta_1}{4} + \frac{\Delta_2}{4}\right)\right]\frac{r_s}{r_\oplus}\,\Omega_\oplus \quad (13)$$

Correction terms due to curvature and drag are apparent. Hence if we can

measure $\Delta\omega$, Ω' and $\Omega_\oplus$ to a precision corresponding to a part in 10^9 to

10^{10} of the earth rate, we can sort out the "extra rotation" predicted

by general relativity. We now consider the precision to which we can

measure these three quantities.

The Sagnac frequency difference $\Delta\omega$ can be used to measure rotation

rate in two different types of experiments depending on whether the laser

is placed inside or outside of the ring cavity. The "locking" problem

associated with internally driven ring lasers has been the subject of

recent investigations [9], and these devices can now operate down to a rate

of 10^{-5} to 10^{-6} of earth rate using a ring of 1m diameter. Note that ring

radii larger than this are difficult to work with as the laser tends to go

multi-mode in that case. However, very interesting new developments in

passive Sagnac interferometry, in which the laser is removed from the

cavity, allow large ring diameters and also avoid the locking problem.

The following is a quote from the Paper [10] of Ezekiel et al., "With a

10 m by 10 m cavity and a 4 watt stabilized argon laser it should be

possible to reach a sensitivity of $10^{-10}\Omega_\oplus$".

The rotation rate Ω' can be measured to a high precision by observing

the time it takes for the table to turn through 2π. We can measure the time

to essentially "arbitrary" precision. The error in this measurement comes

from the angular uncertainty associated with the determination that the

table has turned through 2π. However, this too can be accomplished to a

very high precision by several optical techniques. For example, we can

mount a mirror on the turntable and by means of an autocollimator determine

that the mirror has returned to its original position to within $\sim 10^{-4}$ sec of arc [11]. This corresponds to an error $\delta\Omega'$ of about $10^{-10}\Omega'$. By mounting

several mirrors on the turntable, we could monitor [12] Ω' quasi-continuously.

Earth rate, $\Omega_{\oplus}$, is known to an amazing accuracy from lunar ranging [13]

and very long baseline radio astronomy [14]. We are grateful to Dr. Bender

for his comments [15] on the earth rate problem and for the following clear

statement, "0.1 millisec [accuracy of measurements of earth's period] over

1 day seems realistic or better if there is little high-frequency noise in

UT1." This translates into an error in $\Omega_{\oplus}$ of magnitude $\delta\Omega \approx (\delta T/T)\Omega_{\oplus}$, and

since $T \lesssim 10^5$ sec this implies an error in the measurement of $\Omega_{\oplus}$ to a part

in 10^9 to 10^{10}.

For a discussion of the techniques of radio astronomy applied to $\Omega_{\oplus}$,

see the review by Councilman [14]. Table 1 of that paper lists the

expected limitation on earth rotation measurements to be $\lesssim 0.001$ sec of

arc. This implies an error in earth rate of $\approx (\delta\phi/2\pi)\Omega_{\oplus}$, i.e., again an

error of between 10^{-9} and 10^{-10} earth rate.

In conclusion it appears that state of the art technology applied

to the experiment of Eq. (12); could provide a new measurement of γ. The

frame dragging experiment [16] as implied by (13) would require a measure-

ment of $\Omega_{\oplus}$ to a part in 10^{10}. At present $\Omega_{\oplus}$ is known to $\sim 10^{-9}$ to 10^{-10}

of earth rate. However, in view of recent accomplishments in the measure-

ment of $\Omega_{\oplus}$ an extension of current technology to a sensitivity $10^{-10}\Omega_{\oplus}$

seems likely. Clearly such precision measurements will call for imaginative

experimentation, but the problem area [17] is potentially rich in both

fundamental and applied payoff.

Acknowledgments

Much of this work was carried out during a recent visit to the

University of Warsaw Institute for Theoretical Physics. We gratefully

acknowledge their hospitality. The author wishes to thank I. Bialynicki-

Birula, D. Brill, B. Carr, J. Cohen, H. Hill, K. Just, G. Moore, E. Power, H. Rund, and K. Thorne, for helpful and stimulating discussions concerning the analysis. The experimental situation has been clarified and refined in conversations with P. Bender. P. Franken, H. Hill, T. Hutchings, S. Jacobs, W. Lamb, P. Meystre, V. Sanders, R. Shack, J. Small, H. Walther, W. Wing and J. Wyant.

This work was supported in part by Wright-Patterson AFB, U. S. Air Force, and in part by the Air Force Office of Scientific Research (AFSC), United States Air Force, and the Army Research Office, United States Army.

References

1. C. Misner, K. Thorne, and J. Wheeler, Gravitation, (Freeman, 1973).

2. For a detailed discussion of experimental relativity see Ref. 1. In this context we mention especially the works of: R. Dicke, C. Everett, W. Fairbank, H. Hill, K. Nordtvedt, R. Pound, G. Rebka, I. Shapiro, and C. Will. The recent paper of A. Baillet and J. Hall, P.R.L. 42, 549, (1979) is a beautiful example of modern optics applied to the study of space-time.

3. For a comprehensive review of the Sagnac effect see E. Post, Rev. Mod. Phys. 39, 475 (1967). The ring laser is discussed in Chap. 12 of "Laser Physics" by M. Sargent, M. Scully, and W. Lamb, (Addison-Wesley 1974). For a first principles theory of ring laser physics see L. Menegozzi and W. Lamb, Phys. Rev. A8, 2103, (1973). For a summary of ring laser physics and applications the article by Aronowtiz in "Laser Applications", Vol. 1 (Academic Press, N.Y.), 1971, is especially recommended.

4. In an interesting paper V. Braginsky, C. Caves and K. Thorne [Phys. Rev. D 15, 2047, (1977)] have suggested use of a high Q microwave cavity as a device to detect frame dragging. We thank Prof. Thorne for calling this paper to our attention. An essential difference between that

paper and the present work is that our ring laser is mounted on a
rotating platform. This notion is central to the analysis and
proposed experiments.

5. C. Will and K. Nordtvedt, Jr., Astrophys J. __177__, 757, (1972)

6. Wei-Tou Ni, Astrophysics J. __176__, 769, (1972). For example, the curvature
    parameter γ assumes the value 1 in general relativity and $(1+\omega)/(2+\omega)$
    in Brans, Dicke, Jordon theory where ω is the coupling constant of
    Dicke. The frame dragging parameters (Δ_1, Δ_2) assume the values $(1,1)$
    in general relativity and $[(10+7\omega)/(14/7\omega),1]$ is Brans, Dicke,
    Jordan theory.

7. In this paper, we take the rotation axis of Ω' to be parallel to
    the earth's axis.

8. Note that this implies we have "calibrated" our instrument so that
    we "know" $\Delta\omega_o/\Omega_\oplus$ to $\sim$ a part in 10^9.

9. See for example: M. Scully, V. Sanders and M. Sargent, Optics Lett.
    __3__, 43, (1978); and R. Cahill and E. Udd, Optics Lett. __4__, 93, (1979).
    Note that we can measure $\Delta\omega$ to a much higher precision than $10^{-6}\Omega_\oplus$
    since the device is rotating at $\Omega'\sim$1Hz and such a rotation produces
    a bias which keeps the ring laser from locking.

10. S. Ezekiel et. al, "Laser Inertial Rotation Sensors", SPIE Vol. __157__,
    69, (1978).

11. R. Shack private communication.

12. This may be accomplished in several ways; moreover, by averaging over
    many cycles we can dramatically reduce the effects of random noise.

13. P. Bender et. al., Science __182__, 228 (1973).

14. C. Counselman, Ann. Rev. Astro. and Astrophys., __14__, 197, (1976).

15. P. Bender private communication. A single measurement of $\Delta\omega$ would take
    $\sim$ 10 minutes and $\Omega_\oplus$ could be monitored in similar intervals.

16. Space limitations require that several technical aspects of the experimental arrangement have been omitted from the present discussion e.g. motion of the earth around the sun, earth wobble, etc. Inclusion of these, and other details, presents no real problem and will be discussed elsewhere.

17. For another problem in which off-diagonal elements of the metric tensor are important, see the paper on synchronization in noninertial systems by J. Cohen and H. Moses, P.R.L. $\underline{39}$, 1641 (1977). We are grateful to Professor Cohen for calling this work to our attention.

High Resolution Spectroscopy of Atomic Hydrogen

A.I. Ferguson[1], J.E.M. Goldsmith, T.W. Hänsch, and E.W. Weber[2]

Department of Physics, Stanford University, Stanford, CA 94305, USA

1. Introduction

Spectroscopy of hydrogen, the simplest of the stable atoms, has played an important role in the development of atomic theory and quantum mechanics [1]. In recent years, tunable lasers and coherent light techniques have led to dramatic improvements in spectral resolution, opening new opportunities for the measurement of fundamental constants and for stringent tests of quantum electrodynamic calculations. Much progress has been achieved since the previous conference in this series [2]. Nonetheless, laser spectroscopy of atomic hydrogen continues to hold exciting challenges, and future refined experiments may well lead to some surprising fundamental discovery.

2. High Resolution Spectroscopy of the Balmer-α Line

The prominent red Balmer-α or H_α line, subject of innumerable classical investigations, was the first hydrogen line to be studied by Doppler-free laser spectroscopy. By applying the method of saturated absorption spectroscopy with a pulsed tunable dye laser to a Wood-type hydrogen gas discharge, it became possible to resolve single fine-structure components in the optical spectrum. An absolute wavelength measurement of the strong $2P_{3/2}$ – $3D_{5/2}$ component provided a new, tenfold improved value of the Rydberg constant [3]. For more than four years this remained the only Doppler-free measurement of the Rydberg, even though the result fell somewhat outside the error limits of earlier conventional measurements.

During the past year, new studies of the Balmer-α line have been completed in our laboratory at Stanford, which take advantage of recent developments in instrumentation and coherent light techniques to achieve much improved resolution and accuracy.

2.1 Polarization Spetroscopy Yields a New Rydberg

Polarization spectroscopy [4] takes advantage of the fact that small changes in light polarization can be detected with higher sensitivity than changes in light intensity, at least at visible wavelengths. Polarization

---

[1] Lindemann Fellow, Present address: Department of Physics, University of St. Andrews, North Haugh, St. Andrews, Scotland.

[2] Max-Kade Fellow, Present address: Physikalisches Institut der Universitat Heidelberg, Philosophenweg 12, D-6900 Heidelberg, West Germany.

spectroscopy makes it possible to monitor the nonlinear interaction of two counterpropagating monochromatic light beams in a gas at lower atom density and lower light intensity than are required for saturated absorption spectroscopy, thus alleviating pressure broadening, power broadening, and related problems.

Polarization spectra of the Balmer-α line [5], recorded with a cw dye laser in a mild He-H_2 discharge, exhibit an at least fivefold improvement in resolution over earlier saturation spectra (Fig. 1). The excited n=2 hydrogen atoms were produced in a 5.5 cm diameter hot cathode dc discharge tube, filled with a mixture of 15% H_2 in He. The tube could operate down to lower pressure and current density than the Wood's tube previously used, and the axial electric field in the positive column was almost an order of magnitude smaller. Since the gas mixture was predominantly monatomic helium, dissociation of the hydrogen caused only a relatively minor change in pressure. Thus it was possible to study the effects of current and pressure separately, and to determine any necessary corrections.

The line components in Fig. 1 have been recorded as derivatives of dispersion-shaped resonances from the gyrotropic birefringence of the hydrogen gas which is induced by a circularly polarized saturating beam. The excellent signal-to-noise ratio made it feasible to measure the absolute wavelength of the weak but narrow $2S_{1/2} - 3P_{1/2}$ fine structure component. Only transitions from the F=1 hyperfine level of the 2S state give a strong signal, since atoms with angular momentum F=0 cannot be oriented. A near-coincident I_2 line (the i-th hyperfine component of the $^{127}I_2$ B-X R(73) 5-5 transition) served as an intermediate reference line. It, too, was observed by Doppler-free polarization spectroscopy, and its wavelength was determined relative to the i-th hyperfine component of the $^{127}I_2$ B-X R(127) 11-5 transition at 632.8 nm, using a novel interferometer calibration technique.[6]

A new Rydberg value, R_∞ = 109 737.314 76 (32) cm^{-1}, has been determined from the measured wavelength. It is in agreement with and three times more accurate than that obtained by the first Doppler-free measurement (see Fig. 2). Another twofold improvement can be obtained by a more precise wavelength

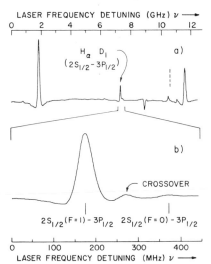

Fig. 1 (a) Polarization spectrum of H_α. The dashed line indicates the position of the iodine reference line (b) Expanded portion of polarization spectrum showing the $2S_{1/2} - 3P_{1/2}$ fine structure component (Ref.5)

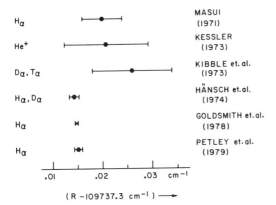

Fig. 2 Comparison of recent Rydberg measurements

measurement of the nearby iodine reference line. An independent though less accurate Rydberg measurement has very recently been reported by Petley and Morris [7], using saturated absorption spectroscopy of the Balmer-α line in a Wood's discharge.

2.2 Anomalous Pressure Effects

A very careful study of systematic effects due to pressure and discharge conditions has been carried out in support of the Rydberg measurement [8,9]. More than 1000 individual spectra have been evaluated in these investigations.

All investigated Balmer-α fine structure components showed a negative shift for pressures of He above 0.1 torr (Fig. 3), in contrast to the

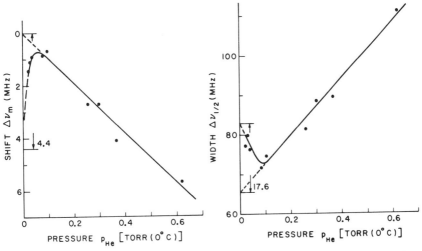

Fig. 3 Pressure shift and pressure broadening of the $2S_{1/2} - 3P_{1/2}$ fine structure component of H$_\alpha$ (Ref. 8)

positive shift reported for the unresolved, blended H_α line. The anomalous negative shifts and the large pressure broadening are correctly given by an impact approximation theory, assuming classical straight trajectories and taking into account only the van der Waals interaction and a special property of excited hydrogen, the large cross section for fine structure transfer collisions [8]. For pressures below 0.1 torr, some of the individual fine structure lines show a strong positive shift and a narrowing of the line width (Fig. 3), which are attributed to a collisional decoupling of the 3P hyperfine structure. These effects are analogs of the collisional narrowing or motional narrowing reported previously in nuclear magnetic resonance, radiofrequency, microwave, and Raman spectroscopy of gases.

2.3 Optical-Radiofrequency Double-Quantum Saturation Spectroscopy

The resolution in the polarization spectra of the H_α line is approaching the natural line width (29 MHz), determined by the short lifetime of the 3P state. Still narrower lines down to 20 MHz width have been observed by using a two-wire transmission line in the discharge tube to apply a radiofrequency field to the atoms, in addition to the laser light, so that optical radiofrequency double-quantum transitions are induced from the $2S_{1/2}$ level to the longer living $3S_{1/2}$ or $3D_{3/2}$ level [10]. The natural width of the $3S_{1/2}$ level is less than 1 MHz, and double-quantum spectroscopy of a collision-free atomic beam should make it feasible to reach such a line width experimentally.

The small hydrogen $3P_{3/2}$ - $3D_{3/2}$ Lamb shift has been measured directly by comparing single- and double-quantum saturation signals. The result, -5.5(0.9) MHz, agrees with theory and confirms that the $3D_{3/2}$ state lies lower than the $3P_{3/2}$ state.

3. Two-Photon Spectroscopy of Hydrogen 1S-2S

Perhaps the most intriguing transition to be studied in atomic hydrogen is the two-photon transition from the 1S ground state to the metastable 2S state, because first order Doppler broadening can be conveniently eliminated by excitation with two counterpropagating light beams, whose Doppler shifts cancel [11], and because the 1/7 sec lifetime of the 2S state implies an ultimate linewidth as narrow as 1 Hz. Unfortunately, however, two-photon excitation of hydrogen 1S-2S requires monochromatic ultraviolet radiation near 243 nm, where there are still no good tunable laser sources available.

The best 1S-2S spectra so far have been recorded about two years ago by C. WIEMAN at Stanford [2], employing a blue single-mode cw dye laser oscillator with nitrogen-pumped pulsed dye laser amplifier chain and lithium niobate frequency doubler. The short pulse duration and some frequency chirping introduced by the dye amplifiers limited the resolution to about 120 MHz (fwhm at 243 nm.). Nonetheless it was possible to measure the hydrogen-deuterium isotope shift of the 1S-2S interval to within 6.3 MHz or one part in 10^5 and to obtain a first qualitative confirmation of the predicted small 11.9 MHz relativistic correction due to the recoil of the nucleus. A comparison of the 1S-2S transition with the n=2 to n=4 Balmer-β line yielded an experimental value for the ground state Lamb shift of hydrogen, accurate to better than 0.4%.

34

3.1 Multi-Pulse Spectroscopy

Several approaches promising much improved resolution have been explored since then. In particular it has been demonstrated in experiments with sodium vapor that it is feasible to record Doppler-free two-photon spectra with a coherent train of picosecond pulses emitted by a synchronously pumped mode-locked cw dye laser [12]. A resonant signal is obtained whenever the frequencies of two oscillating modes add up to the atomic excitation frequency, and this signal can be strong, because many pairs of modes contribute simultaneously to the excitation. The spectral resolution can be comparable to that of a single-mode cw laser, and the actively controlled mode-spacing can provide an accurate calibration scale for the measurement of large frequency intervals.

Using presently available instrumentation, it is relatively easy to produce a continuous train of picosecond pulses in the blue spectral region with peak powers reaching a few hundred Watts [13]. But unfortunately this power level is still too low for efficient second harmonic generation. The requirement to match both phase velocities and group velocities of fundamental and second harmonic wave dictates the use of very short crystals (length typically less than 1 mm), and the small nonlinear optical coefficients of crystals suitable for this spectral region limit the conversion efficiency to less than one percent at best. Nonetheless, it may be feasible to observe multi-pulse two-photon transitions in atomic hydrogen in this way.

3.2 CW Spectroscopy of 1S-2S

We are confident that it will be possible to observe hydrogen 1S-2S two-photon transitions even with low-power cw ultraviolet radiation, generated by nonlinear sum frequency mixing of two different single-mode lasers [14]. We are presently experimenting with a setup as shown in Fig. 4., and although no signal has yet been observed, a discussion of this scheme can serve to illuminate some of the technical difficulties and to estimate what should be possible in the near future.

While known nonlinear optical crystals do not permit 90 degree phase-matched second harmonic generation down to 243 nm, ultraviolet radiation of this wavelength can be produced as the sum frequency of a blue krypton ion laser and a yellow rhodamine 6G dye laser in a crystal of ammonium dihydrogen phosphate (ADP), cooled close to its Curie temperature [14]. Using frequency-stabilized single mode lasers of 600 and 200 mW power, respectively, focused to a spot diameter of 100 μm inside a 5 cm long ADP crystal, we obtain tunable cw ultraviolet radiation of about 50 μW. The primary power could clearly be much increased by the use of a ring-cavity dye laser or by inserting the summing crystal inside one or both laser cavities. But the ultraviolet power seems limited by some damage mechanism in the crystal, which leads to beam deterioration and power loss after about five minutes of operation even at the present low intensities. This damage may be associated with the formation of color centers due to heavy metal impurities in the crystal material and can perhaps be alleviated by more careful crystal selection.

The ultraviolet intensity can be boosted by coupling the crystal output beam into an enhancement cavity, formed by two dielectric mirrors and electronically locked to resonance. A standing wave field of about 1 mW circulating power is thus easily produced inside the observation chamber.

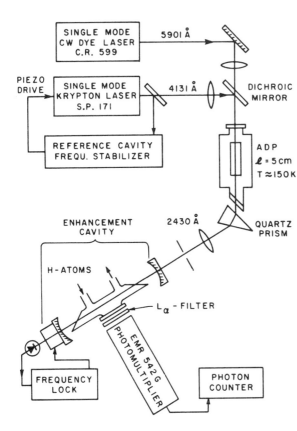

Fig. 4 Apparatus for cw
two-photon spectroscopy
of hydrogen 1S-2S.
Ultraviolet radiation
near 243 nm is generated
in a cooled ADP crystal
as the sum frequency of
a yellow dye laser and a
blue krypton ion laser.
An enchancement cavity
around the observation
chamber provides a
standing wave field of
about 1 mW circulating
UV power

Ground state hydrogen atoms from a Wood-type gas discharge [11] are sent by
flow and diffusion into the chamber, and the collision-induced vacuum
ultraviolet emission of the laser-excited hydrogen 2S atoms is monitored
through a MgF₂ window by a solar-blind photomultiplier with Lyman-α
interference filter.

The signal expected under these conditions is rather small, since the
two-photon excitation cross section is not enhanced by any near resonant
intermediate state [15]. For a laser intensity $I = 2.5$ W/cm² corresponding to a beam waist diameter of about 200 μm, and at a line width $\Delta\nu = $
4 MHz, the transition rate per atom at resonance is of the order $\Gamma = 3.3$
10^{-6} $I^2/\Delta\nu = 5 \cdot 10^{-6}$ sec^{-1}. In a He-H₂ (10%) mixture of 0.1 torr total
pressure with 15% of the H₂ molecules dissociated, the density of ground
state hydrogen atoms is 10^{14} cm^{-3}, and about $2 \cdot 10^5$ atoms per second
will be excited in an interaction volume of 1 cm length. At a detection
solid angle of 0.02×4π sterad, a filter transmission of 16%, and a
photocathode efficiency of 20% the expected signal is about 120 counts per
second, if the loss of Lyman-α photons due to resonance trapping and
quenching can be kept negligible.

In preliminary experiments with pure hydrogen, a background of about 50
counts per second has been registered, much above the multiplier dark count

rate of one per second. Such a background makes it difficult, though hopefully not impossible, to detect the 1S-2S signal. We are presently exploring ways to reduce the background, and we are considering the use of optional pulsed dye amplifiers for one or both primary beams in order to find the resonance quickly.

3.3 Practical Resolution Limits

Although our present laser sources have linewidths on the order of 1 MHz, more sophisticated servo controls could reduce this width to a few kHz. But other sources of line broadening, in particular pressure broadening and transit broadening, have to be overcome before such a resolution can be approached in the 1S-2S two-photon spectrum.

The dominant causes of collision broadening are 2S-2P fine structure transfer and electron spin depolarization [9]. For encounters with He buffer gas atoms, the respective collision cross sections are 1.35×10^{-14} cm^2 [16] and $(4 \pm 2) \times 10^{-15}$ cm^2 [9]. At room temperature and at a He pressure of 0.1 torr, the resulting pressure broadening amounts to approximately 1.4 MHz (fwhm at 243 nm). Cooling near the temperature of liquid helium [17] and lowering the pressure to 0.01 torr can reduce this broadening to about 15 kHz.

The short transit time of the light hydrogen atoms moving across the 200 µm beam waist results in a line broadening of about 2 MHz at room temperature or 230 kHz at 4 Kelvin. Near longitudinal interaction of the ultraviolet light with a cooled hydrogen beam [17] over a 1 cm long intersection would reduce transit broadening to less than 5 kHz.

The relativistic transverse Doppler effect which is not cancelled by the use of counterpropagating beams will produce a line broadening and shift of the order 100 kHz at room temperature or 1.5 kHz near 4 Kelvin. Further reduction may be possible with the help of radiation cooling [18].

Light shifts [19] are a negligible problem, by comparison. At an ultraviolet intensity of 2.5 Wcm^{-2}, the calculated a.c. Stark shift of the 1S-2S transition is only of the order 10 Hz.

3.4 Possible Future Measurements of Fundamental Constants and Tests of QED

In summary then it appears technically quite feasible to observe the 1S-2S two-photon transition with a line width of a few tens of kHz or a resolution approaching one part in 10^{11}. What can be learned by measuring the center frequency to this or better accuracy? It will certainly be possible to extract a better value of the Rydberg constant. However, the uncertainty of the electron/proton mass ratio (0.4 ppm) limits the achievable accuracy to about two parts in 10^{10}. On the other hand, if the 611 GHz H-D isotope shift of the 1S-2S transition is measured to better than 500 kHz, an improved value of the electron/proton mass ratio and hence the Rydberg can be determined [2]. Uncertainties in the fine structure constant and the nuclear size corrections impose error limits of about 0.08 ppm for the electron/proton mass ratio and four parts in 10^{11} for the Rydberg that can be obtained in this way.

Obviously, neither a measurement of the 1S-2S frequency nor of the isotope shift can provide a very stringent test of quantum electrodynamic theory, because we are free to adjust some fundamental constants until the calculations agree with the observations. But once these measurements and adjustments are made, we are able to predict the frequencies of all other optical transitions in hydrogen with comparable accuracy, and any precise comparison of two different transition frequencies becomes a very interesting test of theory. For instance, a comparison of the 1S-2S transition with one of the visible Balmer lines can be used to confirm the predicted 8149.43 ± 0.08 MHz Lamb shift of the 1S ground state [20], and some of the technical advances discussed here make it appear quite feasible to surpass even the accuracy of radiofrequency measurements of the 2S Lamb shift in this way. Another very interesting test would be a comparison of the 1S-2S transition with transitions to or between highly excited hydrogen Rydberg states. In this way one might detect, for instance, some small deviations from Coulomb's law which may exist within atomic dimensions and which may have escaped detection in the past.

Work supported by the National Science Foundation under Grant No. NSF-9687 and by the U.S. Office of Naval Research under Contract N00014-78-0403.

References

1. T.W.Hänsch, A.L.Schawlow, and G.W.Series, Sci. Am. 240, 94 (1979)
2. C.Wieman and T.W.Hänsch, in "Laser Spectroscopy III," J.L. Hall and J.L Carlsten, eds., Springer Series in Optical Sciences Vol. 7, Berlin, Heidelberg, New York, 1977, p 39
3. T.W.Hänsch, M.H.Nayfeh, S.A.Lee, S.M.Curry, and I.S.Shahin, Phys. Rev. Letters 32,. 1336 (1974)
4. C.Wieman and T.W.Hänsch, Phys. Rev. Letters 36, 1170 (1976)
5. J.E.M.Goldsmith, E.W.Weber, and T.W.Hänsch, Phys. Rev. Letters 41, 1525 (1978)
6. J.E.M.Goldsmith, E.W.Weber, F.V.Kowalsky, and A.L.Schawlow, Applied Optics, accepted for publication
7. B.W.Petley and K.Morris, Nature 279, 141 (1979)
8. E.W.Weber and J.E.M.Goldsmith, Phys. Letters 70A, 95 (1979)
9. E.W Weber, Physical Review, to be published
10. E.W.Weber and J.E.M.Goldsmith, Phys. Rev. Letters 41, 940 (1978)
11. T.W.Hänsch, S.A.Lee, R.Wallenstein and C.Wieman, Phys. Rev. Letters 34, 307 (1975)
12. J.N.Eckstein, A.I.Ferguson, and T.W.Hänsch, Phys. Rev. Letters 40, 847 (1978)
13. J.N.Eckstein, A.I.Ferguson, T.W.Hänsch, C.A.Minard, and C.K.Chan, Opt. Comm. 27, 466 (1978)
14. T.W Hänsch, in "Tunable Lasers and Applications," A. Mooradian, T. Jaeger, and P. Stokseth, eds., Springer Series in Optical Sciences, Vol. 3, Berlin, Heidelberg, New York, 1976, pp. 340
15. Y.Gontier, and M.Trahin, Phys. Letters 36A, 463 (1971)
16. S.R.Ryan, S.J.Czuchlewski, and M.V.McCusker, Phys. Rev. A16, 1892 (1977)
17. S.B.Crampton, T.J. Greytak, D.Kleppner, W.D.Phillips, D.A. Smith, and A.Weinrib, Phys. Rev Letters 42, 1039 (1979)
18. T.W.Hänsch and A.L.Schawlow, Opt. Comm. 13, 86 (1975)
19. P.F.Liao and J.E.Bjorkholm, Phys. Rev. Letters 34, 1 (1975)
20. G.Erickson, J. Chem. Phys. Ref. Data 6, 831 (1977)

Direct Frequency Measurement of the 260 THz (1.15 μm) ^{20}Ne Laser: And Beyond

D.A. Jennings, F.R. Petersen, and K.M. Evenson

Time and Frequency Division, National Bureau of Standards, Boulder, CO 80303, USA

ABSTRACT

Absolute frequency measurement has been extended to the visible spectrum with the measurement of the strong 1.15 μm laser line in ^{20}Ne at 260 THz and lines in iodine at twice this frequency. The 260 THz frequency was synthesized in nonlinear crystals of $CdGeAs_2$ and $AgAsS_3$ from stabilized CO_2 lasers and the 1.5 μm laser line in ^{20}Ne. The visible frequencies were synthesized by generating the second harmonic of the 260 THz radiation with a $LiNbO_3$ crystal. The absolute frequencies of ten hyperfine components of $^{127}I_2$ near 520 THz were measured.

INTRODUCTION

Since the frequency measurement by Hocker et al. [1] in 1967 of the 890 GHz line of the HCN laser the progress in laser frequency measurement has been steady (See Fig. 1). The frequency measurement of the 10 THz line of H_2O

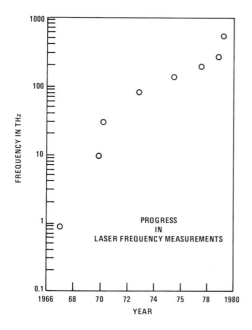

Fig. 1 Progress in laser frequency measurements.

39

in 1970 by Evenson et al. [2] and the 1973 frequency measurement of the CO_2 laser at 30 THz opened the way for the 88 THz (3.39 µm) frequency measurement of the CH_4 stabilized Ne line by Evenson, et al. [3]

Since the 88 GHz measurement, progress toward higher frequencies has been somewhat slower due to the falloff in the sensitivity of the point-contact MIM diode used in the measurements. Nevertheless, in 1974 the 148 THz frequency of Xe was measured [4], and then, in 1977 the 197 THz Ne radiation was also measured [5]. Efforts to reach 260 THz and the visible were still elusive. During the past year, however, a change in experimental techniques has resulted in attainment of both of these goals.

The 260 THz Measurement

In principle, the measurement of the 260 THz (1.15 µm) laser line in ^{20}Ne is straightforward as can be seen in Fig. 2. The addition of two CO_2 laser lines to the previously measured 197 THz laser radiation of ^{20}Ne synthesizes the required frequency of 260 THz within the 1.5 GHz pass band of the RF amplifier used.

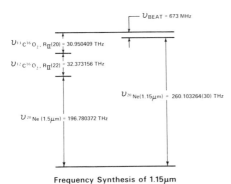

Frequency Synthesis of 1.15µm

Fig. 2 Synthesis of the 1.15 µm ^{20}Ne radiation.

The actual synthesis of 260 THz was achieved by using the quadratic nonlinear susceptibility in crystals to mix the known laser frequencies in the following manner. A crystal of $CdGeAs_2$ was used to sum the $R_{II}(20)$, $^{13}C^{16}O_2$ and the $R_{II}(22)$, $^{12}C^{16}O_2$ laser frequencies. The 63 THz output frequency (4.7 µm) was then summed with the 197 THz frequency of a ^{20}Ne, 1.5 µm laser in a crystal of Ag_3AsS_3 (proustite). This synthesized radiation (260 THz) was combined with the 260 THz ^{20}Ne, 1.15 µm laser radiation, and the difference frequency (i.e., the beat frequency) was detected on a fast photovoltaic Ge diode. The resulting beat frequency was amplified and measured with a spectrum analyzer.

The experimental setup is shown in Fig. 3. The CO_2 reference lasers were stabilized to the saturated absorption in CO_2, [5] and the CO_2 power lasers were frequency offset locked. The 1.15 µm, ^{3}He-^{20}Ne laser was frequency offset locked to a Lamb-dip stabilized 1.15 µm, ^{20}Ne laser [6]. The 1.5 µm, ^{3}He-^{20}Ne laser was manually set to the center of its gain curve. The basic lasers have been described elsewhere with the exception that mirrors and gas fills were changed to enhance laser performance for each particular frequency [3,7].

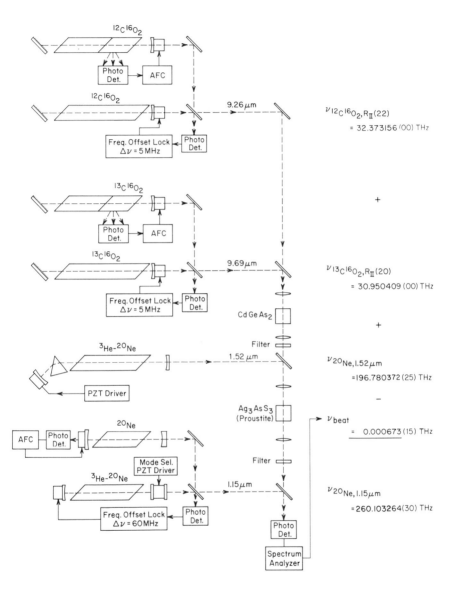

Fig. 3 A block diagram of the 260 THz experimental setup with the final result.

Therefore, the frequency of the ^{20}Ne, 1.15 μm laser was:

$$\nu_{^{20}Ne,\ 1.15\ \mu m} = \nu_{^{12}C^{16}O_2,\ R_{II}(22)} + \nu_{^{13}C^{12}O_2,\ R_{II}(20)}$$
$$+ \nu_{^{20}Ne,\ 1.52\ \mu m} - \nu_{beat},$$

where

$$\nu_{^{12}C^{16}O_2,R_{II}(22)} = 32.373\ 156(00) \quad THz\ [8],$$

$$\nu_{^{13}C^{16}O_2,R_{II}(20)} = 30.950\ 409(00) \quad THz\ [9],$$

$$\nu_{^{20}Ne,\ 1.52\ \mu m} = 196.780\ 372(25)\ THz\ [10],\ and$$

$$\nu_{beat} = 0.000\ 673(15)\ THz.$$

Thus,

$$\nu_{^{20}Ne,\ 1.15\ \mu m} = 260.103\ 264(30)\ THz.$$

This number is in agreement with the frequencies derived from wavelength measurements in the spectra tables [11] and a recent, more accurate wavelength measurement [12]. The uncertainty in the ^{20}Ne, 1.15 μm frequency comes from the uncertainties in the ^{20}Ne, 1.52 μm frequency (25 MHz) and the determination of line center (15 MHz).

Previous frequency measurements up to 197 THz (1.52 μm) have utilized the tungsten-nickel, point-contact diode as the nonlinear element for synthesis and detection. Several unsuccessful attempts were made to use this diode in the measurement of the 260 THz frequency prior to the measurement described above. After this frequency had been successfully measured with the nonlinear crystals (i.e. the beat frequency was known) the 4.73, 1.52, and 1.15 μm radiations were again focused on the point-contact diode in another attempt to use this device. All rectified signals were of the order of 1 mV; the polarity of the 1.52 and 1.15 μm signals was opposite to that of the 4.73 μm signal. A search was made at the known beat frequency with diode impedances from a few hundred ohms to several thousand ohms. The results were unsuccessful.

Once the 260 THz ^{20}Ne line is measured it is in principle a simple matter to double this frequency and thereby synthesize a known frequency in the visible, 520 THz. In a joint experiment with the National Research Council in Ottawa, Canada ten hyperfine transitions in $^{127}I_2$ near 520 THz were measured by comparison with the known frequency of the Lamb-dip stabilized pure ^{20}Ne laser at 260 THz. The yellow-green light at 520 THz, generated (in the NRC laser) [13] by intracavity doubling in lithium niobate of 260 THz radiation from a He-Ne discharge, was servo-locked to individual hyperfine components of $^{127}I_2$ observed in saturated absorption, and their frequencies were determined simply by measurement of the beat frequencies of the two radiations at 260 THz.

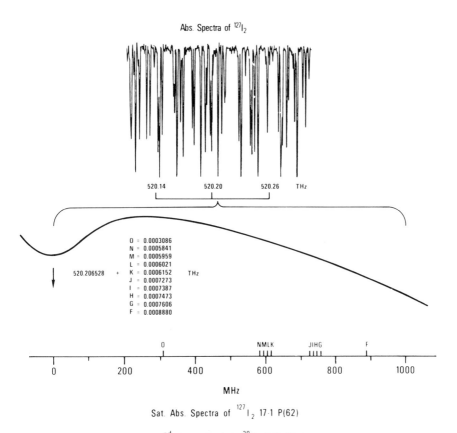

Abs. Spectra of $^{127}I_2$

520.14   520.20   520.26   THz

O = 0.0003086
N = 0.0005841
M = 0.0005959
L = 0.0006021
520.206528   +   K = 0.0006152   THz
J = 0.0007273
I = 0.0007387
H = 0.0007473
G = 0.0007606
F = 0.0008880

0   NMLK   JIHG   F

0   200   400   600   800   1000

MHz

Sat. Abs. Spectra of $^{127}I_2$ 17-1 P(62)

at the 2nd Harmonic of the ^{20}Ne 260 THz Laser

Fig. 4 Results of the $^{127}I_2$ hyperfine frequency measurement at 520 Hz.
Absorption spectra of $^{127}I_2$ from the work of GERSTENKORN et al. [16]

There is a substantial overlap of the doubled radiation and the strong
P(62) line in the 17-1 band of $^{127}I_2$ (See Fig. 4). Fifteen hyperfine
components (labeled a to o in order of decreasing frequency) are expected
in this line, and ten of these (f to o) were observed in saturated absorption within the laser tuning range. The lowest frequency component, o, is
well separated from the others and provides by far the best signal-to-noise
ratio because the background absorption from the other iodine components is
relatively small and because it occurs near the peak of the laser output
power. The components had a full width at half height of 2 MHz for an
iodine pressure of 4 Pa. With 2 MHz frequency modulation (at 1.8 kHz), the
laser could be servo-locked to the zero-crossing of the amplitude modulation at 5.4 kHz that occurs at the center of each component (third harmonic
locking). The infrared laser was locked in turn to each hyperfine component for the frequency measurements described below, and thus was at half
the frequency of the hyperfine line.

The frequency measurements were done simply by combining the 260 THz
beams from the two lasers on a high speed photodiode. The beat frequency

43

was displayed on a spectrum analyzer and measured with an adjustable marker oscillator and counter. The frequency, f_α, of the α-component is given by

$$f_\alpha = 2 [f_{Ne} + (f_\alpha/2 - f_{Ne})].$$

Six determinations of the frequency difference between the o-component and the Lamb-dip were made with a readjustment of the mirrors of the NBS laser for symmetrical Lamb-dip between each one. The standard deviation of six such settings was about 0.1 MHz. Systematic errors due to asymmetry of the modulation envelope were estimated to be less than about 1% of the 45 MHz full width of the Lamb-dip. Other sources of error were significantly less than this. The mean value is

$$f_o/2 - f_{Ne} = 154.3 \text{ MHz},$$

and the estimated 1-σ error is 0.5 MHz. The frequency of the o-component of P(62), 17-1 band of $^{127}I_2$ is thus

$$f_o = 2 [f_{Ne} + (f_o Y/2 - f_{Ne})] = 520\ 206\ 837 \pm 60 \text{ MHz}.$$

A preliminary measurement of the wavelength of this component gave $\lambda = 576\ 294\ 758 \pm 6$ fm [14], from which we calculate $f_o = c/\lambda = 520\ 206\ 811 \pm 6$ MHz [15]. The agreement between the above values for the frequency is satisfactorily within the error limits. In addition, the laser was locked to each of the components (n to f) and the beat frequencies between $f_o/2$ and f_{Ne} were measured. The results are shown in Fig. 4. The uncertainty in each of these beat frequencies is also 0.5 MHz.

This extension of absolute frequency measurements to the visible paves the way for highly accurate measurements in this portion of the electromagnetic spectrum. The rather large error limit on f_α is due to the free running 197 THz He-Ne laser used in the measurement of the Lamb-dip stabilized ^{20}Ne, 1.15 μm laser. In view of the reproducibility of this Lamb-dip stabilized laser, an improved determination of its frequency, f_{Ne}, can be combined with the above value of $f_\alpha - 2f_{Ne}$ to decrease the uncertainty of these iodine frequencies by about two orders of magnitude.

The Frequency Chain to the Visible

Fig. 5 illustrates the entire chain of frequency measurements which link the frequency at 520 THz to the Cs frequency standard. Fourteen lasers and six klystrons were used in seven steps, each terminated by a laser actively stabilized to a Doppler-free absorption line when possible. Unfortunately, such a stabilization technique does not exist for the 2.03 μm Xe and 1.52 μm Ne lasers. Consequently, the principal uncertainty in the 520 THz measurement results from the uncertainties in these two frequencies. With this chain, a significant improvement can therefore be realized by connecting the 88 THz and 260 THz lasers in a single step. This measurement can be done in a straightforward way by simultaneously observing beats between the 88 and 148 THz lasers, the 148 and 197 THz lasers, and the 197 and 260 THz lasers. Unfortunately, this measurement would require eleven lasers and one klystron in a major experimental effort.

Therefore, the chain shown in Fig. 6 is proposed as an alternative technique for connecting the $^{127}I_2$ stabilized 520 laser to the Cs frequency standard. The multiplication factor of 48020 is accomplished with six lasers and five klystrons, a significant reduction from the previous chain.

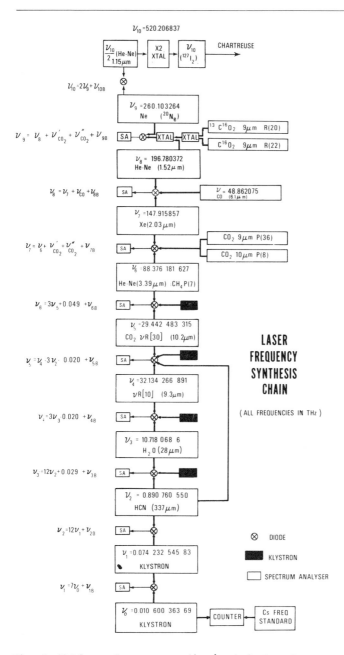

$\nu_{10} = 520.206837$

$\nu_{10} = 2\nu_9 + \nu_{10B}$

$\nu_9 = \nu_8 + \nu'_{CO_2} + \nu''_{CO_2} + \nu_{9B}$

$\nu_8 = \nu_7 + \nu_{CO} + \nu_{8B}$

$\nu_7 = \nu_6 + \nu'_{CO_2} + \nu''_{CO_2} + \nu_{7B}$

$\nu_6 = 3\nu_5 + 0.049 + \nu_{6B}$

$\nu_5 = \nu_4 - 3\nu_2 - 0.020 + \nu_{5B}$

$\nu_4 = 3\nu_3 - 0.020 + \nu_{4B}$

$\nu_3 = 12\nu_2 + 0.029 + \nu_{3B}$

$\nu_2 = 12\nu_1 + \nu_{2B}$

$\nu_1 = 7\nu_0 + \nu_{1B}$

CHARTREUSE

$\frac{\nu_{10}}{2}$ (He-Ne) 1.15 μm — X2 XTAL — ν_{10} ($^{127}I_2$)

$\nu_9 = 260.103264$ Ne (^{20}Ne)

$^{13}C^{16}O_2$ 9 μm R(20)
$C^{16}O_2$ 9 μm R(22)

$\nu_8 = 196.780372$ He-Ne (1.52 μm)

$\nu = 48.862075$ CO (6.1 μm)

$\nu_7 = 147.915857$ Xe(2.03 μm)

CO$_2$ 9 μm P(36)
CO$_2$ 10 μm P(8)

$\nu_6 = 88.376\ 181\ 627$ He-Ne(3.39 μm) : CH$_4$ P(7)

$\nu_5 = 29.442\ 483\ 315$ CO$_2$ νR[30] (10.2 μm)

LASER FREQUENCY SYNTHESIS CHAIN

(ALL FREQUENCIES IN THz)

$\nu_4 = 32.134\ 266\ 891$ νR[10] (9.3 μm)

$\nu_3 = 10.718\ 068\ 6$ H$_2$O (28 μm)

$\nu_2 = 0.890\ 760\ 550$ HCN (337 μm)

$\otimes$ DIODE
KLYSTRON
SPECTRUM ANALYSER

$\nu_1 = 0.074\ 232\ 545\ 83$ KLYSTRON

$\nu_0 = 0.010\ 600\ 363\ 69$ KLYSTRON — COUNTER — Cs FREQ STANDARD

Fig. 5 Old laser frequency synthesis chain from X-band to visible.

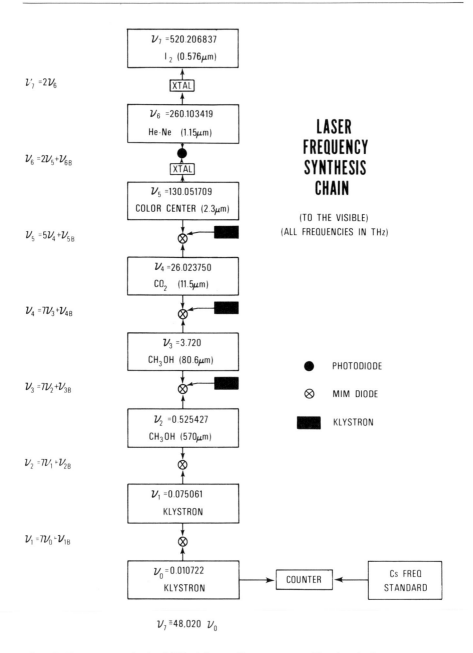

Fig. 6 New proposed simplified laser frequency synthesis chain.

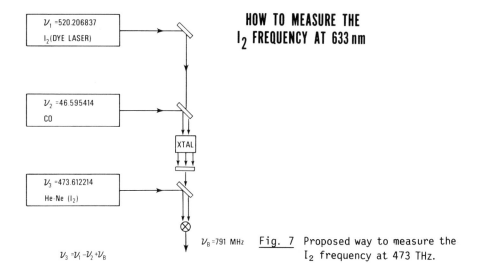

HOW TO MEASURE THE
I$_2$ FREQUENCY AT 633 nm

ν_1 =520.206837
I$_2$(DYE LASER)

ν_2 =46.595414
CO

XTAL

ν_3 =473.612214
He-Ne (I$_2$)

ν_B =791 MHz

$\nu_3 = \nu_1 - \nu_2 + \nu_B$

Fig. 7 Proposed way to measure the I$_2$ frequency at 473 THz.

All lasers in the chain exist, and each multiplication step has been demonstrated to be experimentally feasible with the nonlinear devices indicated in the diagram. This chain should have a measurement capability with a fractional frequency uncertainty between 10^{-10} and 10^{-11} for the I$_2$ transition at 520 THz.

This stabilized frequency at 520 THz appears to be an ideal reference from which to synthesize other standard frequencies in the visible spectrum. It is anticipated that these frequencies will again be synthesized by mixing the radiation from the 520 THz laser with an appropriate infrared laser in a nonlinear crystal such as AgGaS$_2$. This crystal has been produced with a wide, low absorption, transmission band (α<0.5 cm^{-1} for 0.5<λ <10 μm) [16] and has been used to up-convert 10.6 μm radiation into the green spectral range [17]. As an example of this method, Fig. 7 shows a possible way to measure the frequency of the I$_2$ stabilized 0.633 μm Ne laser.

Summary

In summary, we have demonstrated the technique of visible frequency measurements that are directly related to the CS frequency standard. In the future we would hope to use this technique to generate a number of standard frequencies in the visible.

References

1. L. O. Hocker, A. Javan, and D. Ramachandra Rao, Appl. Phys. Lett. 10, 147 (1967).
2. K. M. Evenson, J. S. Wells, L. M. Matarrese, and L. B. Elwell, Appl. Phys. Lett. 16, 159 (1970).
3. K. M. Evenson, J. S. Wells, F. R. Petersen, B. L. Danielson, and G. W. Day, Appl. Phys. Lett. 22, 192 (1973).

47

4. D. A. Jennings, F. R. Petersen, and K. M. Evenson, Appl. Phys. Lett. $\underline{26}$, 510 (1975).

5. K. M. Evenson, D. A. Jennings, F. R. Petersen, and J. S. Wells, Laser Spectroscopy III, J. L. Hall and J. L. Carlsten, eds. (Springer-Verlag, Berlin 1977), pp. 56-68.

6. J. H. Hall, IEEE, J. Quantum Electron., $\underline{QE4}$, 638 (1968).

7. K. M. Evenson, J. S. Wells, F. R. Petersen, B. L. Danielson, G. W. Day, R. L. Barger, and J. L. Hall, Phys. Rev. Lett. $\underline{29}$, 1346 (1972).

8. F. R. Petersen, D. G. McDonald, J. D. Cupp, and B. L. Danielson, "Laser Spectroscopy", edited by R. G. Brewer and A. Mooradian (Plenum, New York, 1974), pp. 171-191.

9. Charles Freed, A. H. M. Ross, and Robert G. O'Donnell, J. Mol. Spectrosc. $\underline{49}$, 439 (1974). Quoted error is a result of recent measurements with stabilized lasers (saturated absorption). Private communication with Charles Freed.

10. The frequency of the 1.52 µm He-Ne laser was remeasured and the corrected value is listed. In the previous measurement (K. M. Evenson, D. A. Jennings, F. R. Petersen, and J. S. Wells, in Laser Spectroscopy III, J. S. Hall and J. L. Carlsten, eds. (Springer-Verlag, Berlin 1977), pp. 56-68.), a higher frequency mode of the 1.52 µm laser was not detected, the CO frequencies were not directly measured, and the 1.5 µm laser discharge tube was filled to a total pressure of 324 Pa (with He:Ne=10:1) instead of the 648 Pa used in the present experiment. The new measurements, made with the $^{12}C^{16}O$, $P_{18}(14)$ line only, revealed the following corrections: (1) Mode$_v$ +71 MHz; (2) directly measured frequency of $^{12}C^{16}O$, $P_{18}(14)$ [$\nu^{12}C^{16}O$, $P_{18}(14)$ = 48.862 075(3) THz] + 11 MHz; and (3) pressure shift [+65(9)kHz/Pa] + 21 MHz. The resultant frequency is $\nu^{20}Ne$, 1.5 µm = 196.780 372(25) THz.

11. Charlotte E. Moore, "Atomic Energy Levels", Vol. I, NSRDS-NBS35, National Bureau of Standards, 1971 (U.S. GPO, Washington, D.C. 20402).

12. The wavelength of 1.15 µm is from J. L. Hall and S. A. Lee, private communication, using the techniques of J. L. Hall and S. A. Lee, Appl. Phys. Lett. $\underline{29}$, 367 (1976).

13. G. R. Hanes, N.R.C. Ottawa, Canada, paper in preparation.

14. The authors wish to thank Dr. K. H. Hart of NRC, Ottawa, for this wavelength value.

15. Using the value c = 299 792 458 M/s recommended in BIPM, Comptes Rendus des Séances de la Conf. Gen. des Poids et Mesures, 15th, p. 103, 1975.

16. S. Gerstenkorn and P. Luc, "Atlas du Spectre d'Absorption de la Molécule d'Iode," Centre National de la Recherche Scientifique, Paris, 1978.

17. H. Matthes, R. Viehman, and N. Marschall, Appl. Phys. Lett. $\underline{26}$, 237 (1975).

18. Tien-Lac Hwang and S. E. Schwarz, Appl. Phys. Lett. $\underline{31}$, 99 (1977).

Transverse Resonance-Radiation Pressure on Atomic Beams and the Influence of Fluctuations

J.E. Bjorkholm, R.R. Freeman, A. Ashkin, and D.B. Pearson

Bell Telephone Laboratories, Holmdel, NJ 07733, USA

The development of tunable laser sources has led to a resurgence of interest in various effects caused by resonance-radiation pressure on atoms and ions. Our immediate interest in this subject was initiated by the independent proposals of LETOKHOV [1] and of ASHKIN [2] for using radiation pressure to construct optical traps for neutral atoms. The intent of both proposals is to confine and cool neutral atoms in these optical traps. Recently it was demonstrated that ions stored in electromagnetic traps can be cooled using the radiation pressure of nearly-resonant light [3]. We are currently investigating the transverse dipole force of resonance-radiation pressure; an understanding of this type of force is crucial to the construction of optical traps.

Recently we made the first direct observation of the effects of the optical dipole force on free atoms [4]. Those experiments demonstrated focusing and defocusing forces exerted on a beam of neutral sodium atoms by the transverse radiation pressure associated with a superimposed and copropaga- ting light beam from a cw dye laser. The forces and the resulting effects were made large by tuning the frequency of the laser to be near the sodium resonance transition. The work presented here was aimed at determining the degree to which the atomic beam could be focused using this basic technique. The atomic beam focal spot size is, in some instances, determined by input characteristics of the atomic beam such as longitudinal velocity distribu- tion and initial divergence. However, an important role can also be played by fluctuations of the radiation-pressure forces which occur because of the quantized nature of light. These fluctuations serve to heat the atoms; this heating can also limit the atomic beam focal-spot sizes that can be obtained. Fluctuations of the radiation pressure can also be a limiting factor in the design of optical traps [5,6]. In general, fluctuations can severely limit the applicability of resonance-radiation pressure if they are not properly compensated for.

A diagram of the basic situation we consider is shown in Fig.1. An atom having a velocity v along the z-axis (the longitudinal direction) is illuminated by a Gaussian laser beam of frequency ν, also propagating along the z-axis. The transverse intensity distribution of the light is

$$I(r) = I_0 e^{-2r^2/w^2}$$

where w is the spot size of the laser beam. The forces of resonance- radiation pressure can be classified into two types and, for a weakly

49

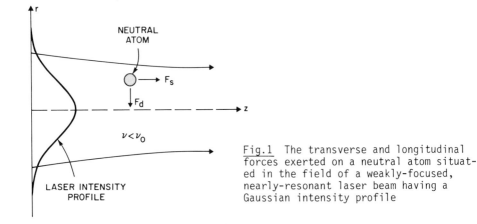

Fig.1 The transverse and longitudinal forces exerted on a neutral atom situated in the field of a weakly-focused, nearly-resonant laser beam having a Gaussian intensity profile

focused laser beam as being considered here, they are mainly transverse and longitudinal in character, as shown in Fig.1.

The focusing effects arise primarily from the transverse dipole resonance-radiation pressure [7,2]. This force arises from stimulated scattering of the light by the atoms and it is the same as the force exerted on an induced dipole situated in an electric-field gradient. Thus it is proportional to, and in the direction of, the optical intensity gradient. The dipole force is written as

$$\vec{F}_d = -\frac{1}{2} \alpha \vec{\nabla} |E|^2, \qquad (1)$$

where $\vec{E}$ is the optical electric field and α is the atomic polarizability which, for a simple two-level atom, is

$$\alpha = \frac{\lambda^3 \Delta\nu_n}{32\pi^3} \frac{\Delta\nu}{\Delta\nu^2 + \Delta\nu_n^2/4} \frac{1}{1 + p} . \qquad (2)$$

In (2), $\Delta\nu = (\nu' - \nu_0)$ where $\nu' = (1-v/c)c/\lambda$ is the Doppler-shifted frequency of the optical field, ν_0 and $\Delta\nu_n$ are the frequency and natural linewidth of the resonance transition of the atom, and p is a saturation parameter. Explicitly, p is given by

$$p = \frac{I}{I_s} \frac{\Delta\nu_n^2/4}{\Delta\nu^2 + \Delta\nu_n^2/4} \qquad (3)$$

where $I = |E|^2 c/8\pi$ is the optical intensity and $I_s = 2\pi^2 \Delta\nu_n h\nu/\lambda^2$ is the saturation intensity for the two-level atom. In the limit $p \gg 1$, the dipole force reduces to $h\Delta\nu \vec{\nabla}I/I$; this shows that, with sufficient laser power to insure that p remains large, the dipole force can be arbitrarily increased by increasing $|\Delta\nu|$. For a weakly focused TEM_{00} (Gaussian) mode laser beam the force is essentially transverse to the direction of propagation of the light. When the light is tuned below the atomic resonance frequency, the atoms are attracted to the high intensity regions of the light beam, while for tunings above resonance the atoms tend to be expelled from the light.

Spontaneous resonance-radiation pressure arises from spontaneous scattering of the light by the atoms and it exists even in uniform plane-waves [8]. The average spontaneous force is in the direction of the light propagation and its magnitude is given by

$$F_s = \frac{h}{2\lambda\tau} \quad \frac{p}{1+p} \, , \tag{4}$$

where $\tau = 1/2\pi\Delta\nu_n$ is the natural lifetime of the excited state of the atom. The average spontaneous force is simply the rate at which photon momentum is absorbed by the atom; for $p \gg 1$ it saturates to a maximum value of $h/2\lambda\tau$. Thus in many situations the dipole force can greatly exceed the spontaneous force. The effects of the spontaneous force have been observed in numerous experiments [9]. The role of spontaneous scattering in our experiment is twofold. First, the average spontaneous force accelerates the atoms in the direction of light propagation; this acceleration is rather unimportant to an understanding of our results. Secondly, spontaneous scattering occurs in a discrete and random fashion. The recoil experienced by an atom in a single scattering event can be in any direction; only the average recoil is in the direction of the light propagation. This random scattering serves to "heat" the atoms in the transverse and longitudinal directions, as recently discussed elsewhere [5,6,10]. We will present a discussion indicating that, under some conditions, it is transverse heating due to spontaneous scattering which determines the degree to which an atomic beam can be focused using resonance-radiation pressure. Spontaneous decay of excited atoms can also lead to fluctuations in the dipole force and consequently to another source of heating for the atoms. For our experimental conditions, however, the heating due to the random nature of the spontaneous force was dominant [11]. It should be emphasized that heating caused by fluctuations of the radiation pressure forces is to be distinguished from the heating or cooling due to the velocity dependence of the average spontaneous force resulting from Doppler shifts of moving atoms [12,3].

A schematic diagram of our experimental setup is shown in Fig.2. Light from a continuously-tunable, single-mode cw dye laser was superimposed upon an effusive beam of neutral sodium atoms ($T \sim 500°C$) by reflection off a 3 mm thick dielectric-coated mirror having a 230 μ diameter hole in it. Typical laser power superimposed on the atoms was 40 mw and the light was linearly polarized. A 75 cm focal-length lens focused the light to a spot size $w_0 = 75 \mu$ at a nominal distance of 25 cm from the mirror. A moveable hot-wire detector was used to measure the atomic beam profile in the focus of the laser beam. The detector was insensitive to light incident upon it and its resolution was determined by an aperture of 50 μ diameter mounted in front of the hot iridium wire. Strong focusing or defocusing effects were obtained when the laser was tuned to within several GHz of the sodium D_2 resonance transition at 5890A° for which $\Delta\nu_n = 10$ MHz. The corresponding value for I_s is about 19 mW/cm^2. As a cautionary note for the application of the various formulas given here, it must be realized that under most situations the sodium atom is not a good approximation to the idealized two-level atom.

51

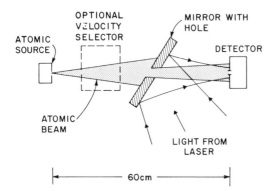

Fig.2 Schematic diagram of the experimental setup. The transverse scale is greatly magnified relative to the longitudinal scale; for instance, the diameter of the hole in the mirror is 250μ

Examples of the atomic beam profiles we measured under different conditions are shown in Fig.3. Curve (a) shows the beam profile in the absence of light; the peak beam current is normalized to 1. For profile (b) the laser is tuned about 1 GHz below resonance for atoms at the peak of the velocity distribution ($\sim$9 x 10^4 cm/sec); the on-axis beam intensity is increased to a normalized value of 5.8 and the diameter of the focused atomic beam (FWHM) is 68 μ, a factor a 6 reduction. Clearly the actual atomic beam diameter is smaller and the peak intensity higher as the experimental result is a convolution of the actual profile with the 50 μ aperture on the detector. Profile (c) shows the results obtained when the laser was tuned about 1 GHz above resonance. In this case the atoms are strongly expelled from the light. In principle the beam profiles are cylindrically symmetric; it isn't immediately obvious that the areas under the three curves represent the same total number of atoms in the beam. Numerical integration verifies that the numbers are in fact the same to within 10 percent. The symmetry of the beam profiles we measured with the light applied was crucially dependent upon precise overlap and alignment of the laser and atomic beams. Effects due to the input laser beam not being perfectly Gaussian and due to distortion of the laser mode caused by the hole in the mirror also may cause asymmetries in the atomic beam profiles.

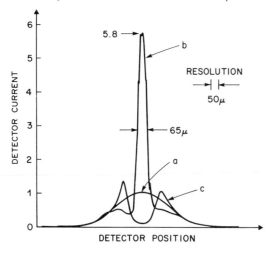

Fig.3 Atomic beam current measured by detector as a function of its transverse position; a, no light; b, laser tuned $\sim$1GHz below resonance for focusing; c, laser tuned $\sim$1GHz above resonance for defocusing. Peak beam current is approximately 5 x 10^8 atoms/sec

A simple calculation carried out for a two-level atom illustrates that heating caused by the fluctuations of spontaneous scattering can limit the spot size of the focused atomic beam. The transverse dipole force experienced by the atoms can be obtained from a transverse potential energy given by [2]

$$U(r) = \frac{1}{2} h \Delta\nu \, \ln[1+p(r)] \qquad (5)$$

where $p(r) = p_0 \exp(-2r^2/w^2)$ and p_0 in the on-axis value of the saturation parameter. For our experimental parameters the well depth in the focus of the laser beam is 1.6×10^{-18} ergs, meaning that an atom with a maximum transverse velocity of 280 cm/sec can be confined in the well. The divergence of our atomic beam is about 3×10^{-4} rad, which corresponds to a maximum initial transverse energy for an atom emerging from the hole in the mirror of about 1.4×10^{-20} ergs. Assuming no change of transverse energy as the atoms propagate along the light, confinement in the transverse potential well implies that the spot size of the atomic beam in the focus of the light would be about 11 μ in diameter, much smaller than we observe.

Now consider the additional transverse energy acquired by the atoms due to the heating caused by spontaneous scattering. Assuming that the spontaneous scattering is isotropic (for simplicity), a random walk analysis indicates that, for an atom having zero initial transverse velocity, the probability density that its transverse velocity is v_t after N scattering events is

$$P_N(v_t) = \frac{m\lambda}{h} \left[\frac{3}{2\pi N} \right]^{1/2} \exp \left[-v_t^2 \frac{3m^2\lambda^2}{2h^2 N} \right] .$$

The corresponding rms transverse velocity is

$$v_t^{rms} (N) = \frac{h}{m\lambda}(N/3)^{1/2} \sim 1.7(N)^{1/2} \text{ cm/sec.}$$

For our experimental parameters and for the most probable atomic velocity, we estimate that an atom undergoes $N \sim 1200$ scattering events in travelling along the laser beam. At the focus, then, we have $v_t^{rms} \sim 60$ cm/sec which corresponds to a transverse energy of 6.6×10^{-20} ergs. This is about a factor of 5 more than the initial transverse energy of the beam and it corresponds to an atomic beam focal spot size of about 25 μ. This simple analysis indicates that the effects of heating by spontaneous scattering are significant.

To more rigorously verify the importance of fluctuations we used a computer to numerically calculate atomic trajectories and the related beam profiles; a detector resolution of 50 μ was included in the computation. In the absence of light the atomic trajectories were taken to be straight lines emanating from a point source; the various rays were weighted to closely approximate the actual beam profile. The average dipole and spontaneous forces were exactly accounted for. Figures 4a and 4b show the beam profiles calculated for a laser power of 40 mW and $\Delta\nu = -1$ GHz. For Figure 4a the fluctuations in the spontaneous force were not included and the calculated

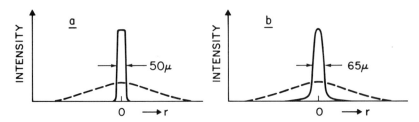

Fig.4 Computed atomic beam profiles for 40 mW of laser power and $\Delta\nu$ = -1GHz; a, spontaneous heating not included; b, spontaneous heating included. The dashed curves show the beam profile in the absence of light. The vertical scales for all curves are arbitrary and comparisons between the various amplitudes should not be made.

atomic beam diameter is clearly much smaller than the detector resolution, in disagreement with our measurements. For Figure 4b the fluctuations of the spontaneous force were modeled using a random number generator and by keeping track of time and the rate of spontaneous scattering. The beam profile shown is the result of averaging over 40 such calculations. Clearly the atomic beam spot size has been greatly increased by the inclusion of spontaneous scattering; the computed beam profile is similar to the experimental profile b in Fig.3. This computation demonstrates that the random nature of spontaneous scattering can limit the degree of atomic beam focusing.

Assuming spontaneous scattering to be an important limitation, smaller spot sizes could be obtained if it were possible to reduce N while keeping the dipole focusing forces constant. With increased laser power this can, in fact, be done. For instance, in the limit p>>1, we have $F_d \propto$ power/$\Delta\nu$ and $N \propto$ power/$\Delta\nu^2$. Consequently, if the laser power and $\Delta\nu$ are increased by the same factor m, the dipole forces remain approximately constant while N is reduced by the factor 1/m. Consider 1W of laser power and $\Delta\nu$ = -25GHz; for this case N∿100 and the additional transverse energy imparted by spontaneous heating is less than the initial transverse energy in the beam. The resulting atomic beam spot diameter indicated by the simple model is about 11 μ (as compared with the approximately 50 μ presently achieved); further improvements could be made by decreasing the initial atomic beam divergence. A computer beam profile for this case verifies the conclusions of this simple analysis. Thus it appears that atomic beam focal spot sizes much smaller than obtained in our present experiments should be readily attainable; we hope to verify this shortly using the increased power available from ring-type cw dye lasers.

In conclusion, we have experimentally demonstrated that a beam of neutral sodium atoms can be focused to a spot diameter of approximately 50 μ using the transverse dipole resonance-radiation pressure exerted by a 40 mW laser beam. Simple analysis shows that in some cases the spot sizes are limited by the random fluctuations of the spontaneous radiation pressure; with 1W of laser power, spot sizes less than 10 μ should be attainable. Our discussion makes it clear that the effects of heating by spontaneous scattering can have important detrimental effects in other applications of resonance-radiation pressure on atoms, such as the slowing or guiding of atoms. As discussed recently [6], consideration of heating effects is of paramount importance in the design of optical traps for neutral atoms.

Acknowledgement

The authors are indebted to J. P. Gordon for many useful discussions.

References

1. V. S. Letokhov, V. G. Minogin, and B. D. Pavlik, Opt. Commun. 19, 72 (1976).

2. A. Ashkin, Phys. Rev. Lett. 40, 729 (1978).

3. D. J. Wineland, R. E. Drullinger, and F. L. Walls, Phys. Rev. Lett. 40, 1639 (1978); W. Neuhauser, M. Hohenstatt, P. E. Toschek, and H. Dehmelt, Phys. Rev. Lett. 41, 233 (1978).

4. J. E. Bjorkholm, R. R. Freeman, A. Ashkin, and D. B. Pearson, Phys. Rev. Lett. 41, 1361 (1978).

5. V. S. Letokhov, V. G. Minogin, and B. D. Pavlik, Zh. Eksp. Teor. Fiz. 72, 1328 (1977) [Sov. Phys. JETP 45, 698 (1977)].

6. A. Ashkin and J. P. Gordon, Opt. Lett. 4, 161 (1979).

7. G. A. Askar'yan, Zh. Eksp. Teor. Fiz. 42, 1567 (1962) [Sov. Phys. - JETP 15, 1088 (1962)].

8. A. Ashkin, Phys. Rev. Lett. 25, 1321 (1970).

9. See, for example; O. R. Frisch, Z. Phys. 86, 42 (1933); R. Schieder, H. Walther, and L. Woste, Opt. Commun. 5, 337 (1972); J. L. Picque and J. L. Vialle, Opt. Commun. 5, 402 (1972); A. F. Bernhardt, D. E. Duerre, J. R. Simpson, and L. L. Wood, Appl. Phys. Lett. 25, 617 (1974); J. E. Bjorkholm, A. Ashkin, and D. B. Pearson, Appl. Phys. Lett. 27, 534 (1975).

10. J. L. Picque, Phys. Rev. A 19, 1622 (1979).

11. J. P. Gordon, private communication.

12. T. W. Hansch and A. L. Schawlow, Opt. Commun. 13, 68 (1975).

Deflection of a Na Beam by Resonant Standing Wave Radiation

E. Arimondo, H. Lew, and *T. Oka*

Herzberg Institute of Astrophysics, National Research Council of Canada
Ottawa, Ontario, K1A OR6, Canada

1. Introduction

In 1933 KAPITZA and DIRAC published a paper titled "The Reflection of Electrons from Standing Light Waves" in which they proposed a new experiment [1]. Fig. 1 shows a picture from their paper.

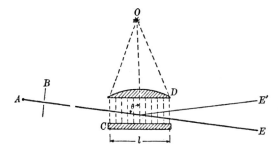

Fig.1 A picture from the paper of Kapitza and Dirac [1]

An electron beam AE is reflected by standing waves between a lens D and a mirror C. This is the opposite to a normal optical diffraction experiment or X-ray diffraction in that matter is reflected by radiation.

The force involved in the Kapitza-Dirac effect is very small because it is caused by the Compton scattering which is a second order process. (See [2] for a summary of experiments). It is obvious that if we replace the electron beam with an atomic beam and use resonant standing wave the interaction is of first order and a much more efficient deflection of beam can be observed in spite of the larger mass of the atoms. Here we report such an experiment.

Because of the dual nature of both atoms and radiation, the experiment can be looked at in various ways. If everything goes ideally we can look at it as a diffraction of a plane matter wave by a grating of electromagnetic field. Or we can look at the experiment as deflection of particles by photon momentum. We wish to use this latter photon momentum picture (and come to the former later in Section 7) as was also done by Kapitza and Dirac when they estimated the magnitude of reflection.

2. Photon Momentum Picture

First let us recall the Na deflection experiments by a travelling wave [3]. In those experiments an unfocussed resonant travelling wave ω_+ is applied to the atomic beam. The absorption of the radiation $\omega_+\uparrow$ pushes an atom to the +x direction while a spontaneous emission pushes a molecule in a random fashion. By alternately repeating an absorption $\omega_+\uparrow$ and a spontaneous emission, the average position of an atom is pushed to the direction of +x. Here the induced emission $\omega_+\downarrow$ does not help the deflection because it pushes the atom back to the -x direction. Therefore we have to wait for the time of spontaneous emission which for Na is 1.63×10^{-8} sec.

The situation is different if we use standing wave radiation which we can look at as being composed of two travelling waves ω_+ and ω_-. Here an emission induced by ω_- pushes an atom in the same direction as the absorption by ω_+ does. The stimulated processes $\omega_+\uparrow$ and $\omega_-\downarrow$ push an atom to the +x direction and $\omega_+\downarrow$ and $\omega_-\uparrow$ to the -x direction. Therefore the magnitude of the deflection depends on how these four stimulated processes are combined.

Three typical cases are considered,

(a) The maximum deflection in the +x direction would occur if $\omega_+\uparrow$ and $\omega_-\downarrow$ are repeated and that in the -x direction if $\omega_+\downarrow$ and $\omega_-\uparrow$ are repeated, The total deflection of the beam would then be proportional to the number of stimulated processes $n = \Omega t$ where $\Omega = \mu E/h$ is the Rabi flopping frequency and t is the transit time. We will call this case "coherent pushing".

(b) The minimum deflection would occur if an atom keeps interacting only with one standing wave.

(c) If in the alternating absorption and emission processes it is random whether ω_+ or ω_- is involved, then the net deflection results from a random walk and has a Gaussian shape with a half width which is proportional to $\sqrt{n}$ We will call this case "random pushing".

The average momentum transfer p for the three cases is expressed as

$$p = n\cdot\Delta p \qquad \text{for coherent pushing}$$
$$p = \sqrt{n}\cdot\Delta p \qquad \text{for random pushing}$$
$$\text{and} \quad p = \tfrac{1}{2}\Delta p \qquad \text{for a travelling wave,}$$

where $\Delta p = h\nu/c = \hbar k$ is the unit photon momentum.

For a typical experimental arrangement (100 mWatts focussed to 0.1 mm radius and an atomic beam size of 0.1 mm radius), the Rabi frequency $\Omega \sim 1$ GHz and the transit time $t \sim 2\times10^{-7}$ sec so that the number of stimulated processes $n = \Omega t \sim 200$. Therefore there is an order of magnitude difference between the three cases.

The speed that a Na atom at rest acquires after absorbing one photon of the Na D line is ∿ 3 cm/sec. Thus after a coherent pushing a Na atom would acquire in the above estimate a speed of ∿ 6 m/sec, which is the speed of the world's best marathon runners.

Obviously we are novices in this field but we had not started from nothing. We have previously observed velocity-tuned three-photon processes [4] shown in Fig. 2.

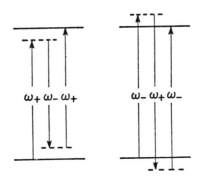

Fig.2 Velocity-tuned three-photon processes, an example of coherent pushing

Here, in order to match the resonant condition, a molecule is forced to interact alternately with ω_+ and ω_- and thus a coherent pushing is achieved. This process can occur in general for $2\ell+1$ photons. Our experiment was initiated out of curiosity on what will happen if ℓ is very large.

Note also the transverse Stern-Gerlach experiment done in the rf region [5]. The present experiment is an optical analogue of this experiment.

3. Experimental

The essential elements of our apparatus can be seen in Fig.3.

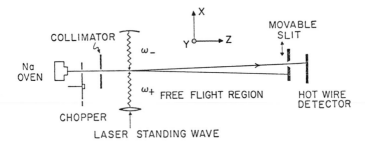

Fig.3 Experimental arrangement

The Na beam was produced by heating Na metal to 380°C in an iron oven with a hole of 0.1 mm. The collimator aperture was 30 cm from the oven and the opening was adjusted by micrometers attached to the four blades of the collimator such that the beam size at the ionizing wire was 0.25 mm vertically; horizontally the beam was two to three times wider. The slit in front of the ionization wire detector was 0.1 mm wide in the vertical direction and its vertical position was swept by a motor for scanning the beam profile. The detector was a hot wire ionizer of 0.125 mm tungsten grown from $W(CO)_6$. The Na ions were collected on an electron multiplier after a crude mass spectrometer separated them from the background potassium ions from the wire. The Na beam was chopped at 30 Hz with a toothed wheel driven by a synchronous motor. The output from the electron multiplier was processed with a lock-in amplifier and displayed on a recorder.

A stabilized Spectra Physics 580A dye laser provided single mode resonant radiation with a power of up to 100 mWatts. The tuning of the frequency was done by observing the Na D_2 line fluorescence from the atomic beam. The laser radiation was focussed on the atomic beam with a lens of 10 cm focal length at a position 5 cm from the collimator slit and reflected by a concave mirror. A crude adjustment of the laser beam to the atomic beam was done by seeing fluorescence from the atomic beam. The final adjustment was done in the following manner. The position of the atomic beam detector was fixed at the centre of the undeflected atomic beam and the laser beam position was adjusted so that a minimum atomic beam intensity was reached. This was based on the assumption that the deflection is maximum at the correct alignment.

The free flight distance after the interaction region was 75 cm.

4. Travelling Wave Experiment

First we repeated the travelling wave experiment done by the group of Dr. Walther [3] in order to check the performance of our apparatus. Very efficient deflection was observed, as shown in Fig.4.

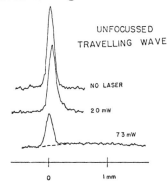

UNFOCUSSED
TRAVELLING WAVE

NO LASER

20 mW

73 mW

0   1 mm

Fig.4 Deflection of a Na beam by resonant travelling wave radiation

59

The half width at half maximum of the Na beam without laser
radiation was ∿ 0.13 mm. The asymmetry of the Na beam profile
due to radiation pressure is obvious for 20 mWatts of unfocussed
laser power. When 73 mWatts of laser power was applied, Na
atoms with F=2 (which are resonant to the radiation) were
completely deflected, leaving only Na atoms with F=1. The
latter give the undeflected profile with the intensity of 3/8 as
expected from statistics.

5. Standing Wave Experiment

For the standing wave experiment the laser beam was focussed on
the atomic beam and reflected by a concave mirror. A typical
result is given in Fig.5.

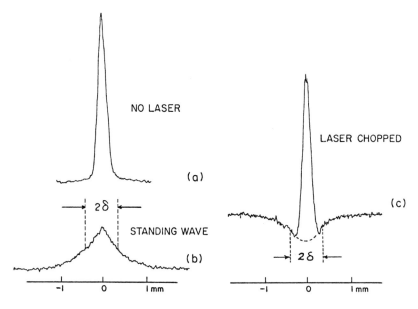

Fig.5   Na beam intensity profiles in the standing
wave experiment

(a) Undeflected beam (hwhm ∿ 0.13 mm)

(b) Symmetric deflection of the Na beam
    due to resonant standing wave
    radiation. The Rabi frequency was
    ∿ 2.1 GHz

(c) Laser-chopped signal which gives a
    difference between (a) and (b)

In these experiments the laser frequency was fixed at the
maximum of the Na D_2 line and the slit in front of the detector
wire was swept. Because of the large Rabi frequency of ∿ 2 GHz,
the setting of the laser frequency was not critical. Fig. 5(a)

60

shows a trace of the Na beam without laser radiation. Fig.5(b)
shows the symmetric deflection of the beam due to the standing
wave. The large reduction of the peak and the broadening is
clearly seen. When the reflecting mirror is blocked so that
the atomic beam interacts with a focussed travelling wave, a
trace similar to Fig.5(a) with a very small shift was obtained
but broadening was not noticeable.

The results in Figs. 5(a) and 5(b) are the "raw" data in which
contributions from the portion of the Na beam which was not
affected by the laser beam are also included. In order to
exclude these, the laser radiation was chopped (at 30 Hz) and the
signal was processed in a lock-in amplifier. This is equivalent
to taking a difference between the traces (a) and (b) in Fig.5.
The result is shown in Fig.5(c). The deflection 2δ of the beam
was measured from Fig.5(c) by fitting a Gaussian curve to the
lobes of the trace as shown by the dotted line.

In the other set of experiments, the movable slit in front of
the detector was set at the maximum of the undeflected Na beam
and the frequency of the dye laser was swept. The result is
shown in Fig.6.

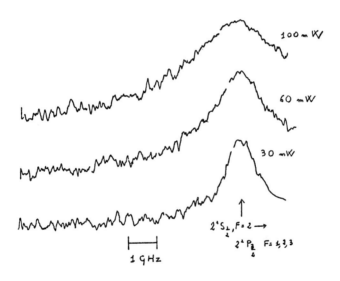

Fig.6 Spectroscopy using Na beam deflection

These results were used to determine the Rabi frequency.

The Na beam deflection 2δ in Fig.5 has been measured as a
function of dye laser power from 5 to 100 mWatts. The results
are plotted in Fig.7.

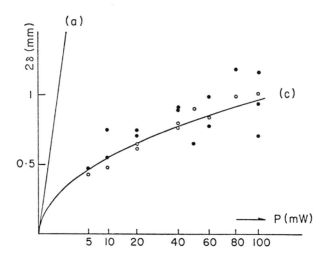

Fig.7 Dependence of the deflection 2δ (fwhm) on the
radiation power. Black circles represent "raw"
values without correction. White circles are
corrected values. The straight line (a)
represents calculated values for coherent
pushing whereas the curve (c) represents random
pushing.

The black circles represent values which were directly obtained
from the measurements. The scatter is mainly due to unavoidable
slight misalignment of the laser radiation with respect to the
atomic beam for different runs. If we normalize the set of
measurements in a given run by adjusting one value of the set to
the theoretical curve, we obtain values represented by the white
circles in Fig.7. The theoretical curves for the coherent pushing (a) and random pushing (c) are also shown in Fig.7. In
drawing the latter the laser electric field was estimated from
the power broadening shown in Fig.6. It is seen that both the
absolute value of the deflection and the power dependence agree
with the model of random pushing.

6. Hyperfine Structure

One complication arises due to the fact that the Na atom has two
hyperfine levels with F=2 and 1 in the ground state and the
optical transitions starting from these levels are separated by
$\nu_1 - \nu_2 \sim 1.77$ GHz. Therefore if we set the laser frequency to the
transition from the F=2 level, the other transition from the F=1
level is off-resonant. The deflection of the Na atom in the F=1
level is then estimated from $n = (\mu_1 E/h)^2 [(\nu_1 - \nu_2)^2 + (\mu_1 E/h)]^{-1/2} t$
rather than from $(\mu_2 E/h)t$. For small laser power such that
$\mu_1 E/h \ll |\nu_1 - \nu_2|$, the F=1 atoms are not deflected and produce a
sharp peak with an intensity of $\sim 3/8$ of the original beam at
the centre of profile. This residual component does not affect

our measurement of 2δ when the laser chopped picture of Fig.5(c) is used.

For higher powers the deflection of the F=1 atoms becomes significant and causes the lobes of the profile in Fig.5(c) to be less Gaussian (i.e. sharper). This portion was excluded in determining 2δ. For very high power (e.g. 100 mWatts) the F=1 Na atoms are also efficiently deflected.

Further complications due to possible optical pumping between the F=2 and F=1 levels have not been considered here.

7. Theory

The theory for the transfer of momentum to atoms by a resonant standing wave has been actively studied recently by KAZANTSEV et al. [6], STENHOLM et al. [7], and Cook and Bernhardt [8,9]. Instead of the intuitive photon momentum picture so far used in this paper all the previous theoretical papers use a classical treatment for the interaction between the standing wave field and the atomic beam.

To the extent we understand it, Kazantsev's theory is based on the fact that an atom in resonant radiation field ε has a potential energy of $-\mu\varepsilon$. In a standing wave $\varepsilon = E\sin kx$, therefore the momentum transfer is

$$\Delta p_x = Ft = -t\partial U/\partial x = \mu Ekt\cos kx .$$

Using $k = \omega/c$, we see that for x=0 the coherent pushing is realized. Such calculations explain the result of the transverse Stern-Gerlach experiment [5] but differs both in orders of magnitude and the field dependence from what we observed. One problem is that the dimension of the atomic beam ($\sim$ 0.1 mm) is much larger than the wavelength λ. If the atoms hit the nodes and abdomens of the standing waves randomly then the momentum variation is better estimated by

$$(\Delta p)^2/2m = \mu E .$$

This gives the field dependence observed by us but not the magnitude of deflection. Also here the momentum transfer is independent of the transit time t.

Cook and Bernhardt [8] have solved equations of motion

$$\dot{C}_+(p) = (i\Omega/4)[C_-(p-\hbar k) + C_+(p+\hbar k)]$$
$$\dot{C}_-(p) = (i\Omega/4)[C_+(p-\hbar k) + C_+(p+\hbar k)] .$$

Compared with the normal equations of motion for a two-level system, their equations are different in that they contain the transfer of momentum $p\pm\hbar k$ explicitly. After solving the equations using rather elegant mathematics they obtained the momentum transfer probability as

$$W(p,t) = \sum_{h=-\infty}^{+\infty} J_n^2(\Omega t)\delta(p-n\hbar k) \quad .$$

This equation means that the probability of the momentum transfer $p = n\hbar k$ is equal to $J_n^2(\Omega t)$ where J_n is the Bessel's function. This result is identical to the Fraunhofer diffraction of a plane wave by a sinusoidal plane grating. Here we see clearly the wave nature of an atom. One atom "sees" the whole grating and is diffracted.

Since the value of $J_n^2(\Omega t)$ is largest at $n \sim \Omega t$, Cook and Bernhardt theory again gives coherent pushing. It is also interesting to note that since $J_n^2(\Omega t)$ is not zero even for $n > \Omega t$, momentum transfers with more than the number of stimulated processes are allowed.

In obtaining the above results, Cook and Bernhardt neglected the Doppler shift due to momentum transfer. When this effect is included they obtained a maximum deflection which is proportional to $\sqrt{E}$ in agreement with our observation but is larger by a factor of 5.

8. Experimental Problems

The discrepancy between theory and experiment is also caused by the non-ideal experimental set up. For example, in order to treat the Na atom as plane wave, the beam has to be completely collimated in the z direction so that $\Delta p_x=0$ initially. For our atomic beam of $\Delta p_x/p \sim 3\times10^{-4}$, the "size" of the atomic wave will be $\Delta x \sim \hbar/\Delta p_x \sim 100$ Å and still smaller than the wavelength of radiation. Thus the atom cannot see the standing wave as a grating.

The standing wave is also not an ideal one because of the focussing. Ideally a narrowed parallel beam will be closer to the standing wave model used in theory. The probability of spontaneous emission is much less than that of stimulated processes in our experiment but is still not negligible and part of the randomness may be ascribed to it.

Finally we note that in order to test the lensing effect of standing wave predicted by Cook [9] or the even finer diffraction "pattern" of the beam, an atomic beam system with much higher collimation is needed.

We have profited from suggestions by two anonymous referees and discussions with A. Bambini, A.F. Bernhardt, M. Bloom, R.J. Cook and U. Fano, about the interpretation of our experimental results.

References

1. P.L. Kapitza and P.A.M. Dirac, Proc. Cambr. Phil. Soc. $\underline{29}$, 297 (1933).

2. H. Schwarz, Phys. Lett. $\underline{43}$A, 457 (1973).

3. O.R. Frisch, Z. Physik $\underline{86}$, 42 (1933);
   R. Schieder, H. Walther, and L. Wöste, Opt. Commun. $\underline{5}$, 337 (1972);
   J.L. Picqué and J.L. Vialle, Opt. Commun. $\underline{5}$, 402 (1972).

4. S.M. Freund, M. Römheld, and T. Oka, Phys. Rev. Lett. $\underline{35}$, 1497 (1975);
   J. Reid and T. Oka, Phys. Rev. Lett. $\underline{38}$, 67 (1977).

5. M. Bloom, E. Enga, and H. Lew, Can. J. Phys. $\underline{45}$, 1481 (1967).
   R.J. Hill and T.F. Gallagher, Phys. Rev. A $\underline{12}$, 451 (1975)

6. A.P. Kazantsev, JETP $\underline{63}$, 1628 (1972), JETP Lett. 17, 150 (1973), JETP $\underline{66}$, 1599, $\underline{67}$, 1660 (1974), Lett. $\underline{21}$, 158 (1975);
   A.P. Botin and A.P. Kazantsev, JETP $\underline{68}$, 2075 (1975);
   A.P. Botin, A.P. Kazantsev, and V.S. Smirnov, JETP $\underline{71}$, 122 (1976);
   B.L. Zhelnov, A.P. Kazantsev, and G.I. Surdutovich, Kvant. Elektr. $\underline{4}$, 893 (1977);
   A.P. Kazantsev, Uspekhi. Phys. Nauk $\underline{124}$, 113 (1978);
   G.A. Delone, V.A. Grinchuk, A.P. Kazantsev and G.I. Surdutovich, Opt. Commun. $\underline{25}$, 399 (1978).

7. S. Stenholm, J. Phys. B $\underline{7}$, 1235 (1974);
   S. Stenholm and J. Javanainen, Appl. Phys. $\underline{16}$, 159 (1978);
   S. Stenholm, V.G. Minogin and V.S. Letokhov, Opt. Commun. $\underline{25}$, 107 (1978);
   S. Stenholm, Phys. Reports $\underline{C}$43, 151 (1978).

8. R.J. Cook and A.F. Bernhardt, Phys. Rev. A $\underline{18}$, 2533 (1978).

9. R.J. Cook, Phys. Rev. Lett. $\underline{41}$, 1788 (1978).

Laser Cooling of Ions Bound to a Penning Trap

R.E. Drullinger and D.J. Wineland

Time and Frequency Division, National Bureau of Standards
Boulder, CO 80303, USA

Abstract

MgII ions which are confined in a room temperature Penning trap have been cooled to $\leq$ 0.5 K by scattering photons which are nearly resonant with the $3s^2S_{1/2} \leftrightarrow 3p^2P_{3/2}$ transition. The magnesium loaded into the trap has a natural isotopic abundance consisting of $\sim$ 80% ^{24}Mg, and $\sim$ 10% each ^{25}Mg and ^{26}Mg. The ^{24}Mg is radiatively cooled and it subsequently cools the ^{25,26}Mg by long range coulomb collisions. This allows the use of a "cooling ion" being used in conjunction with a more complex species of spectroscopic interest.

Experiments using two lasers are reported where one laser is fixed in frequency and provides the cooling source and the other laser is swept in frequency. By monitoring the scattered light from the second laser, an optical spectrum of MgII ions is obtained. Because the Doppler width of the cooled ions is reduced, the three isotopic lines are clearly resolved.

Introduction

In the past few years there has been increasing interest in the use of near-resonant photon scattering to cool a collection of atoms, ions or molecules. This interest is motivated in part by the practical need to reduce first- and second-order Doppler shifts in ultra-high-resolution spectroscopy and in part by the esthetic appeal of controlling the positions and velocities of a collection of particles to within the limits imposed by quantum fluctuations. The results of recent experiments show that this control may be close at hand.

Current interest in the possibility of cooling began with independent proposals to use near-resonant laser radiation to reduce the temperature of a gas of neutral atoms [1] or ions which are bound in an electromagnetic "trap" [2]. This method of cooling has subsequently been incorporated into the interesting schemes for trapping of particles using near-resonant optical fields [3]. The first demonstration [4] of cooling using the basic technique described here was made for a slightly modified situation; specifically, the magnetron motion of an electron in a Penning trap was "cooled" by a technique called motional sideband excitation [4,5], which is formally equivalent to the laser cooling of atoms. Cooling of ions bound in an electromagnetic trap was more recently demonstrated; lasers have been used to cool MgII ions stored in a Penning trap [6] and BaII ions in a RF trap [7].

The Cooling Process

To illustrate the cooling concept in simple form, consider a gas of absorbers which possess a resonant electric dipole transition in some convenient spectral region. Now suppose that these absorbers are irradiated with monochromatic, directed radiation tuned near but slightly lower than their "rest" resonant frequency. The absorbers which are moving against the radiation field and which Doppler shift it into resonance will scatter (absorb and reemit) photons at a rate higher than those absorbers which are not at resonance. For each scattering event, the absorber receives a momentum impulse. For an absorber which is moving against the radiation field, this impulse retards its motion, resulting in a cooling effect. The average momentum transferred to the absorber by the reemitted photon is zero because of the randomness of the photons' directions.

The momentum of a visible photon is approximately 10^{-22} g-cm/s, while that of an absorber with mass of 25 a.m.u. and temperature of 300 K is about 10^{-18} g-cm/s. For this case it can be seen that about 10^4 scattering events per absorber would be required to substantially cool the gas. In a practical cooling experiment on free atoms [1] it would be desirable to irradiate the absorbing gas from all sides with radiation that covered the entire lower half of the Doppler profile. The experimental requirements on the photon source can be substantially reduced, however, if the absorbers are bound to a laboratory fixed apparatus. An ion in an electromagnetic bottle such as a Penning trap can serve as such an absorber.

To illustrate the cooling process in this case, consider an absorber which is harmonically bound and constrained to move along the x direction. We assume that its velocity is given by $v_x = v_0 \cos 2\pi\nu_v t$ where ν_v is its vibrational frequency and that the natural linewidth of the optical transition (full width at half maximum = $\gamma/2\pi$) is less than ν_v. As shown in Fig. 1, the spectrum of such an absorber, when observed in the laboratory x direction, contains the central resonance line with first-order-Doppler-effect generated sidebands separated by ν_v having relative intensities $J_m^2(v_0\nu_0/c\nu_v)$ where J_m is the Bessel function of order m. If we irradiate this absorber with photons of frequency $\nu_L = \nu_0 + m\nu_v$, the frequency of the scattered photons occurs nearly symmetrically about ν_0 at the frequencies ν_0, $\nu_0 \pm \nu_v$, $\nu_0 \pm 2\nu_v$ . . . . Therefore, although photons of energy $h(\nu_0 + m\nu_v)$ are absorbed, on average photons of energy approximately equal to $h\nu_0$ are reemitted; when m is negative, this energy difference causes the kinetic energy of the absorber to decrease by $|mh\nu_v|$ per scattering event. In the experiments described below the natural linewidth is greater than the sideband frequency ($\gamma/2\pi \gg \nu_v$) and the observed spectrum of the bound ion is simply the usual Doppler broadened line; however, the above conclusion is still valid.

The rate at which this cooling can be achieved has been worked out in some detail elsewhere [3,5,6,8]. We can get a feeling for it by remembering from above that approximately 10^4 scatterings are needed to cool a room temperature absorber of 25 a.m.u. Again, for our visible absorber with a fully allowed transition, we could achieve a scattering rate of 10^7 sec^{-1} at resonance with a radiant flux of approximately 10 mW/cm^2, implying cooling time as short as 1 ms.

COOLING (HEATING) OF BOUND RESONANT ABSORBER

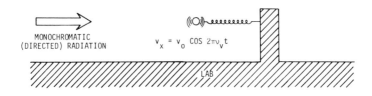

SPECTRUM OF ABSORBER IN LAB FRAME

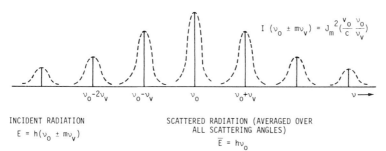

$$I(\nu_0 \pm m\nu_v) = J_m^2(\frac{v_0}{c}\frac{\nu_0}{\nu_v})$$

INCIDENT RADIATION
$E = h(\nu_0 \pm m\nu_v)$

SCATTERED RADIATION (AVERAGED OVER
ALL SCATTERING ANGLES)
$\bar{E} = h\nu_0$

Fig. 1 One dimensional cooling model. In the time domain picture (top of figure) atoms absorb only when moving towards the incident radiation which is tuned below the atom's rest frequency (ν_0). In a spectrum picture (bottom of figure) atoms absorb photons of energy $h(\nu_0 + m\nu_v)$ and emit photons of average energy $h\nu_0$. When m < 0 the energy deficit causes the atoms to cool.

The minimum temperature attainable by the cooling process is determined by a competition of the damping rate on the vibrational energy due to the laser cooling and the heating effect due to recoil [3,6,7,8]. In the limit where the trap vibration frequencies are much less than the absorber natural width ($\nu_v \ll \gamma/2\pi$) this leads to a one dimensional kinetic energy of $\hbar\gamma/4$ [3,7,8] which is the order of 10^{-3} K for our mass 25 absorber with $A = \gamma = 2.6 \times 10^8$/s.

Experimental Apparatus

For the demonstration of the cooling process just outlined, the MgII ion was chosen since it has a simple alkali-like spectrum with no nuclear spin for the common isotope (M=24), is easily loaded into the Penning trap and has its first resonance line at 280 nm (a wavelength which can be generated by frequency doubling a cw dye laser).

The trap is a copper Penning trap with characteristic dimensions [9,10] $r_0 = 1.64z_0 = 0.63$ cm which is typically operated at $V_0 = 7$ V and $B_0 \cong 1$ T. The trap is enclosed in a high vacuum envelope at room temperature ($P < 1.3 \times 10^{-8}$ Pa $\cong 10^{-10}$ Torr) allowing observed thermalization times due to collisions with the background gas of as long as 30 min and storage times of greater than one day. Mg is emitted from an oven and ions are formed by an electron beam coincident with the trap axis. Because the oven is about 0.5 cm from the trap, the temperature of the trap, and therefore the background gas is elevated above room temperature ($\sim$ 350 K).

The motion of an ion in a Penning trap can be resolved into 3 components. In the trap z axis direction (the direction of the B field and the trap symmetry axis) the ion executes simple harmonic motion in the electrostatic potential well ($\nu_z \cong 200$ KHz). In the x-y plane, perpendicular to the B field, the ion path is a paracycloid which is composed of a circular cyclotron motion ($\nu_c' \cong 625$) KHz superimposed on a circular magnetron motion ($\nu_m \cong 30$ KHz). The axial motion and the cyclotron motion are thermal and strongly coupled through the long range coulomb collisions of ions in the trap (coupling time $\cong 1$ ms). However, the magnetron motion is not thermal and to a high degree not coupled to the other two motions.

In these experiments, the cooling beam traverses the trap in the x-y plane and cools primarily the cyclotron motion, which in turn cools the z-axis motion through coulomb collisions.

Two lasers have been used in the experiment to date, both are cw single frequency dye lasers pumped by an Ar$^+$ laser and operating at $\cong$ 560 nm on rhodamine 110 dye. One laser is a commercial, standing wave, cavity stabilized device capable of scanning over 30 GHz. The second is a home-made ring type which can be locked via a saturated absorption to a hyperfine line in the I_2 spectrum near the half harmonic of the MgII resonance. Each of these lasers has a linewidth of approximately 1 MHz and with 3 W of pump power the standing wave laser has an output of approximately 250 mW while the ring laser has an output of approximately 500 mW.

This visible light is frequency doubled to the uv resonance of MgII via a 90° phase matched AD*P crystal having a power conversion k ($P_{out} = kP_{in}^2$) of the order of 10^{-3}/W, hence 100 mW input results in 10 μW at the second harmonic. When these two dye lasers are used together, their respective beams pass parallel through separate sections of the same doubling crystal and are then independently focused into the trap through a hole in the ring electrode.

Results

In previous cooling experiments [6], approximately 5 x 10^4 ions were loaded into the trap (density $\sim$ 2 x 10^7/cm^3). The thermal motion of these ions in the trap axial direction induces image currents in the trap endcaps which are analyzed to give a signal proportional to N · T, where N is the number of ions in the trap and T is their temperature [9,10]. Since the time for cooling is much less than the ion storage time and since observations confirm that the laser does not eject ions from the trap, N can be assumed to be constant. Therefore, this signal provides a direct measurement of ion temperature.

Figure 2 shows a plot of this signal versus time when the cooling radiation is applied for a fixed detuning $\nu_o - \nu_L \cong 2$ GHz. The ions had

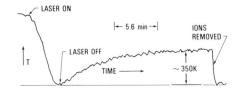

Fig. 2 Ion temperature versus cooling for a fixed radiation frequency $(\nu_0 - \nu_L \cong 2\ \text{GHz})$

previously been heated above the ambient temperature by the laser. When the laser was tuned to the low frequency side of the $M_J = -1/2 \leftrightarrow M_J = -3/2$ transition a temperature of < 40 K was achieved. The cooling radiation was then switched off and the ions allowed to rethermalize. For this plot, the background gas pressure in the trap was intentionally increased to give a shorter rethermalization time.

Since this image current analysis gave an upper temperature bound of only 40 K, while the actual temperature was thought to be much lower, the experiment was reconfigured to collect and count the photons which were scattered into an f/4 cone back toward the laser. In this way the scatter rate versus laser frequency can be used to give a direct measure of the resonance transition linewidth and hence the temperature. Figure 3 shows such a curve taken with only one laser. Here a small load of ions (≲ 500) is first cooled by slowly sweeping the laser in toward line center from the low frequency side. Then, for the linewidth trace, and to minimize the heating that occurs as the laser sweeps out through the high frequency side of the line, the laser is stepped back in frequency and rapidly swept through the line. The resulting line is asymmetric but the low frequency side of the line shows very little Doppler broadening and has a halfwidth which corresponds to a temperature of less than 0.5 K. The broadening due to the natural linewidth of 43 MHz [11] has been taken into account. This may be an upper bound on the cyclotron-axial temperature since the measured width of the line is largely due to the spreading effect of the magnetron motion. That is, as the cloud rotates around the z axis at the magnetron frequency, different portions of the laser beam (∼ 100 μm dia) interact with different velocities due to the magnetron motion, giving rise to a spread in frequencies of approximately 70 MHz. The "temperature" of this magnetron motion for the smaller ion clouds of approximately ≲ 500 ions (cloud diameter in x, y plane ≅ 200 μm) can be approximated by $kT_m = \langle \tfrac{1}{2}m\omega_m^2 r_m^2 \rangle$ giving the result $T_m \cong 0.25\ \text{K}$.

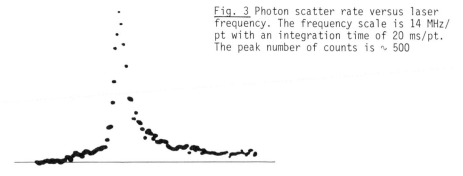

Fig. 3 Photon scatter rate versus laser frequency. The frequency scale is 14 MHz/pt with an integration time of 20 ms/pt. The peak number of counts is ∼ 500

At higher scatter rates one must be aware that the axial temperature may be elevated above that of the cyclotron temperature [8]. This is because although the axial and cyclotron degrees of freedom are thermalized by coulomb collisions, the axial motion is also independently heated by recoil in the emission process. For the experimental conditions described here, however, this effect should be negligible.

The isotopic composition of the magnesium loaded into the trap is natural, containing 80% ^{24}Mg and 10% each of ^{25}Mg and ^{26}Mg. The cooling is usually done on the resonance line of ^{24}Mg but the ^{25}Mg and ^{26}Mg are simultaneously cooled by coulomb collisions. The resonances of ^{25}Mg and ^{26}Mg lie in the high frequency wing of ^{24}Mg and to observe them by resonant light scattering causes the ^{24}Mg to become heated and subsequently heat the entire cloud. To overcome this problem, the more powerful ring dye laser is used to provide a cooling beam to radiate the low frequency side of the ^{24}Mg line and keep the entire sample cold while the lower power standing wave dye laser is tuned through the spectrum. The resulting plot is shown in Fig. 4. ^{24}Mg and ^{26}Mg have zero nuclear spin and yield single lines while ^{25}Mg has nuclear spin I = 5/2 and should exhibit six lines. However, the nuclear hyperfine components are optically pumped such that nearly all of the ^{25}Mg ions are in the M_I = -5/2 state. Hence ^{25}Mg is also observed as a single line. The relative positions of the observed lines depend on the isotope shifts and (in the case of ^{25}Mg) the hyperfine coupling constants [12].

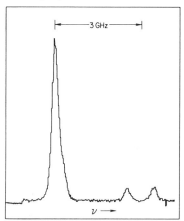

Fig. 4 Photon scatter rate versus laser frequency showing the three isotopes of MgII in the trap. The sample is maintained cold during the scan by irradiating the low frequency side of the strong line with a second laser. From left to right the three lines are due to the $(3s^2S_{1/2}$, M=-1/2) $\leftrightarrow$ $(3p^2P_{3/2})$ transitions in the ^{24}Mg, ^{25}Mg, and ^{26}Mg isotopes. At 300 k the Doppler width is approximately equal to the spacing between the ^{24}Mg and ^{25}Mg isotopes ($\sim$ 3 GHz).

Summary

We have demonstrated laser cooling of MgII bound to a Penning trap and have achieved a temperature of $\lesssim$ 0.5 K. This corresponds to a reduction of the second order Doppler effect of approximately 1,000 relative to experiments conducted at room temperature. Since the rethermalization rates for ions in the trap are very long, and since this cooling laser is not also used for confinement, it can be switched off to eliminate various perturbing effects while some transition in the cooled ions is interrogated under ultra-high resolution. We have shown that cooling ^{24}Mg also cools by coulomb collisions the ^{25}Mg and ^{26}Mg ions which are in the trap. This provides the opportunity to study spectrally complicated species by simultaneously loading a coolable ion (i.e., an ion with (1) no transition bottlenecks which prevent the repeated scattering needed for cooling and (2) resonant transitions in convenient spectral regions).

Acknowledgments

The authors wish particularly to thank J. C. Bergquist for the design and construction of the ring dye laser used in part of this experiment. We acknowledge the help of many useful discussions and suggestions from Drs. Bergquist, Wayne M. Itano and F. L. Walls. We also acknowledge partial support by the Office of Naval Research.

References

1. T. W. Hänsch and A. L. Schawlow, Opt. Commun. 13, 68 (1975).

2. D. J. Wineland and H. Dehmelt, Bull. Am. Phys. Soc. 20, 637 (1975).

3. See, for example, J. E. Bjorkholm, R. R. Freeman, A. Ashkin, and D. B. Pearson, Phys. Rev. Lett. 41, 1361 (1978), A. Ashkin, Phys. Rev. Lett. 40, 729 (1978), V. S. Letokhov, V. G. Minogin, and B. D. Pavlik, Zh. Eksp. Teor. Fiz. 72, 1328 (1977), Sov. Phys. JETP 45, 698 (1977), A. P. Kazantsev, Usp. Fiz. Nauk. 124, 113 (1978), Sov. Phys. Usp. 21, 58 (1978), A. Ashkin and J. P. Gordon, Opt. Lett. 4, 161 (1979).

4. R. S. Van Dyck, Jr., P. B. Schwinberg, and H. G. Dehmelt, in New Frontiers in High-Energy Physics, Kursunoglu, Perlmutter, and Scott eds., (Plenum, New York, 1978) p. 159.

5. H. G. Dehmelt, Nature 262, 777 (1976), D. Wineland and H. Dehmelt, Int. J. Mass Spectrometry and Ion Phys. 16, 338 (1975), erratum 19, 251 (1976).

6. D. J. Wineland, R. E. Drullinger, and F. L. Walls, Phys. Rev. Lett. 40, 1639 (1978).

7. W. Neuhauser, M. Hohenstatt, P. Toschek, and H. Dehmelt, Phys. Rev. Lett. 41, 233 (1978).

8. D. J. Wineland and Wayne M. Itano, Submitted to Phys. Rev. A.

9. H. G. Dehmelt and F. L. Walls, Phys. Rev. Lett. 21, 127 (1968); H. Dehmelt, in Advances in Atomic and Molecular Physics, edited by D. R. Bates and I. Esterman (Academic, New York, 1967, 1969), Vols. 3 and 5; D. J. Wineland and H. G. Dehmelt, J. Appl. Phys. 46, 919 (1975).

10. R. A. Heppner, F. L. Walls, W. T. Armstrong, and G. H. Dunn, Phys. Rev. A 13, 1000 (1976).

11. W. W. Smith and A. Gallagher, Phys. Rev. 145, 26 (1966).

12. M. F. Crawford, F. M. Kelly, A. L. Schawlow, and W. M. Gray, Phys. Rev. 76, 1527 (1949).

Preparation, Cooling, and Spectroscopy of Single, Localized Ions

W. Neuhauser, M. Hohenstatt, and P.E. Toschek

Institut für Angewandte Physik der Universität
D-6900 Heidelberg, Fed. Rep. of Germany
and

H.G. Dehmelt

Physics Department, University of Washington
Seattle, WA 98195, USA

Certainly the most desirable conditions of experiments, to which an atomic physicist could ever aspire under the viewpoint of clarity and simplicity, include the preparation of a single, well-localized atomic particle. The realization of this concept, which requires some kind of entropy reduction or "cooling" of the particle [1, 2], would dramatically simplify the interpretation of experiments, e.g. in quantum statistics, and in studies of collective interactions with electromagnetic fields. Moreover, spectroscopic investigations of that particle could particularly benefit from the absence of most spectral broadening mechanimus: Doppler effect of 1st and 2nd order, transit-time broadening, and collisional broadening are eliminated [3]. Since only natural broadening is left over, this concept permits the detection of optical lines of unprecedented sharpness, and it offers an approach towards the ultimate atomic time and length standard.

Recently, we have succeeded in confining 10 to 20 barium ions in a miniaturized RF quadrupole field configuration, to observe their resonance fluorescence visually, photographically, and photoelectrically, and to demonstrate the effect of optical cooling [4,5, see also 6]. When averaged over time intervals longer than the RF period, the oscillating electric quadrupole field generates a harmonic quasi-potential for particles within a certain range of specific charges [7, 8]. In the experiments (Fig. 1), the fluorescence from an ion cloud of 50 um Ø was excited by a few mW of CW dye laser light at λ = 493 nm, in resonance with the Ba^+ $^2S_{1/2}$ - $^2P_{1/2}$ transition (a-b), but detuned by some 500 MHz to the low-frequency wing of the Doppler-broadened line. A second CW dye laser beam at λ' = 650 nm also irradiated the cloud in order to release ions from the metastable $^2D_{3/2}$ state (c) back into the excitation-emission cycle. The frequencies of the lasers were conrolled by visual observation of resonance fluorescence from a He-Ba discharge on the respective lines. The ions were generated *in situ*, i.e. in the centre of the potential well, by electron impact upon an atomic beam that emerged from a small barium oven at a background pressure of less than 10^{-8} mbar.

Here we report upon a series of improved experiments which include the preparation and cooling of *single ions*. These localized particles are genuine mono-ion oscillators at rest in space and driven by the light fields or, alternatively,

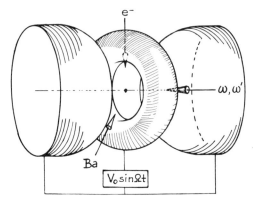

Fig. 1 Ion trap configuration. The internal diameter of the ring electrode is 0.7 mm

part of an ionic quasi-molecule whose negatively charged core consists of the macroscopic trap.

The scheme for the detection of the resonance fluorescence is shown in Fig. 2. One of the eye pieces of a binocular microscope (8 x 25 maximum magnification) is used for the visual, the other one for the photographic detection. Into the latter channel, a 3-step image intensifier is inserted. A third optical channel, originally equipped for microphotography, is converted into the channel for photoelectric detection. Its part of the fluorescence light forms an image of the ion cloud and impinges upon the cathode of a photo-multiplier. In the image plane, a diaphragm suppresses much of the stray light. The PM signal is detected by a photon counter, D/A converter, and XY recorder.

The weak, unfocussed electron beam ($U_e \simeq 220$ V) generates ions in the initially empty trap at a rate of 1 or 2 per min-

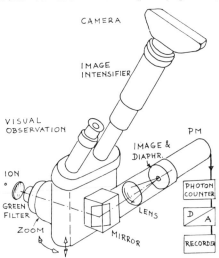

Fig. 2 Detection scheme for ionic resonance fluorescence

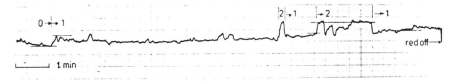

Fig. 3 Resonance fluorescence signal *vs.* time. 0, 1, or 2 ions in trap

ute. The corresponding increments of the resonance fluorescence are visually easily perceivable and give rise to steps of the recording of the fluorescence signal *vs.* time (Fig. 3). Sometimes the process of trapping takes a few seconds, since an ion which moves in a larger orbit is inefficiently cooled. Accordingly, the corresponding steps of the recorder trace are washed out. Occasionally, clouds of two or more ions heat up, increase in size, and reduce the recorded signal; one of the ions may even escape from the trap (Fig. 3, right). In good recordings, clear steps are identified up to clouds of four or five ions (Fig. 4). As a final check, the red light is blocked. The fluorescence cycle is interrupted, and a base line signal is rendered.

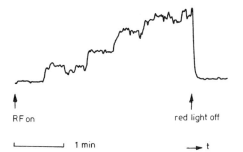

Fig. 4 Resonance fluorescence signal *vs.* time. Accumulation of four ions in the trap

For taking photographs of ion clouds as they appear on the screen of the image intensifier, the electron beam is turned off, when a preselected ion number has accumulated in the trap. With I, $I' \simeq 300$ μW, $\omega_{ab}-\omega \simeq 500$ MHz, $\omega_{bc}-\omega \simeq 300$ MHz, the photograph of Fig. 5 was taken. It shows the images of three ions, two ions, and a *single ion* (exposure time 10 min each, film: Ilford HP5). The cloud diameters, 5 to 8 μm, correspond to a temperature below 10 K. Throughout the exposures, the ion clouds were visually observed. The observation of the single ion revealed a unique peculiarity: the cloud occasionally contracts to a size below the microscopic resolution ($\simeq 0.5$ μm), and appears as a point-like bright light source, whose motional temperature is below 0.1 K.

After some improvements of the frequency stability of the lasers had been accomplished, direct photomicrographs - without applying the image intensifier - have been taken through the photographic channel (Fig. 2, "mirror" replaced by camera).

75

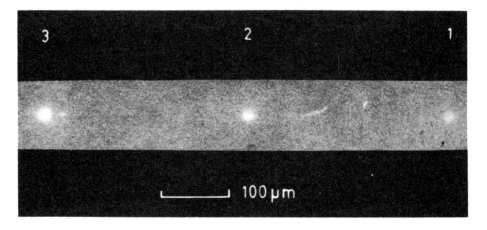

100 µm

Fig. 5 Images of 3, 2 ions and 1 ion, taken from the screen
of an image intensifier

During the 30-min. exposure, the ion was stored in a 15 V deep
pseudo-potential well. Its depth was determined by resonant RF
heating of the ions at the axial secular oscillation frequency
$\bar{\omega}_z$ [5]. The conditions of strong cooling existed for 10 to 12
min, whereas for the rest of the exposuretime the cooling was
incomplete due to minor frequency and power fluctuations of
the light (Fig. 6; astronomical film, type Kodak 103 a-F).
Accordingly, the picture shows a bright point-like spot superimposed upon a diffuse background cloud some 8 µm in size. In
a 5.4-V deep well, complete cooling was achieved for the entire
17-min exposure time (s. Fig. 7). The corresponding ion temperature estimated from the cloud size was smaller than 2.5 K.

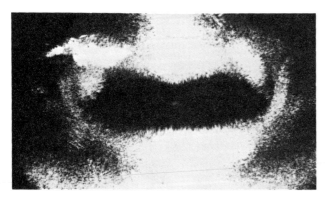

Fig. 6 Direct photograph of resonance fluorescence from *single*
ion (centre of picture): 10 um cloud corresponds to T<15 K,
superimposed spot < 1µm corresponds to T<0,1 K (exp.time: 30
min, well depth 15 V)

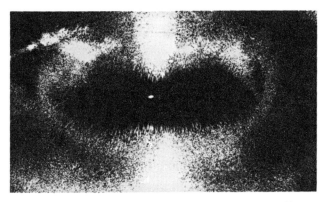

Fig. 7 Direct photograph of resonance fluorescence from *single* ion (centre of picture): 8-μm cloud corresponds to T<2.5 K (exp. time 17 mm, well depth 5.4 V)

It seems evident that further improvements of the stability of the laser frequencies *and* output powers will result in reliably localized, quasi-immobile ions whose residual small oscillation amplitude leaves the interaction with the light field in the Lamb-Dicke regime (orbit size < $\lambda/2$), and the bulk of the resonance fluorescence in the carrier line of the frequency-modulated spectrum.

In another experiment, a narrow diaphragm corresponding to 20 μm $\emptyset$ in the object plane was used (see Fig. 2). Clouds of 5 to 10 ions were irradiated by an additional RF field at $\bar{\omega}_z$. The corresponding heating of the cloud causes an expansion and a concomitant reduction of the fluorescence signal (Fig. 8). With the frequency of the red laser stationary on the low-frequency wing of the corresponding line, some hundred MHz off the centre, and the frequency of the blue laser far off the resonance line centre, $\omega_{ab} - \omega > \Delta \omega_D$, a dim cloud appears whose size is 100 μm, or more. When ω is tuned across the resonance line, the cloud contracts and expands again according to the spectral variation of the cooling power [6]. The fluorescence

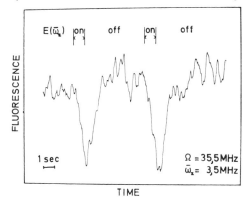

Fig. 8 Resonance fluorescence after small diaphragm. RF heating at $\bar{\omega}_z$ blows ion cloud up and decreases signal

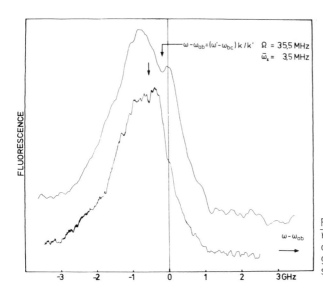

Fig. 9 Resonance fluorescence after small diaphragm *vs.* tuning of green laser across resonance line

signal increases and decreases, respectively, and gives rise to an asymmetric "cooling spectrum" (Fig. 9). When the condition $\omega_{ab} - \omega = (\omega_{bc} - \omega')(K/K')$ is fulfilled, a Doppler-free saturation resonance appears.

This experiment indicates that there seem not to exist any insurmountable complications which prevent this spectroscopic technique from being applied to a single ion.

This work was supported by the Deutsche Forschungsgemeinschaft.

References:

1. H.G. Dehmelt, Nature <u>262</u>, 777 (1976)

2. D.J. Wineland, H.G. Dehmelt, Bulletin APS <u>20</u>, 637 (1975)

3. H.G. Dehmelt, P.E. Toschek, Bulletin APS, <u>20</u>, 61 (1975)

4. W. Neuhauser, M. Hohenstatt, P.E. Toschek, and H.G. Dehmelt, Phys. Rev. Letters <u>41</u>, 233 (1978)

5. W. Neuhauser, M. Hohenstatt, P.E. Toschek, and H.G. Dehmelt, Appl. Phys. <u>17</u>, 123 (1978)

6. D.J. Wineland, R.E. Drullinger, and F.J. Walls, Phys. Rev. Letters <u>40</u>, 1639 (1978)

7. W. Paul, O. Osberghaus, and E. Fischer, Forschungsber. d. Wirtsch.- u. Verkehrsministeriums NRW, Nr. 415 (1958)

8. E. Fischer, Z. Physik <u>156</u>, 1 (1959)

Two and Three Level Atoms/
High Resolution Spectroscopy

Theories of Laser Bandwidth Effects in Spectroscopy

J.H. Eberly

Department of Physics and Astronomy, University of Rochester
Rochester, NY 14627, USA

1.   Introduction

The theory of lineshapes in spectroscopy is a well-developed subject,
treated in too many texts, monographs and reviews to justify individual
citations here. My purpose in this short note is to summarize very recent
progress in one corner of the field, the corner concerned with non-perturbative effects associated with finite-bandwidth laser excitation. These
effects can be encountered easily in laser spectroscopy, whenever the laser
power is sufficient to deplete the initial state of a transition. They give
rise to theoretical problems of lineshape that are just beginning to be at-
tacked. Only the simplest of these has been solved. In this talk I will
only deal with what I will call the "width change" question. That is, to
what extent is the observable spectroscopic linewidth changed if the exciting
light has a finite bandwidth of its own?

In the next Section I have sketched a brief reminder of the answer to this
question in the perturbative limit. In the third and fourth Sections some
exact results, without perturbative assumptions, are presented for a number
of special cases. The final Section contains some applications of the exact
results and a few comments are made about unsolved problems.

2.   Perturbation Theory

It is sufficient for our purposes to
consider the excitation of a single allowed transition by a quasi-monochromatic light field near to resonance. The
notation is explained by the first
figure. The upper level has been given
a finite width due, for example, to the
possibility of spontaneous decay or the
existence of collisions. An ensemble
of atoms with such transition could
be characterized by a finite distribution of values of ω_0, as is the case
with Doppler broadening.

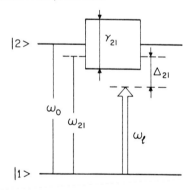

Fig. 1   The model absorber

Schrödinger's equation for this problem has the form:

$$i \, \dot{a}_1 = \tfrac{1}{2} \, \Omega(t) \, a_2 \; .$$

(1a)

$$i \, \dot{a}_2 = \Delta_{21} a_2 + \tfrac{1}{2} \, \Omega(t) \, a_1 \, , \tag{1b}$$

where a_1 and a_2 are the time-dependent amplitudes of states $|1\rangle$ and $|2\rangle$, and $\Omega(t)$ is the Rabi frequency for the transition:

$$\hbar \, \Omega_{12}(t) = 2 \, \underline{d}_{12} \cdot \underline{\varepsilon} \, E(t). \tag{2}$$

Here $\underline{\varepsilon}$ is the polarization and $E(t)$ is the complex envelope of the quasi-monochromatic light field $\underline{E} = \underline{\varepsilon} \, E(t) \exp [-i\omega_\ell t] + c.c.$, and $\underline{d}_{12}$ is the transition dipole matrix element.

If the atom is initially in its ground state, then the expression for the probability of a transition to the second level is:

$$|a_2(t)|^2 = \int_0^t dt' \int_0^t dt'' \, [-\tfrac{1}{2} i \Omega(t')][\tfrac{1}{2} i \Omega^*(t'')]$$
$$\times \exp i\Delta_{21}(t'-t'') \, a_1(t') \, a_1^*(t''). \tag{3}$$

This expression should be averaged over the band of final states associated with level 2. Figure 1 shows γ_{21} to be the width of this level.

If the laser pulse is sufficiently long, on the order of nanoseconds or longer, then a significant part, perhaps all, of the laser's bandwidth can be expected to be due to stochastic fluctuations in $\Omega(t)$. Such fluctuations are predicted by fundamental laser theory, but non-fundamental causes originating in mechanical, thermal, and electronic noise are often dominant. Clearly, $a_1(t)$ and $a_2(t)$ are thus really themselves stochastic variables, due to the action of $\Omega(t)$, as shown in (1). If, as we assume, the fluctuations in $\Omega(t)$ are rapid on the scale of an observation time, then the observable probability of excitation into level 2 is the average of (3) over the ensemble of fluctuations.

In the spirit of perturbation theory we put $a_1(t') \approx a_1^*(t'') \approx 1$ in (3). Then it is straightforward to carry out the average of (3) over the band of final states in $|2\rangle$ and over the ensemble of laser fluctuations. Later comparisons with non-perturbative results will be easier if we characterize the density of final states by the function:

$$\rho(\omega_{21}) = \frac{\gamma_{21}/2\pi}{(\omega_{21}-\omega_0)^2 + (\gamma_{21}/2)^2} \tag{4}$$

and the laser fluctuations by the second order correlation:

$$(\!(\, \Omega(t') \, \Omega^*(t'') \,)\!) = \Omega_0^2 \, e^{-\frac{1}{2}\gamma_\ell |t'-t''|} \tag{5}$$

where $\hbar \, \Omega_0 = 2 |\underline{d}_{21} \cdot \underline{\varepsilon}| E_0$, and $(c/2\pi) E_0^2 = I_0$, the light intensity.

These functions (4) and (5) can be regarded simply as convenient models embodying a finite level width γ_{21} and a laser bandwidth (inverse correlation time) γ_ℓ , but it will be recognized that they have some fundamental significance as well.

Let us define $P_2(t)$ to be the observable transition probability:

$$P_2(t) = \int d\omega_{21} \, \rho(\omega_{21}) \, (\!(\, |a_2(t)|^2 \,)\!) \, . \tag{6}$$

Then the perturbative limit is easily found to be:

$$P_2(t) = \tfrac{1}{4} \Omega_0^2 \frac{1}{\mu} [t - \frac{1 - e^{-\mu t}}{\mu}] + c.c., \tag{7}$$

where

$$\mu = \tfrac{1}{2} (\gamma_{21} + \gamma_\ell) - i\Delta , \quad \Delta \equiv \omega_0 - \omega_\ell. \tag{8}$$

Eq. (7) does not show a linear growth of probability with time, and so perturbation theory does not in general predict a constant transition rate, as is well known. In the limit of very short times, such that $(\gamma_{12} + \gamma_\ell)t \ll 1$, one finds quadratic behavior:

$$P_2(t) \rightarrow \tfrac{1}{4} \Omega_0^2 t^2 , \tag{9}$$

which is the beginning of the first Rabi inversion cycle. Every transition is completely coherent in the beginning ($\omega_0 \gg t^{-1} \gg \gamma_{21} + \gamma_\ell$). In the limit of sufficiently long times ($t^{-1} \ll \gamma_{12} + \gamma_\ell$) one finds the result

$$P_2(t) \rightarrow \tfrac{1}{4} \Omega_0^2 \frac{\gamma_{21} + \gamma_\ell}{\Delta^2 + \tfrac{1}{4} (\gamma_{21} + \gamma_\ell)^2} \times t. \tag{10}$$

All of the limits $\Delta \gg \gamma_{21} + \gamma_\ell$, $\gamma_{21} \gg \Delta$, γ_ℓ, and $\gamma_\ell \gg \Delta$, γ_{21} give text-book expressions that are more easily recognized if the Rabi frequency is rewritten in terms of the light intensity and the square of the transition dipole moment. Then (10) becomes:

$$P_2(t) \rightarrow [\sigma(\Delta)\Phi] \times t , \tag{11}$$

where $\hbar\omega_\ell\Phi = (c/2\pi) E_0^2$ and $\sigma(\Delta)$ is the usual dipole cross section, except that the laser bandwidth contributes to the cross section linewidth:

$$\sigma(\omega_\ell - \omega_0) = \frac{4\pi^2\omega_\ell |\underline{d}_{12} \cdot \underline{\epsilon}|^2}{\hbar c} \frac{(\gamma_{21} + \gamma_\ell)}{(\omega_\ell - \omega_0)^2 + \tfrac{1}{4} (\gamma_{21} + \gamma_\ell)^2} . \tag{12}$$

Either (7) or (12) can be understood to show, in perturbation theory, that a finite laser bandwidth simply adds itself to the usual atomic linewidth, given the simple model (5) of laser fluctuations. Next I will try to suggest some of the difficulties in treating the same model non-perturbatively, and some exact results of that treatment.

3. Excitation by a Phase-Fluctuating Laser, Exact Results

In the absence of perturbation theory (3) becomes more difficult to solve because $a_1 = 1$ is no longer permitted as an approximation. It is necessary to consider the time evolution of $a_1(t)$ as well as of $a_2(t)$, and this leads to the optical BLOCH equations for the density matrix elements of the two-level atom. These equations can be written:

$$(d/dt) \sigma_{21} = -(\gamma_{od} - i\Delta) \sigma_{21} + \tfrac{1}{2} i\Omega(t) w \tag{13}$$

$$(d/dt) \sigma_{12} = -(\gamma_{od} + i\Delta) \sigma_{12} - \tfrac{1}{2} i\Omega(t) w \tag{14}$$

$$(d/dt) \; w = -\gamma_d(1 + w) + i\Omega(t) \; \sigma_{21} + c.c., \tag{15}$$

where $w = \sigma_{22} - \sigma_{11}$, and γ_d and γ_{od} are diagonal and off-diagonal relaxation rates arising from spontaneous emission or collisions.

The difficulty of solving sets of equations such as (13) - (15) is readily apparent. Eq. (13) shows that σ_{21} is coupled to Ωw. From (15) we see that Ωw is coupled to $\Omega^2 \sigma_{21}$, and so on. This produces a decorrelation problem in the general case, and requires an approximation such as $((\Omega^2 w)) \cong ((\Omega^2)) \; ((w))$ in order to obtain a closed set of equations. It is remarkable that there is a model for $\Omega(t)$ that is reasonably realistic and permits exact solution of (13) - (15) without any intermediate decorrelations. Apparently BURSHTEIN [1] was the first to obtain $((w(t)))$ exactly for this problem. BURSHTEIN'S model, $\gamma_d = \gamma_{od} = 0$, and $\Omega(t)$ has the form $\Omega(t) = \Omega_0 \exp[i\phi(t)]$, and $\phi(t)$ executes rapid jumps at random times, remaining constant between jumps, such that $\Omega(t)$ undergoes a slow decorrelation as in (5). Then BURSHTEIN [1] showed that $((w(t)))$ satisfies exactly a third order linear differential equation. Its solution hinges on the roots of a cubic polynomial $p(z)$:

$$p^B(z) = (z + \gamma_\ell) \; [z(z + \gamma_\ell) + \Omega_0^2] + z\Delta^2 . \tag{16}$$

To understand the meaning of this result, $p(z)$ should be compared with the TORREY [2] polynomial $p^T(z)$ obtained from (13) - (15) for a non-fluctuating, purely monochromatic field:

$$p^T(z) = (z + \gamma_{od})[(z + \gamma_d)(z + \gamma_{od}) + \Omega_0^2] + (z + \gamma_d) \; \Delta^2. \tag{17}$$

This comparison makes evident what can be shown in detail,[3] that the effect of phase fluctuations in this model has no effect on the solution for the atomic inversion other than the substitutions:

$$\gamma_d \rightarrow \gamma_d \quad , \quad \gamma_{od} \rightarrow \gamma_{od} + \gamma_\ell . \tag{18}$$

This means that the width-change question has a simple answer for a phase-noisy laser. The entire effect is to add γ_ℓ to the existing absorption linewidth, exactly as in the perturbative case (12). Since the inversion $w(t)$ is not directly affected by the phase of the light field in any event, this result is perhaps not surprising, and maybe even disappointing because it is so conventional.

However, it is much less easy to anticipate that the __emission linewidth__ is an entirely different matter. In that case (18) is not correct and must be replaced[3,4] by

$$\gamma_d \rightarrow \gamma_d + \gamma_\ell \quad , \quad \gamma_{od} \rightarrow \gamma_{od} + 2\gamma_\ell . \tag{19}$$

Thus, in emission, even the diagonal relaxation rate is exaggerated by the existence of phase fluctuations. AGARWAL [3] and WODKIEWICZ [3] have discussed in detail the modifications of the standard Torrey polynomial (17), and has given substitution rules analogous to (18) and (19), appropriate to a number of different experimental situations that need not be discussed in detail here.[5] The important observation that follows from these results [1,3,4] is that even in the intuitively simple case of a laser bandwidth due to random phase jumps with slow phase diffusion there is not a unique answer to the width change question.

4.   Excitation by a Laser with Intensity Fluctuations

All laser light is subject to intensity fluctuations, and it is important
to determine their effect on questions of lineshape. Even the basic "width
change" question is difficult to answer in this case. BURSHTEIN [1] and
WÓDKIEWICZ [3] have found cases that can be solved exactly, but each of
them is hampered by unrealistic assumptions. In BURSHTEIN'S example exact
resonance is required, so there is no possibility of attacking the linewidth
question, and in WÓDKIEWICZ'S case the complex electric field amplitude is
required to be delta-correlated:

$$\Omega(t) = \Omega_0 + \delta\Omega(t) \tag{20}$$

with $((\delta\Omega(t) \, \delta\Omega^*(s)))$ = $2\Gamma_1\delta(t-s)$ and $((\delta\Omega(t) \, \delta\Omega(s)))$ = $2\Gamma_2\delta(t-s)$ The
Hermitean character of Ω requires Γ_1 to be real, and satisfy $\Gamma_1 \geq |\Gamma_2|$.

Nevertheless WÓDKIEWICZ [3] shows that the problem is still characterized
by a cubic polynomial,

$$p^W(z) = (z + \tfrac{1}{2}\,\Gamma_1)\,[(z + \tfrac{1}{2}\,\Gamma_1)(z + \Gamma_1) + \Omega_0^2] +$$
$$+ (\Delta^2 - \tfrac{1}{4}\,|\Gamma_2|^2)\,(z + \Gamma_1) - \tfrac{1}{4}\,\Omega_0^2\,(\Gamma_2 + \Gamma_2^*) \ . \tag{21}$$

There is no obvious connection between (21) and the Torrey polynomial (17),
except in the more or less well-known limit $\Gamma_2 = 0$. One notes in (21) that
even when $\Gamma_2 = 0$ both diagonal and off-diagonal relaxation processes are
affected by Γ_1. This is in line with one's expectations, that intensity
fluctuations have a direct impact on both atomic populations and phases.
However, there is clearly no substitution rule such as (18) or (19) and no
simple general answer to the width-change question, in strong contrast to
the phase-noise case.

We have recently examined, using Monte Carlo techniques, the BURSHTEIN-
WÓDKIEWICZ two-level problem. We have treated a model that permits arbitrary
detuning and arbitrarily large or small γ_ℓ, although still restricted by
the exponential form of (5). No decorrelations are introduced but the
ensemble averages are evaluated numerically. Some analytic results are possible in special cases. Our procedure is the following. Eqs. (13) - (15)
are integrated. The field is chosen to have a phase that jumps by π at
random times and the intensity also jumps randomly from one value to another,
the possible values for $|\Omega|$ coming from a normalized distribution $p(\Omega)$.
The average jump interval is γ_ℓ^{-1}. In Fig. 2 some representative results
are shown for the solution for $((w))$.

Three qualitative remarks are suggested by one's intuition, and are illustrated in the solutions shown in Fig. 2. First, in the weak-field and
long-time limits the results of perturbation theory should be valid, and
Fig. 2a verifies this, showing a linear growth of population in the excited
state at exactly the rate[7] $((|\Omega|^2)) /\gamma_\ell$. Second, *homogeneous* relaxation
should always occur for $\gamma_\ell t > 1$ because the exciting field is itself coherent only for $t < \gamma_\ell^{-1}$.   All of the graphs of Fig. 2 are consistent
with this. Exponential decay at exactly the rate γ_ℓ is shown clearly
in Fig. 2b.

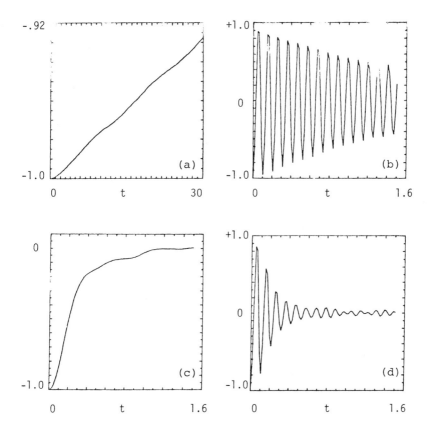

Fig. 2   The average inversion $\langle\!\langle\, w(t)\,\rangle\!\rangle$ in the presence of both phase
jumps and intensity fluctuations. The time scale is in units of the
laser's coherence time $2\gamma_\ell^{-1}$. In the top two examples the effective
relaxation time is equal to the laser's coherence time, but in the
lower two examples it is shorter, due to the intensity fluctuations.

Third, under some conditions a new kind of *inhomogeneous* relaxation
should be observable. This is because the dynamical behavior of every
member of the statistical ensemble is described by the Rabi solution
$w(t) = -\cos\Omega t$. The average of $w(t)$ over a distribution $p(\Omega)$ of values
of Ω will lead to decay of $\langle\!\langle\, w(t)\,\rangle\!\rangle$ at a rate determined by the width of
this distribution and quite independent of γ_ℓ. Both Figs. 2c and 2d show
this effect. In each case the parameters are the same as for the figures
immediately above them, except that the width of $p(\Omega)$ has been increased

from 0.001 γ_ℓ to 5 γ_ℓ. The relaxation rate in Figs. 2c and 2d is clearly not γ_ℓ, and may be shown to be 5 γ_ℓ.

The existence of significant Rabi oscillations even when inhomogeneous decay is dominant over homogeneous decay (laser-induced free-induction decay?), as shown in Fig. 2d, is possible only when $p(\Omega)$ has a relatively narrow peak far from $\Omega = 0$. This is the situation for a moderately well stabilized laser far above threshold. A Gaussian $p(\Omega)$ centered at $\Omega = 0$ has been considered by AVAN and COHEN-TANNOUDJI [3] for resonance fluorescence and by ZOLLER [8] for multiphoton ionization. In this case the difference between $((\Omega^4))$ and $((\Omega^2))^2$ is so slight that coherent effects are inhomogeneously damped out within one Rabi cycle.

5.   Applications and Summary

In the preceding Sections I have used results of model calculations to show that there is not a universal rule that can be used to predict the consequences of finite-laser-bandwidth excitation simply by knowing the consequences of weak or monochromatic laser excitation. In view of the specialized character of the models used it is desirable that their predictions be examined experimentally. To a limited extent this has been done,[9] without conclusive results. As an example, to show some of the advantages and weak points of these theoretical models, I will mention briefly only the last of Refs. 9.

The solid dots and dashed curves in Fig. 3 show experimental observations[10] and an old theoretical prediction[11] for 2-photon ionization of cesium. The theory assumes a zero bandwidth laser. The solid curves show the results of a calculation[12] using a finite-bandwidth phase-noisy laser. It is apparent that the noisy-laser theory can give much better agreement with the data. What is not apparent in Fig. 3 is that expression (5) is involved in this agreement in a crucial way. An exponential correlation function such as (5) implies a Lorentzian spectrum for the laser. It is highly unlikely that the Ar ion laser used in the experiments had a Lorentzian spectrum, although it is the extreme far wings (>100 γ_ℓ) of the spectrum that are significant here, and the shape of the far wings appears not to be known.

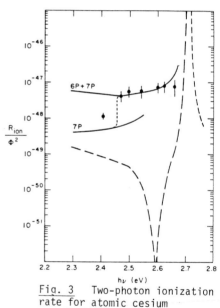

Fig. 3   Two-photon ionization rate for atomic cesium

In closing, without commenting further on the need for more experiments of the type in Refs. [9], it is well to emphasize the question of (5), because it contains the weak point of theoretical work to date. All models that permit fully non-perturbative treatments of absorption appear to be based either on correlation function (5), or on delta-function correlations. All of these are valuable in permitting analytic and parametric estimates of laser bandwidth effects, but none of them correspond to a laser with a realistic spectral shape. This remains an important defect in the theory of laser spectroscopy, one that I hope can be dealt with successfully in the near future.

Acknowledgments

It is a pleasure to acknowledge discussions with G.S. Agarwal, K. Wódkiewicz, and J.J. Yeh. This work was partially supported by the U.S. Department of Energy under Contract EY-76-S-03-1118.

References

1. A.I. Burshtein, Sov. Phys. JETP 21, 567 (1965) and 22, 939 (1966)
2. H.C. Torrey, Phys. Rev. 76, 1059 (1949)
3. G.S. Agarwal, Phys. Rev. A 18, 1490 (1978), and K. Wódkiewicz, Phys. Rev. A 19, 1686 (1979)
4. G.S. Agarwal, Phys. Rev. Lett. 37, 1383 (1976); J.H. Eberly, ibid., 37, 1387 (1976); H.J. Kimble and L. Mandel, Phys. Rev. A 15, 689 (1977); P. Avan and C. Cohen-Tannoudji, J. Phys. B 10, 155 (1977); and P. Zoller, J. Phys. B 10, L321 (1977)
5. A case not discussed by Wódkiewicz [3] but of recent interest is that of a two-level system capable of ionization. Then Eqs. (13) - (15) must be supplemented by a fourth equation that allows the upper level population to decay at a rate proportional to laser intensity. When only phase fluctuations are present this case is also exactly solvable. See J.H. Eberly and S.V. ONeil, Phys. Rev. A 19, 1161 (1979), and references therein.
6. J.J. Yeh, T.A. Nguyen, and J.H. Eberly (unpublished); see also T.A. Nguyen, Thesis, University of Rochester 1979
7. Only 1000 time histories were averaged to produce the graphs in Fig. 2, so some finite-ensemble variations are evident. The rate of growth in Fig. 2a is just that predicted by (10) in the relevant limits, $\Delta = \gamma_{21} = 0$. See also J.R. Ackerhalt and J.H. Eberly, Phys. Rev. A 14, 1705, Appendix
8. P. Zoller, Phys. Rev. A 19, 1151 (1979)
9. P. Agostini, A.T. Georges, S.E. Wheatley, P. Lambropoulos, and M. Levenson, J. Phys. B 11, 1733 (1978); B.R. Marx, J. Simons, L. Allen, ibid., 11, L273 (1978); P.B. Hogan, S.J. Smith, A.T. Georges and P. Lambropoulos, Phys. Rev. Lett. 41, 229 (1978); and L. Armstrong and J.H. Eberly, J. Phys. B (in press)
10. E.H.A. Granneman and M.J. Van der Wiel, J. Phys. B 8, 1617 (1975)
11. M.R. Teague, P. Lambropoulos, D. Goodmanson, and D.W. Norcross, Phys. Rev. A 14, 1057 (1976)
12. L. Armstrong, Jr., and J.H. Eberly, J. Phys. B (in press)

Interaction of Two Resonant Laser Fields with a Folded Doppler Broadened System of I_2

R.P. Hackel and S. Ezekiel

Research Laboratory of Electronics, Massachusetts Institute of Technology
Cambridge, MA 02139, USA

For many years there has been much attention given to the treatment of the interaction of electromagnetic radiation with three-level systems. Theoretical calculations have considered homogeneous and inhomogeneous systems [1-6] in both folded and cascade configurations and experimental data have been obtained in atomic beams [7] and gas cells [8,9]. In this paper, we present high resolution experimental studies of the interaction of two laser fields with a folded Doppler broadened system in I_2.

The experimental arrangement used is shown in Fig.1. A single frequency argon laser which acts as the pump is held fixed within the 21 Doppler broadened $X^1\Sigma \rightarrow B^3\Pi$ (43,0) R(15) hyperfine transitions at 5145Å. A single frequency tunable dye laser, used as the probe is scanned over the $X^1\Sigma \leftarrow B^3\Pi$ (43,11) R(15) hyperfine transitions at 5828Å. Both lasers are frequency stabilized to independent Fabry-Perot cavities [10]. The beam dimensions of the two lasers are closely matched and the combined beams are collimated before traversing the 80 cm long I_2 cell. After three passes through the cell to enhance sensitivity, the probe beam is separated from the pump and reflected back upon itself. In this manner both forward and backward scattering, resulting from copropagating and counterpropagating beams, respectively,

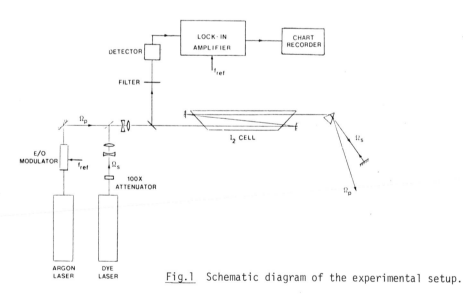

<u>Fig.1</u> Schematic diagram of the experimental setup.

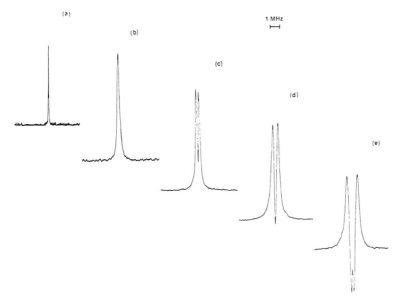

Fig.2 Forward scattered hyperfine lineshape for peak pump intensity of (a) 7.7 mW/cm^2, (b) 68 mW/cm^2, (c) 330 mW/cm^2, (d) 1.0 W/cm^2, (e) 4.7 W/cm^2. Probe intensity 3.1 mW/cm^2, P = 2 mTorr. Vertical scale is arbitrary.

can be observed in one scan. In order to record the probe lineshape, the pump amplitude is modulated by the electrooptic crystal at 2 kHz and the probe is synchronously demodulated by the lock-in amplifier.

Using this setup, this folded Doppler broadened I_2 system was studied as a function of pump intensity for both forward and backward scattering. The lineshape for one of the forward scattered hyperfine lines is shown in Fig.2 for various pump intensities and a weak probe field. Fig.2a displays the extremely narrow resonances obtained with weak pump (7.7 mW/cm^2) and weak probe (3.1 mW/cm^2) fields. The width of these lines can be much narrower than the width of the upper level [9]. With increase in pump intensity, the lineshape first broadens (Fig.2b) and then splits (Fig.2c), a manifestation of the ac Stark effect. For pump intensities greater than about 1 W/cm^2, the lineshape exhibits absorption near line center (Fig.2d), with an added splitting appearing within the absorption feature at a pump intensity of about 4.7 W/cm^2 (Fig.2e).

The backward scattered lineshape as a function of pump intensity is shown in Fig.3, with the corresponding forward scattered lineshape displayed for comparison. The weak field line in Fig.3a appears broader and smaller for the backward scattered case as indicated by the vertical scale factor. Increase in the pump intensity first broadens (Fig.3b) and then splits the backward scattered lineshape (Fig.3c). However the splitting is first observed at a higher pump intensity ($\sim$410 mW/cm^2) than for the forward scattered line which appears at about 100 mW/cm^2. Further increase in pump intensity increases the magnitude of the splitting as shown in Fig.3d and 3e.

89

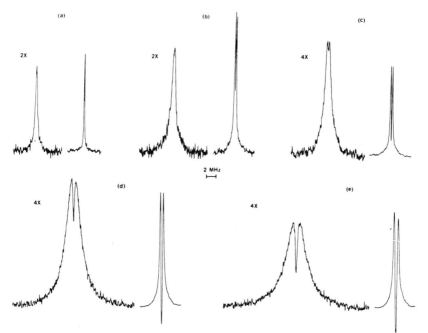

Fig.3 Backward scattered hyperfine lineshape (left) with corresponding forward scattered lineshape (right). Peak pump intensity (a) 27 mW/cm², (b) 200 mW/cm², (c) 410 mW/cm², (d) 1.1 W/cm², (e) 3.2 W/cm². Probe intensity 1.3 mW/cm², P = 2.5 mTorr. Vertical scale factor between corresponding forward and backward scattered lines is indicated.

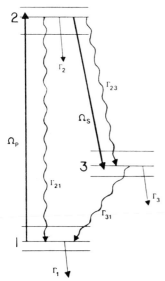

The theory for the interaction of two resonant laser fields with a three-level system has been presented by many authors using a density matrix method [1-6]. To explain our results, we have used a similar approach, modelling the I_2 system as the folded Doppler broadened three-level system illustrated in Fig.4. The pump laser frequency, Ω_p, is close to the ω_{21} transition frequency and the probe frequency, Ω_s, is likewise close to ω_{23}[11]. The decay rate of level i to level j is given by Γ_{ij} (the total decay rate of level i being Γ_i) with the relaxation rate for optical coherence $\gamma_{ij} \geq 1/2$ [$\Gamma_i + \Gamma_j$]. Since only electric dipole transitions are considered, the Γ_{31} rate shown is zero. With a classical description of the driving fields, the usual dipole approximation for the interaction Hamiltonian and the rotating wave approximation, the equations of motion for the density matrix elements are given by:

Fig.4 Schematic diagram of I_2 three-level system in a folded configuration.

$$\dot{\rho}_{11} = \frac{i}{2} (\Omega_R \rho_{21} - \Omega_R^* \rho_{21}^*) - \Gamma_1 \rho_{11} + \Gamma_{21} \rho_{22}$$

$$\dot{\rho}_{22} = -\frac{i}{2} (\Omega_R \rho_{21} - \Omega_R^* \rho_{21}^*) + \frac{i}{2} (\Omega_R' \rho_{32} - \Omega_R'^* \rho_{32}^*) - \Gamma_2 \rho_{22}$$

$$\dot{\rho}_{33} = -\frac{i}{2} (\Omega_R' \rho_{32} - \Omega_R'^* \rho_{32}^*) - \Gamma_3 \rho_{33} + \Gamma_{23} \rho_{22}$$

$$\dot{\rho}_{21} = [i(\Omega_p - \omega_{21}) - \gamma_{21}] \rho_{21} + \frac{i}{2} \Omega_R' \rho_{31} - \frac{i}{2} \Omega_R^* (\rho_{22} - \rho_{11})$$

$$\dot{\rho}_{32} = [-i(\Omega_s - \omega_{23}) - \gamma_{32}] \rho_{32} - \frac{i}{2} \Omega_R \rho_{31} - \frac{i}{2} \Omega_R'^* (\rho_{33} - \rho_{22})$$

$$\dot{\rho}_{31} = [i(\Omega_p - \Omega_s - \omega_{31}) - \gamma_{31}] \rho_{31} - \frac{i}{2} \Omega_R \rho_{32} + \frac{i}{2} \Omega_R'^* \rho_{21}$$

(1)

with $\Omega_R = \mu_{12} E_p/h$ and $\Omega_R' = \mu_{23} E_s/h$ the Rabi frequencies for the pump and probe transitions, respectively.

If the assumption of a weak probe field is made (i.e. $\Omega_R' \ll \Gamma_{ij}$) then steady state solutions for the density matrix elements can be readily found. The quantity of interest is the absorption coefficient for the probe beam, given by:

$$\alpha_s = -\frac{1}{I_s} \frac{dI_s}{dz} = -\frac{h\Omega_s}{I_s} \Omega_R' \, \text{Im} \, \rho_{32}$$

(2)

where I_s is the power density of the probe beam. In the weak probe field limit, α_s can be expressed as

$$\alpha_s = \frac{h\Omega_s \Omega_R'}{I_s} \left\{ \frac{(N_1^\circ - N_2^\circ) \, \Omega_R^2}{4\Gamma_c \, (\Delta_p^2 + \gamma_{21}^2) + 2\Gamma_B \gamma_{21} \Omega_R^2} \times \right.$$

$$\times \, \text{Re} \left[\frac{-2\Gamma_A \gamma_{21}(i[\Delta_p - \Delta_s] - \gamma_{31}) + \Gamma_c(i\Delta_p + \gamma_{21})}{(-i\Delta_s - \gamma_{32})(i[\Delta_p - \Delta_s] - \gamma_{31}) + \Omega_R^2/4} \right]$$

$$\left. + (N_3^\circ - N_2^\circ) \, \text{Re} \left[\frac{i(\Delta_p - \Delta_s) - \gamma_{31}}{(-i\Delta_s - \gamma_{32})(i[\Delta_p - \Delta_s] - \gamma_{31}) + \Omega_R^2/4} \right] \right\}$$

(3)

where

$$\Gamma_A = \Gamma_1(\Gamma_3 - \Gamma_{23})$$

$$\Gamma_B = \Gamma_3(\Gamma_1 + \Gamma_2 - \Gamma_{21})$$

$$\Gamma_C = \Gamma_1\Gamma_2\Gamma_3$$

$$\Delta_p = \Omega_p - \omega_{21}$$

$$\Delta_s = \Omega_s - \omega_{23}$$

and N_i^o is the zero field population of level i. The second term in Eq. (3) is related to the linear susceptibility, whereas the first term, proportional to the pump intensity, is related to the third order susceptibility which is the observable quantity in our setup.

The above result (3) assumes a single velocity group, specifically the v = 0 group. For a collection of molecules in thermal equilibrium with a most probable velocity $v_0 = \sqrt{2kT/m}$, the absorption coefficient must be integrated over the Maxwellian velocity distribution

$$\alpha_s(v) = \int_{-\infty}^{+\infty} \alpha_s' \frac{\exp(-v^2/v_0^2)}{v_0\sqrt{\pi}} \, dv \tag{4}$$

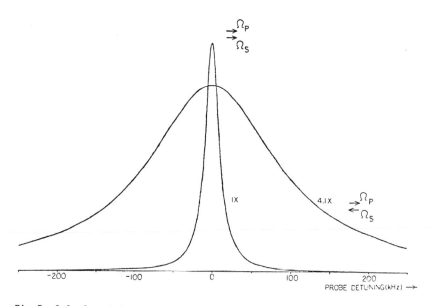

Fig.5 Calculated forward and backward scattered lineshape for a weak pump field (Ω_R = 1 kHz) for Doppler broadened I_2 system.

where α_s is derived from α_s above by replacing $\Delta_p \rightarrow \Delta_p - \Omega_p$ v/c and $\Delta_s \rightarrow \Delta_s - \varepsilon\Omega_s$ v/c with ε = +1 for forward scattering and -1 for backward scattering.

For the three-level I_2 system under consideration, the integration over the velocity distribution was carried out numerically for both forward and backward scattering. The resultant lineshapes are shown in Fig.5 for a weak pump field (Ω_R = 1 kHz) and Fig.6 for a strong pump field (Ω_R = 1 MHz). The values of the parameters used in the calculations of the lineshapes of Figs.5 and 6 are $\Gamma_1 = \Gamma_3$ = 5 kHz, Γ_2 = 100 kHz, Γ_{21} = 0.74 kHz, Γ_{23} = 0.16 kHz, $\gamma_{ij} = (\Gamma_i + \Gamma_j)/2$ and Ω_s/Ω_p = 0.8828 corresponding approximately to the I_2 levels studied in our present setup.

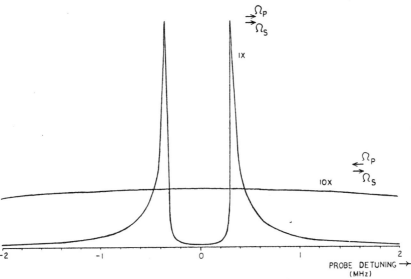

Fig.6 Calculated forward and backward scattered lineshape for a strong pump field (Ω_R = 1 MHz).

For the weak field case the calculated linewidth, Γ_s, simplifies to [2,5].

$$\Gamma_s \approx \Gamma_1 + (1 - \varepsilon\Omega_s/\Omega_p) \Gamma_2 + \Gamma_3 \tag{5}$$

For our I_2 system the Ω_s/Ω_p ratio is approximately 0.9 and Γ_2 is the largest relaxation parameter. Therefore, the difference between forward and backward scattered linewidths is significant as shown in Fig.5, where Γ_s is 198 kHz for backward scattering as compared with 21 kHz for forward scattering. The experimentally observed lineshapes for weak fields (Fig.3a) are consistent with the calculated ones in Fig.5.

For the strong pump field case (Fig.6), the calculated forward scattered lineshape is still much larger and narrower than the backward scattered one. Moreover, the forward scattered lineshape displays a splitting which is absent for backward scattering. In addition, no absorption is observed

93

near line center, even for larger values of the pump intensity. The intensity levels for the calculated lineshape of Fig.6 correspond roughly to the experimental lineshapes in Fig.3d. The presence of a splitting in the observed backward scattered line and the absorption feature near line center in the forward scattered line are inconsistent with the calculated lineshapes. Effects due to optical pumping, the finite transit time, and higher order processes in I_2 are presently under investigation in an attempt to explain these discrepancies.

Many interesting applications have been suggested for this two-step excitation scheme [5,9]. For example, Doppler-free spectroscopy of thermally unpopulated levels in the ground electonic state can be performed using such a two-step process. In addition, with a strong pump field the matrix elements associated with individual pump transitions can be determined from the splitting of the probe lines. This is illustrated in Fig.7, which shows a scan over several (43,11) R(15) hyperfine transitions for both forward and backward scattering with a pump intensity of 0.4 W/cm². (The forward/backward lines are indicated by an F/B subscript.) Another unique application is the measurement of relaxation parameters of individual energy levels. The relaxation rate of the upper level, level 2, can be deduced from corresponding forward and backward scattered linewidths [5] as indicated in Eq.(5). In order to determine the relaxation rates for levels 1 and 3, it is necessary to have one additional measurement, namely the linewidth of the single-step transition between levels 1 and ·2.   Finally, the extremely narrow resonances generated by this resonant two-step process have important applications as secondary frequency standards [9].

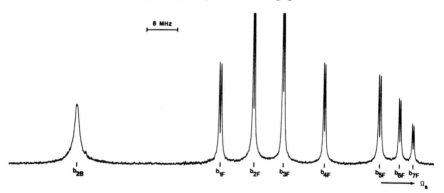

Fig.7 Typical scan over several (43,11) R(15) hyperfine lines for a 0.4 W/cm² pump intensity. Probe intensity 3.4 mW/cm², P = 2 mTorr. Forward/backward scattered lines are marked with an F/B subscript.

This research was supported by the National Science Foundation, grant number PHY77-07156 and the Joint Services Electronics Program at the Massachusetts Institute of Technology.

References

1. M.S. Feld and A. Javan, Phys. Rev. 177, 540 (1969).
2. T. Hänsch and P. Toschek, Z. Phys. 236, 213 (1970).
3. B. Feldman and M.S. Feld, Phys. Rev. A5, 899 (1972).
4. B.R. Mollow, Phys. Rev. A8, 1949 (1973).

5. I.M. Beterov and V.P. Chebotaev, in Progress in Quantum Electronics, edited by J.H. Sanders et. al., (Pergamon, Oxford, England, 1974), Vol. 3, Part 1. This article contains an extensive list of references up to 1973.

6. C. Delsart and J.C. Keller, J. Physique 39, 350 (1978) and references therein.

7. J.-L. Picque and J. Pinard, J. Phys. B9, L-77 (1976); H.R. Gray and C.R. Stroud, Opt. Comm. 25, 359 (1978); S. Ezekiel and F.Y. Wu, in Multiphoton Processes, edited by J.H. Eberly and P. Lambropoulos (Wiley, New York, 1978) pp. 145-56.

8. A Schabert, R. Keil and P. Toschek, Appl. Phys. 6, 181 (1975); P. Cahouzac and R. Vetter, Phys. Rev. A14, 270 (1976); C. Delsart and J.C. Keller, Opt. Comm. 15, 91 (1975); Opt. Comm. 16, 388 (1976); J. Phys. B9, 2769 (1976); B. Wellegehausen and H.H. Heitman, Appl. Phys. Lett. 34, 44 (1979).

9. R.P. Hackel and S. Ezekiel, Phys. Rev. Lett. 42, 1736 (1979).

10. L.A. Hackel, R.P. Hackel and S. Ezekiel, Metrologia 13, 141 (1977); F.Y. Wu and S. Ezekiel, Laser Focus, 13, 78 (1977).

11. Throughout this paper all frequencies, linewidths and relaxation rates will be expressed in circular frequency units.

Compensation of Doppler Broadening by Velocity-Dependent Light-Shifts

S. Reynaud, M. Himbert, J. Dupont-Roc, and C. Cohen-Tannoudji

Ecole Normale Supérieure and Collège de France, 24, rue Lhomond
F-75231 Paris Cedex 05, France

1. Introduction

When a 2-level atom is irradiated by a quasi-resonant light beam, having a frequency ω_L close to the atomic frequency ω_0, its energy levels a and b are shifted. Perturbation theory shows that the so called light-shift ε_a of the lower level $\underline{a}$ (which is equal to $-\varepsilon_b$) is proportional to the light-intensity I and inversely proportional to the detuning $\omega_L-\omega_0$, provided that this detuning is not too small.

$$\varepsilon_a = - \varepsilon_b \sim I/(\omega_L - \omega_0) \tag{1}$$

Light-shifts have been studied several years ago, before the advent of lasers, in optical pumping experiments using ordinary light sources [1][2]. By choosing a convenient polarization for the light beam, it was possible to shift two Zeeman sublevels of the ground state by different amounts and to detect this effect by a shift of the magnetic resonance curve between these two sublevels. As a consequence of the long relaxation times in the ground state, this resonance curve is very narrow and light-shifts as small as one hertz have been easily measured.

Now, with laser sources, light-shifts have considerably increased, from a few Hz to several GHz [3], and they can be observed directly on optical transitions. Actually, they introduce a limitation in the accuracy of high resolution spectroscopic measurements and they have to be avoided or, at least, carefully controlled.

On the other hand, during the last few years, new physical effects using light-shifts have been proposed and studied. For example, atomic beam deflection experiments using transverse dipole forces, and discussed in this meeting [4], may be interpreted in terms of position-dependent light-shifts. An intensity gradient $I(\vec{r})$ of the laser beam introduces a $\vec{r}$-dependent light-shift $\varepsilon_a(\vec{r})$ of the ground state. When the detuning is large enough, $\varepsilon_a(\vec{r})$ appears as a potential energy for the atom, giving rise to dipole forces, $-\vec{\nabla}\varepsilon_a(\vec{r})$, which can deflect an atomic beam.

In this paper, we discuss another effect using velocity-dependent light-shifts for compensating the Doppler broadening of an optical transition. We will briefly outline the principle of such a scheme which is analyzed in more details in references [5] and [6] (see also [7]). Experimental evidence for the narrowing mechanism has been obtained recently [8]. We present here additional experimental results obtained on [20]Ne and which exhibit important polarization effects.

Let us finally mention that other mechanisms for producing velocity dependent energy shifts, which could lead to a compensation of the Doppler broadening, have been recently suggested. They use quadratic Stark shifts in crossed static electric and magnetic fields [9], or motional interactions of spins with static electric fields in polar crystals [10].

2. Discussion of the narrowing mechanism

Consider an atom moving, in the laboratory frame, with a velocity v towards a laser beam with frequency ω_L (Fig.1).

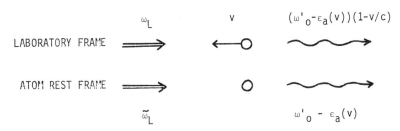

LABORATORY FRAME

ATOM REST FRAME

Fig.1 - *Laser and emission frequencies in laboratory and atom rest frames.*

The light-shifts induced by such a laser irradiation have to be evaluated in the rest frame of the atom where the laser frequency is Doppler shifted from ω_L to $\tilde{\omega}_L = \omega_L (1 + \frac{v}{c})$. The v-dependence of $\tilde{\omega}_L(v)$ results in a v-dependent detuning $\tilde{\omega}_L(v) - \omega_0$ between the apparent laser frequency $\tilde{\omega}_L(v)$ and the frequency ω_0 of the transition a-b. It follows that the light-shift ε_a of $\underline{a}$ which depends on this detuning is also v-dependent.

Suppose now that we observe, in the same direction as the laser, the light spontaneously emitted from a third level $\underline{c}$ to $\underline{a}$ by an atom excited in $\underline{c}$ (for example by a discharge). The frequency ω'_0 of transition a-c is completely off-resonance with ω_L, so that level $\underline{c}$ is not perturbed by the laser. In the rest frame, the emitted frequency is equal to ω'_0 corrected by the light shift $\varepsilon_a(v)$ of $\underline{a}$, i.e to $\omega'_0 - \varepsilon_a(v)$. Coming back to the laboratory frame introduces the well-known Doppler factor $(1 - \frac{v}{c})$ since the atom is moving away with a velocity v.

Now, the basic idea discussed in this paper is to try to achieve a compensation between the v-dependence of the light-shifted internal frequency $\omega'_0 - \varepsilon_a(v)$ and the v-dependence of the emission Doppler factor $(1 - \frac{v}{c})$, in order to have, in the laboratory frame, all atoms emitting at the same frequency in the forward direction.

Such a compensation condition is easily found to be

$$\varepsilon_a(v) = - \omega'_0 \, v/c \tag{2}$$

Before describing how it can be achieved, let's first discuss some important characteristics of this new scheme.

First, the narrowing is not due to a population effect. We are looking at a spontaneous emission signal which is independent of the population of the final state a and which originates from a level c which is not perturbed by the laser. Levels a and b could even be empty.

Second, there is no power broadening of the Doppler free line. When the compensation condition is fulfilled, all atoms emit at the same frequency. The width of the line is just the homogeneous width γ. These features clearly distinguish this effect from others, such as fluorescence line narrowing [11] [12] where the laser irradiation creates in the velocity distribution of atoms a population hole or a population peak which is power broadened.

Finally, and this is perhaps the most interesting point, this effect is highly anisotropic. If one looks at the light emitted not in the forward but in the backward direction, the Doppler emission factor changes from $(1 - \frac{v}{c})$ to $(1 + \frac{v}{c})$. If the Doppler broadening is compensated for one direction of emission, it is clearly doubled for the opposite one. Such a scheme could therefore provide an atomic medium emitting with the homogeneous width γ in one direction and with twice the Doppler width Δ in the opposite one. Forward-backward asymmetries are well known effects in laser spectroscopy of 3-level systems [12] but, here, such an asymmetry could be particularly important because of the large difference between γ and Δ.

3. How to get a light-shift proportional to v ?

Two possibilities are represented on Fig.2. One of them (Fig.2a) considers a 2-level system with frequency ω_0 irradiated by two co-propagating laser beams with frequencies $\omega_0 + \delta$ and $\omega_0 - \delta$. In the second scheme (Fig.2b), which is simpler experimentally, we have a single laser beam ω_0, which is σ-polarized and which excites a $J = 0$ to $J = 1$ transition in a static field, so that the two Zeeman sublevels b_+ and b_-, which are coupled to the laser, are detuned by $+\delta$ and $-\delta$ (δ is the Zeeman splitting).

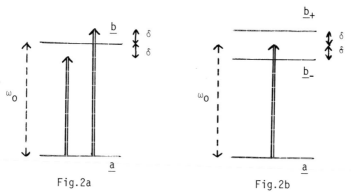

Fig.2a   Fig.2b

For a $v = 0$ atom, we have in both cases two opposite detunings, so that the two light-shifts produced by the two lasers in the first case, by the virtual excitation of the two Zeeman components in the second one, balance. It follows that $\varepsilon_a(v=0) = 0$.

On the other hand, when v is different from zero, the Doppler shift of the laser frequencies introduce unequal detunings and, consequently, unbalanced light-shifts. This shows that $\varepsilon_a(v)$ is an odd function of v which depends on the laser intensity I and on the detuning δ.

From now on, we will consider only the second scheme (Fig.2b). The laser intensity I being fixed (generally at its highest possible value), we have to find if it is possible to adjust δ so that the linear term in the expansion of $\varepsilon_a(v)$ coincides with $- \omega'_0 v/c$. We have also to understand the effect of higher order terms (in v^3, v^5, ...) and to determine under what conditions they can be neglected.

4. Calculation of the compensation condition and of the emission spectra

It turns out that the light intensities required for achieving the compensation condition may be quite large,so that a non-perturbative treatment of the atom-laser coupling is necessary [13].

There are actually three relevant states of the atom-laser system which are strongly coupled. Let $|a, n\rangle$ be the state corresponding to the atom in a in presence of n laser photons. We take its unperturbed energy equal to 0 so that the corresponding perturbed energy will represent the light-shift ε_a. The atom in a can absorb one laser photon and jump into one of the two Zeeman sublevels b_+ or b_- of b, so that we have also to consider the states $|b_+, n-1\rangle$ and $|b_-, n-1\rangle$ with unperturbed energies respectively equal to $\delta - u$ and $- \delta - u$ where δ is the Zeeman splitting and $u = \tilde{\omega}_L - \omega_0 = \omega_0 v/c$ the detuning of the laser with respect to ω_0 in the rest frame, proportional to v. The atom-laser coupling V is characterized by the matrix elements

$$\langle a, n | V | b_+, n-1 \rangle = \omega_1/2 \tag{3}$$

where ω_1 is a Rabi frequency equal to the product of the atomic dipole by the laser electric field. We suppose $\omega_1 \gg \gamma$ so that we can neglect the damping processes in a first step and diagonalize the matrix

$$\begin{pmatrix} 0 & \omega_1/2 & \omega_1/2 \\ \omega_1/2 & \delta-u & 0 \\ \omega_1/2 & 0 & -\delta-u \end{pmatrix} \tag{4}$$

The eigenvalues of (4) give the three perturbed energies versus u, i.e. versus v (in particular, the light-shift of a is not just the sum of the two light-shifts associated with the coupling to $|b_+, n-1\rangle$ alone, and $|b_-, n-1\rangle$ alone, as it would be in a perturbative treatment). One gets in this way an exact expression for the linear term (in v) of $\varepsilon_a(v)$ leading, for the compensation condition (2), to the following exact relation

$$\omega_1^2/(\omega_1^2 + 2\delta^2) = \omega'_0/\omega_0 = \tilde{s} \tag{5}$$

between ω_1^2 (proportional to the laser intensity), δ, and ω'_0/ω_0.

The diagonalization of (4) gives also three perturbed states $|i, n\rangle$ (i=1,2,3), which all contain admixtures of $|a, n\rangle$, so that the emission spectrum $\underline{c} \rightarrow \underline{a}$

is actually a triplet corresponding to the three transitions $|c, n> \rightarrow |i, n>$. These three lines have a simple physical interpretation in the perturbative regime. We have first the Doppler free line corresponding to the spontaneous emission from c to the light-shifted level a (Fig.3a), and then two Raman sidebands corresponding to inverse Raman processes where the atom starting from c spontaneously emits one photon and absorbs one laser photon to end in sublevels b_+ or b_- which are also light-shifted (Fig.3b).

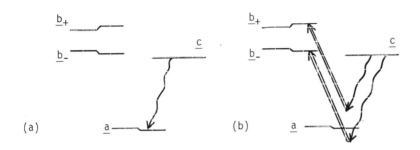

Fig. 3 - *Perturbative interpretation of the three lines emitted from c*

Because of the Maxwellian distribution of velocities, the diagonalization of (4) must be done for a range of values of v corresponding to the Doppler width Δ of a - b.

If the light intensity is very large, so that

$$\omega_1 >> \Delta \qquad (6)$$

we can neglect the curvature of the perturbed energy levels associated with (4) when v varies within the Doppler width. In such a case, we have, when condition (5) is fulfilled, a complete compensation of Doppler broadening : all atoms emit the central line at the same frequency ω'_0. The theoretical spectrum obtained in these conditions (Fig.4) exhibits a central narrow peak at ω'_0 with the homogeneous width γ. The compensation of Doppler broadening does not occur for the two Raman sidebands which are located at a distance $\pm \omega_1/\sqrt{2s}$ and which remain Doppler broadened. The weights of the lines are respectively equal to s/2, 1-s, s/2. They are comparable, so that the height of the narrow peak is much larger.

On the other hand, if

$$\Delta >> \omega_1 >> \gamma \qquad (7)$$

we cannot neglect the curvature of the energy levels versus v within the Doppler width. In such a case, we have only a partial compensation of Doppler broadening, occurring for the fraction of the velocity distribution where the linear approximation is valid. The spontaneously emitted frequencies are redistributed into a central symmetric narrow peak (which represents the contribution of atoms having a light shift linear in v), and two sharp-edged sidebands (see for exemple the theoretical curve B = 42 G of Fig.5).

If one increases the light intensity and, simultaneously, changes the Zeeman splitting δ in order to maintain (5), the height of the central peak would increase without any broadening, whereas the two sidebands would move away and become Doppler broadened.

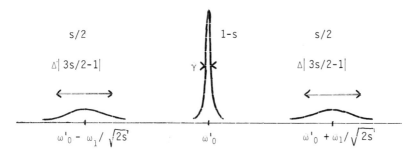

Fig.4 - *Theoretical emission spectrum in the case of a complete compensation of Doppler broadening.*

5. Experimental results

The first experimental evidence for the compensation of Doppler broadening by velocity-dependent light-shifts has been obtained on ^{20}Ne atoms [8].

The three levels a, b, c are respectively the $1s_3$, $2p_2$ and $2p_{10}$ levels (Paschen notations). The output of a c.w. dye laser is tuned to the wavelength 6163 Å of the a - b transition and is focussed after a σ polarizer inside a d.c. discharge cell containing 1.5 torr of ^{20}Ne and put in a static magnetic field. The π-polarized fluorescence which is emitted from the $m = 0$ sublevel of $2p_{10}$, at 7439 Å, is isolated by color and interference filters and spectrally analyzed with a confocal Fabry-Perot interferometer. The laser power was of the order of 150 mW and the waist w in the discharge cell of the order of 400μ, which leads to values of ω_1 of the order of 200 MHz, small compared to Δ ($\Delta \sim 1100$ MHz), but large compared to γ ($\gamma \sim 30$ MHz). We are therefore in a regime of partial compensation of Doppler broadening.

Fig. 5 shows, on the left part, the recorded emission spectral profiles for increasing values of the static magnetic field. They are in good agreement with the corresponding computed curves represented on the right part. In zero magnetic field, we have actually a 2-level system since only a superposition of b_+ and b_- is coupled to the laser. We get the well known Autler-Townes doublet [14] observed on the spontaneous emission from c. When the magnetic field is increased, a narrow structure appears in the center of the spectrum, reaches a maximum around B = 42 G, and then broadens. For B = 42 G, ω_1 and δ satisfy the compensation condition (5) and one gets the Doppler free line discussed above with the two sharp-edged sidebands.

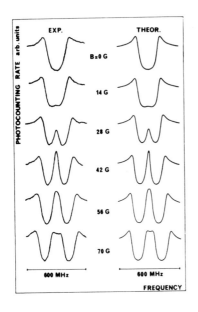

Fig.5 - *Experimental and theoretical emission spectral profiles on the $2p_{10} \rightarrow 1s_3$ transition of ^{20}Ne (laser excitation of the $2p_2 - 1s_3$ transition)*

More recently, we have studied other transitions of ^{20}Ne (Fig.6) for which the Zeeman structure appears now in the lower level $\underline{a}$ ($1s_4$) of the $\underline{a} - \underline{b}$ transition ($1s_4 - 2p_3$) which is coupled to the laser. With this new energy level scheme, it is now the upper state $\underline{b}$ which undergoes a light shift ε_b linear in v, so that the compensation of Doppler broadening can be achieved on the inverse Raman spontaneous line starting from $\underline{c}$ ($2p_{10}$, m =0), whereas the two spontaneous transitions $\underline{c} - a_+$ and $\underline{c} - a_-$ give rise to two sidebands which remain Doppler broadened or sharp-edged.

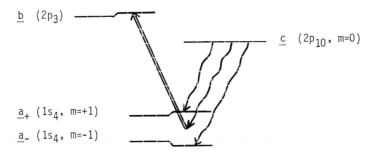

Fig.6 - *Energy level scheme and spontaneous direct and Raman transitions corresponding to the spectra of Fig.7.*

Note that we have now the possibility of choosing at the detection a σ polarization either parallel or perpendicular to the one of the laser beam. The emission rate on the three lines $|c, n> \rightarrow |i, n>$ is proportional to $|<i, n| \vec{\varepsilon} \cdot \vec{D} |c, n>|^2$ ($\vec{\varepsilon}$ is the detection polarization, $\vec{D}$ the atomic dipole operator). The emission spectra are therefore different on these two polarizations.

Fig.7a shows the comparison between experimental [15] and theoretical spectra when the detection polarization is parallel to the laser one. In zero magnetic field, no structure appears. The reason is that this detection polarization connects c to a linear superposition of a_+ and a_- which is just orthogonal to the one coupled to the laser. The central Doppler free line appears, when the field increases, with a better contrast than on Fig.5. The compensation condition corresponds to a field around 114 G [16]. The results obtained when the detection polarization is orthogonal to the laser one are represented on Fig.7b. The zero field spectrum exhibits now the Autler-Townes doublet. But, for this polarization, a perturbative treatment shows that the inverse Raman process vanishes for v = 0, because of a destructive interference between the two possible paths via a_+ and a_- (Fig.6) This is why the height of the central narrow peak is now smaller than the one of the two sidebands.

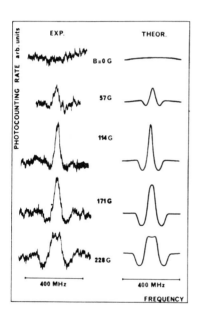

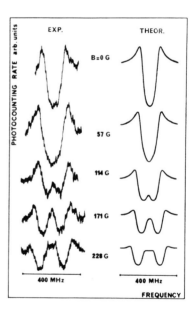

(a)   (b)

Fig.7 - *Experimental and emission spectral profiles on the $2p_{10} \rightarrow 1s_4$ transition of ^{20}Ne (laser excitation of the $2p_3 - 1s_4$ transition). The σ detection polarization is either parallel (a) or perpendicular (b) to the laser one.*

103

6. Conclusion

In conclusion, we have proposed a new scheme for compensating the Doppler broadening by velocity dependent light-shifts and we have obtained experimental results giving good confidence in the theoretical analysis of such a scheme.

Up to now, we have used moderate laser intensities (150 mW) and moderate focalisations ($w = 400\mu$). Commercial ring lasers can give now powers higher by a factor 10, even still higher if one works inside the laser cavity. The laser beam could be also focussed to smaller waists if one uses a second laser beam for probing the $\underline{c}$ -$\underline{a}$ transition rather than a spontaneous emission signal.

Consequently, it does not seem hopeless to obtain, even with c.w. lasers, values of ω_1 of the order of, or larger than Δ. In such a case, one could achieve a nearly complete compensation of Doppler broadening leading to forward-backward asymmetries as large as 10 or 100. This would open the way to various interesting applications such as reduction of threshold in laser media, ring lasers, directed Doppler free superradiance, Doppler free coherent transients, non-reciprocal devices...

References

[1] J.P. Barrat and C. Cohen-Tannoudji, J. Phys. Rad 22, 329,443 (1961)
     C. Cohen-Tannoudji, Ann. de Phys. 7, 423, 469 (1962)
     M. Arditi and T.R. Carver, Phys. Rev. 124, 800 (1961)

[2] W. Happer and B.S. Mathur, Phys. Rev. 163, 12 (1967)
     J. Dupont-Roc, Thesis, Paris (1971)
     W. Happer, Progress in quantum electronics, vol.1 (Pergamon Press, New-York, 1971)
     C. Cohen-Tannoudji and J. Dupont-Roc, Phys. Rev. A5, 968 (1972)

[3] E.B. Aleksandrov, A.M. Bonch-Bruevich, N.N. Kostin and V.A. Khodovoi, JETP Lett. 3, 53 (1966)
     A.M. Bonch-Bruevich, N.N. Kostin, V.A. Khodovoi and V.V. Khromov, Sov. Phys. JETP 29, 82 (1969)
     P. Platz, Appl. Phys. Lett. 14, 168 (1969) 16,70 (1970)
     R. Dubreuil, P. Ranson and J. Chapelle, Phys. Lett., 42A, 323 (1972)

[4] See the contribution of A. Ashkin and J.E. Bjorkholm in this volume

[5] C. Cohen-Tannoudji, Proceedings of the 2nd International Symposium on Frequency Standards, Copper Mountain 1976, published in Metrologia, 13, 161 (1977)

[6] C. Cohen-Tannoudji, F. Hoffbeck and S. Reynaud, Opt. Commun. 27, 71 (1978)

[7] The connection between light-shifts and narrow structures appearing in high resolution laser spectroscopy of three-level systems is discussed by P.E. Toschek in Proceedings of Les Houches Session XXVII (1975), edited by R. Balian, S. Haroche and S. Liberman (North Holland, Amsterdam, 1977)

P.F. Liao and J.E. Bjorkholm, Phys. Rev. Lett. 34, 1 (1975) suggest also the possibility of line narrowing by light-shifts in two-photon Doppler free experiments although they observe only a broadening

[8] S. Reynaud, M. Himbert, J. Dupont-Roc, H.H. Stroke and C. Cohen-Tannoudji, Phys. Rev. Lett. 42 756 (1979)

[9] D.M. Larsen, Phys. Rev. Lett. 39, 878 (1977)

[10] R. Romestain, S. Geschwind and G.E. Devlin, Phys. Rev. Lett. 39, 1583 (1977)

[11] T.W. Ducas, M.S. Feld, L.W. Ryan, N. Skribanowitz and A. Javan, Phys. Rev. A5, 1036 (1972)

[12] M.S. Feld in Fundamental and Applied Laser Physics, Proceedings of Esfahan Symposium, edited by M.S. Feld, A. Javan and N.A. Kurnit (John Wiley, New York, 1971) p. 369

[13] See for example C. Cohen-Tannoudji and S. Reynaud in "Multiphoton Processes" edited by J.H. Eberly and P. Lambropoulos, p. 103 (John Wiley, New York, 1978)

[14] P. Cahuzac and R. Vetter, Phys. Rev. A14, 270 (1976)
A. Shabert, R. Keil and P.E. Toschek, Appl. Phys. 6, 181 (1975) and Opt. Comm. 13, 265 (1975)
C. Delsart and J.C. Keller, Opt. Commun. 16, 388 (1976) and J. Phys. B9, 2769 (1976)

[15] The reabsorption of the fluorescence light in the cell is more important on the $2p_{10}$ to $1s_4$ transition than on the $2p_{10}$ to $1s_3$ one. We have been therefore obliged to reduce the Neon pressure to 0.3 Torr, which explains why the signal to noise ratio is lower than on Fig.5

[16] When the Zeeman structure exists in the lower level a (scheme of Fig.6) calculations similar to the ones of section 4 lead to a slightly different compensation condition $2\delta^2/(\omega_1^2 + 2\delta^2) = s$.

Multiple Coherent Interaction in Optical Separated Fields

V.P. Chebotayev

Institute of Thermophysics, Siberian Branch of the USSR Academy of Sciences
Novosibirsk-90, USSR

Various methods of superhigh resolution spectroscopy have been elaborated for each range of electromagnetic waves. The γ - range uses Mössbauer spectroscopy, the microwave one the method of separated fields, the method of accumulating bulb, and Dicke resonance. In the optical band there have been recently elaborated new methods based on the use of Doppler free non-linear resonances. The methods based on the use of spectroscopy of saturated absorption and two-photon resonances [1] are widely used. The optical resonance widths may be decreased by increasing the time of particle-field interaction.

Starting with 1974 we have been developing a new method for the optical band, the method of separated fields that enables us to increase the time of coherent interaction of particles with field. We were not first who tried to use the method of separated optical fields. However it was impossible to use this method in the optical band in the same way as it was employed by Ramsey for the microwave range [2]. Figure 1 schematically represents the principal idea of the method in the microwave range. A beam of particles interacts with two fields. After resonance interaction with the first field a particle has a dipole moment at the transition frequency ω_{21} equal to

$$d_1 = d_{12} G \tau e^{-i\Omega t_1 + \varphi_1} e^{-i\omega_{21}t}/2 + c.c. \qquad ,$$

where d_{12} is the matrix element of the dipole moment, $G = \dfrac{id_{21}E}{\hbar}$, E is the field amplitude, τ is the interaction time. The energy absorbed in the second field periodically depends on the difference of the dipole moment phase and of the field phase

$$\mathcal{E} = \frac{\hbar\omega}{2} (|G|\tau)^2 \cos(\Omega T + \varphi_1 - \varphi_2) \qquad (1)$$

where $T = L/u$ is the transit time of a particle, $\Omega = \omega - \omega_{21}$ is the field frequency detuning relative to resonance, $\varphi_2 - \varphi_1$ is the phase difference between the fields. In the interaction of an ensemble of particles averaging is related only to the time of flight of particles between the fields. This results in rapid damping of absorption oscillations with an increase of the detuning frequency. In the optical band we

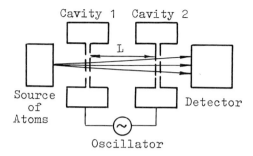

Cavity 1 Cavity 2

L

Source
of
Atoms

Detector

Oscillator

Fig.1 Schematic diagram of
the method of separated
fields in a microwave range

observe a new effect which does not permit a direct use of the
method of separated fields (MSF) in optics: in optics field
dimensions are usually larger than a wavelength, and a Doppler
effect becomes important. Coherent properties of an individual
particle will manifest themselves if they are reflected in
macroscopic medium polarization, that is, of great importance
in the optical band is the macroscopic medium polarization
transfer. Let us consider two optical fields (Fig.2 a, b)
interacting with the particle beams of finite divergence. A
dipole moment of the particles going out of the point z_1 is

identical for all particles and determined by the field phase
in the point z_1. However if the distance between fields is

sufficiently large, the phase difference between the second
field and the dipole moment depends not only on the transit
time of a particle and the phase difference between fields
but also on the particle velocity projection. In Fig.2 a it
is seen that the phase of dipoles in the fields has a "Doppler"
phase angle equal to kv_zT. Inspite of the fact that individual

particles transfer coherence, absorption of the ensemble of
particles associated with the polarization transfer is zero.
The similar results may be obtained in considering the dipole

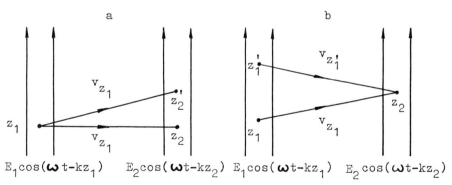

a   b

$E_1\cos(\omega t-kz_1)$ $E_2\cos(\omega t-kz_2)$ $E_1\cos(\omega t-kz_1)$ $E_2\cos(\omega t-kz_2)$

Fig.2 Interaction of particles with two separated optical
fields. a) $z_2-z_2'= \Delta z_2 \gg \lambda$; b) $z_1-z_1'= \Delta z_1 \gg \lambda$.

107

moment that is introduced by particles into the point z_2 of
the second field after the interaction with the first field.
The dipole moment phases depend on the field phases in the
points z_1, z_2, z_3, and so on. If $\Delta z \gg \lambda$, an average dipole
moment in the point z_2 is zero. Thus, observation of the effects
connected with coherence transfer requires elimination of the
Doppler effect influence.

Over the last years we have proposed various methods to
obtain resonances in separated fields based on the nonlinear
interaction of particles with time and spatially limited optical
fields: a three-beam system for two-level atoms [3], a two-level
system based on the use of two-photon absorption in the standing-wave field [4], a three-level system based on the use of
processes like stimulated Raman scattering (SRS) in three-level
systems [5]. A short review of these methods is given in [6].
Even the first experiments gave resonances with the width of
about 10^{-11} [7, 8]. It is evident that the method has great
potentialities from the standpoint of attainment of resonances
with widths of about 10^{-13} to 10^{-14}.

The above methods use various physical ideas for eliminating
the Doppler effect which were not used in the method of separated fields in the microwave range. Moreover, in the course of
investigations of this method in the optical band there have
been discovered new phenomena associated with coherent radiation that are closely related to such known phenomena as photon
echo, superradiation, induction decay, and wave front reversal.
The totality of phenomena that occur in spatially and time separated fields under resonance conditions underlies the method
of separated fields.

We can dwell on some properties of the method in the optical
band. The following phenomena are of interest to us:
1. macroscopic polarization transfer
2. coherent radiation and its properties
3. SRS - processes in separated fields
4. resonant transients.

Most of these phenomena are connected with optical polarization transfer that is of prime importance for the optical band.
The polarization transfer is critical both in attainment of
resonances and in coherent radiation for two-level atoms and
processes of coherent Raman scattering. Let us consider this
phenomenon in detail in accordance with [3, 9].

1. Macroscopic Polarization Transfer and Coherent
   Radiation in Separated Fields

1.1 Traveling Waves

To make a better analysis of elementary processes let us consider a beam of two level atoms interacting with plane traveling and standing waves. The distance between them is L. Let

the beam divergence be so small that $kv\theta < 1/\tau$. This means
that all particles interact with the field and, hence, homogeneous saturation takes place.

We are interested in the processes near the line center.
So we shall assume that $\Omega\tau \gg 1$. After the interaction with the
first traveling wave the dipole moment is

$$d_1(t) = d_{12}\, a_1^*\, b_1\, e^{-i\omega_{21}t} + c.c.,$$

where a_1 and b_1 are the amplitudes of probability of finding
a particle on the upper and lower levels. After the interaction
with the second beam in the second order of perturbation theory
the probability amplitudes are

$$b_2 = b_1 + a_1 G_2 \tau e^{-i\Omega t_2}\cos(kz_2 + \varphi_2) - b_1\frac{(|G_2|\tau)^2}{2}\cos^2(kz_2 + \varphi_2)$$

$$a_2 = a_1 + b_1 G_2^* \tau e^{i\Omega t_2}\cos(kz_2 + \varphi_2) - a_1\frac{(|G_2|\tau)^2}{2}\cos^2(kz_2 + \varphi_2).$$

The second terms correspond to the second order of perturbation
theory which complies with one-photon transitions. The terms
quadratic in G_2 correspond to two-photon processes. The pertinent dipole moment is

$$d_2(t) = d_1(t) + d_2(t, G_2) + d_2(t, G_2^2) + c.c.$$

The dipole moment nonlinear in G_2 contains two terms. The
first one is connected with one-quantum processes, the second
with two-quantum processes. A very important point is that the
two-quantum processes result in no changes of the dipole moment
phase but change its influence. The one-quantum processes are
responsible for a jump of the dipole moment phase that is equal
to

$$\Delta\varphi = 2k(z_2 - z_1) - 2\Omega T.$$

It can be easily seen that at the distance L from the second
beam the dipole moment phase of a particle will fully coincide
with the field phase. The phase jump $\Delta\varphi$ contains a spatial and
time phases. At the distance L from the second beam the dipole
moment phase of a particle is

$$\varphi = kz_1 + 2k(z_2 - z_1) + 2\Omega T + \omega_{21}T.$$

It is seen that the dipole moment phase of a particle will
coincide with the field phase in the point z_3, that is, all
particles will lock in spatial and time synchronism with a wave.
Note that the time jump of the phase 2 T cancels the phase
difference between the dipole and the field at frequency detuning Ω . So interference effects and frequency dependence of

absorption or coherent radiation are not observed. The mechanism of elimination of the Doppler effect in separated traveling waves is closely related in its nature to a photon echo. The time phase cancellation explains why the resonance effects are not observed in numerous experiments on photon echo. There is no polarization transfer in the interaction of oppositely traveling waves.

2.2 Standing Wave

Let us consider the case where the second field is a standing wave. There are two mechanisms of elimination of the Doppler effect. The first one corresponds to the interaction with traveling waves. (It should be remembered that a standing wave may be represented as a sum of two oppositely traveling waves). We discussed this mechanims above. Qualitatively new peculiarities arise in the interaction of standing waves. The phenomena that occur in this case are typical for a gas only and associated with the two-photon process of absorption and emission from oppositely traveling waves. It is evident that these processes may occur only in the line center where the particles interact simultaneously with both waves.

The phase jump of the dipole moment in the interaction with oppositely traveling waves is (Fig.3)

$$\Delta \varphi = \pm 2kz_2.$$

Since before the interaction the dipole moment phase of a particle was kz_1, after the interaction it will be $-2kz_2+kz_1$. At the distance x=2L from the first beam $2z_2-z_1=z_3$ and the dipole moment phase is $-kz_3$. When considering similarly the other points one can easily notice that the polarization wave propagation occurs in the direction that is opposite to the direction of the first traveling wave.

The difference from the case of traveling waves is that the phase jump $\Delta \varphi = -2kz_2$ necessary for polarization transfer arises at the expense of the two-photon process of absorption and emission of oppositely traveling waves. At the distance 2L the phase of the field and of the particle dipole moment is shifted by a value of $2\Omega T$. It is the circumstance that is responsible for the appearance of resonance phenomena.

The macroscopic polarization transferred at a distance may be observed in absorption of a probe wave and coherent radiation in separated fields (CRSF). Let us discuss the latter in detail. The principal features of CRSF are as follows:
1. Coherent radiation has a resonance character whose manifestation depends on the conditions of observation.
2. Coherent radiation has properties of continuous superradiation. Its intensity is proportional to the square of particle density. When working with an inverted beam one can observe other phenomena inherent to superradiation.

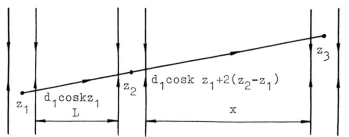

$E_1\cos kz_1\cos\omega t$ $E_2\cos kz_2\cos\omega t$ $E_3\cos kz_3\cos\omega t$

<u>Fig.3</u>   Interaction of particles with three separated optical fields

3. The wave front reversal of a traveling wave occurs in the system.
4. Coherent radiation is sensitive to changes in the dipole moment phases during drift.
5. Coherent radiation associated with stimulated processes arises at the field frequency rather than at the transition one.
6. Since coherent radiation arises with a delay with respect to an exciting field, the system operates as a coherent delay line or a system with record and restoration of an optical image.
7. The radiation is very sensitive to deflection of particles in the region of drift under the action of various physical factors (gravitational, electric and magnetic fields, collisions, recoil effect and others).
8. Under the action of fields with different frequencies polarization appears at combination frequencies.
9. In strong standing-wave fields the polarization transfer occurs at the distances multiple to L.
Let us consider some properties of the method of separated fields.

2. Resonance Phenomena in Separated Fields

Resonance phenomena that are associated with interference ones may occur only in the interaction of standing waves. In the standing wave at the two-photon process of absorption and emission of oppositely traveling waves the cancellation of the Doppler shift only occurs. So the polarization phase for the particle drift time T differs from the field phase at the distance 2L by a value of $2\Omega T$. The occurrence of interference phenomena considerably depends on beam properties, field configurations and other experimental conditions. This results in occurrence of so-called Ramsey fringes in observation of the probe wave absorption. Another situation takes place in the CRSF. Let us consider as an illustration resonance phenomena when the first field is a traveling wave, the second is a standing wave (Fig.4). Let the particle beam that transfers polarization be monokinetic. It is a good analog of the interaction of the system with two pulses. As we can see, in this case the coherent radiation arises in two directions. In the

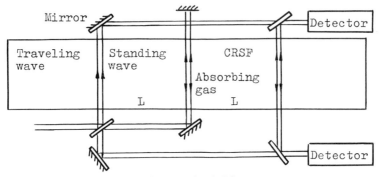

Fig.4 CRSF in separated optical fields

same direction the radiation phase coincides with the field phase, in the opposite direction the radiation phase differs from the field phase by a value of $2\Omega T$. However this phenomenon is not observed if the radiation intensity is dependent on frequency. In the first case the radiation will be observed within the limits of a Doppler width, in the second within the limits of $1/\tau$. Now let us change the scheme of experiment and observe an intensity by heterodyning the CRSF with the radiation of a pumping laser. It is evident that in the first case the signal intensity is dependent only on the constant phase difference of the CRSF and of the pumping laser radiation and independent of the frequency detuning Ω. In the second case with varying Ω the change in the signal resembling Ramsey fringes occurs. If the first field is a standing wave, the CRSF will occur in both directions. Unlike the first case each direction has two waves that are due to the interaction of traveling and standing waves. In this case an intense interference arises between two CRSF components connected with traveling and standing waves, intensity beats arise in each direction.

A good analog of spatially separated fields with a monokinetic beam is time separated fields. Here all particles interact with the field in the same time interval and have identical phases. Fig.5 shows the dependence of the CRSF intensity in the standing-wave field in SF_6. An interference pattern is well defined.

In obtaining resonances in spatially separated fields actual beams have velocity spread. The time phase jump depends on the atomic velocity. This results in blurring the interference pattern. Modulation becomes less contrast. Simultaneously an interference pattern appears in the back CRSF, which is a consequence of addition of radiation with various phases dependent on the transit time.

2.1 Destruction of the Interference Structure and
     Attainment of Resonances with a Radiation Picture

The interference pattern makes it difficult to gain spectro-

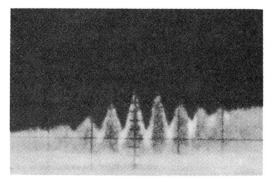

Fig.5 Intensity change of the coherent radiation with the laser frequency tuning. The time between pulses is T=0,4 μs

scopic information connected with the presence of a fine line structure. There is a simple way to reduce and probably to eliminate its influence. It is based on the change of the delay time between fields. In this case with detuning Ω oscillating terms will periodically depend on T. It is easier to vary T in the time separated fields. Averaging the dependence of the signal over the delay time as a function of frequency we obtain an ordinary resonance curve with a width equal to a linewidth.

As we have already seen, the signal intensity in the time T is described by the expression

$$I = I_0(1 + \cos 2\Omega T)e^{-\gamma T}.$$

Averaging over T we obtain

$$I = I_0(1 + \frac{\gamma^2}{\gamma^2 + \Omega^2}).$$

In spatially separated fields T may be changed by varying the distance between beams and the beam velocities. Ion beams may be used here as their velocity is readily changed. It is more complicated to change velocity of atomic or molecular beams. Here a particle velocity can be changed with the aid of light pressure effects. By choosing temperatures of various beams one can considerably decrease oscillations and destroy Ramsey fringes. For example, the second oscillation amplitude decreased by a factor of 10 in comparison with the use of one beam. Fig.6 shows the resonance shape for the beams with two different distributions of particles over velocities.

2.2 Particle Scattering

In the interaction of particles with the standing-wave field particle scattering occurs due to a recoil effect. This may result in variation of properties of resonances of separated fields. There appears an effect of resonance distribution. However its mechanism differs from the mechanism of splitting of saturation resonances. A more detailed consideration of the influence of the recoil effect on the CRSF properties is made in [10].

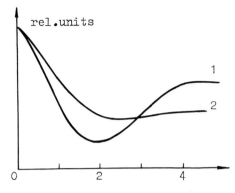

rel.units

Fig.6 Resonance in excited particles arising from the interaction of a particle beam with T_1(curve 1) and a two-temperature beam with T_1 and T_2 (curve 2) with separated fields. $T_2/T_1 = 5.2$.

$\frac{\Omega L}{v_0}/_{T=T_1}$

Since the recoil should be taken into account in the case where Δ is compared with the resonance width $\gamma \sim T^{-1}$ that is considerably less than the transit linewidth $1/\tau$, $\Delta \ll 1/\tau$. This means that the splitting of absorption and radiation line contours may be neglected. However owing to the recoil effect the particle velocity changes (Fig.7). It is important for us that in the interaction of an atom with the first wave the recoil may be neglected. The occurrence of the CRSF requires the ensemble of polarized atoms with random velocity spread to arise in the medium after the interaction with the first wave. The presence of recoil does not change this assumption. The variation of the phase for the time of interaction with the wave τ may be neglected, as it is $\Delta\tau \ll$ 1.

In the interaction with the second field there takes place an additional change of the particle velocity due to the recoil effect. During drift between the second and third fields the particles suffer an additional Doppler shift equal to 2 $\Delta v_z kT$, where Δv_z is the change of the particle velocity in photon absorption (emission). This shift may be cancelled by varying the field phase by a value equal to $2\Omega T = 2 \Delta v_z k T$. Hence $\Omega = \pm \Delta v_z k$, i.e., the resonance is split.

An interesting situation arises at the distances $x_n=(2n-1)L$. The coherent radiation occurs in the orders that are even in the second-wave field. This means that the even number of photons participates in the interaction and, hence, there are processes where an atomic velocity is not changed at all ($\Delta v = 0$). Here the condition of phase cancellation is not varied by allowing for the recoil. These are the processes that are responsible for the resonance component in the line center. A general theory of the CRSF effect that simultaneously allows for the influence of recoil and strong field has been elaborated in [10]. It should be emphasized that the component in the line center arises not only in the CRSF line but also in absorption resonances in separated fields [11].

If a particle in the region of drift is affected by some forces, the particle deflection produces an extra phase shift.

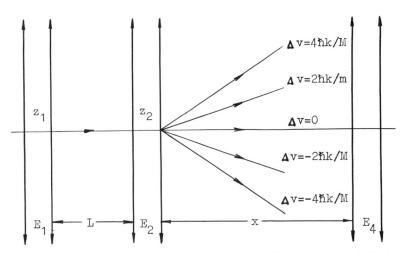

Fig.7 Interaction of particles with three separated fields in consideration of a recoil effect

It may be compensated by frequency detuning. With the extension of the region of drift of about 50 cm the system described becomes very sensitive. For example, the gravitational field curves particle trajectories, which results in the resonance shift by a value of about 1 kHz (for particles with $M = 10$, $T = 300°K$ the field axis is vertical). The systems with ion beams may be very sensitive to electric and magnetic fields. In the electric field of about 10^{-8} V/cm with the distance between fields of 50 cm the frequency shift for an ion with $M = 10$ is about 1 kHz. The magnetic sensitivity is high too.

2.3 Strong Fields

Since the phenomena under consideration are due to the action of time-limited fields, one can observe the phenomena that are physically similar to the action of the $\pi/2$ pulse. These phenomena are more or less clear. We shall dwell on qualitatively new phenomena which arise in the standing-wave fields and are due to polarization transfer. In consideration of the effects in higher orders the polarization transfer occurs at the distances 3L, 4L only in the standing-wave fields. The polarization transfer at the other distances may be also explained by using the model of phase jumps. The dipole moment that arises in the particle after the interaction with the field contains the signal "n" corresponding to the number of the order of perturbation theory $d \sim |G_1|^n \cos^n kz_1$.

The interaction with the second field must be considered in even orders of perturbation theory. The dipole moment of interest is $\sim \cos^n kz_1 \cos^m kz_3$. The orders of perturbation theory must be combined so that at the distance kL from the first beam the dipole moment has the first spatial harmonic. After the interaction with two fields the dipole moment will contain the term:

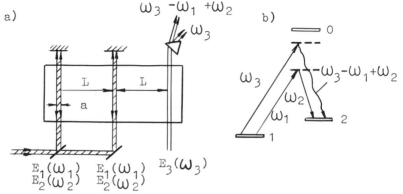

Fig.8 Scheme for observing coherent Raman scattering of three-level atoms (a). Layout of transitions in a three-level system (b).

$$k(nz_1 - mz_2).$$

Since the coordinate z at the distance x = kL from the first beam is

$$z_3 = z_1 + k(z_2 - z_1).$$

It is evident that the perturbation theory orders, i.e., the numbers m, n, and ℓ , should be combined so that the equality may take place

$$nz_1 - mz_2 + z_1 + \ell (z_2 - z_1).$$

The polarization can be produced at the distance 4L from the first beam, if n is equal to 5, m = 4. This case is of interest in obtaining resonances with no recoil.

3. SRS in Separated Fields

Here we shall discuss a new phenomenon, coherent Raman scattering (CRS) in separated fields [5]. It arises in the approximation that is linear in an incident field. Its physical nature

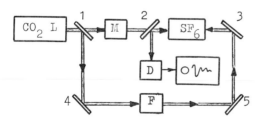

Fig.9 Experimental scheme for observing nonstationary resonance effects: 1-5 – mirrors, M – modulator, F – filter, D– detector.

strongly differs from the processes in SRS. This phenomenon may be observed in an absorbing medium and, hence, is not associated with the process of amplification.

Under the action of two fields resonant to the $1 \longrightarrow 2$ transition the polarization is produced at the distance x=2L from the first beam (Fig.8). If there is a field at the frequency ω_2 in this region and Raman transitions are allowed, the coherent Raman scattering is expected to occur at the frequencies $\omega \overset{+}{-} \omega_{12}$.

If a one-photon transition between levels 1 and 2 is forbidden, coherence between these levels may be produced by acting upon the system with two fields with frequencies ω_1 and ω_2, the difference between them is equal to the frequency $\omega_{12} = \omega_1 - \omega_2$.

This consideration is related to spatially separated fields. Similar phenomena may be observed in pulsed fields. In this case in the time 2T after the first pulse one can observe a signal at combination frequencies which may be interpreted as an effect of CARS-echo.

This method may be employed in studying coherent relaxation processes when one-photon transitions are forbidden. This may be important in studying condensed media.

4. Nonstationary Resonance Effects

The method of separated fields is underlain by the variation of the dipole moment phase of a particle in the nonlinear interaction of the other fields that permits elimination of the Doppler effect. In this case we neglect the induced dipole moment decay. Let us see what will happen to the induced dipole moment in the presence of other fields. In other words, unlike early studies into free induction decay, we are interested in induction decay in the presence of other fields. The first observation of this phenomenon was reported in [12].

At the time $t_1=0$ a gas cell is affected by the traveling-wave pulse with duration τ that polarizes the medium. After that the medium is affected by the standing-wave field, and there appears damping radiation in the opposite direction (this may be interpreted as the reversal of a wave front of the induction). The radiation frequency is located in the line center. Interference of the radiation with the traveling component of the standing wave results in radiation pulsation.

Figure 9 gives the schematic representation of the experiment on observation of this phenomenon. The radiation of a CO_2 laser passes through the cell with SF_6 and through the modulator M. After the pulse action there is the standing-wave field in the cell which is formed by mirrors and by

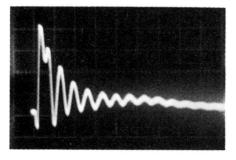

modulator M. The radiation is examined in the direction opposite to the direction of the radiation pulse propagation.
Fig.10 shows the signal of beats at the detuning frequency.

The physical picture of occurrence of the radiation may be
briefly described as follows. After the interaction with a
traveling wave there arises the distribution of decaying dipoles over velocities described by the expression

$$d^{(1)} = -ie^{ikz}|d_{21}|^2 h^{-1} E \tau e^{-(\Gamma - i\omega_{21} + ikv)t}$$

The oppositely traveling wave interacting with the polarized medium produces spatial inhomogeneity of the medium with
the period $T/2$, which is associated with the decaying dipoles.
The second component of the standing wave interacting with the
spatially inhomogeneous medium produces a polarization wave
in the opposite direction. The radiation that is due to the
polarization undergoes beats with the first component of the
standing wave. It is interesting to note that the radiation
in the line center results from averaging over the particle
ensemble. Let us trace this phenomenon in more details. The
population associated with the interaction of polarized particles with the first component of the standing wave is

$$n_1 - n_2 = |d_{21}|^2 h^{-2} E_1^* E \tau \exp -(\Gamma - i\Omega + ikv)t /(\Gamma - \gamma - i\Omega + ikv) + c.c.$$

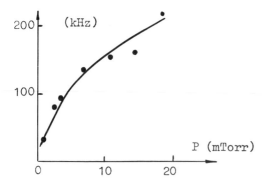

The second component of the standing wave produces the dipole moment of particles

$$d^{(nl)} = \frac{i|d_{21}|^4}{\hbar^3} E_2 E_1^* E \tau e^{-\Gamma t} \frac{e^{-i(\omega_{21}+kv)t} - e^{-i(\omega_{21}-kv)t}}{2ikv(\Gamma - \gamma - i\Omega + ikv)} e^{-ikz}$$

The method is useful in studying collisional shifts and broadening. Fig.11 shows the dependence of the decay rate on particle density. This dependence is nonlinear due to the influence of scattering at small angles. The similar phenomena were observed in studying the nonlinear dependence of collisional resonance broadening [13] and in experiments with photon echo [14].

References

1. V.S.Letokhov, V.P.Chebotayev: Principles of Nonlinear Laser Spectroscopy (Izd. Nauka, Moscow, 1975)
2. N.F.Ramsey: Molecular Beams (Oxford University Press, New York, London, 1956)
3. Ye.V.Baklanov, B.Ya.Dubetsky, V.P.Chebotayev: Appl.Phys. 9, 171 (1976)
4. Ye.V.Baklanov, B.Ya.Dubetsky, V.P.Chebotayev: Appl.Phys. 11, 201 (1976)
5. V.P.Chebotayev: Vestnik MGU 19, 159 (1978)
6. V.P.Chebotayev: In Coherence in Spectroscopy and Modern Physics, ed. by F.T.Arecchi, R.Bonifacio, and M.O.Scully (Plenum Press, New York, 1978) p. 173
7. S.N.Bagayev, A.S.Dychkov, V.P.Chebotayev: Pisma ZhETF 26, 591 (1977)
8. J.C.Bergquist, S.A.Lee, J.L.Hall: Phys.Rev.Lett. 38, 159 (1977)
9. V.P.Chebotayev: Appl.Phys. 15, 213 (1978)
10. Ye.V.Baklanov, B.Ya.Dubetsky, V.M.Semibalamut: ZhETF 76, 482 (1979)
11. B.Ya.Dubetsky: Kvant.Elektronika 3, 1258 (1976); B.Ya.Dubetsky, V.M.Semibalamut: Kvant.Elektronika 5, 176 (1976)
12. L.S.Vasilenko, M.N.Skvortsov, V.P.Chebotayev: Pisma ZhTF 4, 1120 (1978)
13. S.N.Bagayev, Ye.V.Baklanov, V.P.Chebotayev: Pisma ZhETF 16, 15 (1972); S.N.Bagayev, A.S.Dychkov, V.P.Chebotayev: Pisma ZhETF 29, No. 9 (1979)
14. R.G.Brewer, R.L.Shoemaker: Phys.Rev.A 6, 2001 (1972).

High Resolution Spectroscopy of Calcium Atoms

J.C. Bergquist, R.L. Barger, and D.J. Glaze

Time and Frequency Division, National Bureau of Standards
Boulder, CO 80303, USA

Abstract

The recent results on saturated absorption optical interference spectroscopy of calcium are presented. The photon recoil splitting of the Ca $^1S_0$ - $^3P_1$ intercombination line at 657 nm has been fully resolved. Linewidths as narrow as 3 kHz half width half maximum intensity (HWHM) are reported for radiation beams spatially separated by up to 3.5 cm. Second order Doppler is shown to be the present limitation to the accuracy of this technique. Methods are discussed which could lead to an optical wavelength/ frequency standard with an accuracy of better than 10^{-14}.

Introduction

Saturated absorption with spatially separated optical fields [1-5] permits extremely high resolution with good signal-to-noise ratio and can be accomplished with simple, easily achievable, optical systems. Use of this optical Ramsey interference method with highly stabilized dye lasers and long lived transitions [6-8], such as the $^1S_0$ - $^3P_1$, intercombination resonance lines of Ca and Mg can now lead to resolution of the photon recoil doublet. This partially removes the accuracy limitation previously imposed by unresolved recoil components in the visible spectrum. We summarize our investigations of the Ca $^1S_0$ - $^3P_1$ line at 657 nm using the above methods with which we have obtained linewidths as narrow as 3 kHz HWHM and completely resolved the recoil doublet which is split by 23.051 kHz. Also we discuss techniques which could lead to an optical wavelength/ frequency standard for this line with an accuracy of better than 10^{-14}. Fractional frequency stability for such a calcium stabilized dye laser oscillator is calculated to exceed 10^{-16}.

The Ca transition [6-7] is very suitable for measurements of the highest level of accuracy, such as those in metrology, in spectroscopy and in various relativity experiments. It has a long lifetime of 0.39 ms (natural linewidth of 410 Hz), small electromagnetic field shifts of about 10^8 Hz/T^2 (1 Hz/G^2) and 1 Hz/(V/cm)^2 for the $\Delta m_J = 0$ transition. Its transition probability allows an optimum laser power of less than 1 mW. ^{40}Ca possesses a nondegenerate ground state and has zero nuclear spin giving no hyperfine structure. A complete resolution of the recoil peaks and full knowledge of the distributed velocity contribution to the Ramsey fringes of each recoil component should permit precise definition of line center.

Basic Concepts

The theory of photon recoil effects in saturation spectroscopy has been discussed by Kol'chenko, et al. [9] and more thoroughly by Bordé [2,10]. Conservation of 4-momentum (momentum and energy) requires that, in the absorption process, the atom absorbs the energy and momentum of the photon, thereby changing not only the atom's internal energy, but also its motional energy. Similarly, in the emission process the atom must change its kinetic and internal energy to provide for the emitted photon's energy and momentum. If there are no other energy sources available, then the kinetic energy of recoil must be provided by the absorbed or emitted photon. This requires that the frequency of the absorbed photon be blue shifted relative to the Doppler shifted natural resonance frequency of the atom and, correspondingly, that the emitted photon be red shifted. In linear spectroscopy only the absorption or emission process is observed, with the result that the frequency shift is not directly discernable (in fact it could be very easily included in the definition of the two level atom state energy difference). Saturation spectroscopy includes both emission and absorption aspects, thus, with sufficient resolution, the counterrunning probe wave will disclose the effect of the recoil frequency shift as a doublet Lamb dip structure, i.e., there will be two frequencies of anomalously high transmission for the probe wave. One occurs at the frequency where the atom's ground state density is reduced by the power wave interacting with the same set of absorbers. The second transmission peak comes at the frequency where both the power and probe waves interact with the same excited state velocity class absorbers. The two peaks are symmetrically displaced from the location of the ordinary Bohr frequency with the total frequency splitting being $\delta v = hk^2/4\pi^2 m$, where h is Planck's constant, k is the wave vector amplitude, and m is the atomic mass. For ^{40}Ca, this recoil splitting amounts to 23.051 kHz for the $^1S_0 - {}^3P_1$ intercombination line at 6573 Å.

Until now the only instance of photon recoil resolution in this optical region has been that of Hall, et al. [11], with the methane saturation absorption peaks at the longer 3.39 μm wavelength. The resolution represented in the methane experiment of $\sim 2 \times 10^{-11}$ came with continuous excitation (one zone) throughout the interaction region with many hours of integration time and high quality, large aperture optics. In fact, in most spectroscopy experiments, the radiation field extends more or less uniformly throughout the volume in which move the absorbers to be studied. Ramsey [12] was the first to point out that this may not be the most advantageous method in which to apply the oscillating driving field. He observed that useful spectroscopic information could be obtained if the amplitude and phase of the radiation field were non-uniform throughout the region. Of particular value is the arrangement that allows independent phase evolution of atoms and radiation field between interactions of atom and field. A simple scheme to produce this field on, field off, field on effect is to intercept a beam of atoms with spatially separated radiation fields. This gives an absorption profile dependent on the absorber's natural transition frequency with spectral width determined by the time between the radiation regions, rather than the transit time through each region. To help visualize this, we note that the interaction of absorber and field in the first radiation beam produces a coherent superposition of upper and lower states. This results in a dipole which oscillates at the natural resonance frequency in the field free region between the radiation zones. The effect of the second light field depends on the phase of the radiation relative to the absorber's oscillations, so that the absorbers

121

passing through this field will either be further excited or returned to the ground state by stimulated emission. Thus, an interference results which produces line narrowing since the quantum absorption transition probability is dependent not only on the frequency of the driving field, but also on the phase evolution difference of quantum system and field between zones. The phase evolution difference is proportional to the interzone transit time.

For detailed discussions of saturated absorption optical Ramsey interference with separated oscillatory fields the reader is referred to theoretical discussions of Baklanov, et al. [1] and Bordé [2] and to the treatments of Bergquist, et al [3-5]. But for completeness in this paper let us remind the reader of the important features. For three equally spaced, parallel, standing wave radiation fields, transversely crossed by a mono-velocity atomic beam, the line profile is a complex combination of linear and nonlinear terms. First there is a broad Doppler pedestal given by the first order linear absorption term. Superimposed on this is a Lorentzian Lamb dip associated with the third order nonlinear terms which carry only single zone resolution in either the preparation or probing process or both. Note, then, that the low resolution, nonlinear field-atom interactions can occur in any one of the three zones, or any two, or even all three to produce our "single zone" Lamb dip (the low resolution term from interactions in all three zones is a factor of two narrower than the other terms). Finally, superimposed on this is the sharp Ramsey fringe pattern produced by the previously described interference experienced by those atoms which nonlinearly interact with all three zones. The sharp interference fringe results only if the nonlinear, atom-field interactions have high resolution in both the preparation and probing process. Bordé [2] has shown that the fringe pattern or oscillatory part of the signal for equal relaxation constants $\gamma_{ij} = \gamma_{ii} = \gamma$ is an exponentially damped cosine proportional to

$$(1/v^2)\exp\left[-(\omega-\omega_0')^2\, a^2/v_r^2\right] \cos\left[(\omega-\omega_0')(2v_x L/a^2-\gamma)a^2/v_r^2\right]$$

where a is the laser mode radius, L the common zone separation, $v^2 = (v_r^2 + v_z^2)$, $v_r^2 = v_x^2 + v_y^2$, and $\omega_0' = \omega_0(1 -v^2/2c^2 \pm \hbar k/2mc)$ to a good approximation. In the expression for ω_0', we have included not only the shift in the natural resonance frequency due to the recoil terms, $\pm \hbar k/2mc$, but also due to the second order Doppler term, $-v^2/2c^2$. In the limit that the second order Doppler frequency shift is much smaller than the detuning frequency, $\Delta\omega = \omega - \omega_0$, at which $\Delta\omega(v_x L/v_r^2) = \pi$, we can ignore this shift, and velocity averaging for a Maxwell-Boltzmann beam velocity distribution will then produce a strongly damped cosine pattern consisting of a primary peak at line center plus one or two smaller side peaks, similar to those described by Ramsey for linear RF excitation with separated oscillatory fields. We will return to a fuller discussion of this point later. If the radiation standing waves are composed of equal intensity counter propagating waves, variation of the relative cavity phases across the three zones produces a variation of fringe intensity (from positive through zero to negative) but no asymmetry, and hence, no shift in the fringe center, a particularly appealing characteristic for accurate measurements. If the cavity phase condition is nonstationary then the fringes will wash away with averaging time. The opposition of two cat's-eye retroreflectors to produce three or more radiation zones intrinsically holds the cavity phase condition constant [3,4]. This is due to the cat's-eye's high insensitivity to thermal and mechanical fluctuations and to small angle variations of the input laser beam. Finally, if properly focused, the cat's-eyes give fringes of maximum positive amplitude for all velocity classes. We are unaware of any other arrangement which automatically (i.e., without servoing) holds constant the initial phase condition.

There are important advantages of the separated oscillatory field method as compared to the cw, single zone interrogation method. Higher resolution can be obtained without loss of signal-to-noise ratio, whereas in the cw case, higher resolution is inversely proportional to S/N ratio (best case) because increasingly fewer atoms contribute to the increasingly narrower resonance. Note the immediate consequence to frequency standards applications, since $\sigma_\tau \propto 1/(Q \cdot S/N)$. Of course, this also directly impacts spectroscopy with the possibility of real time data acquisition as opposed to many hours of averaging. Secondly, the wavefront flatness is required only over the mode diameter 2a, rather than three zone interaction length of 2L. And finally, power broadening and shift are minimized.

Experimental Apparatus

The very high resolution possible in this experiment consequently demands a dye laser spectrometer of unprecedented frequency characteristics. Figure 1 shows a block diagram representation which highlights some of the important features of our Ca dye laser spectrometer as well as the Ca beam and interrogation method. Much of our fast-stabilized dye laser system has been previously described [13]. The dye laser is stabilized to the midpoint of a transmission fringe in an external optical Fabry-Perot cavity of extremely high finesse. The cavity length is controlled to provide long term stability and frequency tuning capability. Two important improvements have been made in the system [6]. Firstly, the short term linewidth of our laser has been reduced to approximately 800 Hz rms with an improved high frequency servo amplifier with gain to 5 MHz [14]. Secondly, we have reduced the long term drift to less than 2 kHz/hr by controlling the servo cavity length by means of a first derivative line center lock to the few MHz wide saturated absorption line in an external calcium cell. In the present setup the line center lock was obtained by directly frequency modulating the laser at 300 kHz. This modulation frequency is chosen so that the FM sidebands are within the cell saturated absorption linewidth, to give the first derivative signal, but largely outside the narrower beam saturated absorption line [15]. The modulation index is chosen to essentially only produce the first order sidebands at 300 kHz, leaving the majority of the power in the carrier. A dc scan voltage introduced at the integrator sweeps the laser frequency over the atomic beam line. The dc scan is calibrated using 3.39 μm He-Ne lasers by measuring the beat frequency between a local oscillator locked to the servo cavity, and a methane stabilized laser. Poor finesse of the cavity mirrors at the 3.39 microns presently limits the calibration accuracy to about 10 percent.

Adding sidebands to the laser adds complexity to the correct line analysis. First, three equally spaced laser frequencies (our carrier plus the two sidebands) produce five equally separated saturated absorption lines on the beam Doppler profile (provided the beam Doppler is sufficiently large). These five features may or may not overlap depending on the ratio of the inverse single zone transit time to the modulation frequency. In our case this ratio is approximately one, which results in only

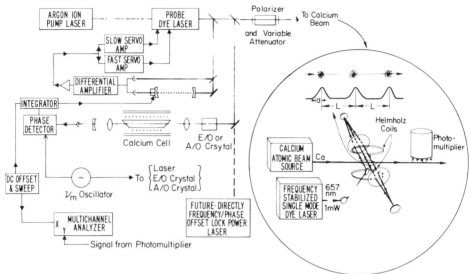

Calcium Dye Laser Spectrometer

Figure 1

slightly overlapping lines. This presents a difficulty in correctly subtracting out the baseline. The second aspect is that there are three
separate and distinct contributions to the central "resonant" saturated
absorption line. There is the usual contribution from the $v_z \sim 0$ absorbers
which are excited by the carrier and subsequently probed by the carrier.
But there are two other class of $v_z \not\sim 0$ absorbers which also contribute to
the central feature. One of these classes sees the blue shifted sideband
as the pump and the red shifted one as the probe. In a symmetrical way,
the oppositely signed v_z class sees the red shifted sideband as the pump
and the blue shifted as the probe. Interestingly, these all contribute in
a coherent way to the sharp Ramsey fringe structure for each recoil peak.
Thus, there are now two possibilities to use the full beam Doppler to
further enhance the signal-to-noise ratio while maintaining full resolution
in the saturated absorption Ramsey fringe method. The first is to make the
radiation zones narrow so that the acceptance angle $\sim \lambda/3a$ matches the beam
Doppler (wavefront curvature and the corresponding confocal parameter are
the limitations here); the second is to simply add sidebands to the laser
to fully cover the beam Doppler [16].

We also show the possibility in Fig. 1 to either phase modulate with
electro-optic crystal or to frequency modulate with an acousto-optic crystal, both of which are exterior to the laser. This allows the addition of
sidebands to line center lock, but we now are able to choose whether or not
to modulate that part of the laser light sent to the atomic beam. We have
made preliminary investigations with both crystals. Perhaps the most
elegant method [4] is to build a second low power reference laser which
could be locked to one of the magnetically shifted, $\Delta m_i = 1$, $^1S_0 - ^3P_1$
lines in the Ca absorption cell. The more powerful probe dye laser could

be phase or frequency offset locked from the stable reference laser with a tunable RF oscillator which would provide a precise frequency scale. More generally, the reference laser could be stabilized to any of a number of possible atomic lines which are nearby the line(s) to be studied. For example, iodine nearly fills the entire visible spectrum with sharp transitions to which a laser could be locked. This would permit an extremely precise, stable, tunable dye laser spectrometer of general versatility.

In the circled inset in Fig. 1, we show a three dimensional representation of the atomic beam and interaction region. A Ca atomic beam from a resistively heated oven sequentially interacts with three equally spaced and parallel standing wave light beams from our stabilized dye laser. The atomic beam is collimated to give a Doppler beam width of approximately 1.7 MHz HWHM and has a density of about $10^8/cm^3$ at the laser excitation region. Typically, the laser power in the interaction region is on the order of 1 mW focused to a beam spot radius of approximately 0.15 cm. To form the three standing wave radiation fields, we use the duo cat's-eye reflector which was described above. The common spatial separation of the light fields is variable from 0 to $\sim$ 3.5 cm. A transverse magnetic field of a few times 10^{-4} Tesla (a few Gauss) is superimposed on the interaction region which splits off the $m_j = \pm 1$ components. Furthermore, with light linearly polarized in the direction of the applied magnetic field, we excite only the $m_j = 0$ to $m_j = 0$ transition, which has no first order Zeeman effect. For signal detection we use a 5 cm diameter cathode photomultiplier located 20 cm downstream from the excitation region and 1 cm from the beam. With this arrangement we estimate that we collect about one percent of the total fluorescence photons. The signal is recorded on a multichannel analyzer with typical signal averaging times of a few minutes.

Results

The photon recoil doublet structure produces two partially overlapping fringe patterns as is shown in Fig. 2. The outer curve is the atomic beam fluorescence profile, with HWHM of 1.7 MHz, showing the single zone saturation dip with HWHM of 175 kHz. The two solid inner curves are the bottom of the single zone dip greatly expanded to show the observed Ramsey patterns for 2L = 3.5 and 7 cm, with the positions of the recoil components indicated by the vertical dashed lines. These curves were obtained by recording first derivative signals, with a narrowbanded S/N of about 10, and then integrating them with the multichannel analyzer to improve the apparent S/N. The separate fringe patterns for the two recoil peaks are shown as dashed curves above each experimental curve. The fringe intensity is a few percent of the total signal. The intensity ratio of the recoil peaks is approximately one, but the scatter in this ratio for our data is so large at present that no meaningful comparison can be made with the expected ratio of 0.998:1.0 [9].

Our measured value for the recoil splitting for a power density of about 10 mW/cm^2 is $\delta\nu$ = 23.6 ± 2 kHz, where the error of about 10 percent is mostly due to the poor absolute calibration of the dc frequency scan mentioned above. By holding this scan constant and obtaining relative measurements of the splitting versus power, we have obtained a possible indication of the contraction of the recoil splitting with power predicted by Bordé [2,10]. This is a light intensity shift of each recoil component toward line center caused by higher order coherent processes contributing

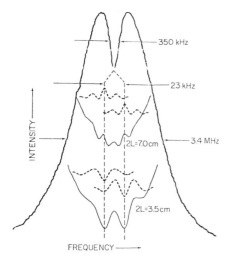

Figure 2 Observed fluorescence profiles. Outer curve: Atomic beam
Doppler profile showing "single"
zone saturation dip. Inner curve:
Expansions of bottom of single zone
dip (solid curves) showing Ramsey
fringes for 2L = 3.5 and 7 cm.
Fringe patterns for each photon recoil component shown with dashed
curves. Note fringe sign inversion
between the two resolution cases.
See text for discussion.

extra intensity to the inner side of each line. Bordé calculates the shift
to be, in first approximation, $\Delta = (\mu E/2h)^2/(f \cdot \delta v)$ (Hz) for the case of
continuous excitation and well resolved recoil peaks. In this expression μ
is the electric dipole moment (for Ca, $\mu = 4.8 \times 10^{-20}$ esu-cm) and E is the
electric field amplitude. For three-zone Ramsey excitation, this contraction should be reduced by a factor approximately equal to 2a/L, since the
atom experiences the electric field for only the same fraction, 2a/L, of
its time in the interaction region. With the electric field averaged over
the Gaussian mode and diluted by 2a/L, the predicted contraction for 2L = 7
cm is $\Delta = -0.14$ kHz/(mW/cm²). Experimentally, the recoil splitting versus
power for 2L = 7 cm and power densities up to 17 mW/cm² shows a contraction
of $\Delta = -0.2$ (±0.2) kHz/(mW/cm²). Although the experimental error at present is large, this result suggests agreement with Bordé's theory for
continuous excitation attenuated in this way.

The power induced shifts for the two recoil peaks are equal, to first
order, but opposite in sign, and thus the center of the line remains unshifted. It should be possible to line center lock to both peaks simultaneously, using modulation sidebands for instance, to give an average
frequency for the centroid which is power independent to first order.
Also, one might use identical circular polarizations to obtain the $m_j = 0$
$\leftrightarrow m_j = \pm 1$ crossover resonance which has only one recoil peak, and hence
no light-induced shift [2]. Thus, we do not believe this shift will be a
serious problem for accurate frequency measurements.

The predominant limit to the accuracy of a frequency measurement for
this line is the second order Doppler shift, 1.7 kHz for the most probable
velocity. The second order Doppler shift has been an important limitation
to all present frequency standards and requires careful velocity measurements to reduce this systematic offset. Similarly for calcium, a very
careful determination of each velocity class contribution to the signal
would be necessary to reduce this systematic offset below 10^{-14}.

126

It is an interesting feature, peculiar to the Ramsey interference method, that the line center position becomes highly sensitive to resolving power when the magnitude of the resolution, $\Delta v/v$, approaches that of the second order Doppler shift, $-\frac{1}{2} v^2/c^2$ [17]. This can be easily understood when one remembers that the resonant frequency of each recoil peak for a particular velocity class, v, is shifted by the second order Doppler term, $-\frac{1}{2} v^2/c^2$, but that the interfringe spacing is proportional to $v/2L$ (actually, to $v_r/2L$, but $v_r \sim v$ for saturation spectroscopy signals). We already have seen that velocity averaging washes out the fringe structure away from the recoil shifted line center, because each cosine, with period dependent on v, only contributes coherently at line center, but adds with random phase elsewhere, thus washing to zero in the wings. But, recall that this result obtains with fractional resolution much lower than the second order Doppler shift. When the resolution improves to where the interference fringe spacing begins to approach the same order of magnitude as the second order Doppler shift, then there is no frequency, or "line center", where all velocity contributions add coherently. Rather, there is an arbitrary frequency position, in general blue shifted from the recoil resonance frequency, which receives the majority coherent build up. This position depends critically on the interzone spacing and on the atomic beam velocity distribution. Contrast this to the single zone, cw excitation method where the second order Doppler shifted resonance is also highly sensitive to the velocity distribution but largeley independent of resolving power. However, even for sufficiently long lived systems, the single zone method has an ultimate resonance width limited by the quadratic Doppler broadening (for the multiple velocity case) independent of potentially better spectrometer resolving power, whereas the multiple zone interference method can yield artificially narrow resonances. Depending on the combination of velocity distribution and interzone spacing, it is possible to produce linewidths that are sub second order doppler and, perhaps, even subnatural in some cases.

Returning now to Fig. 2, note that the fringe pattern for each recoil peak in the high resolution case (2L = 7 cm) is inverted from that of the lower resolution case of 2L = 3.5 cm. We had earlier attributed this inversion to cavity phase shift between radiation zones. However, a velocity integration which includes the quadratic Doppler shift and no cavity phase shift gives inverted fringes for the 2L = 7 cm spacing. In fact, already at 3.5 cm there is a small asymmetry due to the quadratic effect.

It is clear that the task of determining the unshifted line center, or the accuracy and reproducibility has increased in complexity. But the method of optical Ramsey fringes applied to long lived atoms, such as Ca, is not hopeless, and there remain a number of potentially useful schemes. One could pulse the radiation field to select one, or at most a few, velocity classes in order to precisely determine the second order Doppler shift (also provides a means to determine the velocity distribution). Only those atoms with velocities, $v = nL/\tau$, where τ^{-1} is the pulse repetition rate and n is an integer, will interact with all radiation fields. However, this method is very expensive in signal-to-noise ratio since most atoms do not contribute to the signal. Additionally, the single velocity atomic beam signal is a slightly damped cosine function, which presents problems in determining and locking to line center.

Another alternative would be radiation pressure cooling of all velocity components with a broadband laser frequency red shifted from the natural resonance frequency of a suitable cooling line. For both Ca and Mg, the $^1S_0 - ^1P_1$, line, at 4228 Å and 2853 Å, respectively, is a potential candidate. A possibly better cooling mechanism for Ca would be to simultaneously optically pump on the three $4^3P_{0,1,2} - 5^3S$ transitions near 600 nm. In either case, it should be possible to thermally cool Ca by a factor of 10^2 with $\sim 5 \times 10^4$ scattering events in an approximately 1 cm interaction zone immediately outside the oven. The velocity would be reduced by a factor of 10 and the quadratic effect by a factor of 100. This would give a second order shift of only 17 Hz for the most probably velocity, permitting a clear resolution of the natural linewidth of 410 Hz. The three zone interaction region would follow the cooling region in a way such that the cooling process does not take place in the interaction region (a mirror with a hole in it would serve to bring in the longitudinal cooling beam). A common interzone spacing of 1.5 cm would fully resolve the recoil splitting and give a linewidth of 750 Hz HWHM. The fluorescence decay length is also reduced form approximately 32 cm to 3.2 cm, enhancing collection efficiency. As described in detail elsewhere, it would be possible to further improve the signal-to-noise ratio by photon amplification of the excited $^3P_1$ state to overcome detection and collection efficiencies [6]. The useful limit to S/N ratio could then be determined by the shot noise in atomic beam intensity. A conservative estimate of 10^4 for the interference fringe S/N ratio would give a pointing precision, or fractional frequency stability, of 3×10^{-16} at one second for our 750 Hz line. This precision should make it possible to study and correct systematic errors to better than 10^{-14} and result in a very high accuracy frequency standard in the visible spectrum.

In conclusion, we have for the first time obtained ultra high resolution with optical Ramsey interference fringes in a long lived atomic system. We have fully resolved the recoil splitting of the saturated absorption signal in the Ca, $^1S_0 - ^3P_1$ intercombination line at 657 nm. We have introduced a locking scheme which greatly extends the useful stability of a dye laser spectrometer. Methods have been proposed which should permit a visible wavelength frequency standard, based on this Ca transition, which should exhibit a frequency stability of $3 \times 10^{-16} (\tau)^{-\frac{1}{2}}$ and a reproducibility exceeding 1×10^{-14}. This precision should allow improvements in experiments involving measurements of frequency offsets, such as gravitational red shift.

The authors acknowledge useful contributions from others: In particular, we thank Jan Hall for continued advice and discussions. We have received beneficial input from Siu Au Lee, Jürgen Helmcke, and Christian Bordé.

References

1. Ye. V. Baklanov, B. Ya. Dubetsky and V. P. Chebotayev, Appl. Phys. 9, 171 (1976).
2. C. J. Bordé, Proc. of 1977 Laser Spectroscopy Conference, in Laser Spectroscopy III (Springer-Verlag, 1977), p. 121, and references therein.
3. J. C. Bergquist, S. A. Lee and J. L. Hall, Phys. Rev. Lett. 38, 159 (1977).
4. J. C. Bergquist, High Resolution Optical Spectroscopy: Optical Ramsey Fringes; Neon Isotope and Hyperfine Studies (PhD thesis, University of Colorado, Boulder, Colorado, 1978).
5. J. C. Bergquist, S. A. Lee and J. L. Hall, Proc. of 1977 Laser

Spectroscopy Conference, in Laser Spectroscopy III (Springer-Verlag, 1977), p. 142.

6.   R. L. Barger, J. C. Bergquist, J. C. English, and D. J. Glaze, Appl. Phys. Lett., to be published.

7.   R. L. Barger, Proc. of 1973 Laser Spectroscopy Conference, in Laser Spectroscopy (Plenum Press, New York, 1974), p. 273.

8.   R. L. Barger, T. C. English and J. B. West, in Proc. 29th Annual Symp. Frequency Control (Electronic Industries Assoc., 2001 Eye St. N.W., Washington, DC 20006, 1975), p. 316.

9.   A. P. Kol'chenko, S. G. Rautian and R. I. Sokoloviskii, Soviet Phys. JETP 28, 986 (1969).

10. C. J. Bordé, C. R. Acad. Sc. Paris 283B, 181 (1976).

11. J. L. Hall, C. J. Bordé, and K. Uehara, Phys. Rev. Lett. 37, 1339 (1976).

12. N. F. Ramsey, Molecular Beams (Oxford Univ. Press, Amen House, London E.C., 1956), Ch. V.

13. R. L. Barger, J. B. West and T. C. English, Appl. Phys. Lett. 27, 31 (1975).

14. J. Helmcke, J. L. Hall and S. A. Lee, to be published.

15. J. C. Bergquist, to be published. This technique for optical frequencies is similar to the RF techique recently introduced for hydrogen masers. See F. L. Walls and D. Howe, in Proc. 32nd Annual Symp. Frequency Control (Electronic Industries Assoc., 2001 Eye St. NW, Washington, DC 20006, 1978) (in press).

16. This possibility has been more fully explored by J. L. Hall for the case of single zone saturation spectroscopy, private communication.

17. R. L. Barger, to be published; and J. L. Hall, private communication. Also see article by S. A. Lee, J. Helmcke and J. L. Hall in this volume.

18. H. Hellwig, S. Jarvis, Jr., D. Halford, and H. E. Bell, Metrologia 9, 107 (1973).

19. T. W. Hänsch and A. L. Schawlow, Opt. Commun., 13, 68 (1975). D. J. Wineland and W. M. Itano, to be published.

High Resolution Two-Photon Spectroscopy of Rb Rydberg Levels

S.A. Lee[1], J. Helmcke[2,3], and J.L. Hall

Joint Institute for Laboratory Astrophysics
National Bureau of Standards and University of Colorado
Boulder, CO 80309, USA

In this paper we report on our investigation of the interesting features, and problems, associated with ultrahigh resolution two-photon spectroscopy, using a frequency controlled dye laser to study the 5^2S-n^2S transitions in rubidium. The optical Ramsey technique of separated interactions is used to reduce transit broadening. Narrow resonances of 17 kHz full width half maximum are observed.

The use of counterpropagating laser beams to eliminate Doppler broadening in two-photon transitions by now is a well-known and standard method in laser spectroscopy. In principle, transitions between long-lived levels, such as the ground and metastable states, should lead to extremely narrow resonances, and various authors have proposed using these two-photon resonances as possible frequency standards in the optical region [1-5]. In practice, however, the resolution obtained in two-photon spectroscopy has thus far been limited either by the laser line width, or by the short lifetime of the states being investigated. Important aspects of high resolution two-photon spectroscopy, such as the transit-limited line shape, or small level shifts, have remained almost unexplored. The reasons may be summarized as a lack of suitable laser sources, and the lack of experimentally attractive candidate transitions.

The major disadvantage of two-photon transitions between long-lived states is the extremely low transition strength. Large laser intensity is needed and even so, two-photon signals usually require long periods of signal averaging (e.g., Bi) [6,7]. This puts rather stringent requirements on the laser source: reasonable power output, narrow spectral width, and good long term frequency stability.

If the laser width may be neglected, transit broadening becomes the main limiting factor in two-photon line widths. The line shape of two-photon absorption, including the effects of the finite transit time of the atoms through a Gaussian laser field and the natural broadening of the levels, has been studied in detail by BORDE [8], and by BAKLANOV and DUBETSKII [9]. Their treatment has recently been extended [7] to the

[1] Present address: Argonne National Laboratory, Argonne, Illinois.
[2] Present address: Physikalisch-Technische Bundesanstalt, Braunschweig, West Germany.
[3] Supported in part at JILA by a NATO fellowship (1977-78).

experimental situation in an atomic beam, where the excited atoms are detected at a distance downstream from the interaction region. Due to the decay of the excited atoms in flight, the detected signal will show not only a reduction in strength, but also a broadening in width, since the faster atoms have a higher probability of reaching the detector before decaying. Taking this decay into account, the Doppler-free two-photon transit-limited line shape may be obtained by the following integral over all possible velocities v in the atomic beam [7]:

$$I_{det} \sim \int_0^\infty dV \cdot V \exp\left(-V^2 - \frac{G}{V} - \frac{\xi^2}{4V^2}\right) \quad . \tag{1}$$

Here, $V = v/u$, where $u = (2kT/M)^{1/2}$; $G = \gamma\ell/u$ is a decay parameter, where γ is the population decay rate of the upper state, and ℓ is the detector to interaction distance; $\xi = (2\omega-\omega_0)w_0/u$ is the detuning parameter, where ω and ω_0 are the laser and atomic transition frequencies, respectively, and w_0 is the laser mode radius. Figure 1 shows the resulting line shape for different values of G. For G = 0, i.e., detecting the atoms at the interaction region, the transit width $\Delta\omega$ (FWHM at the laser frequency scale) is 1.25 u/w_0 in agreement with Ref. 9.

The line shape calculated with G = 0.29 is found to be in good agreement with our recent experiment on the Bi $6p^3$ $^4S_{3/2}$ to the metastable $6p^3$ $^2P_{3/2}$ two-photon transition [7].

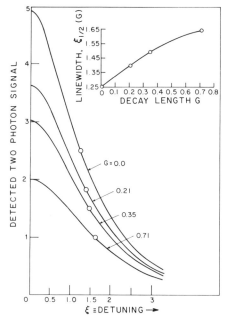

Fig. 1. Transit-limited two-photon line shape, calculated from Eq. (1). The full profile is symmetric around $\xi = 0$. In-flight decay between excitation and detection is parametrized by G. Inset shows line width increasing effect of inflight decay.

To reduce transit broadening, BAKLANOV et al. [10] proposed using two spatially separated laser beams. The resulting fringe structure, arising from the interference between two separate interactions of an atom with the radiation field, will have a width determined by the atomic transit time between the two light beams. This is the optical analog of the well-known Ramsey fringes in microwave and rf spectroscopy. The first radiation zone creates a coherence between the atomic lower and upper states. In the region between the laser beams, the atomic polarization precesses freely, and at a rate which is, in general, different from the radiation field. The effect of the second radiation zone depends on the phase of the radiation relative to the phase of the atom, thereby producing interference dependent on the interzone transit time T. The fringe pattern is of the form $\cos(2\omega-\omega_0)T$. When the fringes are integrated over the velocity distribution, the side fringes tend to be averaged out, leaving only the central fringe.

The total phase of the Ramsey signal also includes a contribution from the relative phase difference between the laser fields themselves in the two zones [11]. This "cavity phase" introduces a displacement of the fringe system. Thus it is important to hold this "cavity phase" fixed in an experiment, otherwise the random fluctuations will wash out the fringe pattern. We note that in a folded standing wave configuration the phase factors cancel exactly, and if the laser beams are matched, the fringe pattern will be symmetric and unshifted. Ultimately, at very high resolution the second-order Doppler effect gives an interesting distortion -- as well as shift -- of the Ramsey fringe pattern. (See below.)

In the experimentally appropriate limit that the interzone separation d is much greater than the size of each zone, the Ramsey fringes have the form [10,11]

$$I_{det} \sim \int_0^\infty dV \cos \left(\frac{z}{V} + 2\pi\ yV\right)\left[\frac{1}{V^2}\right][\exp\left(-\frac{G}{V}\right)][V^3 \exp(-V^2)] \qquad (2)$$

where as before $V = v/u$ and the decay parameter $G = (\gamma/u)[\ell+(d/2)]$. The cosine term arises from the Ramsey interference between excited state amplitudes produced in the two separate interaction regions. The detuning parameter is $z = (2\omega-\omega_0)d/u$, where d is the interzone separation. The second-order Doppler shift is $yV = (ud\omega_0/4\pi c^2)\cdot V$, normalized to the Ramsey period, $\delta\omega_R = \pi v/d$. The $[1/V^2]$ term arises from integration of the interaction through the laser mode, as may be shown with the diagrammatic method of BORDE [11]. The next factor represents exponential decay in the distance between excitation and detection regions. Finally the contributions of each velocity group need to be weighted with the appropriate beam velocity distribution. Figure 2A shows a Ramsey profile computed from Eq. (2) using conditions appropriate to our high resolution experiment $(\omega_0/2\pi) = 1.0 \times 10^{15}$ Hz, d = 0.5 cm, u = 3.13 × 10^4 cm/sec, G = 2.0, y = 0.0087). Figure 2B shows the effect of increasing the interzone spacing by a factor 10 to d = 5.0 cm (y = 0.087) at which point the nominal Ramsey fringe spacing just equals the expected lifetime-limited line width (~3 kHz at the laser frequency). Some fringe distortion is evident due to the influence of the second-order Doppler shift.

In the future we may expect to observe Ramsey fringes of even narrower spectral widths in atomic systems with longer lifetimes. For example, the Bi $^2P_{3/2}$ metastable level mentioned earlier has a 4 msec

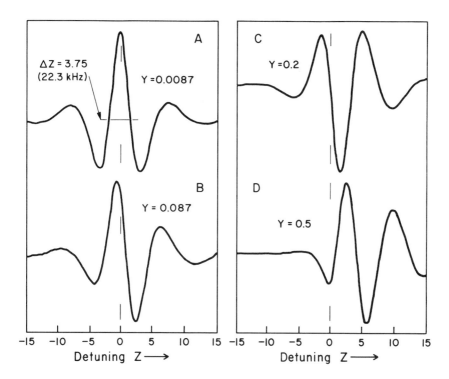

Fig. 2. Calculated Ramsey fringe profiles in limit of negligible interaction time in laser beams. All curves correspond to detection downstream at 2 decay lengths (G=2), and differ in the ratio, y, of second-order Doppler redshift to the Ramsey spacing. A) Corresponds to experimental conditions of Fig. 7B, y = 0.0087; B) y = 0.087; C) y = 0.2; D) y = 0.5. Note that the fringe-like structure can appear on the blue side of the rest frequency.

estimated lifetime, leading potentially to 20 Hz wide two-photon resonances. However from a usual atomic beam source, the thermal velocity $u \sim 3.5 \times 10^4$ cm/sec leads to a second-order Doppler shift of 1.3 kHz. Such domination of the Ramsey line width by the quadratic Doppler shift gives rise to interesting spectral profiles. We may appreciate this effect physically from the following kind of arguments [22].

For a given atomic velocity v the Ramsey pattern contains, in addition to the central fringe near $\omega = \omega_0/2$, a number of side fringes of laser frequency spacing $\Delta\omega_{Ramsey} = \pi(v/d) \cdot m$ where m is the side-fringe order number. On the red side of line center this linear side-fringe frequency shift and the second-order Doppler shift are additive, both increasing with atomic velocity v. On the blue side of line center, however, the two contributions to the fringe offset have opposite signs. Thus with low but increasing velocities the side fringe resonance position moves first linearly to higher frequency. With further velocity increase the resonance position becomes stationary as the linear fringe offset and second-order Doppler shift approximately cancel. Ultimately at higher velocities, the second-order shift dominates and the side fringe components again move to the red. Computer simulation under very high resolution conditions leads to "fringe-like" wiggles on the blue side of true line center superimposed on a nearly structureless background. Figure 2(C,D) shows two representative curves for G = 2 and y = 0.2 and 0.5. As the shape and position of these wiggles are highly sensitive to the assumed conditions, it is amply clear that some type of velocity-selection technique will be essential in this ultrahigh-resolution domain. Alternatively, the observed profiles may be used to infer the actual velocity distribution.

Turning now to experimental aspects, two-photon Ramsey fringes have been observed in Na 3S-4D transition, in the time domain by SALOUR and COHEN-TANNOUDJI [12] using two coherent pulses, and by CHEBOTAYEV and coworkers [13] using two spatially separated laser beams. As the short lifetime of the 4D state ($\tau \sim$ 50 nsec) is really the main limiting factor in these experiments, they could not bring out the high resolution aspect of the Ramsey technique. We report now on an investigation of ultrahigh resolution two-photon spectroscopy using the Ramsey technique and a frequency-controlled dye laser to study the 5^2S-n^2S transitions in rubidium.

The choice of rubidium has the following rationale: The Rydberg levels of Rb can be reached by two photons of a Rhodamine 6G dye laser; and more importantly, they have reasonably long lifetimes and are easy to detect. For Rydberg levels, the lifetime varies $\sim (n^*)^3$, where n^* is the effective principal quantum number. Thus for $n^* \sim$ 30, the lifetime is ~25 μsec, giving a two-photon line width of 3 kHz (in the laser frequency scale), which is adequate for our purpose. In addition, the Rydberg levels are easily ionized by an electric field, and so the two-photon transitions may be detected with high efficiency and sensitivity by field ionization and positive ion counting.

In our experiment a beam of Rb atoms is produced from a resistively heated oven (T ~ 200°C), and is collimated to approximately 1 mm in diameter at the interaction region. Figure 3 shows a schematic of the atomic beam apparatus. To increase the circulating laser flux, we use a resonant cavity inside the vacuum chamber. This slightly folded cavity (see Fig. 3) produces a power buildup ~30 and has two separate waists, of 155 μm

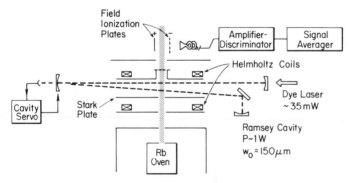

Fig. 3. Rb atomic beam apparatus for two-photon Ramsey fringes.

134

radius, displaced along the atomic beam axis. The use of careful input
mode matching and a high-finesse, non-degenerate cavity insures that its
internal fields accurately represent a spatially matched standing wave
laser field and so have the proper amplitude and phase conditions for a
clear study of the optical Ramsey effect. The atom beam crosses the laser
beams at right angles, and the excited atoms are field ionized at a distance of 1.5 cm downstream from the interaction region. The positive ions
are accelerated into the cone of a channeltron electron multiplier. The
multiplier's output pulses are stored in a signal averager, which is swept
synchronously with the laser frequency. The detection region is enclosed
in a metal box to shield its electric field from the interaction region.
A pair of parallel plate electrodes in the interaction region allow us to
apply a small, uniform electric field to study the dc Stark effect. We
note that for S levels, the Stark effect introduces a red shift in the
transition frequency but does not produce splitting. By observing the
Stark shift as the polarity of the applied electric field is reversed,
we conclude that stray electric field in the interaction region is less
than 0.15 V/cm, giving a frequency shift of less than 10 kHz. A small
Helmholtz coil pair is used to generate ~20 G of magnetic field for
Zeeman studies.

A schematic of the frequency-controlled dye laser spectrometer [14]
is shown in Fig. 4. The dye laser is basically a three-mirror astigma-
tically compensated cavity, with a birefringent tuner and two etalons for
single frequency operation. The dye cavity optical length may be varied
by applying voltages to an intracavity ADP phase modulator, and to a
piezo-driven mirror. For frequency control, a part of the dye laser output
is mode matched into a high-finesse, invar-spaced cavity. The laser frequency is servo-controlled to the side of a transmission fringe of the

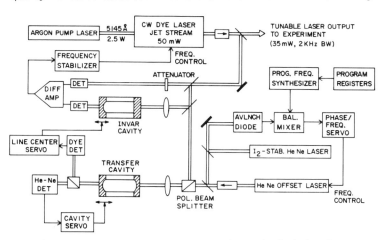

Fig. 4. Narrow-bandwidth frequency-offset-locked dye laser spectrometer.
Lock of dye laser on side of "invar cavity" fringe narrows dye laser line
width. Long-term frequency control is via first derivative lock to the
transfer cavity which is similarly controlled to the HeNe offset laser.
The offset frequency of this laser from the I$_2$-stabilized laser is locked
to the programmable frequency synthesizer. A multichannel signal averager
and the frequency synthesizer are swept synchronously.

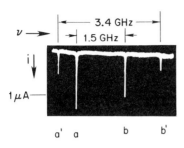

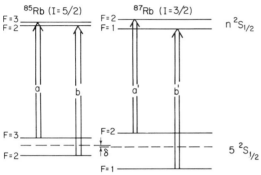

Fig. 5. Doppler-free two-photon peaks in an ionization cell filled with natural Rb. Sweep time 4 sec.

cavity, using the usual fast differencing technique. With an improved amplifier system and an intra-dye-laser-cavity ADP phase modulator to provide fast response, the dye laser spectral width is narrowed to 2 kHz RMS. The long term stability and frequency information is obtained by servo-controlling the invar cavity length so that the dye laser is locked on the center of the transmission fringe of a quartz-spaced transfer cavity. The length of the transfer cavity is servo-controlled to a 633 nm HeNe local oscillator laser, which is in turn frequency offset locked [15] to an $^{129}I_2$ stabilized HeNe laser. Digital scanning of the dye laser frequency is accomplished by the frequency synthesizer in the frequency-offset-locked loop. The system is capable of a continuous scan of 900 MHz, at 1 kHz step resolution. The limit on the scan range is due to the gain profile of the HeNe laser. The use of a separate transfer cavity greatly improves the long term stability and frequency reproducibility of the dye laser, particularly against spurious frequency shifts introduced by very small angular drifts in the direction of the dye laser output beam (which degrade the mode-matching into the invar cavity). Drift rate of the absolute optical frequency was <1 kHz/min under favorable conditions. The useful dye laser output for the experiment is ~35 mW, at 2 kHz spectral width.

We note that strong realtime two-photon 5S-nS transitions in Rb may be obtained with a simple electrostatically shielded, space-charge limited ionization cell [16,17]. In Fig. 5 we show a spectrum of the 5S-32S transition in natural rubidium. As illustrated in the lower part of the figure, the two larger central components arise from the more abundant ^{85}Rb and are separated in laser frequency by one half of the 3.03573 GHz ground state hyperfine splitting. The outer two peaks are from ^{87}Rb, and are separated by one half of its 6.78468 GHz ground state hfs. The upper state hfs is negligible at this resolution level. Unfortunately, condi-

tions in a cell are rather complex. Stray electric fields induce Stark
shifts, and pressure shifts and broadenings of ~1.6 GHz/Torr and (2.6±1.0)
GHz/Torr are observed for the 5S-32S transition [17].

In an atomic beam the collision effects are negligible. When the
spectrum was obtained with high resolution, each of the lines described
above showed a complex structure. This was found to be caused by the
presence of a small magnetic field (~1 G) in the apparatus, which caused
Zeeman splitting of the transitions. To investigate the Zeeman effect
more clearly, a magnetic field of about 20 G was applied. For ^{85}Rb, I =
5/2, giving F = 3 or 2. The F = 3-3 transition consists of 11 components, whereas in the F = 2-2 transition the central component is missing.
See Fig. 6. The effect may be explained by the following simple model.
The ^{2}S levels have a total angular momentum $\vec{F} = \vec{I} + \vec{S}$, where I and S are
the nuclear and electronic spin angular momentum, respectively. The spin
Hamiltonian may be expressed as

$$H = A(\vec{I} \cdot \vec{J}) + g\mu_B \vec{S} \cdot \vec{B} - g_I \mu_N \vec{I} \cdot \vec{B} \quad ,$$

where g, g_I are respectively the electronic and nuclear g-factors, and
μ_B and μ_N are the electronic and nuclear magnetons. The selection rules
for transitions between Zeeman levels for ^{2}S states were discussed by
BLOEMBERGEN et al [18]. For small magnetic fields such that $g\mu_B B \ll A$,
the selection rules are $\Delta F = 0$, and $\Delta m_F = 0$, assuming the spin orbit coupling may be neglected in the intermediate P states. The Zeeman levels
may be labeled by the quantum numbers F and m_F. We note that in a
hydrogenic atom, the hyperfine constant A varies as $(n^*)^{-3}$. For ^{85}Rb,
$A_{5S} = 1011.9$ MHz, giving $A_{32S} \sim 200$ kHz. Thus the weak field criterion,
$B \ll A/g\mu_B$ (~0.1 G) is not satisfied even in the earth's magnetic field.
For a field of 20 G which we used, the upper level is in the Paschen-Back
region.

In the strong field domain, where the levels may be labeled by their
magnetic quantum numbers m_S and m_I, the Zeeman energy of the levels is
simply $A m_I m_S + g\mu_B m_S B - g_I \mu_N m_I B$. The selection rules are $\Delta m_I = 0$, $\Delta m_S = 0$.

In our case we have $A_{32S} \ll g\mu_B B \ll A_{5S}$. Consider the transitions
from the F = 3 hyperfine level only. The lower and upper states have
different spin wave functions, and the projection of the lower m_F levels

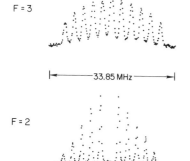

F = 3

|←———33.85 MHz———→|

F = 2

Fig. 6. Zeeman spectra of
32S ← 5S two-photon transition in Rb atomic beam. Note
the absence of the central
$|m_F| = 3$ component in the
F = 2 spectrum.

137

onto the upper $m_S m_I$ levels give two different transition frequencies for each m_F value, except for m_F = +3 or -3 which are pure states. These latter two transitions contribute to a single line at the zero field F=3-3 position. Thus 11 transitions are possible, in agreement with experiment. Similarly, the F=2-2 transition can be shown to have 10 Zeeman components and to be missing the central component (which derives from $m_F = \pm3$), in agreement with Fig. 6.

These Zeeman patterns can be used to obtain the hyperfine constant of the upper level. The frequency interval of adjacent transitions belonging to the m_S = +1/2 group is different from the interval of the m_S = -1/2 group, the difference being $A_{32S}/2$. Here the factor of 2 is introduced so that the frequency scale is in terms of the laser frequency. We measured A_{32S} = (0.23±0.07)MHz. A precise knowledge of the magnitude of the magnetic field is not needed to deduce the excited state hyperfine structure constants but it would allow determintion of the g factors.

We note that laser stepwise excitation in conjunction with rf resonance technique has been very successful in measuring the hyperfine structure of nS levels in alkali atoms, where $n \lesssim 10$ [19]. The extension of the double resonance technique to higher n values proves to be difficult, due to the rather poor signal-to-noise ratio in observing fluorescence. The present method has a comparable frequency resolution but with extremely good S/N. Typical time spent in obtaining a Zeeman pattern is 1 min. Thus our method should be very useful in measuring hyperfine parameters of high Rydberg levels.

Another interesting type of possible measurement with the m_F-state-resolved spectrum is the determination of the excited state static polarizabilities. For example when even small voltages were applied to the electric field plates mentioned earlier, rather dramatic frequency shifts were observed. For the 5S-32S (F = 3, m_F = -2) transition a quadratic Stark frequency shift rate of -620 kHz/(V/cm)^2 was measured. Presumably more complicated behavior would occur in the nD levels where a tensor polarizability term can also be non-zero.

To investigate the optical Ramsey effect, the ^{85}Rb 5S-32S, F = 3 line was resolved into its 11 components with a 20 G magnetic field. The central Zeeman component ($m_F = \pm3$) was chosen for its strength, and also because this line is least perturbed by magnetic field inhomogeneity in the interaction regions. The velocity of the beam is u ~ 3×10^4 cm/sec. With an upper state radiative lifetime of ~25 μsec, the population decay length is ~0.75 cm.

Figure 7A shows the Ramsey feature obtained with two spatially separated laser beams. The fringe contrast is ~35%. The signal was obtained in one sweep, at 1 sec/channel, and 5 kHz step size (at 633 nm). The two cavity waists, of mode radius 155 μm, were separated by approximately 1.5 mm. As expected, the side fringes rapidly wash out, due to the velocity distribution in the atomic beam. The estimated experimental fringe full width of ~40 kHz was somewhat larger than the expected 30 kHz width in this early experiment. (The dye laser width was ~5 kHz when this spectrum was taken.) The large background in Fig. 7A is the transit broadened signal from those atoms which interact with only one radiation zone. The full width, according to calculations described above, is 390 kHz. But with a detection to interaction length of 1.5 cm, corresponding to a decay parameter G = 2, the width is broadened to ~630 kHz. The ex-

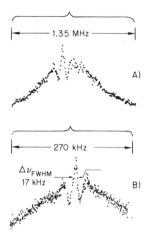

Fig. 7. Two-Photon Ramsey Fringes. Frequency increases to the left. A) Waist separation $\approx$ 1.5 mm. Single zone background resonance width ~600 kHz, Ramsey fringe width ~ 40 kHz. Dwell time 0.25 sec per point. B) High resolution, waist separation $\approx$ 4.2 mm. Single sweep at 1.06 kHz/channel. Dwell was 0.35 sec per point.

perimentally observed width is ~600 kHz. On the peak the ion counting rate is ~10 kcps.

Figure 7B shows the narrowest Ramsey fringes that we have obtained, with an interzone separation of 0.42 cm. The fringe full width of 17 kHz is almost entirely due to the interzone transit time [20]. Further increase of the separation unfortunately led to a severe reduction in the Ramsey signal. (We note that as the separation is increased, fewer atoms will be intercepted by both zones. If the oven aperture is too large the reduction in fringe contrast will be thus quadratic in the separation.) Furthermore, at n = 32 the natural decay is beginning to take its toll.

In the spectra we obtained, the Ramsey fringe pattern always appears to be displaced relative to the background peak. (The central Ramsey fringe is to the high frequency side of the background peak.) See Fig. 7A for example. These data were obtained on the magnetic field-independent $|m_F|$ = 3 transitions so that magnetic field inhomogeneities can play no role in producing the somewhat asymmetric shape. In principle asymmetric Ramsey fringes could be caused by a residual phase difference between the radiation zones. However the high finesse and accurate mode-matching of our cavity makes this explanation unlikely. A more tantalizing cause may be the reduction in the ac Stark shift with the Ramsey method: the background peak is shifted the full amount, while the shift of the Ramsey peak is diluted by a factor proportional to the one-zone to interzone transit time ratio. At our wavelength, the Rb 5S level is pushed upwards in energy by the ac Stark effect by 40 Hz/(W/cm^2), when the 5 to 10P levels contributions are taken into account. The 32S level shift is very difficult to calculate accurately. We note that the largest radial matrix elements occur for nearby nP states, but their nearly symmetric locations relative to 32S give rise to strong cancellation via the $(E_{32S}-E_{nP}\pm\hbar\omega)^{-1}$ denominators in the summation which appears in the ac Stark shift formula. Only very weak matrix elements connect to lower levels (~5P) where the energy denominators are small. Thus we believe that the dominant effect is an upward displacement of the 5S level and a consequent redshift of the two-photon transition frequency.

In conclusion, we have explored the two-photon optical Ramsey effect at high resolution using transitions to rubidium Rydberg levels. A Ramsey fringe as narrow as 17 kHz was easily observed [20]. We are anxious to test the ideas that the ac Stark shift is responsible for our displaced Ramsey resonances and that the shift is diluted by a factor $F \sim w_0/d$ relative to the transit-limited background profile. Certainly higher laser power with still narrower line width would be helpful and presumably ring laser techniques will be attractive in future work. However, as Fig. 7A illustrates, reasonably large shifts are possible even with our modest laser power. It is clear that a quantitative understanding of the intensity shift and its reduction via the "Ramsey dilution effect" are prerequisite to the use of two-photon Ramsey spectroscopy for optical frequency standards and precision spectroscopy. A possibly more serious challenge is presented by the distortion of the Ramsey profiles due to the second-order Doppler effect. Perhaps radiative cooling or other techniques can be used to significantly reduce the kinetic temperature, or perhaps supersonic expansion can be used to narrow the velocity distribution. Another clear area of interest is the influence of thermal fields [21] in depopulating -- and more interestingly -- in shifting the Rydberg levels.

One of us (JLH) wishes to thank R. L. Barger for communication of his unpublished results and for useful discussion of the second-order shifts. Numerous discussions with C. J. Borde on the subject of line shapes are also gratefully acknowledged. The work has been sponsored in part by the Office of Naval Research, the National Science Foundation, and the National Bureau of Standards.

References

1. L. S. Vasilenko, V. P. Chebotayev and A. V. Shishayev, JETP Lett. 12, 113 (1970); Ye. V. Baklanov and V. P. Chebotayev, Opt. Comm. 12, 312 (1974).
2. B. Cagnac, G. Grynberg and F. Biraben, J. Phys. (Paris) 34, 845 (1973).
3. T. W. Hänsch, S. A. Lee, R. Wallenstein and C. Weiman, Phys. Rev. Lett. 34, 307 (1975).
4. D. E. Roberts and E. N. Fortsan, Phys. Rev. Lett. 31, 1539 (1973).
5. P. L. Bender, J. L. Hall, R. H. Garstang, F. M. J. Pichanick, W. W. Smith, R. L. Barger and J. B. West, Bull. Am. Phys. Soc. 21, 599 (1976).
6. O. Poulsen, J. L. Hall, S. A. Lee and J. C. Bergquist, JOSA 68, 697 (1978).
7. O. Poulsen, J. L. Hall, S. A. Lee and J. C. Bergquist, in preparation.
8. C. Borde, C. R. Acad. Sci. (Paris) 282B, 341 (1976).
9. Ye. V. Baklanov and B. Ya. Dubetskii, Sov. J. Quant. Electr. 8, 51 (1978).
10. Ye. V. Baklanov, V. P. Chebotayev and B. Ya. Dubetsky, Appl. Phys. 11, 201 (1976).
11. C. Borde, C. R. Acad. Sci. (Paris) 284B, 101 (1977).
12. M. M. Salour and C. Cohen-Tannoudji, Phys. Rev. Lett. 38, 757 (1977).
13. V. P. Chebotaycv, A. V. Shishayev, B. Ya. Yurshin and L. S. Vasilenko, Appl. Phys. 15, 43 (1978).
14. J. Helmcke, S. A. Lee and J. L. Hall, to be published.
15. J. L. Hall, IEEE J. Quant. Electr. QE4, 638 (1968).

16. K. C. Harvey and B. P. Stoicheff, Phys. Rev. Lett. 38, 537 (1977).
17. S. A. Lee, J. Helmcke, J. L. Hall and B. P. Stoicheff, Opt. Lett. 3, 141 (1978).
18. N. Bloembergen, M. D. Levenson and M. M. Salour, Phys. Rev. Lett. 32, 867 (1974).
19. S. Svanberg, in Laser Spectroscopy III, Proceedings of the Third International Conference, Jackson Lake Lodge, Wyoming, USA (J. Hall and J. Carlsten, Eds., Springer-Verlag, 1977), pp. 183-194.
20. Although the width of the observed Ramsey feature can be brought into agreement with the prediction of Eq. (2), an unrealistically small value of G is implied. After the apparatus is modified to reduce the uncertainty in G caused by fringing fields near the Rydberg atom quench region, it will be interesting to pursue a quantitative comparison with theory.
21. T. F. Gallagher and W. E. Cooke, Phys. Rev. Lett. 42, 835 (1979); E. J. Beiting, G. F. Hildebrandt, F. G. Kellert, G. W. Foltz, K. A. Smith, F. B. Dunning and R. F. Stebbings 1979, J. Chem. Phys. 70, 3557 (1979); and S. Haroche, paper in this Conference.
22. These ideas were developed by R.L. Barger in collaboration with one of us (J.L.H.).

High Resolution Saturation Spectroscopy with CO_2 Lasers. Application to the ν_3 Bands of SF_6 and OsO_4

Ch.J. Bordé, M. Ouhayoun, A. van Lerberghe, C. Salomon, S. Avrillier
Laboratoire de Physique des Lasers, Université Paris-Nord, Avenue J.B. Clément
F-93430 Villetaneuse, France

C.D. Cantrell
Theoretical Division, University of California, Los Alamos Scientific Lab.
Los Alamos, NM 87545, USA, and

J. Bordé
Department of Chemistry, University of Chicago, 5735 S. Ellis Avenue
Chicago, IL 60637, USA, and
Laboratoire de Physique Moléculaire et d'Optique Atmosphérique, Bâtiment 221
Campus d'Orsay, F-91405, France

I - INTRODUCTION

Saturation spectroscopy of vibration-rotation transitions started in 1967 with the first observations of molecular Lamb dips in H_2O and CO_2 lasers [1, 2]. One year later the inverted Lamb dip technique of LEE and SKOLNICK was beautifully applied to a vibration-rotation line of CH_4 by J.L. HALL and R. BARGER [3]. With the large output power available from CO_2 lasers it appeared very quickly that the absorbing gas could be removed outside the laser resonator with a much greater flexibility in the experiments. For example it became possible to use the saturation chopper method in which only the signal corresponding to the non-linear resonances was amplified and recorded [4]. Many spectra corresponding to numerous molecules were obtained by this technique but most experiments used SF_6 as the absorbing gas.

For these early spectra the linewidth was limited to a few hundred kilohertz and the bandwidth of the laser could not be extended beyond 100 MHz. No absolute frequency calibration was available and the interpretation of such complicated spectra appeared almost impossible. The emphasis was therefore more on applications to optical frequency standards than to spectroscopy. Since these early days of saturation spectroscopy the resolving power, accuracy and bandwidth of saturation spectrometers has increased significantly.

As early as 1968, J.L. HALL introduced the idea of a frequency offset-locked spectrometer to study line shapes [3]. The frequency stability of a reference laser slaved to a saturation peak can be transferred to a second laser locked with a tunable frequency offset from the reference laser [5]. With this technique the first magnetic hyperfine structure of a vibration-rotation line was resolved in 1972 [6] and by paying great attention to all broadening mechanisms and especially transit-time broadening the half-width of the CH_4 hyperfine components was finally reduced below one kHz in 1974. The recoil splitting was then clearly resolved and precise hyperfine splittings were obtained [7]. These were in good agreement with the splittings calculated for the lower state from magnetic resonance data but suggested a spin-vibration interaction in the excited state [8]. As we shall see this spin-vibration interaction is presently the best interpretation for our measurements of the magnetic hyperfine structures of SF_6 lines [9]. Thus saturation spectroscopy appears as a powerful tool to study hyperfine interactions in the excited state as well as to confirm or extend magnetic resonance data in the ground state. Other demonstrations of this capability can be found in the various studies where a dependence of the quadrupole coupling constant with vibration was observed e.g. in NH_3 [10] or OsO_4 [11]. We shall see how spectacular this effect can be in OsO_4 .

The interest of saturation spectroscopy is not limited to hyperfine structure. A number of other recent developments make it possible to use saturation spectroscopy to obtain very accurate spectroscopic constants for a full vibration-rotation band. The first of these developments has been the measurement of optical frequencies with point contact diodes up to the very near infrared. For example the CH_4 3.39 μm line and the numerous CO_2 laser lines have been measured directly in units of the frequency given by a Cesium clock [12]. These measurements and the possibility to beat lasers (and klystrons) together turns infrared spectroscopy from a wavelength spectroscopy into a frequency spectroscopy. The second important point for spectroscopy in the 10 μm spectral region is the development of high pressure waveguide CO_2 lasers which extends the tuning range of these lasers up to several hundred MHz around each vibration-rotation line of CO_2 [13]. This increases the number of coincidences with absorption lines by a large factor and it becomes possible to sample many lines of the P, Q and R branches of a vibration-rotation band. At present this number of lines is not quite large enough to go in one step from a previously unresolved band structure to the final assignment of the observed saturation peaks. Some intermediate resolution spectroscopy is necessary and this gap is very nicely filled by semiconductor diode spectroscopy and by Fourier transform spectroscopy. Saturation spectroscopy only provides the final step in which the jump in accuracy locks the spectroscopic constants to their ultimate value. We shall illustrate such a chain with the typical cases of SF_6 and OsO_4 but we shall first describe the present stage of development of the saturation spectrometer on which these results were obtained.

II - THE FREQUENCY-CONTROLLED SATURATION SPECTROMETER

A simplified overall schematic diagram of the spectrometer is shown in Fig.1 : The CO_2 (or N_2O) laser sources are mounted on a massive (16 T) vibration-isolated concrete slab. Two broad-band waveguide lasers have been added to an original set-up comprising two highly stable low-pressure lasers. A detailed

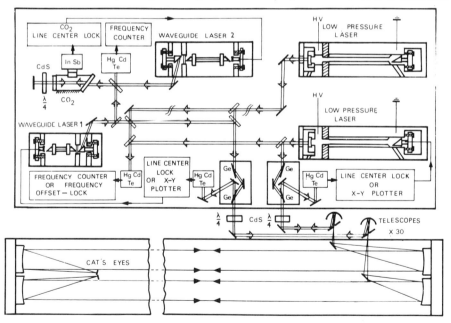

Figure 1

143

description of these various lasers and of their performance can be found in references [13,14]. For current work only two lasers are used simultaneously, one acting as the frequency reference and the other as the measurement source. The beams from the selected pair are independently expanded by 30 cm diameter telescopes in a single large absorption cell (diameter 70 cm, length 18 m) to reduce transit-time broadening and curvature induced shifts. They are reflected by cat's eyes retroreflectors to generate two independent standing waves. The return beams are steered towards HgCdTe detectors by CdS quarter-wave plates and Germanium polarizers. If necessary auxiliary absorption cells (not represented in Fig. 1) can be used to lock the reference laser to a saturation peak under pressure conditions different from those used in the large cell or with a different gas. The absolute frequency reference is obtained by locking one of the waveguide lasers to a CO_2 saturated fluorescence peak monitored with an InSb detector at 4.3 μm[15]. Two other HgCdTe detectors are used to get the beat frequency between any pair of lasers. A view of this spectrometer can be seen through the courtesy of Dr. J.L. HALL, in the Oct. 78 issue of Science. The whole set-up offers a lot of flexibility in the choice of operating modes :

- the first step in the spectroscopy of a new molecule is to record the "landscape" (broad-band spectrum) with the free-running waveguide laser to locate the resonances. Spectra of this type are displayed on Fig.2 and 8. A resolving power of 20 to 40 kHz is typically obtained for this first operation:
- the line positions are then measured by locking a conventional laser to a reference line and the waveguide laser to the resonance to be measured. The accuracy of these measurements has been limited to ± 5 kHz (for isolated lines) mainly by baseline tilts created by the laser output profile, by the gas linear absorption or by optical feedback. This situation has been greatly improved recently through the use of third-harmonic frequency locks;

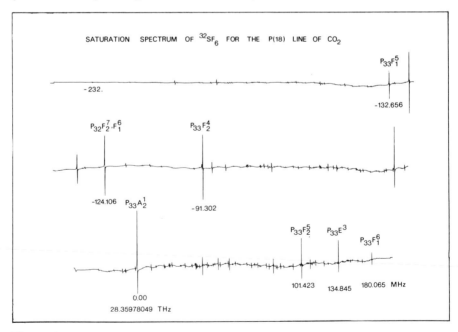

Figure 2

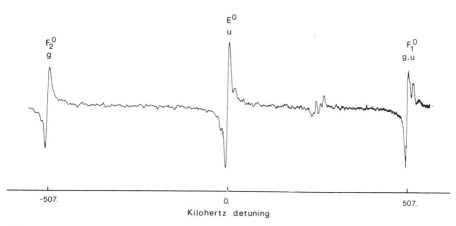

SF$_6$ Q$_{38}$ CLUSTER at 28.412582452 THz

E^0_u

F$^0_{2g}$

F$^0_{1g,u}$

-507.   0.   507.

Kilohertz detuning

Figure 3

- in the last step, the absolute frequency of the reference line is measured
by comparison with the CO$_2$ saturation peak giving a final absolute accuracy
better than ± 50 kHz;
- to study hyperfine structures or line shapes it is essential to slave the
measurement laser to the reference laser with a tunable frequency offset.
Figures 3 and 4 are two examples of spectra obtained in this way. The resolution of the apparatus is presently limited to peak-to-peak widths of the
order of 3 kHz (for heavy molecules like OsO$_4$) by imperfect focussing of the
cat'eyes retroreflectors. Indeed good agreement is found between the observed
line shape and the theoretical one predicted by introducing wave-front curvature in the formulae of reference [16]. A careful interferometric adjustement
of the cat'eyes is necessary to obtain the final increase in resolution
leading to results comparable to the methane work at 3.39 μm.

III - APPLICATION TO THE ROTATIONAL FINE STRUCTURE OF THE ν_3 BAND OF ^{32}SF$_6$
AND DETERMINATION OF A NEW SET OF SPECTROSCOPIC PARAMETERS

To illustrate the applicability of our spectrometer to fine structure studies
we shall now emphasize the example of the ν_3 band of SF$_6$. The procedure described previously was used to record and measure the absolute frequencies of
more than one hundred SF$_6$ resonances falling within the tuning ranges of the
P(12) to P(20) lines of CO$_2$. This corresponds to a nice coverage of the full
vibration-rotation band since high and low J values are sampled in the P and R
branches as well as a large number of Q lines. Fig. 2 shows the example of
part of the P$_{32}$ and P$_{33}$ manifolds in coincidence with the P(18) CO$_2$ laser line.
Similar spectra corresponding to the P(14) and P(16) CO$_2$ laser lines have already been published in references [13,17] and complete detailed spectra will
be found in [18].
    The determination of spectroscopic constants from our saturation-spectros-
copic data fell into three stages : (a) provisional assignment; (b) determination of preliminary constants given a provisional assignment; (c) assignment
of additional lines and final determination of constants. Although the SF$_6$
spectrum near the CO$_2$ laser lines from P(12) to P(22) has been assigned
within the precision allowed by Doppler-limited spectra [19-21], the relative

145

positions of the many lines separated by less than one Doppler width were not
firmly determined in that work. In the vicinity of CO_2 P(14) and P(18) the
assignment of our spectra was clear from previous work [19-21]. However, this
was not the case near CO_2 P(12) or P(20). As is well known [20,22], the relative positions of lines in the Q branch of a spherical-top molecule are largely determined by the ratio v/g of only two of the spectroscopic parameters
of BOBIN and FOX [23], so that the parameters found previously by saturation
spectroscopy with a small tuning range [24] were adequate for an assignment
of many of the lines observed near CO_2 P(16) in this work. The same set of
parameters, however, gave deviations of more than 10 MHz for many of the lines
observed near CO_2 P(12) and P(20) over our broader range of tuning. For the
purpose of assigning the SF_6 lines near CO_2 P(20), two classes of lines were
considered to permit firm assignment : (i) isolated lines in the Doppler-limited spectra [19] and (ii) lines with a characteristic pattern of splitting,
and for which the splittings may be approximately determined from previous
parameters [19-21]. A preliminary set of Bobin-Fox parameters was determined
by a least-squares fit to the provisionally assigned lines near CO_2 P(12)
through P(20); the Bobin-Fox [23] expression for the transition frequencies
was modified by the inclusion of fixed off-diagonal corrections determined by
a full matrix diagonalization using the previous parameters [19-21]. Next, a
full diagonalization of the vibration-rotation Hamiltonian [25,26] for the v_3
mode of SF_6 using the new (preliminary) parameters was found to give good
agreement with the positions of the remaining unassigned lines near CO_2 P(20).
A new set of off-diagonal corrections, and the frequencies of additional
lines now considered to be firmly assigned, were employed in a new least-squares
fit to obtain an improved set of Bobin-Fox parameters, which were then employed in another full diagonalization. This procedure is nearly the same as
was previously used in the analysis of Doppler-limited SF_6 spectra [19-21].
In our work the values of B_0 and D_0 were held fixed, equal to the values determined by Raman spectroscopy [29], in order to provide enough equations to
determine the Hamiltonian parameters [30] from the spectroscopic parameters
[23]. In previous work [19-21], ζ_3 was held fixed.
The spectroscopic parameters determined in our work, and their standard
deviations, are given below in cm^{-1}

$$m = 947.9763307(62) \qquad n = 5.581731(39) \times 10^{-2}$$
$$p = -1.615414(89) \times 10^{-4} \qquad q = 1.236(47) \times 10^{-8}$$
$$s = -6.77(23) \times 10^{-11} \qquad t = -5.8(9.0) \times 10^{-14}$$
$$v = -6.9876(46) \times 10^{-5} \qquad w = -0.9(1.3) \times 10^{-11}$$
$$g = -2.45621(32) \times 10^{-5} \qquad h = -1.910(12) \times 10^{-9}$$
$$k = -1.54(11) \times 10^{-11} \qquad u = 0.4(1.0) \times 10^{-11}$$

The standard deviation of the fit is 328 kHz. The agreement between prediction and observation is good for CO_2 P(14) through P(20) but near CO_2 P(12),
there is room for further work. We expect that further improvement in line
assignments and accuracy of the spectroscopic parameters will follow from the
consideration of hyperfine interactions and from the determination of the
tensor-splitting parameters of the ground state to be discussed below.

IV - MAGNETIC HYPERFINE STRUCTURE OF THE SF_6 LINES : EVIDENCE FOR A SPIN-
VIBRATION INTERACTION

On the basis of magnetic resonance studies [29] it could be predicted that
SF_6 would exhibit magnetic hyperfine structures comparable to the 3.39 μm CH_4
line with splittings of the order of a few kilohertz which could be resolved

Figure 4

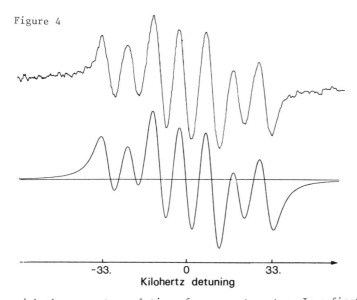

-33.　　　　0　　　　33.
Kilohertz detuning

with the present resolution of our spectrometer. In a first study of these
structures three lines with symmetry A_2 close to the $P(14)$, $P(18)$ and $P(20)$
CO_2 line centers were of particular interest since in their case only the
scalar terms in the Hamiltonian need to be considered [30]. As an example
Fig. 4 shows the structure of the R_{28} A_2^o line of $^{32}SF_6$ at 28.46469125 THz.
The spectrum exhibits seven resolved hyperfine components. This is consistent
with the values $I=1$ and $I=3$ predicted for the total nuclear spin allowed for
A_2 rovibronic levels [31]. The $I=1$ triplet almost coincides with the three
central components of the $I=3$ septuplet. The lower curve in the figure is a
theoretical spectrum which has been obtained by considering only the scalar
spin-rotation interaction $W_{SRS} = -h\,c_a\,\vec{I}.\vec{J}$. The coupling constant c_a has
been fixed in the lower level to the value given by OZIER, YI and RAMSEY [29] ;
$c_a = -5.27$ kHz. The only adjustable parameters are the value $c_a + \Delta c_a$ of this
coupling constant in the upper level and the width of the Lorentzian used as
line shape. The heights of the Lorentzians are deduced from the general formulae of intensities for saturation spectroscopy [32]. Apart from some minor
line shape differences the agreement between both curves is very good for
each case but leads to very different values of Δc_a for the three lines :
$\Delta c_a = -0.165$ kHz for R_{28} A_2^o ; $\Delta c_a = +0.125$ kHz for P_{33} A_2^1 ; $\Delta c_a = +0.068$ kHz
for P_{59} A_2^3 (at 28.3062526 THz). A much more satisfactory model is obtained
by adding a spin-vibration term to the Hamiltonian [8]. The scalar part is:
$W_{SVS} = h\,A\,\vec{I}.\vec{\ell}$ where $\vec{\ell}$ is the vibrational angular momentum corresponding
to the triply degenerate mode ν_3 . The expectation values of this Hamiltonian
which may be obtained from the projection theorem

$$< \vec{I}.\vec{\ell} > = < \vec{I}.\vec{J} > < \vec{\ell}.\vec{J} >/< J^2 >$$

are also proportional to the expectation values of a pure $\vec{I}.\vec{J}$ Hamiltonian
with a proportionality factor which varies from line to line and has opposite
signs for P and R lines. We understand why the previous theory worked so well
for single lines by introducing an effective phenomenological Δc_a which we
can now write as a combination of a residual spin-rotation δc_a and of a term
resulting from the spin-vibration interaction :

Figure 5

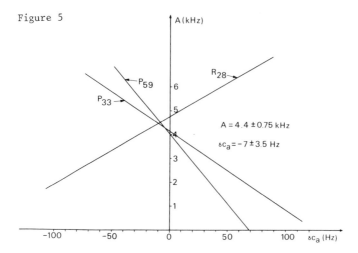

$$\Delta c_a = \delta c_a - A[(2J'+1)\Delta J+1]/2J'(J'+1) \quad \text{with} \quad \Delta J = J'-J$$

For each A_2 line there is a redundancy between δc_a and A but as illustrated in Fig. 5 the three A_2 structures together are only compatible with a single value of A and a single reasonably small value of δc_a .

Systematic studies of the other lines that can be reached with the wave-guide CO_2 laser should bring us a more complete picture of these hyperfine interactions in the near future.

V - DIRECT OPTICAL OBSERVATION OF THE GROUND STATE OCTAHEDRAL SPLITTINGS OF SF_6 AND VIOLATION OF THE $\Delta C = 0$ SELECTION RULE

For some time we have been puzzled by additional resonances in the spectrum which look like hot lines but appear exactly in the center of the interval between main resonances. A typical example is the resonance between the E^o and F_1^o components of the Q(38) triplet of Fig. 3. The positions of these peaks strongly suggest that they are crossover resonances but this explana-tion runs into two difficulties :
1) - this implies that the $\Delta C = 0$ selection rule is violated;
2) - some of the crossover peaks are conspicuously missing : for example there is no resonance in the center of the Q(38) $E^o - F_2^o$ interval. A tenta-tive explanation that resolves these difficulties is that owing to the small value of the centrifugal distortion constant t_{044} in the ground vibrational state, hyperfine interactions may mix substantially some vibration-rotation states having different point group symmetry types but the same overall pa-rity (if we neglect possible parity-violating effects).

Hyperfine operators are tensorial products of nuclear spin operators with vibration-rotation operators both having the same symmetry species C in order to give a total Hamiltonian of symmetry A_{1g} . The symmetry types C can be A_{1g} (scalar spin-rotation) E_g (tensor spin-rotation and tensor spin-spin) and F_{2g} (tensor spin-spin) [30]. They will mix vibration-rotation states having symmetry types C' and C" such that $C \times C' \times C" \supset A_{1g}$.

The nuclear spin wavefunctions have only certain definite symmetry types in O_h : A_{1g} , A_{2g} , A_{2u} , E_g , F_{1u} , F_{2u} , F_{2g} . The consequence of the absen-

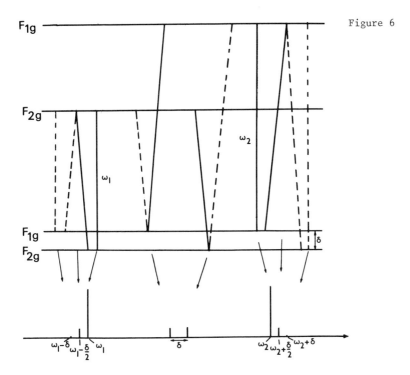

Figure 6

ce of some symmetry types in this list (no E_u , no A_{1u} , no F_{1g}) combined with the Pauli principle (the total wavefunction must be A_{2u}) is that one of the two parities cannot exist for the vibration-rotation levels of some symmetry types [33] : in the lower vibrational level A_1 and F_1 vibration-rotation levels can be u or g but A_2 and E levels may only be u and F_2 levels may only be g . The consequence, for instance in the case of the Q(38) F_2+E+F_1 cluster, is that the hyperfine interactions can mix together E and F_1 states or F_1 and F_2 states but not E and F_2 , hence the missing crossover.

Furthermore these crossover resonances have a structure which can be understood from Fig. 6 where we consider the simple case of F_1 and F_2 clusters.

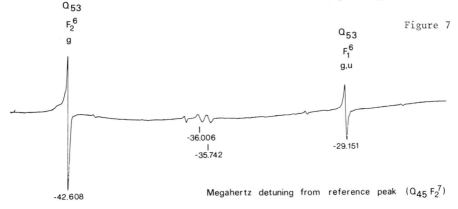

Q_{53}
F_2^6
g

Q_{53}
F_1^6
g,u

Figure 7

-36.006
-35.742

-29.151

-42.608

Megahertz detuning from reference peak (Q_{45} F_2^7)

149

We expect a crossover splitting which reflects exactly the ground state splitting. We observe many examples of such doublets and one example is given in Fig. 7. From the observed splitting one may infer for the first time a value for the tensor centrifugal distortion constant t_{044} for SF_6. A precise value requires the full diagonalization of the vibration-rotation + hyperfine Hamiltonian. Such a program is presently being carried out but we can already give a preliminary estimate for this constant of the order of 6 Hz.

VI – FINE AND HYPERFINE STRUCTURE STUDIES IN THE ν_3 BAND OF OsO_4

Another spherical top of great theoretical and practical interest for spectroscopy, photochemistry and optical frequency standards is OsO_4. The perfect match between the ν_3 band of this molecule and the 10.4 µm band of CO_2 has already stimulated many laser spectroscopy studies [11]. The essential intermediate resolution step has been taken by R.S. McDOWELL and coworkers using both Fourier transform spectroscopy and laser diode spectroscopy [34]. Each CO_2 laser line from P(8) to P(24) and from R(6) to R(24) hits many resonances of various isotopic species. The Q-branch itself can be sampled with the N_2O laser lines. Since the hyperfine structures are either absent or can be well resolved we have there a full grid of high quality secondary optical frequency standards in this spectral region. We have made systematic measurements of the absolute frequencies of a large number of lines corresponding to the main isotopes found in the natural mixture : $^{192}OsO_4$, $^{190}OsO_4$, $^{189}OsO_4$. The detailed assignment and derivation of new spectroscopic constants is presently underway. As an example Fig. 8 shows the spectrum obtained for the P(14) CO_2 laser line which is in close coincidence with a hot line of $^{192}OsO_4$ but also reaches the P_{46} A_2^2 line of the same isotopic species as well as the P_{49} A_2^3 and A_1^3 components of $^{189}OsO_4$ and some as yet unidentified lines of $^{190}OsO_4$. The interference fringes on the spectrum come from the insufficient

SATURATION SPECTRUM OF OsO_4 FOR THE P(14) LINE OF CO_2 Figure 8

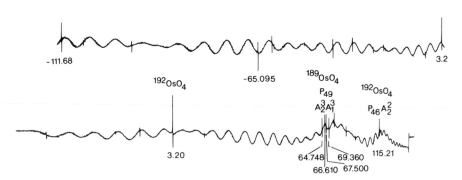

FREQUENCY OFFSET FROM THE CO_2 LINE CENTER (MEGAHERTZ)

150

optical isolation in this particular experiment. The P_{49} A_2^3 and A_1^3 lines show up as a quadruplet. The only possible origin for this additional splitting is through an electric quadrupole interaction. KOMPANETS and coll. [11] have already reported such structures but with much smaller values for the splittings. To understand this difference one needs to consider the expected hyperfine level structure illustrated on Fig. 9 (For negative eqQ's).

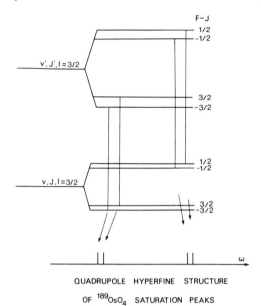

QUADRUPOLE HYPERFINE STRUCTURE

OF ^{189}OsO$_4$ SATURATION PEAKS   Figure 9

The ^{189}Os nucleus has a spin 3/2 (and a quadrupole moment Q) which gives rise to a 4-fold degeneracy in each level. This degeneracy is lifted by the quadrupole interaction since vibration-rotation interactions induce a field gradient q_J at the osmium nucleus. The selection rule $\Delta F = \Delta J$ results in two doublets of equal spacing with a frequency distance roughly equal to half the difference between the quadrupole coupling constants $\Delta(eqQ)$. In the ground vibrational state the only source for q_J is the centrifugal distortion and this explains why the splittings that we observe for fundamentals are much larger than those corresponding to hot lines.

The next unexpected feature is that the observed pattern at higher resolution than Fig. 8 differs from a pure quadrupolar spectrum : for P lines instead of two equally spaced doublets for each rotational line we get a single resonance on the lower frequency side and a barely resolved doublet on the high frequency side (see Fig. 10). This can be understood if we introduce the next class of hyperfine interactions, namely magnetic hyperfine interactions. For high J values and negligible δc_a we may write in first approximation the corresponding shift for the X = F-J component : $\Delta\nu_M(X) = (A-c_a)X \Delta J$. If the difference between the spin-vibration and spin-rotation coupling constants is positive, then for P-branch lines ($\Delta J = -1$) the splitting between the two $|X| = 3/2$ components will be reduced by a quantity three times larger than that corresponding to the $|X| = 1/2$ components which explains the observed pattern. A confirmation of this interpretation has been obtained by the

151

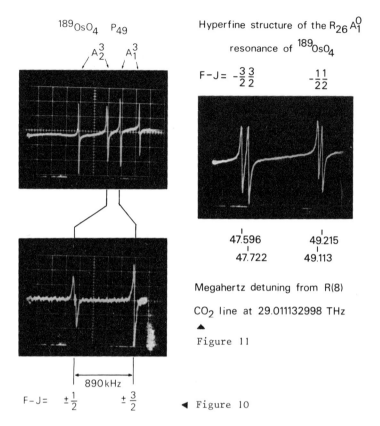

$^{189}OsO_4$ P_{49}

A_2^3 A_1^3

Hyperfine structure of the $R_{26} A_1^0$

resonance of $^{189}OsO_4$

$F - J = -\frac{3}{2}\frac{3}{2}$ $-\frac{1}{2}\frac{1}{2}$

47.596 49.215
47.722 49.113

Megahertz detuning from R(8)

CO_2 line at 29.011132998 THz

▲

Figure 11

890 kHz

$F - J = \pm\frac{1}{2}$ $\pm\frac{3}{2}$

◀ Figure 10

recent observation of the patterns for a large number of R-branch lines. In this case $\Delta J = +1$ and the effect of the magnetic interactions is to split further the two components of each doublet by a quantity again three times larger for the $|X| = 3/2$ doublet than for the $|X| = 1/2$ one.

We give in Fig. 11 the example of $R_{26} A_1^0$. From this spectrum we can make a first estimate for $A-c_a$ which is of the order of 14 kHz. For this line $\Delta(eqQ) = eqQ(upper) - eqQ(lower) = -3.01\,MHz$ but we should point out that this last quantity varies importantly from line to line as illustrated by the comparison of the splittings between P_{49} and R_{26}. A precise study of this dependence is underway.

As there are other spherical tops (e.g. SiF_4) whose ν_3 band falls within the tuning range of the CO_2 lasers we can conclude that we are just at the beginning of a detailed quantitative understanding of fine and hyperfine interactions in spherical tops. Such a systematic study is of great interest, both from a fundamental point of view, to test and refine the theory of these interactions and specify their role in relaxation processes and from a practical point of view to improve our comprehension of the various experiments where spherical tops are used in conjunction with CO_2 lasers : optical frequency standards, laser induced-chemistry including the well-known laser isotope separation as well as laser pyrolysis [35], passive Q-switching and mode-locking, transient experiments (self-induced transparency, optical nutation, photon echoes, Ramsey fringes), optical isolators and optically pumped far-infrared lasers.

References

[1] M.A.Pollack, T.J.Bridges and A.R.Strnad,Appl.Phys.Lett. 10,182-183(1967)

[2] Ch.J.Bordé and L.Henry,C.R.Acad.Sc.Paris, 265 B, 1251-1254(1967) and
IEEE J.of Quantum Electron. QE-4, 874-880(1968)

[3] J.L.Hall, IEEE J.of Quantum Electron. QE-4, 638(1968),
R.L.Barger and J.L.Hall,Phys.Rev.Lett. 22, 4-8(1969)

[4] Ch.J.Bordé,C.R.Acad.Sc.Paris, 271 B, 371-374(1970)

[5] Ch.J.Bordé and J.L.Hall in Laser Spectroscopy ed. by R.G.Brewer and
A. Mooradian Plenum Press, pp.125-142(1974)

[6] J.L.Hall and Ch.J.Bordé,Phys.Rev.Lett. 30,1101-1104(1973)

[7] J.L.Hall,Ch.J.Bordé and K.Uehara,Phys.Rev.Lett. 37, 1339-1342(1976)

[8] K.Uehara and K.Shimoda,J.of Phys.Soc.Jap. 36, 542-551(1974)

[9] Ch.J.Bordé,M.Ouhayoun and J. Bordé,J.of Mol.Spectrosc. 73, 344-346(1978)

[10] M.Ouhayoun,Ch.J.Bordé and J. Bordé,Molecular Physics, 33, 597-600(1977)

[11] O.N.Kompanets,A.R.Kukudzhanov,V.S.Letokhov,V.G.Minogin and E.L.Mikhailov,
Sov.Phys.-JETP, 42, 15-22(1976) and references therein

[12] F.R.Petersen,D.G.McDonald,J.D.Cupp and B.L.Danielson in Laser Spectroscopy ed. by R.G.Brewer and A.Mooradian, Plenum Press 1974

[13] A.Van Lerberghe,S.Avrillier and Ch.J.Bordé,IEEE J.of Quantum Electron.
QE-14, 481-486(1978) and references therein

[14] M.Ouhayoun and Ch.J.Bordé, Metrologia, 13, 149-150(1977)

[15] C.Freed and A.Javan, Appl.Phys.Lett. 17, 53(1970)

[16] Ch.J.Bordé,J.L.Hall,C.V.Kunasz and D.G.Hummer,Phys.Rev. 14, 236-263(1976)

[17] A.Van Lerberghe,S.Avrillier,Ch.J.Bordé and C.D.Cantrell,JOSA,68,624(1978)

[18] Ch.J.Bordé,M.Ouhayoun,S.Avrillier,A.Van Lerberghe,C.Salomon,C.D.Cantrell,
and J.Bordé, to be published

[19] R.S.McDowell,H.W.Galbraith,B.J.Krohn,C.D.Cantrell and E.D.Hinkley,Optics
Commun. 17, 178-183(1976)

[20] R.S.McDowell, H.W.Galbraith,N.G.Nereson,C.D.Cantrell and E.D.Hinkley,
J.Mol.Spectrosc. 68, 288-298(1977)

[21] R.S.McDowell,H.W.Galbraith,N.G.Nereson,C.D.Cantrell,P.F.Moulton and
E.D.Hinkley,Optics Letters, 2, 97-99(1978)

[22] F.Michelot and J.Moret-Bailly,J.de Physique Lettres,39,L 275-277(1978)

[23] B.Bobin and K.Fox,J. de Physique (Paris) 34, 571-582(1973)

[24] M.Loëte,A.Clairon,A.Frichet,R.S.McDowell,H.W.Galbraith,J.-C.Hilico,
J.Moret-Bailly and L.Henry,C.R.Acad.Sci.(Paris) 285 B, 175-178(1977)

[25] K.T.Hecht,J.Mol.Spectrosc. 5, 355-389(1960)

[26] J.Moret-Bailly,Cahiers Phys. 15, 237(1971)

[27] H.Berger,A.Aboumajd and R.Saint-Loup,J. de Physique Lettres,38, L 373-375
(1977)

[28] A.G.Robiette,D.L.Gray and F.W.Birss,Mol.Phys. 32, 1591-1607(1976)

[29] I.Ozier, P.N.Yi and N.F.Ramsey,J.Chem.Phys. 66, 143-145(1977)

[30] F.Michelot,B.Bobin and J.Moret-Bailly,J.of Mol.Spectrosc. in press(1979)

[31] J.Bordé,J.de Physique Lettres,39, L 175-178(1978)

[32] J.Bordé and Ch.J.Bordé,C.R.Acad.Sci.(Paris) 285 B, 287-290(1977) and J.
Mol.Spectrosc. (in press)1979

[33] H.Berger,J. de Physique,38, 1371-1375(1977)

[34] R.S.McDowell,L.J.Radziemski,H.Flicker,H.W.Galbraith,R.C.Kennedy,N.G.
Nereson,B.J.Krohn,J.P.Aldridge,J.D.King,K.Fox,J.Chem.Phys. 69, 1513-
1521(1978)

[35] Ch.J.Bordé,A.Henry and L.Henry, C.R.Acad.Sci.Paris, 263 B, 619-620(1966)

High Resolution Laser Spectroscopy of Small Molecules

W. Demtröder, D. Eisel, H.J. Foth, G. Höning, M. Raab, and H.J. Vedder

Fachbereich Physik, Universität Kaiserslautern
D-6750 Kaiserslautern, Fed. Rep. of Germany

1. Introduction

The visible absorption spectra of several heavy diatomic molecules and of many polyatomic molecules show a very complex
structure. The spectral line density is often so large that
many absorption lines overlap within their Doppler-width.
When measured with Doppler-limited spectral resolution, these
spectra may therefore appear quasicontinuous, concealing finer details, such as rotational structure or fine- and hyperfine-splittings.

In this paper we report on applications of three different
sub-Doppler techniques to the spectroscopy of some diatomic
and triatomic molecules. These techniques include linear laser spectroscopy in collimated molecular beams with single
mode tunable lasers, nonlinear polarization spectroscopy and
a combination of these two methods with optical-optical double-resonance techniques. The three methods are illustrated
by examples taken from detailed investigations of the molecules NaK, Cs_2 and NO_2.

2. Sub-Doppler Laser Spectroscopy in Molecular Beams

A tunable single mode argon- or dye-laser is crossed perpendicularly with a collimated molecular beam (Fig.1). The un-
dispersed total fluorescence $I_{Fl}(\lambda_L)$ is observed as a function of the laser-wavelength λ_L by photomultiplier PM1. The
resultant excitation spectrum exhibits line profiles which represent a convolution of the homogeneous linewidth and the re-

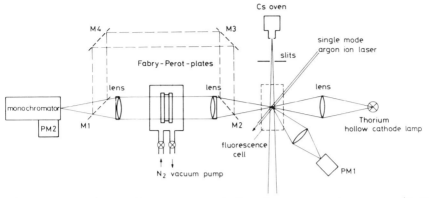

Fig.1 Experimental arrangement for laser spectroscopy in molecular beams

duced Doppler-width $\Delta v = \xi \cdot \Delta_D$ where $\xi \ll 1$ gives the collimation ratio of the molecular beam. Fig.2 shows a section of the Cs_2-excitation spectrum, excited by a single mode argon laser which could be continuously tuned over 10 GHz around $\lambda = 476,5$nm.

For the assignment of the different absorption lines the laser is stabilized onto the center of a line and the fluorescence spectrum, excited at this transition, is recorded by photomultiplier PM2 behind a monochromator. To increase the accuracy of wavelength-measurements, a Fabry-Perot-interferometer is placed in front of the monochromator. Since the Doppler-free excitation allows the selective population of single rovibronic levels (v',J') in the upper electronic state, the fluorescence spectra, excited by each of the excitation lines in Fig.2, consist of single P-R or Q-progressions ($\Delta J = \pm 1$ or 0) which can be readily assigned, at least with respect to the vibrational quantum number v". The rotational quantum number J can

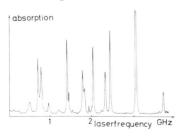

Fig.2 Section of the Cs_2-excitation spectrum around $\lambda = 476,5$ nm

be also determined unambiguously, if several progressions with different J-values are measured with sufficiently high accuracy. These measurements yield the molecular constants and the potential curve of the electronic ground state [1] .

In some molecules, such as NaK, the upper electronic singlet states, accessible by optical excitation from the Σ-ground-state, are perturbed by nearby triplet states. These perturbations cause a mixing of singlet- and triplet wave functions and allow to observe in the fluorescence spectrum simultaneously singlet-transitions and triplet-transitions terminating on the $a^3\Sigma$-ground state (Fig.3). Since the $a^3\Sigma$-potential is mainly repulsive and shows only a shallow van der Waals minimum, the triplet fluorescence spectrum is similar to the fluorescence of an excimer-molecule. It consists of a modulated continuum [2] and some discrete lines (Fig.4). According to the Franck-Condon principle the transition probability $A \propto$ $|\langle\Psi'_{vib}(R).\Psi''_{cont}\rangle|^2$ is proportional to the squared vibrational wavefunction $\Psi_{vib}(R)$ in the upper bound state, and the intensity modulation of the continuous fluorescence reflects the dependence of Ψ'_{vib} on the internuclear distance R as well as the slope of the repulsive part of the $a^3\Sigma$-potential. [3]

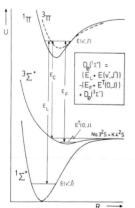

Fig.3 Schematic diagram of singlet- and triplet-transitions in NaK

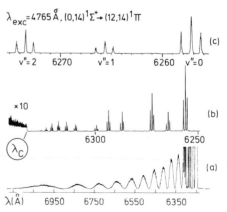

Fig.4 Triplet fluorescence spectrum of NaK. a) medium, b) higher resolution, c) expanded section of b)

Fig.4 shows such a triplet fluorescence spectrum recorded with medium resolution. In the upper trace an extended plot of the discrete spectrum is shown as monitored with higher resolution. From the rotational and vibrational spacings of the triplet lines the well depth of the triplet potential can be determined as 203 ± 4 cm^{-1}. Since the $a^3\Sigma$-potential and the $X^1\Sigma$-potential both dissociate into the same limit Na $^2S_{1/2}$ + K $^2S_{1/2}$ of two ground state atoms, the dissociation limit of the triplet fluorescence also allows to determine the much larger dissociation energy of the $X^1\Sigma$-ground state within a few cm^{-1}! [4] The result of these measurements is $D(X^1\Sigma) = 5269 \pm 6$ cm^{-1}. This represents a substantial improvement in accuracy over other methods which generally determine the dissociation energy by extrapolation from lower bound levels (Birge-Sponer-Plot).

3. Polarization Spectroscopy

Polarization spectroscopy [5,6] is a very sensitive Doppler-free technique which has several definite advantages for high resolution molecular spectroscopy. Fig.5 shows schematically the experimental arrangement: The output from a tunable single mode dye laser is split into a circularly polarized pump beam and a linearly polarized probe beam, which pass into opposite directions through the sample cell, placed between two nearly

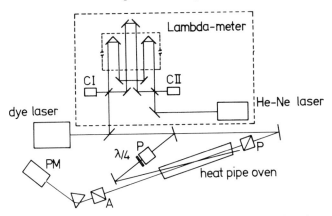

Fig.5 Experimental arrangement for polarization spectroscopy and absolute wavelength-measurement with the lambda-meter

crossed polarizers A and P. If the laser frequency is tuned to
the center of a molecular absorption line (J'→ J") the saturation of the pump transition causes a nonuniform depletion of
the (J", M") sublevels. The saturated M"-population depends on
the transition probability A(J',M'→ J",M") which in turn depends on the polarization of the pump wave. The sample becomes
optically birefringent and alters the polarization of the
transmitted probe wave. The photomultiplier PM behind the analyser A receives a signal, if probe wave and pump wave interact with the same velocity group of molecules in the absorbing
level J". The Doppler-free line-profile of the signal $I(\omega_L)$
as a function of the laser frequency ω_L shows either dispersion
or Lorentzian shape. The line shape is different for molecular
transitions with ΔJ = 0 (Q-lines) or ΔJ = ±1 (P- or R-lines).

Fig.6 illustrates a polarization-spectrum of the P-branch
of Cs_2 at the band head of the 0→0 vibrational band in a
$^1\Pi_u \leftarrow X^1\Sigma_g$-transition. The small dispersion shaped signal
belongs to a Q-line from another band.

Part of the laser beam is sent through a 120 cm long confocal Fabry-Perot-interferometer which provides frequency marks
every 62 MHz. For absolute wavenumber measurements a travelling
Michelson interferometer [7] is used which measures the laser
wavelength stabilized to the center of a molecular line, to

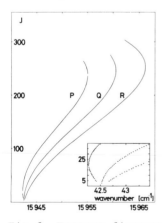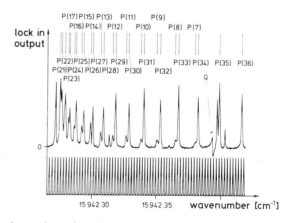

Fig.6 Fortrat diagram of P, Q and R lines of the 0↔0 band in
the $^1\Pi_u \leftarrow X^1\Sigma_g$-transition of Cs_2 and section of the polarization
spectrum of the P-branch.

within 2 x 10^{-4} Å, corresponding to a frequency accuracy of
about 20 MHz. The main uncertainty in the absolute wavelength-
determination is caused by short term fluctuations of the dye
laser (model CR 599) output, probably caused by air bubbles in
the jet or dust particles close to the dye focus. The inherent
accuracy of the wavemeter is better than 5 MHz [8] . Measuring
more than 800 Q-, P- and R-lines in the 0-0-band yields the
Fortrat-diagram of Fig.7. The scattering of the experimental
points is much less than the diameter of the plotted dots in
the expanded insert. A least squares fit to the measured line
position gives the rotational constants of the Cs_2-molecule
in the $X^1\Sigma_g^-$ and the $^1\Pi_u$-state with an accuracy of about
10^{-5} [9] . Besides a recently published work [10] on the same
band with Doppler-limited resolution this is the first complete resolution of all rotational lines in a band in the visible Cs_2-spectrum.

The fact that the proper choice of the pump polarization
allows to select either Q- or P- and R-lines is a definite advantage of polarization spectroscopy since it greatly facilitates the assignment of complex molecular spectra. Combined
with double resonance methods the technique of "polarization
labelling" [11] is an excellent method to investigate the de-
tailled structure of perturbed upper molecular states.

4. Optical-Optical Double Resonance Spectroscopy

This Doppler-free technique is based on the simultaneous interaction of a molecule with two different laser waves, a pump
wave and a probe wave, which pass in opposite directions through
the sample [12] . Assume the pump laser is stabilized onto a
molecular transition $E_i \rightarrow E_k$. Due to optical pumping the lower
level population N_i is partly depleted and the upper level population N_k increases. Chopping of the pump wave intensity
therefore results in corresponding modulations of N_i and N_k.
When the wavelength of the probe wave is tuned across the molecular absorption spectrum, the absorption of the probe beam
will monitor this population modulation whenever the probe

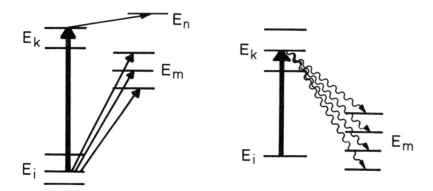

Fig.7 Schematic diagram of optical-optical double resonance, compared with laser induced fluorescence

wavelength coincides with a molecular transition $E_i \rightarrow E_n$ or $E_k \rightarrow E_m$ (Fig.7). The modulation phase is opposite for both cases and phase sensitive detection therefore allows to distinguish between transitions $E_i \rightarrow E_m$ or $E_k \rightarrow E_m$.

Provided the pump transition $E_i \rightarrow E_k$ has been assigned, the common lower level E_i for all probe transitions $E_i \rightarrow E_n$ is known. From the measured wavelengths the level spacings of the upper levels E_n can be immediately obtained. In a way this double resonance technique represents an inversion of the laser induced fluorescence technique. While in the latter method a single upper level is selectively populated and the lower level diagram is deduced from the fluorescence spectrum, in the former method a lower level is selectively depleted and the upper level spectrum is obtained from the modulated absorption spectrum of the probe laser.

Combined with saturation spectroscopy [13] or polarization spectroscopy [11] the OODR-technique allows Doppler-free resolution and substantially reduces the number of lines. This is of particular importance in complex spectra and in case of perturbed upper states. However, if the experiments are performed in gas cells, the wavelength of the single mode pump laser may simultaneously overlap with several Doppler-broadened molecular absorption lines. This causes saturation dips in the population distribution $N_i(v_z)$ of several different levels E_i, where the dips are located at different velocities v_z within the Doppler-profile: Furthermore collisions may transfer these dips to neighbouring levels. Also the fluorescence emitted from the modulated upper level N_k, which terminates at various lower levels will cause a slight population-modulation of these levels. All these effects increase the number of double resonance signals and may impede the unambiguous assignment. In case of complex spectra it is therefore desirable to avoid collisions as well as the simultaneous interaction of several molecular transitions with the pump wave.

A possible solution is the use of a collimated molecular beam. Fig.8 shows the experimental arrangement, where the beams

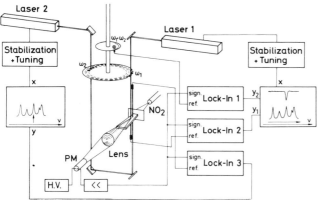

Fig.8   Experimental arrangement for OODR in a collimated NO_2-beam

from two tunable single mode argon lasers cross the molecular NO_2-beam perpendicularly. The two laser beams are chopped at two different frequencies and the OODR-signal is monitored through the laser induced fluorescence at the sum frequency. Fig.9 illustrates the results by showing a small section of the NO_2-excitation spectrum around $\lambda = 488$ nm (lower spectrum).

While the pump-laser is stabilized on a line indicated by the arrow, the probe laser is tuned. The OODR-spectrum (upper part) demonstrates, that two transitions in the lower spectrum share

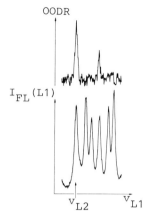 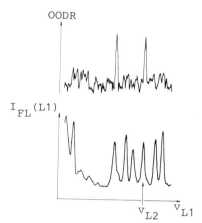

Fig.9 Excitation spectrum of NO_2 around λ = 488 nm (lower trace) and OODR-spectrum (upper trace)

a common lower level. Since in unperturbed spectra the corresponding splitting of the upper levels cannot be due to fine- or hyperfine-splittings (because of the selection rules $\Delta I = 0$ and $\Delta S = 0$ for allowed transitions) the second transitions must be due to perturbations, such as spin-orbit interactions [14].

The application of this OODR-technique to a wider spectral range, covered by a dye laser, will be very helpful to separate unperturbed and perturbed lines and to gain insight in the level structure of perturbed upper electronic states. Such experiments are presently underway in our laboratory.

References
1. G. Höning, M. Czajkowski, M. Stock and W. Demtröder;
   J. Chem. Phys. Sept. 1979 in print
2. E.J. Breford and F. Engelke; Chem. Phys. Lett. 53, 282 (1978)
3. H. Scheingraber and C.R. Vidal; J. Chem. Phys. 66, 3694 (1977)
4. D. Eisel, D. Zevgolis and W. Demtröder; J. Chem. Phys.
   Sept. 1979 in print
5. C. Wieman and T.W. Hänsch; Phys. Rev. Lett. 36, 1170 (1976)
· 6. R.E. Teets, F.V. Kowalski, W.T. Hill, N. Carlson, T.W.Hänsch;
   Proc. Soc. Photo-Optical Instr. Engineers 113; Advances in Laser
   Spectroscopy, San Diego 1977 p. 80 ff
7. F.V. Kowalski, R.E. Teets, W. Demtröder and A.L. Schawlow;
   J. Opt. Soc. Am. 68, 1611 (1978)
8. K. Wickert; Diplomthesis; Kaiserslautern 1979
9. M. Raab, G. Höning, R. Castell and W. Demtröder;
   Chem. Phys. Lett. 1979 in print

10. A.I. Kobyliansky, A.N. Kulikov and L.V. Gurvich; Chem. Phys. Lett. <u>62</u>, 198 (1979)
11. R. Teets, R. Feinberg, T.W. Hänsch, A.L. Schawlow; Phys. Rev. Lett. <u>37</u>, 683 (1976)
12. V.P. Chebotayev: "Three Level Laser Spectroscopy", in: High Resolution Laser Spectroscopy, K. Shimoda, Ed., Topics in Appl. Phys. Vol. 13, (Springer, Berlin, Heidelberg, New York 1976)
13. M. E. Kaminsky, R.T. Hawkins, F.V. Kowalski and A.L. Schawlow; Phys. Rev. Lett. <u>36</u>, 671 (1976)
14. J.C.D. Brand and P.H. Chiu; J. Mol. Spectrosc. <u>75</u>, 1 (1979)

Single Molecular Isotope Absorption-Spectra

U. Boesl, H.J. Neusser, and E.W. Schlag

Institut für Physikalische und Theoretische Chemie der Technischen Universität München, Lichtenbergstraße 4 D-8046 Garching, Fed. Rep. of Germany

Introduction

The task of isotope separation in molecules is facilitated greatly by the availability of sharply tuned lasers; however, this alone is not a sufficient condition for an efficient molecular isotope separation scheme. The spectra of polyatomic molecules are sufficiently complex so that it is all but impossible to predict a priori at what wavelength a particular scheme can function optimally. Without such knowledge the sharply tunable laser is of little use; it contains no knowledge of chemical reactivity, which ultimately must form the basis of any specific separation scheme.

Specific reactivity of molecules can no doubt be induced by tuning sharp laser lines to particular molecular transitions. The wavelengths for these transitions are usually obtained by measuring the molecular absorption spectrum, although this as yet gives no indication of the photochemical activity of a particular transition. Upon isotopic substitution these absorption bands are known to undergo spectral shifts. However, the magnitude of such a shift can in general not be predicted since detailed information about the nature of electronic excited state is in general not available. Hence an experimental method for determining these shifts becomes essential. In the normal absorption spectrum such transitions due to isotopic isomers are usually not seen as they are obscured by the far more preponderant isomer. The question then is the method that can be employed to find the molecular transitions of isotopic isomers. One obvious answer is the "intelligent guess". The second obvious answer is to synthesize each isotopic isomer and then measure its spectrum. A third method is presented in this paper.

Photochemical Isotope Enrichment: General Aspects

Narrow band lasers constitute an interesting tool for highly specific photochemistry. One of the most sensitive applications of this technique is the exploration of the reactivity of isotopic species in excited electronic states. Such processes can also be used as a basis of isotope enrichment [1]. It is of interest, however, to further distinguish various categories of isotopic enrichment. The first category would then consist of the classical methods such as diffusion, the expansion nozzle, or the ultra-centrifuge, all of which separate isotopes on the basis of the fact that the total mass of the species is higher due to the isotopic substitution of atoms. In the photochemical method we have a second category, which offers the additional possibility of separating molecules with the same mass, due to the fact that different isotopes, even though at the same mass, absorb and hence can react at different wavelengths. One can further sub-divide the second category into laser enrichment by photochemical destruction, and into laser enrichment by photochemical synthesis.

Isotope separation often has the further requirement that the isotope has to be in a particular location in a large molecular complex. For example it is here of considerable import not just to prepare ^{15}N in NO_2, but perhaps it is more important to have the heavy nitrogen in a particular large organic-nitrogen compound. The success at achieving ^{15}N-isotopes out of NO_2 is of little direct use for the latter application since synthesis is here quite difficult, and usually of low yield, making this also an expensive pathway. Another important application for laser isotope enrichment is the treatment of partially enriched mixtures. Such synthetically enhanced mixtures, which might still be quite impure, contain the desired isotope at a higher absolute concentration. One such example results from the exchange of hydrogen for deuterium in an acid medium, a method which never yields 100 % of the desired isotopic isomer. Here laser techniques are successfully employed to obtain nearly 100 % of a particular isomer [2]. This latter application, of further cleanup of a previous enrichment scheme is, of course, a main strength of laser isotope enrichment schemes.

The need of such molecular species in small laboratory applications should not be underestimated. Such applications are essential for accurate spectroscopic measurements, such as for force constant determinations or structure determinations. They are also essential for the study of kinetic isotope effects in chemical reactions and thermodynamic processes, as well as in the deter-mination of the mechanism of organic and inorganic reactions. The arsenal of chemists would be greatly enhanced if isotopically specific preparations would be as facile to handle in the laboratory as normal synthetic procedures. A practical small enrichment apparatus in the laboratory would greatly increase the possibil-ities of the synthetic chemist. So far no such scheme is availa-ble for laboratory utilization.

Present schemes of photochemical isotope enrichment consist of two methods:
(1) A particular photochemically active absorption is used by shifting the exciting light slightly to the red or blue, hoping there by to enhance either the light or the heavy isotopes. This scheme can often lead to reasonable enhancement factors. The method is usually trial- and -error since little usually is known or can be calculated about the isotopic shifts in the excited state.

(2) A more systematic method is to perform the necessary spectro-scopy on all the isotopic molecules of photochemical interest. Such comparisons have been performed for the case of formaldehyde $H^{13}CHO$

and $H^{12}CHO$ [3], however the measurement of these spectra has
necessarily demanded the prior availability of these species
in order to be able to measure the spectra in the first place.
In other words one has to synthesize the isomers to evolve a
scheme for separating them. This is a feasable plan when large
commercial separations are to be considered, but it can hardly
be sensible for laboratory scale preparations. If this can be
avoided, it would be of great help to either scheme.

We here wish to present a third, new spectroscopic analytical
method on the basis of which a highly efficient isotope enrichment scheme can be evolved, either in the laboratory, or on
a larger scale. Historically photochemical enrichment antidates
lasers by many years. The first separation is probably the work
of Kuhn and Martin [4] who separated phosgene in 1932 employing
a monochromatic Al sparc and obtained an enrichment factor of
1.09. The classic later work by Gunning et al. [5] on the mercury
isotopes was a further successful example of photochemical isotope
enrichment. Since 1972 such experiments have also been carried out
with the use of lasers.

Enrichment factor and Enrichment yield

Usually the merit of an isotope enrichment scheme is assumed to
be reflected by its

$$\text{Enrichment Factor} = \frac{N_D \ / \ N_H}{N_{Do} \ / \ N_{Ho}},$$

which just compares the ratio of concentrations of the two isomers
after, to that before the start of the enrichment scheme. In
photochemical schemes, even if the spectroscopy of all species
is completely known, it is often difficult to obtain good enrichment, since the molecular absorption bands are rarely completely
separated. The ro-vibronic envelopes of the transitions usually
occupy many wavenumbers. For this reason alone, it is necessary
to find bands which offer an optimum in spectral separation.

Since the enrichment factor does not reflect an elementary step,
it is often possible to influence its value considerably by
enhancing the separation at the expense of the total sample. This
is particularly the case if the bands overlap. If the goal of an
enrichment scheme, however, must also be the attainment of useful
quantities in a relatively short time, here again, as in all synthetic schemes, the question of

$$\text{yield} = N_D \ / \ N_H$$

cannot be ignored. A figure-of-merit could be defined as the
product of these two quantities, which then serves to temper any
overestimations of the enrichment scheme based solely on the
enrichment factor. Furthermore, it must be mentioned that many
laboratory experiments which give an enrichment factor are "in-
principle" schemes which are often difficult, if not impossible
to scale to macroscopic proportions, even to those small amounts

useful in laboratory experimentation. Enrichment schemes in single crystals at very low temperatures are probably still in this category. In very few suggestions for photochemical processes of enrichment has the practical proof of a macroscopic preparation been furnished as well.

Isotope Specific Spectroscopy of Sym-Tetrazine

Of the some dozen molecules so far proposed for photochemical enrichment perhaps one of the best studied is the enrichment of sym-tetrazine. This molecule photodissociates with high quantum yield into N_2 and two molecules of HCN. As such it is an ideal photochemical system. It also presents an interesting challenge in that (b), (c) and (d) in Table 1 all have the same total mass, though (d) is small in natural abundance and can usually be neglected.

Table 1   Isotopic isomers of sym-tetrazine

(a)	AMU 82:	$H_2C_2N_4$	96.3 %
(b)	AMU 83:	$H_2{}^{13}CCN_4$	2.2 %
(c)		$H_2C_2{}^{15}NN_3$	1.4 %
(d)		HDC_2N_4	0.03 %
	AMU $\geq$ 84 ...		0.07 %

But the presence of (b) and (c) alone makes it necessary to have an enrichment scheme which is based not just on the total mass of the molecule. Furthermore, it is not clear whether any non-systematic schemes, such as (1) above would lead to success here. Biasing the laser to the red of any absorption line gives no assurance of absorbing into (b) or (c) or both. The absorption spectrum of Spencer, Cross and Wiberg is also of little help in making any further decisions here [6].
We would like to propose the method in Fig. 1 as a third new method for isotope specific spectroscopy, by which one can decide on an optimal laser frequency for an isotopic enrichment scheme. In the scheme a CW laser is scaned and the original wavelength, the absorption, and the output of a mass spectrometer are monitored. The light is chopped at 5 Hz and the output of the mass spectrometer is switched between preselected masses. Prior to the mass spectrometer is a rapid flow photochemical reactor. By suitably adjusting the scan rate of the laser, the frequency of ligth chopping, the flow rate in the reactor, and the frequency of mass switching in the mass spectrometer one can obtain a mass spectrum synchronous with the absorption of several isotopic species. This spectrum is seen in Fig. 2. The upper trace clearly shows the absorption spectrum (action spectrum) of the heavy isotope. It is seen that the absorption is frequency shifted from that in the light molecule, retaining, however, a similar line shape. From this spectrum an optimal laser frequency for

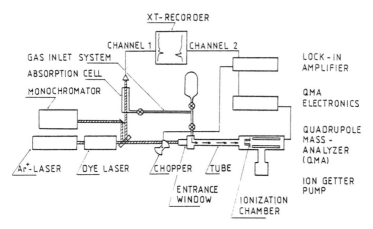

Fig. 1   Experimental setup for isotope selective molecular
spectroscopy in the unseparated natural mixture

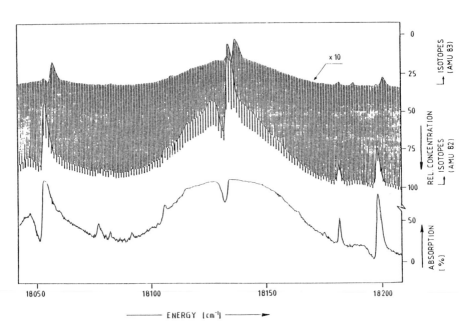

Fig. 2   Isotopeselective Spectra of "light" (middle trace) and
"heavy" (upper trace) sym-tetratine, measured in the
natural isotopic mixture.

an enrichment scheme can be found. It should be emphasized that
this apparatus permits the measurement of both spectra in the
<u>unseparated</u> natural mixture. Hence it is here shown that it is
possible to obtain absorption spectra for developing an enrichment
scheme without the <u>prior</u> availability of pure isotopic species.
It is a spectroscopic method for determining the absorption of a
single isotope at a time within the mixture. More careful investigation of Fig. 2 shows that there is not an one to one
correlation between all peaks in the two spectra. This is expected in view of the fact that several isotopes are present at
the same total mass (Tab. 1). By setting the mass spectrometer
at $^m/_e$ = 29 these isotopic isomers can be distinguished by their
photoproducts.

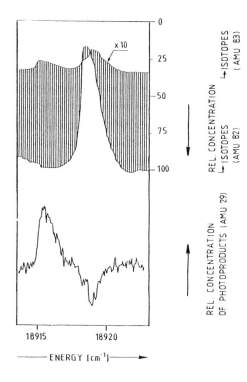

Fig. 3 Isotopeselective
spectra of "light" (middle
trace) and "heavy" (upper
trace) sym tetrazine. In the
lower trace the photoproduct
spectrum of molecules with
mass 29.

In Fig. 3 it is clearly seen that the peak at 18918 cm^{-1} splits
into two peaks such that only the one at 18915 cm^{-1} produces
heavy nitrogen with mass 29. The second peak at 18919 cm^{-1} then
must be due to the ^{13}C isomer. This has been confirmed by mass
spectrometry (see Fig. 3). The interesting feature now is that
we have two classes of frequencies i.e. some which are mapped
one to one in the heavy and the light absorption spectrum, but
others that are mapped two to one. The latter class perhaps
being more the expected result than the former. In either case

169

it is of no small importance to know in which class a particular absorption band falls prior to deciding its suitability
for an enrichment scheme. It is particularly at this point at
which trail- and -error methods are aggrivated by the new
possibilities of different classes of absorption with unkown
splittings. The isotope shifts for a number of transitions in
sym-tetrazine are shown in Table 2, half of which fall into
each class. Hence the origin might be an excellent method of
preparing the heavy compound, but further irradiation with the
X transition (18919 cm^{-1}) is needed to separate the ^{15}N from the
^{13}C isomer. Though the 16 b$_2^2$ transition also presents a differential splitting, this is clearly not as suitable for an isotope
separation scheme as the X transition. The detailed knowledge of
the information in Table 2 for all possible isotopic species is
an almost essential prerequisite for developing any photochemical
enrichment scheme. Here this information is all derived from
measurements on the unseparated mixture.

Table 2   Isotopic shifts in the spectrum of sym tetrazine, measured
           by isotope selective molecular spectroscopy

Band	Excitation Energy of light ST [cm^{-1}] in air	Isotope Shift	
		ST-^{13}C [cm^{-1}]	ST-^{15}N [cm^{-1}]
? (X)	18918.6 +	+0.6	-3.3
?	18888.6	-3.5	-3.5
$6a_0^1$	18837.0	-1.3	-1.3
$16b_2^2$	18430.5 +	+2.5	+2.0
$16b_1^1$	18283.0	+3.0	+3.0
$16a_1^1 16b_1^1$	18198.2	+3.0	+3.0
?	18181.7 +	+6.5	+1.0
origin	18133.9	+3.2	+3.2
$16a_1^1$	18052.9 +	+3.9	+4.9
$16a_2^2$	17970.3	+3.9	?
$6a_1^0$	17396.8 +	+9.3	+6.3
$6a_1^0 16a_1^1$	17317.2 +	+9.1	+6.7

Production of Isotopically Pure Sym-Tetrazine

Using the origin and the X transition some milligram of pure
material can be readily prepared. A typical run is shown in
Table 3 in which the natural, unfortified isotopes are prepared
in high purity with a yield of ca. 40 %. The enrichment factors
achieved, which are in excess of 1000, are so far the highest
published values for a photochemical enrichment scheme. Both the
^{15}N and ^{13}C-isotopic molecules were each prepared macroscopically
with high isotopic purity. The application on this method is,

of course, not limited to isotope enrichment schemes, but more
generally can be applied to the preparation of particular molecular compounds with high purity. It allows the determination
of absorption spectra of minor constituents in mixtures without
a prior separation of the mixture. It can be used to enrich
natural isotopic species, or to clean up a previous isotopic
separation or synthesis.

Table 3 Production of isotopically pure sym-tetrazine isomers

enriched material	ST-^{15}N	ST-^{13}C
initial quantity		
natural ST	155 mg	25 mg
herein contained		
ST-^{15}N	2.2 mg	0.35 mg
ST-^{13}C	3.5 mg	0.55 mg
final enriched quantity	1 mg	0.25 mg
enrichment factor	1077	1308
purity	85 %	91 %

Mass Selective Optical Spectroscopy by Two Photon Ionization

Two or three photon absorption in molecules can lead to
ionization. Such ionization has been employed as a method of
detection of two-photon absorption. Such a method of ionization
might also be of interest as an alternate method for making the
source of a mass spectrometer. This would provide a new method
of photoionization for mass-spectrometry. Suggestions to this
effect have been made by a number of workers [7, 8]. The first
combined optical and mass scanning has been realized for the
case of Alkali metal dimers by Schumacher et al. [9]. Rothe
et al. [10] applied this to the case of Li$_2$ and observed, for
the various lines of an Argon ion laser, differing mass spectra
for the various isotopic species. This is the first case of employing optical methods to enhance certain isotopic species in
a mass spectrometer.

In 1978 Boesl et al. [11] reported on the two-photon ionization
of benzene with a tunable laser source in the ionization chamber
of a mass spectrometer. At a resolution of 0.05 Å of the laser
source the mass spectrum of light and heavy benzene could be
distinguished and separately recorded, from a sample containing
but the natural abundance in a mixture. Here optical methods
enhanced the spectrum of choice beyond that expected from total
photoionization with a broad light source. The emphasis here is
that a highly tunable laser source can tune to one or the other
intermediate level on the way to ionization and hence lead to
optical enhancement. At the same time also Lethokov et al [12]

published work on the photoionization of various larger organic
species employing a fixed nitrogen and fixed hydrogen laser in
the ion source of a mass spectrometer. Furthermore Bernstein et
al. [13] published work on the multiphoton ionization of iodine,
and more recently on benzene, with a tunable laser source in a
mass spectrometer. This latter work focussed on the change in mass
spectral breakdown pattern with laser wavelength.

We here want to report high resolution results which further
emphasize the applicability of optical enhancement in the mass
spectrum.

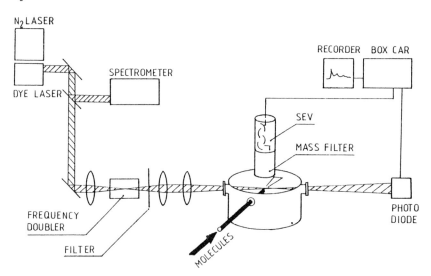

Fig. 4 Experimental setup for spectroscopy with mass selective
two-photon ionization

The apparatus is shown in Fig. 4. Light from a nitrogen pumped
dye laser system (Molectron) is doubled and crossed with a
molecular beam in the ionization region of a mass spectrometer.
The resolution of the system is about 1.5 GHz which means here
that it is Doppler limited. About 10^4 ions are produced in each
laser pulse leading to a strong spectrum. The results are seen
in Fig. 5. The two peaks are clearly separated by 1.65 cm^{-1}. The
envelope represents the 6^1_0 ro-vibronic transition of benzene, with
the rotational structure of the band being well reproduced by a
computer simulation (Fig. 6). This means that no further band-
structure comes from the photoionization step. Fig. 5 also
shows the almost pure absorption due to light benzene or due to
heavy benzene, depending on the setting of the mass selector.

In the first instance the ratio of heavy to light benzene is
0.5 to 100, whereas in the latter case this is turned over
into 960 to 100, i.e. optical enhancement has "tilted" the mass
spectra by three orders of magnitude.

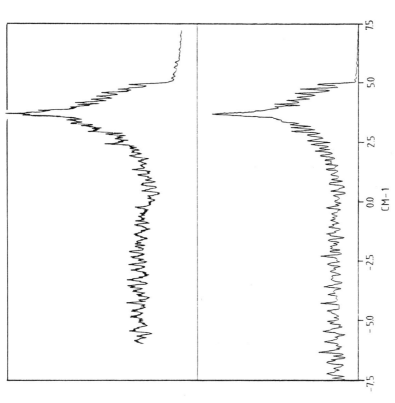

ION CURRENT / AMU 78

CM-1

Fig. 6 Highresolved and masselective intermediate state spectra of two photon ionization of benzene: experimental (upper trace) and theoretical (lower trace) spectrum

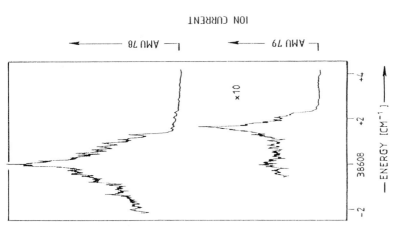

ION CURRENT

AMU 78

AMU 79

×10

— ENERGY [CM⁻¹] —

Fig. 5 Highresolved (1,5 GHz) and mass selective intermediate state spectra of two-photon ionization: C_6H_6 (upper trace) and $13CC_5H_6$ (lower trace)

173

Mass spectrometers with optically tunable ion sources have been of interest since vacuum UV sources where used by Inghram et al. [14] but two-photon ionization now presents the simplification of producing the same energy range with visible and near UV light, as well as working with a well focussed light source. The extreme sharpness of the laser further enables one, via optical enhancement, to bias the mass spectra toward species present in low concentration since the mass spectra will change strongly with the tuned laser frequency. Hence one can now perform two dimensional mass spectra with a laser scan as well as a mass scan. This two dimensional mapping technique has obvious powerful analytical applications.

References

[1] H. Okabe, Photochemistry of Small Molecules, Wiley, New York 1978, Ch. VIII

[2] U. Boesl, Thesis TU München 1978

[3] V.S. Letokhov, C.B. Moore, Kvant. Elektron. (Moscow) 3, 248 and 485 (1976)

[4] W. Kuhn, H. Martin, Naturwiss. 20, 772 (1932)

[5] C.C. McDonald, H.E. Gunning, J.Chem.Phys. 20, 1817 (1952)

[6] G.H. Spencer, P.C. Cross, K.B. Wiberg, J.Chem.Phys. 35, 1925 (1961)

[7] D.L. Feldman, R.K. Lengel, R.N. Zare, Chem.Phys.Lett. 52, 413 (1977)

[8] S.V. Andreyev, V.S. Antonov, I.N. Knyazev, V.S. Letokhov, Chem.Phys.Lett. 45, 166 (1977)

[9] A. Hermann, S. Leutwyler, E. Schumacher, L. Wöste, Chem. Phys.Lett. 52, 418 (1977)

[10] E.W. Rothe, B.P. Mathur, G.P. Reck, Chem.Phys.Lett. 53, 74 (1978)

[11] U. Boesl, H.J. Neusser, E.W. Schlag in Laser-Induced Processes in Molecules, Conference Proceedings Edinburgh 1978, Ed. K.L. Kompa and S.D. Smith, Springer-Verlag Berlin 1979, p.219

[12] V.S. Letokhov et al. Optics Letters 3, 37 (1978)

[13] L. Zandee, R.B. Bernstein, D.A. Lichtin, J.Chem.Phys. 69, 3427 (1978)

[14] H. Hürzeler, M.G. Inghram, J.D. Morrison, J.Chem.Phys. 28, 76 (1958)

[15] see also:
U. Boesl, H.J. Neusser, E.W. Schlag, Chem.Phys.Lett., 61, 57 (1979), Chem.Phys.Lett., 61, 62 (1979), Z.Naturforsch., 33a, 1546 (1978)

High Resolution Coherent Raman Spectroscopy of Gases[1]

A. Owyoung

Division 4214, Sandia Laboratories, Albuquerque, NM 87185, USA

1. Introduction

Although it has been known for well over a decade that the stimulated Raman gain and inverse Raman absorption processes are potentially powerful tools for Raman spectroscopy in the gas phase [1,2], the unavailability of precisely controlled, tuneable, narrow line-width laser sources has severely hindered the effective development of these two forms of stimulated Raman spectroscopy (SRS). Recently, low-power cw laser sources have been employed along with synchronous detection to obtain stimulated Raman gain spectra of the ν_1 fundamental in CH_4 under near Doppler-limited conditions [3]. Although these results have served to illustrate the inherently high sensitivity of the process and demonstrated its ability to yield complex Raman spectra with very high resolution, they also reflect the limitations imposed upon the technique by the use of low-power laser sources.

In the present paper we report on the recent progress made in using a "quasi-cw" laser scheme to obtain significant enhancement in the sensitivity of a stimulated Raman spectroscopy system designed for high-resolution gas-phase studies. The principles underlying the quasi-cw SRS scheme will first be described and its advantages discussed. The experimental implementation of the system is then considered with two sets of examples serving to illustrate its present capabilities. Finally, improvements of the existing system and potential future applications will be briefly outlined.

2. Quasi-CW Stimulated Raman Spectroscopy (SRS)

Both stimulated Raman gain spectroscopy (SRGS) and inverse Raman spectroscopy (IRS) fall into the class of coherent Raman techniques that can offer considerable advantage over conventional spontaneous Raman spectroscopy under the proper conditions. SRGS and IRS differ only in that spectra are obtained in the former case by using a Stokes shifted probe laser to scan the gain profile produced by a pump laser source, whereas, in the latter case of IRS, an anti-Stokes shifted probe laser is used to scan the absorption profile on the opposite side of the spectrum relative to the pump frequency [1,2]. In either case these SRS schemes offer two primary advantages over other alternatives: (1) a strong coherent output signal which scales linearly in species density and input power and (2) direct Raman spectra with resolution limited only by laser linewidth. These two qualities give the SRS techniques particular appeal under conditions where interference from background luminescence is problematical or in situations where very high resolution is required.

---
[1]This work supported by the U.S. Department of Energy

The practical implementation of the cw SRGS system this past year has suggested some promise for the availability of a new tool for use in high-resolution Raman studies in the gas phase; particularly with the introduction of the multiple pass optical cell for enhancing weak signals [3]. Careful consideration of such systems, however, has revealed that they are only practical for studying the strongest Raman transitions because of their modest sensitivity [4] .

It is intuitively reasonable that the limits imposed on existing cw SRS systems would be lifted if input powers could be raised to simply increase the signal level. This is in fact the case, and the signal-to-noise analysis that was previously applied to cw SRS systems is found to be directly applicable to a system incorporating the use of a pulsed laser [5]. These results suggest that the S/N ratio in a shot noise limited system obeys the relation,

$$\frac{S}{N} \propto P_{pump}^{1/2} \, P_{probe}^{1/2} \, (\Delta \nu)^{-1/2} \tag{1}$$

where P_{pump} and P_{probe} are the pump and Stokes or anti-Stokes probe laser input powers respectively, and $\Delta \nu$ is the effective bandwidth of the signal averaging system. Moreover since probe laser amplitude fluctuations are found to be the primary impediment to shot-noise-limited operation of cw systems [4], it is reasonable to take a "quasi-cw" approach wherein a single-mode stable cw laser provides a low-noise probe source and a high-power repetitively pulsed laser system provides the pump source. Such a scheme was previously applied to a SRS study of the Raman susceptibility in liquids [6] and is described diagrammatically in Fig. 1. The probe beam is temporally gated on for a time that is long compared to the high-power pump pulse, yet short enough to result in a low duty cycle, and thus preclude premature saturation of the detector. As the two beams cross in the sample at their common focus, the Stokes gain or anti-Stokes absorption produced by the pump beam is reflected in an amplitude modulation of the probe beam. High-pass filtering results in the elimination of the background probe level thus leaving the pulsed perturbation, S, which obeys the relation

$$S \propto \mathrm{Im}\chi_3^{iijj}(-\omega,\omega,\Omega,-\Omega) \, P_{pump} \, P_{probe} \tag{2}$$

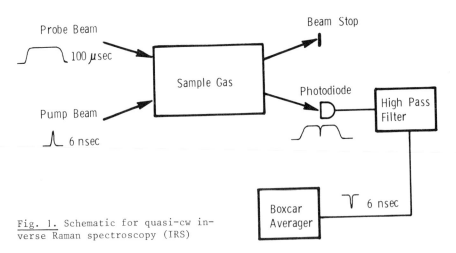

Fig. 1. Schematic for quasi-cw inverse Raman spectroscopy (IRS)

where Ω and ω are the optical pump and probe laser frequencies respectively, and $Im\chi_3^{iijj}$ is the Raman susceptibility [5]. Such an approach thus incorporates both the large gains in sensitivity predicted by (1) using a high-power pump source and the stable noise free advantages of a cw probe. Also, since the pump source is repetitively pulsed, a boxcar averager may be used to provide signal averaging to increase the signal-to-noise even further. As with the cw system, the spectrum is obtained by spectrally scanning either the pump or probe source, which permits the gain or absorption profile to be displayed directly.

An estimate of the S/N improvement to be expected in a quasi-cw SRS system compared to a purely cw system is easily obtained by applying typical operating parameters to Fig. 1. Pump powers which are in excess of 1 MW should be realistic for pulsed systems compared to 1W for a cw system. A conservative power for the gated probe laser is 250 mW whereas the amplitude noise at the detection frequencies used in a cw system gave an upper limit of 0.5 mW for shot noise limited detection,[4]. Finally, the bandwidth of a pulsed system will increase by approximately 10^7 compared to a cw system if one considers pump pulses of 5 nsec duration and a repetition rate of 10 pps. The net result is a S/N improvement of approximately 7×10^3. Clearly, such an improvement in sensitivity would considerably enhance the utility of the SRS technique.

A system such as that described by Fig. 1 may appear to be more complex and difficult to implement than a cw system, but experience has borne out that a quasi-cw apparatus is in fact not only more reliable but simpler to operate. This is attributed primarily to the fact that the pulse-gated detection scheme samples the signal on a 10 nsec time scale where both the effects of thermal lensing and probe laser noise are minimal.

In addition to the obvious improvement in sensitivity available through the quasi-cw approach, several other features make the concept attractive. Although the system should be operable using multi-pass cells similar to those used in cw studies to attain maximum sensitivity, the absence of phase matching requirements also allows a choice of performing a crossed beam experiment as shown in Fig. 1. Here the pump and probe beams only intersect at their foci, giving good spatial resolution, a feature that can be invaluable for diagnostic applications. In addition, since the quasi-cw approach utilizes a short pulsed source, the spectroscopic probe can be timed to perform time resolved measurements of chemical processes, thus suggesting new possibilities for numerous applications.

3. Experimental Discussion

An inverse Raman configuration with a fixed-frequency probe beam and frequency-scanned pump source was selected for feasibility studies of the quasi-cw technique. An IRS configuration should provide slightly better background suppression since an anti-Stokes probe source is used. Also, by fixing the probe beam frequency and scanning the pump beam frequency, a constant optical path is maintained between the probe laser and the detection system. Various regions of the Raman spectrum can thus be accessed without the necessity of adjusting the dispersion system which isolates the probe source from the pump radiation.

3.1 Narrow-Linewidth High-Power Pump Laser Source

The most important feature of the inverse Raman spectroscopy (IRS) apparatus is the high-power tuneable pump laser source that has a linewidth sufficiently narrow to be useful for high resolution gas phase spectroscopy. This system, shown schematically in Fig. 2, uses a frequency-doubled Nd:YAG laser system to pump successive dye amplifiers which are used to amplify the input from a cw dye oscillator. A similar system has recently been reported by DRELL and CHU [7].

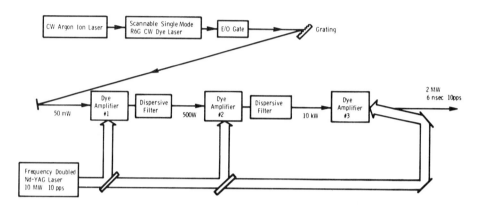

Fig. 2. Narrow linewidth pulsed dye laser system

The cw laser oscillator for the system is a rhodamine-6G single-mode electronically scannable cw dye laser, which supplies a narrow linewidth ($\sim$1MHz) input to the system of $\geq$ 50 mW that is continuously scannable over ranges of 1 cm^{-1} in the spectral region of 575-610 nm. An electro-optic gate shutters the cw output in synchronism with a 10-pps diffraction coupled resonator Q-switched Nd:YAG source, which lowers the average power through the dye amplifiers to prevent heating and thermal blooming.

The first two dye amplifier stages of the system are transverse pumped magnetically stirred dye cuvettes. The stages are isolated by 1200 line/mm gratings used in first order and pinhole spatial filters. Also, a grating between the oscillator and first amplifier prevents backward traveling fluorescence from interfering with the oscillator frequency stabilization system. Amplifier #1 is pumped at a nominal level of $\sim$ 3 mJ resulting in an output of $\sim$ 500W through the first pinhole isolator. Amplifier #2 then provides 10kW to the final amplifier when pumped with $\sim$ 5 mJ from the frequency doubled Nd: YAG source. Both amplifiers operate with a 2 x 10^{-4}M concentration of dye in methanol solution. Using rhodamine 610 and 640 dyes, the wavelength range from 575-610 nm is covered.

The final stage of amplification is supplied by a 25 mm pathlength flowing dye cell, which is longitudinally pumped by a counterpropagating beam from the Nd:YAG laser system to provide maximum energy extraction. A methanol dye solution at concentrations near 4 x 10^{-5}M is used with 50 mJ pump levels resulting in output energies varying from 13 to 17 mJ in 6 nsec pulses at 10 pps.

The final ~ 2 MW output of the laser system exhibits a high degree of directional stability, which is characteristic of the cw oscillator. Also, since the output of amplifier #2 is used to overfill the active volume of the final amplifier, the spatial quality of the beam is characteristic of the Nd: YAG pump source. With careful alignment, transform limited pulses of 75 MHz linewidth (FWHM) are obtained reproducibly. The system operates routinely with a spectral width of 120 MHz when no extraordinary effort is taken to optimize for minimum spectral width.

3.2 Inverse Raman Spectrometer

In Fig. 3 a more detailed schematic is shown of the quasi-cw IRS apparatus that uses the high power pulsed pump source. An argon ion laser operating in a single longitudinal mode at 514.5 nm serves as the primary probe laser source. Alternatively, a cw single-mode ring dye laser is being installed to provide a probe source that will facilitate IRS spectra in the range of $0-2000$ cm^{-1}.

In order to obtain accurate frequency measurements an interferometric cw wavemeter of the type reported by Kowalski, et al., is used to monitor both cw oscillators [8]. The argon ion laser source is also referenced to the flourescence from an iodine-129 cell to assure reproducible frequency settings and both sources are continuously monitored by four temperature stabilized scanning Fabry-Perot interferometers which are used to calibrate the scanning ranges and establish reproducible starting frequencies.

As shown in the figure, a 10 Hz mechanical chopper is used to produce the 100 μsec gate on the cw probe source. A reference signal from this device supplies the trigger to the Nd:YAG dye amplifier pump source and the electro-optic gate, which are both driven in synchronism with the probe pulse. Both pump and probe sources are thus focussed into the gas sample simultaneously where they overlap and the probe experiences a small absorption from the inverse Raman interaction.

At the output of the sample cell the two beams are recollimated and the pump beam terminated by a beam stop. Two Pellin-Broca prisms and a grating pinhole pair provide further isolation of the probe from the pump source. Finally, an EGG Model FND-100 silicon photodiode provides wideband detection of the IRS signal. High-pass filtering and amplification serve to separate the inverse Raman signal from the background probe level and to present the boxcar averager with sufficient signal to operate effectively. Furthermore, a reference signal from a diode monitoring the pump pulse power provides a normalization signal to the boxcar to reduce the effects of pulse to pulse variations in the pump laser. Spectral scanning of the pump source thus causes the inverse Raman absorption profile to be displayed at the boxcar output, which is plotted or stored by a small computer.

4. Experimental Results

Preliminary results obtained using the quasi-cw inverse Raman spectrometer described above have confirmed the capability of the technique for increasing sensitivity well beyond that obtained using cw sources. In addition to the gains which are realized through the increase in pump power, the shot noise limit, which cannot be reached because of probe laser noise in cw SRGS studies, can in fact be realized in the quasi-cw system. This

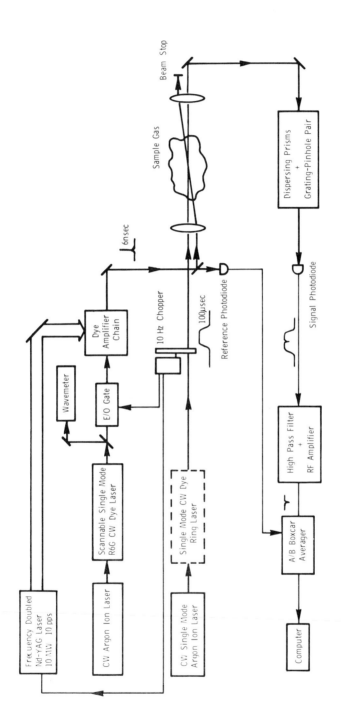

Fig. 3. Experimental configuration for the quasi-cw IRS apparatus

180

is primarily because the probe laser noise spectrum is concentrated at lower frequencies whereas shot noise contributions are uniformly distributed. Thus in the quasi-cw IRS studies, where the detection bandwidth is increased by nearly 10^7 over the cw case, laser noise contributions are not appreciably increased. The net result is that at pump power levels of 2 MW using 6 ns pulses at 10 pps with probe powers as high as 300 mW, the single pass sensitivity realized by the quasi-cw IRS system is over 10^4 times that observed in the cw system [3].

In addition to the increased sensitivity, another advantage found in implementing the new system is the absence of any spurious signals that offset the spectrum and make it difficult if not impossible to ascertain a true zero level. Such offsets were characteristic of the cw system where thermal refractivity changes produced by the pump beam would either offset the base line or cause it to drift as the spectrum was scanned. The absence of such signals in the quasi-cw system considerably simplifies the experimental procedure in addition to providing much cleaner data.

We shall illustrate the present capabilities of the quasi-cw IRS system by two examples where fruitful application has been made. Both will entail full utilization of the spectral resolving capabilities of the system. The former will be a fundamental application to molecular spectroscopy, whereas the latter will illustrate the capability of the system to provide spectra in an environment of strong background luminescence.

4.1 Quasi-cw IRS Spectra of Methanes

The study of the ν_1 fundamental in $^{12}CH_4$ has been a longstanding problem and has often been used to illustrate the resolving power of various Raman techniques [9-11]. Our investigation of this Q-branch spectrum was begun using the cw SRGS system [3]. It became clear, however, that even with the advantage of a multipass cell, the apparatus would be marginally adequate for yielding high quality spectra of the ν_1 fundamental, and that spectra of weaker modes would be out of the question.

In Fig. 4 a quasi-cw IRS spectrum of a portion of the ν_1 fundamental of $^{12}CH_4$ is shown with the experimental parameters used in obtaining the spectrum. A direct comparison of this spectrum with that taken by the cw multipass SRGS technique shows that the cw spectrum, which was taken at 35 torr, evidences a small amount of pressure broadening [3]. In the quasi-cw single pass spectrum, the lines are much more deeply modulated and several lines which were totally blended in the cw spectrum are beginning to appear. Note specifically $Q(5)F_2^o$ and the pairs

$$Q(3)F_1^o - Q(3)F_2^o \; , \; Q(4)F_1^o - Q(4)E^o \; , \; \text{and} \; Q(9)E^o - Q(9)F_1^1 \; .$$

The spectrum of Fig. 4 is unusual in that a positive value for the rotational constant $\Delta\beta = \beta - \beta_o = 0.01075 \text{ cm}^{-1}$ causes the band to unfold toward higher frequencies with the band origin being at $2916.47 \pm 0.01 \text{ cm}^{-1}$. Even more unexpected, however, is the spectrum of the ν_1 fundamental of $^{13}CH_4$ which is shown in Fig. 5. The rather drastic difference in the spectra of the two isotopic species have been attributed to a very small value of $\Delta\beta = 0.00219 \text{ cm}^{-1}$ for the latter species. This is a result of a small change in the value of the excited state rotational constant, which amounts to only 0.16%, but it makes a rather marked contribution to the value of $\Delta\beta$. The line assignments shown in Fig. 5 for $J \leq 11$ have been verified by a least squares fit and the spectral synthesis which is drawn below the experimental spectrum.

181

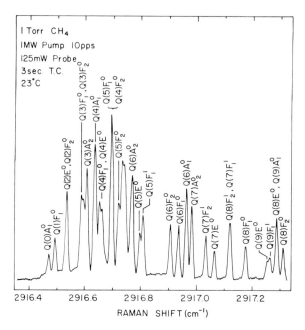

Fig. 4. The ν_1 fundamental for carbon-12 methane

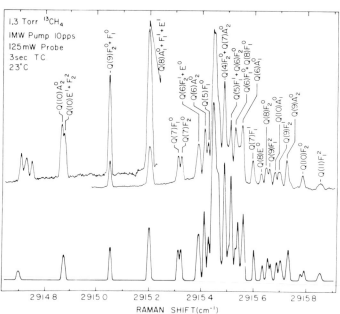

Fig. 5. The ν_1 fundamental for carbon-13 methane (top) is shown with a simulation of the spectrum (bottom)

This fit, which entailed the use of 25 of the experimentally resolved transitions, resulted in a standard deviation of 0.0015 cm^{-1} and the determination of spectroscopic constants [12]. From the figure it is seen that the small value of $\Delta\beta$ has resulted in a strong blending of the manifolds for $J \leq 5$ with the band origin being contained within the strong peak at 2915.43 cm^{-1}. For $J > 5$ the comparatively large size of the tensor splitting results in a spreading of the J manifolds in both directions around the band origin. The -1.04 cm^{-1} isotopic shift of the band origin was entirely unexpected [13] and has been attributed to Coriolis and Fermi interactions with neighboring bands.

For the two isotopic species discussed above, Doppler-limited resolution was essential to the identification of individual transitions. Such is not the case in Fig. 6 where a spectrum of a portion of the ν_1 fundamental of fully deuterated methane, CD_4, is shown with preliminary line assignments [14]. The experimental conditions used for obtaining this spectrum are identical to those used for the previous spectra with the exception of the total pressure, which is 12 torr. The very different appearance of this spectrum from either of the previous isotopic forms is attributed to a much larger positive rotational constant, 0.2 cm^{-1} and to the different nuclear spin. In contrast to previous cases that illustrated the resolving power of the apparatus, this spectrum illustrates a case where an extended tuning range is desirable. The discontinuities in the spectrum show that a 1 cm^{-1} limit is imposed by the continuous scanning capability of the cw dye laser. Yet, in spite of this inconvenience, the resolution required to obtain a spectrum such as this is still beyond the reach of standard Raman techniques.

Although the present system is somewhat limited in its abilities to scan over larger ranges, the limitation is not fundamental. Dye laser systems with considerably extended single mode scanning ranges should soon be available and a computer based digital data acquisition system, which is presently being tested on the apparatus, should solve the problems of rescaling data where piecewise scans are necessary. This latter system will also be a tremendous aid in reading tabulating and calibrating the positions and intensities of lines for the enormous amount of spectral data that is expected to result from a Raman apparatus with the resolution reported herein.

4.2 Quasi-CW IRS Spectra in a Methane-Air Flame

One of the areas for potentially fruitful application of the coherent Raman techniques is in the diagnostics of reacting or combusting gases where background luminescence poses a considerable problem for conventional Raman techniques. CARS methods have undergone extensive consideration for such applications and continue to show much promise for providing very nice spectra in hostile environments [15].

The advantages of the quasi-cw IRS technique in this particular application center not only on its ability to examine spectra with very high sensitivity and resolution, but also in the fact that the linear scaling of the IRS signal with density provides direct Raman spectra that are particularly amenable to analysis. The resulting spectra are thus free of any background contributions or spectral interferences from adjacent lines and virtually unaffected by sample luminescence.

In Fig. 7 we show a Q-branch spectrum of nitrogen in a methane-air flat flame at 1730°K. The J=17 and 18 lines of the second hot band (v=2 → 3) are shown adjacent to the J=44 line of the hot band. A spectrum such as this and

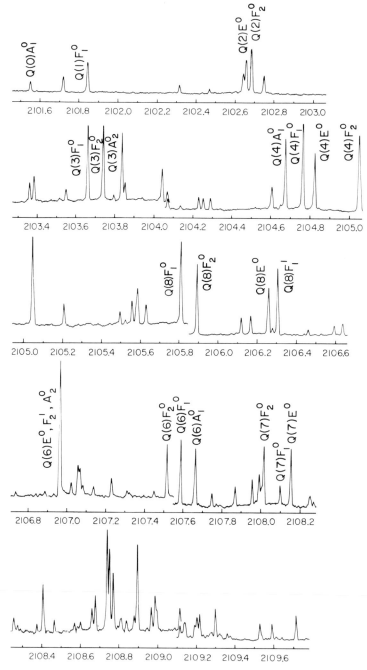

Fig. 6. The ν_1 fundamental for $^{12}CD_4$

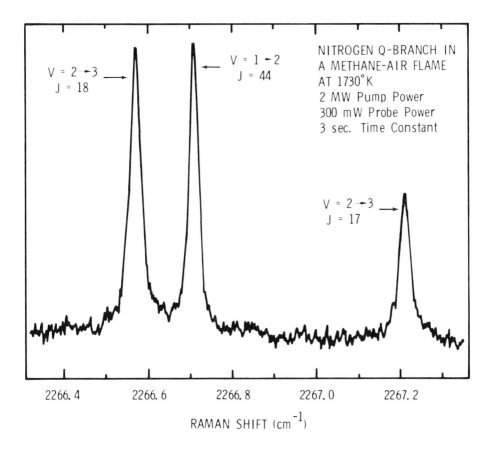

Fig. 7.   Hot band spectra of the nitrogen Q-branch in a methane-air flat flame

similar data involving transitions from the v=0 and v=1 levels should provide
precise means of performing thermometry in the flame.   This example also
illustrates the sensitivity of the technique since the signal levels involved
in this spectrum are over a factor of 200 weaker than those that would be observed in room temperature nitrogen on the J=6 Q-branch line at atmospheric
pressure.   The technique is thus applicable to the monitoring of minor species
such as H_2 and CO, which we have in fact observed at comparable signal levels.
High resolution spectra at elevated temperatures should also provide the capability to obtain much more precise determinations of the molecular constants
for nitrogen.

As a final illustration, we show in Fig. 8 a graph of the measured nitrogen
Q-branch linewidth as a function of J in a methane-air flame at 1730°K and
620 torr, and at room temperature and 225 torr [16].   The collision rate is
nearly identical under the two sets of conditions, yet the linewidth of the
room temperature gas exhibits a much more rapid decay with J than is evidenced
in the hot gas.   Superimposed graphs of the population distributions for the
two cases show the population to be much more sharply peaked and rapidly

185

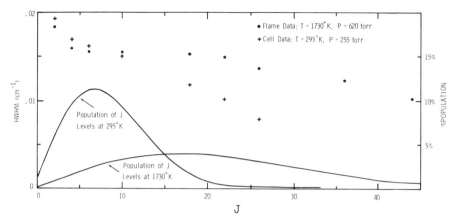

Fig. 8. Graph of linewidth (HWHM) and population distributions for nitrogen
Q-branch transitions vs J at 295°K and 1730°K

decreasing with J in the room-temperature case. This would suggest that phase
interrupting collisions are more likely to occur among species that are somewhat mono-energetic in rotational energy.

We have illustrated how IRS techniques are useful for thermometry and monitoring of minor species in reacting gases. Perhaps even more important, however,
is the fact that high resolution IRS will be able to provide a fundamental data
base of line positions, widths, shapes, and intensities. Such data will be very
useful to interpret lower resolution Raman data.

5. Conclusions

It has been seen that over four orders of magnitude increase in single pass
sensitivity is offered by a quasi-cw IRS system over purely cw approaches.
The quasi-cw system thus provides adequate sensitivity for the fruitful application of the IRS technique to many problems in molecular spectroscopy where
scattering cross-sections had heretofore been too small for such methods to be
useful. Yet the system is still in a rapid state of development and considerable improvements are to be expected in several respects.

Laser sources with larger scanning ranges and a computer based data acquisition system will greatly enhance the convenience of obtaining high quality
spectra. Multipass cells such as the one used in the cw SRGS study will provide more than an order of magnitude more sensitivity [3]. Improvements in the
digital wavemeter should provide absolute frequency measurements to better than
0.001 cm^{-1} rather than the 0.01 cm^{-1} which is the present accuracy. Finally,
instrumental linewidths of less than 10 MHz should be possible with longer pulse
sources [17], thus minimizing the importance of this advantage offered by cw
systems.

The possibilities for diverse applications of the quasi-cw IRS techniques
are many. Fundamental studies of diatomics and more complex spherical tops
will continue. Such ideas as resonantly enhanced IRS, IRS in molecular beams,
and the use of IRS to study the kinetics of photo-induced reactions and photo-

fragmentation are clearly areas in which the technique will find future application.

I wish to gratefully acknowledge the collaboration of my colleagues Drs. Robin S. McDowell, Chris W. Patterson and Larry A. Rahn in various portions of this work. The expert technical assistance of Mr. Randy E. Asbill was also essential to the success of the experimental phases of this work.

References

1. A. Owyoung, C. W Patterson and R. S. McDowell, Chem. Phys. Lett. 59, 156 (1978); and erratum, ibid. 61, 636 (1979).
2. W. J. Jones and B. P. Stoicheff, Phys. Rev. Lett. 13, 657 (1964).
3. P. Lallemand, P. Simova, and G. Bret, Phys. Rev. Lett. 17, 1239 (1966).
4. A. Owyoung, Chemical Application of Nonlinear Spectroscopy, ed. A. B. Harvey, to be published by Academic Press.
5. A. Owyoung, IEEE J. Quantum Electron. QE-14, 192 (1978).
6. A. Owyoung and P. S. Peercy, J. Appl. Phys. 48, 674 (1977).
7. P. Drell and S. Chu, Opt. Comm. 28, 343 (1979).
8. E. V. Kowalski, R. T. Hawkins and A. L. Schawlow, J. Opt. Soc. Am. 66, 965 (1976).
9. A. D. May, M. A. Henesian and R. L. Byer, Can. J. Phys. 56, 248 (1978).
10. M. R. Aliev, D. N. Kozlov, and V. V. Smirnov, JETP Lett. 26, 27 (1977).
11. J. P. Boquillon and R. Bregier, Appl. Phys. 18, 195 (1979).
12. R. S. McDowell, C. W. Patterson and A. Owyoung, to be published.
13. R. S. McDowell, J. Mol. Spec. 21, 280 (1966).
14. A. Owyoung, R. S. McDowell and C. W. Patterson, to be published.
15. W. M. Tolles, J. W. Nibler, J. R. McDonald, and A. B. Harvey, Appl. Spec. 31, 253 (1977).
16. A. Owyoung and L. A. Rahn, to be published.
17. M. D. Levenson, private communication.

Doppler-Free Optogalvanic Spectroscopy

J.E. Lawler, A.I. Ferguson[1], J.E.M. Goldsmith, D.J. Jackson, and A.L. Schawlow

Department of Physics, Stanford University, Stanford, CA 94305, USA

I. Introduction

There has been considerable interest in optogalvanic detection of atomic and molecular transitions since GREEN and his collaborators [1] reported impressive signal to noise ratios using this method. The history of the optogalvanic effect is summarized elsewhere [1,2]. We simply note that several researchers have observed electrical changes in a discharge when it is exposed to radiation corresponding to atomic transitions in the discharge. The potential of this detection method was vastly increased with the development of tunable dye lasers. Optogalvanic detection is different from charged particle detection methods involving space charge limited diodes because of the presence of a plasma with a relatively high electron temperature. This plasma allows one to conveniently observe transitions between excited levels. The spectra of nonvolatile materials are also conveniently observed using optogalvanic detection with sputtering in hollow cathode discharges. The potential analytical application of optogalvanic detection was demonstrated by trace element detection at the 0.1 ppb level [3]. Optogalvanic detection has also been used to lock a dye laser to an atomic transition [4]. Our research at Stanford has concentrated on the development of Doppler-free techniques using this detection method [5,6], and on improvements in the sensitivity of this detection method.

II. Doppler-Free Intermodulated Optogalvanic Spectroscopy

Optogalvanic spectroscopy is based on the detection of an impedance change in a gaseous discharge, produced by irradiation with a laser tuned to an atomic transition occurring within the discharge. The laser perturbs the steady state populations of atomic levels in the discharge; in general, ionization rates of the different levels will be different. Several ionization mechanisms can contribute to the effect. The change in discharge impedance is detected as a change in voltage across a ballast resistor by a lock-in amplifier tuned to the chopping frequency of the laser. A spectrum with ordinary Doppler broadening is obtained by tuning the laser across an optically thin transition.

To generate Doppler-free optogalvanic signals on single photon transitions we use intermodulation, as used in intermodulated fluorescence spectroscopy [7]. Doppler-free intermodulated opto-galvanic spectroscopy (IMOGS) has a similar detection scheme to Doppler limited optogalvanic spectroscopy. In IMOGS, however, the laser beam is split into two

---

[1]Lindemann Fellow, present address: Department of Physics, University of St. Andrews, North Haugh, St. Andrews, Scotland.

components of roughly equal intensity. One beam is chopped at a frequency f_1 and sent through the discharge. The second beam is chopped at a different frequency f_2 and sent through the discharge in the opposite direction. Because the two beams propagate in opposite directions, they interact in general with different velocity groups of atoms under the Doppler profile of a given transition. When the laser is tuned within one homogeneous width of line center, the two beams interact with the same group of atoms. Nonlinearities caused by the two beams acting to saturate the same atoms then give rise to Doppler-free optogalvanic signals at sum (f_1+f_2) and difference (f_1-f_2) modulation frequencies; the sum frequency is conveniently detected with a lock-in amplifier.

We have studied the $2^3P - 3^3D$ transition in ^3He as a demonstration of IMOGS. This transition is of interest because, to our knowledge, no Doppler-free study of it has been reported, and only one limited radiofrequency study of the 3^3D level has been reported in which the $3^3D_3(F=5/2) - 3^3D_3(F=7/2)$ splitting was measured[8]. Furthermore, the structure of this ^3He level is amenable to precise theoretical calculation. Besides being of spectroscopic interest, this transition is of interest for another reason. The kinetics of the He levels involved in this transition are well understood. Transfer rates [9], associative ionization rates [10], and radiative decay rates [11] are well known. We hope to use this information and our measurements of the magnitude of the effect to gain a better understanding of the optogalvanic effect in the rare gas positive column discharge, and ultimately to improve the sensitivity of the detection method.

A schematic diagram of the apparatus used in our first experiment is shown in Fig. 1. The dc discharge cell is a commercial He-Ne laser tube (Jodon Engineering Associates CE(16.7-2CS) with a ^3He partial pressure of 1.62 Torr and a Ne partial pressure of 0.23 Torr. It is operated with a discharge current of 5.5 mA and a tube voltage of 700 V. Tunable laser radiation is provided by an actively stabilized rhodamine 6G dye laser (Coherent Incorporated 599-21). The dye laser beam is split into two beams of roughly equal intensity. One beam is chopped at a frequency of 600 Hz and the other at 840 Hz by the same mechanical chopper, which also provides a reference signal at the sum frequency of 1440 Hz. Changes in discharge tube current are determined by measuring the voltage across a 100 kΩ ballast resistor with a lock-in amplifier tuned to 1440 Hz (PARC HR-8). The dc voltage across the ballast resistor is blocked with a coupling capacitor. The noise current in a 1 Hz bandwidth at 1440 Hz is on the order of 10^{-11} A, consistent with the shot noise limit for the given conditions.

A typical partial scan of the 587.5 nm $2^3P - 3^3D$ multiplet of ^3He is shown in Fig. 2. Only the part of the multiplet connected to the 2^3P_0 level is shown, although a large number of components and crossovers associated with the 2^3P_1 and 2^3P_2 levels can also be observed. The laser power is attenuated to 0.3 W/cm^2 to avoid power broadening. The laser bandwidth is 1 MHz. The observed linewidth of 130 MHz is due primarily to pressure broadening at the ^3He partial pressure of 1.62 torr, and is consistent with previous measurements [12].

Preliminary values for hyperfine splittings of the 3^3D level are given in Table 1, with quoted uncertainties of one standard deviation. The splittings are measured with a precisely calibrated 100 cm confocal interferometer[13]. Calculated values are shown for comparison [14]. We are currently investigating possible systematic errors due to pressure shifts, and attempting to resolve the rest of the multiplet.

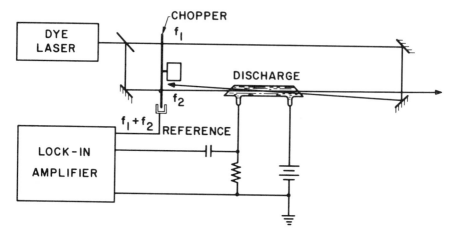

<u>Fig. 1</u> Experimental apparatus for intermodulated optogalvanic spectroscopy

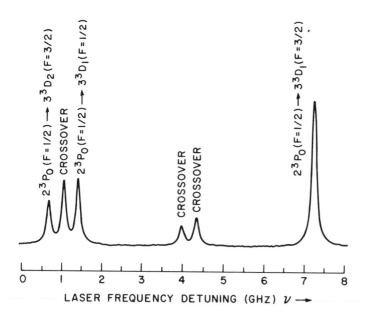

<u>Fig. 2</u> Intermodulated optogalvanic spectrum of part of the ^{3}He 2 ^{3}P – 3 ^{3}D transition at 587.5 nm

	Experimental (MHz)	Theoretical (MHz)
3^3D_2 (F = 3/2) - 3^3D_1 (F = 1/2)	725 ± 4[a]	750[b]
3^3D_1 (F = 1/2) - 3^3D_1 (F = 3/2)	5830 ± 6[a]	5760[b]
3^3D_3 (F = 5/2) - 3^3D_3 (F = 7/2)	55.8 ± 0.8[c]	0[b]

[a]This work.   [b]See Ref. 14.   [c]Ref. 8.

Table I Hyperfine splitting of the ^{3}He 3^3D level

It is interesting to compare the sensitivity of IMOGS to that of Doppler-free saturated absorption spectroscopy. In IMOGS, the signal is detected as a change is the discharge current, rather than a change in probe beam intensity. Both signal and noise are very different in form from those in saturated absorption spectroscopy. The use of a two-level model in the Doppler broadened, weakly saturated limit greatly facilitates the comparison. Although the ^{3}He 2^3P - 3^3D multiplet is not a simple two-level system, we use parameters from the experiment to quantify the comparison. Let n_u and $n\ell$ be the density of atoms in the upper and lower levels, respectively, of the transition under steady state discharge conditions. The change in these densities at line center modulated at the sum of the laser modulation frequencies is

$$\Delta n_u - \Delta n_\ell \equiv 2 \Delta n = (n_u - n_\ell)(\Delta\nu_H/\Delta\nu_I) I_1 I_2/(8 I_{sat}^2) \quad . \quad (1)$$

where $\Delta\nu_H$ and $\Delta\nu_I$ are the homogeneous and inhomogeneous widths, I_1 and I_2 are counterpropagating beam intensities, and I_{sat} is the saturation intensity. For simplicity we have ignored the difference in level degeneracy and assumed that the discharge is optically thin. Equation (1) follows from a standard derivation of the change in absorption or gain[7,15] due to saturation by including a factor, $\Delta\nu_H/\Delta\nu_I$, representing the fraction of atoms in the velocity distribution which interact with the laser beams. Recombination in the positive column of a low pressure glow discharge is negligible, although electrons are lost by ambipolar diffusion to the wall[16]. We assume as an approximation that all excess electrons are collected with no gain. By assuming a "collection efficiency" of unity we are making the simplest approximation possible. Experiments have indicated that the collection efficiency is a function of the sustaining direct current, ballast resistor, and other discharge parameters. Our data on the magnitude of the optogalvanic effect on the He 587.5 nm transition indicates that a collection efficiency of unity is feasible, and that gain may be possible. This approximation allows us to write the signal current as

$$i_s = e V R \Delta n \quad , \quad (2)$$

where e is the electron charge, V is the volume of the positive column perturbed by the laser beams, and R is the difference in ionization rates of the upper and lower levels. The limiting noise in IMOGS is due to fluctuations in discharge current, rather than fluctuations in laser intensity. We observe current fluctuations at audio modulation frequencies, in carefully designed discharge tubes, that are consistent with shot noise. Shot noise provides a fundamental noise limit if each electron emitted from the cathode proceeds to the anode independently. The RMS shot noise current in a bandwidth B due to a current i_{DC} is [17]

$$i_N = (2e \, i_{DC} \, B)^{1/2} \quad . \tag{3}$$

If the electron motion is not independent the current noise can be lower [17]. Thermal noise from the plasma with an electron temperature of several volts also provides a limit. In this discussion the RMS shot noise current will be used as a limit because it corresponds to our observed noise current. The signal-to-noise in IMOGS is given by

$$(S/N)_{IMOGS} = VR \, (n_u - n_\ell) \, \Delta\nu_H \, I_1 \, I_2/(16 \, \Delta\nu_I \, I_{sat}^2 \, (2 \, Bi_{DC}/e)^{1/2}) \quad . \tag{4}$$

In the method of Doppler-free saturated absorption spectroscopy, an intensity modulation ΔI_1 on a laser beam of intensity I_1 is produced by the presence of a chopped saturating beam of intensity I_2. The signal at line center is

$$\Delta I_1/I_1 = \alpha \, \ell \, I_2/(4 \, I_{sat}) \quad , \tag{5}$$

where α is the absorption coefficient of the sample of length ℓ, and I_{sat} is the saturation intensity of the transition. The signal-to-noise ratio in saturated absorption spectroscopy is

$$(S/N)_{SAS} = \alpha \, \ell \, I_2/(4 \, \delta \, I_{sat}) \quad , \tag{6}$$

where δ is the fractional intensity fluctuation within the detection bandwidth.

The ratio of the signal to noise of IMOGS to that of saturated absorption spectroscopy is

$$(S/N)_{IMOGS}/(S/N)_{SAS} = (V/\ell)(R/A)(I_1/h\nu) \, \delta/(2(2B \, i_{DC}/e)^{1/2}) \quad , \tag{7}$$

where A is the Einstein coefficient for the transition, $h\nu$ is the photon energy, and I_{sat} has been approximated as

$$I_{sat} = A \, h\nu \, \Delta\nu_H(n_u - n_\ell)/(2\alpha \, \Delta \, \nu_I) \quad . \tag{8}$$

In the approximation for I_{sat} it is assumed that radiative decay is the primary mechanism restoring steady state discharge populations. For the ^{3}He $2\,^3$P - $3\,^3$D transition, A is $7 \cdot 10^7$ sec^{-1}, [11] and R is approximated as $4 \cdot 10^6$ sec^{-1}, the rate of associative ionization of the $3\,^3$D level at a He partial pressure of 1.62 torr[10]. The rate of associative ionization of the $2\,^3$P level is zero because it is well below threshold. The presence of Ne in the commercial discharge tube is neglected. Helium $3\,^3$D atoms can cheminize ground state Ne atoms whereas He $2\,^3$P atoms cannot, so reactions involving Ne atoms may enhance the opto-galvanic effect. The difference in electron impact ionization rates should in general be included in R. In the experiment, V/ℓ is 0.03 cm^2, I_1 is 0.3 W/cm^2, i_{DC} is 5.5 mA, and B is 1 Hz For typical cw dye lasers, the fractional intensity fluctuation δ is far from the shot noise limit. At the least noisy modulation frequencies we observe a δ of the order of 10^{-4} with a detector bandwidth of 1 Hz. Using these parameters we estimate $(S/N)_{IMOGS}$ / $(S/N)_{SAS}$ = 300.

The analysis of the optogalvanic effect on the He 587.5 nm transition is particularly simple because the dominant decay channel of a $3\,^3$D atom is radiative decay at the laser wavelength. We note that R/A emerges as an ionization efficiency. We have estimated R/A as 5.7% and the collection efficiency as 100%. The quantum efficiency of the optogalvanic detector is the product of the ionization efficiency and the collection efficiency. This quantum efficiency is measurable whenever absorption is detectable. We have measured a quantum efficiency of 4.3% on this transition and quantum efficiencies over 1% on several Ne transitions. Our estimate of the sensitivity is realistic. Substantial improvements in sensitivity may yet be possible.

It is apparent that with favorable reaction rates IMOGS can be more sensitive than saturated absorption spectroscopy. Other spectroscopic techniques have been developed to improve the sensitivity of saturated absorption spectroscopy. The fractional intensity fluctuation δ may be reduced by intensity stabilizing the laser, or its effect may be reduced by a differential detection scheme [18]. Saturated interference spectroscopy [19] and Doppler-free laser polarization spectroscopy [20] have been demonstrated. Each of these techniques has its own set of performance requirements for optics and detectors which must be met if the technique is to provide a substantial improvement in sensitivity over saturated absorption spectroscopy. IMOGS may be most useful in regions of the spectrum where low noise detectors, interferometric quality optics, or high quality polarizers are unavailable. IMOGS is also an extremely simple technique to use, requiring a bare minimum of optical components.

III. Conclusion

We describe intermodulated optogalvanic spectroscopy, a powerful new method of performing Doppler-free saturation spectroscopy in a discharge. Hyperfine splittings of the 3^3D level in ^3He obtained using this method are reported. The high sensitivity which can be achieved with optogalvanic detection offsets the difficulties associated with a discharge system. Spectroscopic pertubrations in a discharge will always be a source of concern in high resolution work, but the perturbations in many cases are not severe. In our initial demonstration of intermodulated optogalvanic spectroscopy we used a visible laser and a rare gas positive column discharge. The technique may be very useful in other regions of the spectrum. The spectra of non-volatile materials may also be observed using this technique in the negative glow region of hollow cathode lamps.

Work supported by tha National Science Foundation under Grant No. NSF-9687 and by the U.S. Office of Naval Research under Contract N00014-78-0403.

References

1. R.B.Green, R.A.Keller, G.G.Luther, P.K.Schenck, and J.C.Travis, Appl. Phys. Letters 29, 727 (1976)
2. W.B.Bridges, J. Opt. Soc. Am. 68, 352 (1978)
3. G.C.Turk, J.C.Travis, J.R.DeVoe, and T.C.O'Haver, Anal. Chem. 50, 817 (1978)
4. R.B.Green, R.A.Keller, G.G.Luther, P.K.Schenck, and J.C.Travis, IEEE J. Quant. Electr. 13, 63 (1977)
5. J.E.Lawler, A.I.Ferguson, J.E.M.Goldsmith, D.J.Jackson, and A.L.Schawlow, Phys. Rev. Letters 42, 1046 (1979)
6. J.E.M.Goldsmith, A.I.Ferguson, J.E.Lawler, and A.L.Schawlow, accepted for publication in Optics Letters
7. M.S.Sorem and A.L.Schawlow, Opt Comm. 5, 148 (1972)

8. J.P.Descoubes, in "Physics of the One and Two-Electron Atoms," F.Bopp and H.Kleinpoppen, eds. (North-Holland 1969), p. 341

9. H.F.Wellenstein and W.W.Robertson J. Chem. Phys. 56, 1072 (1972)

10. H.F.Wellenstein and W.W.Robertson, J. Chem. Phys. 56, 1077 (1972)

11. W.L.Wiese, M.W.Smith, and B.M.Glennon "Atomic Transition Probabilites," U.S.National Bureau of Standards, Ref. Data Series 4 (U.S. GPO, Washington, D.C., 1966), Vol. I, p. 14, p. 128

12. Ph.Cahuzac and R.Damaschini, Opt. Comm. 20, 111 (1977)

13. J.E.M.Goldsmith, E.W.Weber, F.V.Kowalski, and A.L.Schawlow, accepted for publication in Applied Optics

14. M.Fred, F.S.Tomkins, J.K.Brody, and M.Hamermesh, Phys. Rev. 82, 406 (1951)

15. A.Yariv, "Quantum Electronics" (Wiley & Sons, New York, 1975), p. 167

16. M.J.Druyvesteyn and F.M.Penning Rev. Mod. Phys. 12, 87, (1940)

17. J.J.Brophy, "Basic Electronics for Scientists" (McGraw-Hill, New York, 1966), p. 271

18. T.W.Hansch, I.S.Shahin, and A.L.Schawlow, Phys. Rev. Letters 27, 707 (1971)

19. F.V.Kowalski, W.T.Hill, and A.L.Schawlow, Optics Letters 2, 112 (1978)

20. C.Wieman and T.W.Hansch, Phys. Rev. Letters 36, 1170 (1976)

194

Laser Saturation Spectroscopy with Optical Pumping

D.E. Murnick

Bell Laboratories, Murray Hill, NJ 07974, USA

and

M.S. Feld, M.M. Burns, T.U. Kuhl, and P.G. Pappas

Department of Physics and Spectroscopy Laboratory[1]
Massachusetts Institute of Technology
Cambridge, MA 02139, USA

1. Introduction

This contribution is heavily influenced by the antecedents to Doppler-free laser spectroscopy reviewed by our introductory speaker, Professor SERIES.[1] In particular, we have considered the effect of optical pumping as introduced by KASTLER [2] on saturation spectroscopy in atomic systems. Several interesting and important effects can occur in this case. Our primary motivation for these studies has been to devise techniques for efficient nuclear orientation as described in a companion paper [3]. Work related to that presented here has been contributed to these proceedings by SERIES and GAWLIK [4] and has also been carried out by PINARD et al.[5]

As is well known, laser saturation techniques provide powerful means of removing the Doppler resolution limitation of atomic and molecular spectroscopy and attaining narrow, natural-width resonances for high resolution studies.[6] Combining saturation spectroscopy with optical pumping of atomic resonance transitions leads to the following interesting effects: (i) saturation signals of anomalous sign and increased amplitude dependent on laser polarization, (ii) greatly reduced saturation thresholds, and (iii) in certain situations spectral features far narrower than the natural radiative limit. These effects have important practical consequences: (i) implies that weak transitions can exhibit large saturation signals, (ii) that saturation studies can be performed using relatively weak laser sources, and (iii) that heretofore unattainable optical resolution is possible.

Atomic resonance transitions are unique in that their lower levels are stable ground states, and that at pressures below a few torr excited state relaxation occurs predominantly via radiative decay to the ground state. These are also necessary conditions for optical pumping.[7] The simultaneous occurrence of optical pumping and laser saturation leads to the features enumerated above. Here we describe their physical basis and present theoretical predictions and experimental verification in resonance transitions of atomic sodium and barium. Further details are presented elsewhere.[8]

2. Anomalous structure

We have studied Lamb dip [9] and crossover resonances [10] in naturally occurring atomic barium. A single-mode cw dye laser scanned the $^1S_0 \rightarrow {}^1P_1$

[1]Supported in part by U. S. Department of Energy

resonance transitions at 5535Å. A low pressure Ba sample cell (oven temperature 450°C), length L = 1 cm, was irradiated by an intense beam, which
saturated and optically pumped the transitions, and a weak reflected
counter-propagating probe beam. The observed spectra [Fig. 1 (a,b)] are
composed of saturation resonances due to the hyperfine structure of the
naturally occurring isotopes. Three noteworthy features are: the simultaneous occurrence of resonances of opposite signs; the change of sign of
Lamb dips n and o with polarization configuration; and the inverted crossover resonances l and m, [9] which are comparable in intensity to the ^{138}Ba
Lamb dip (a), although the latter would be expected to be 8 times more
intense based on relative abundances and statistical weights.

These features are due to strong optical pumping of the degenerate M_F
sublevels of the odd isotopes, induced by the intense beam, leading to nearly complete orientation of the atomic ground state. Figures 1 (c,d,e)
illustrate this effect for the specific case of the F = 3/2→1/2 hyperfine
component of ^{137,135}Ba. The weak beam probes the modified sublevel populations. Lock-in detection gives the difference between this absorption

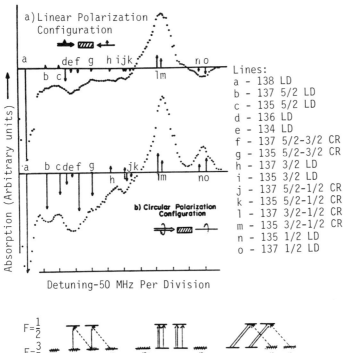

Lines:
a - 138 LD
b - 137 5/2 LD
c - 135 5/2 LD
d - 136 LD
e - 134 LD
f - 137 5/2-3/2 CR
g - 135 5/2-3/2 CR
h - 137 3/2 LD
i - 135 3/2 LD
j - 137 5/2-1/2 CR
k - 135 5/2-1/2 CR
l - 137 3/2-1/2 CR
m - 135 3/2-1/2 CR
n - 135 1/2 LD
o - 137 1/2 LD

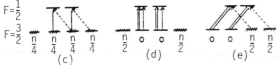

Fig.1   (a,b) Observed probe resonances (dotted lines) and theoretical fits
(arrows) for I_1 = 4 mW/cm^2 and T = 1 μsec. The odd isotope hyperfine
resonances are specified by their upper level F values. F lower = 3/2 in
all cases. (c,d,e) Level diagrams. Relative M_F ground state populations
are indicated. (c) Reference absorption, no optical pumping; (d) linear
polarization configuration. (e) circular polarization configuration.

[Fig. 1 (d,e)] and the absorption without the saturating beam [Fig. 1 (c)]. Depending on polarizations, the probe may sample either depleted [Fig. 1 (d)] or enhanced [Fig. 1 (e)] populations, resulting in reduced or increased absorption, respectively, when appropriate tuning conditions are met. Similar arguments explain the details of the other hyperfine components and the crossover resonances. Quantitative calculations along these lines [8] confirm this discussion [arrows of Fig. 1 (a,b)]. Previous investigators have reported inverted crossover transitions in saturation spectra which are due to pumping between hyperfine levels.[11] Inverted Lamb dips, however, which depend on laser polarization are unique to the optical pumping effect described here.

3. Reduced saturation thresholds

The above explanation assumes complete optical pumping, and thus neglects the influence of ground state relaxation. In any real system, collisions, transit time effects and finite laser linewidth limit the duration, T, of the light-atom interaction, hence the extent of optical pumping. The balance reached between laser optical pumping and relaxation gives rise to a steady state population distribution with a given degree of orientation. This steady state condition can be formulated in terms of a saturation process in which the normal saturation parameter[12]

$$I_S = \hbar\omega/2\sigma\tau \tag{1}$$

is replaced by an optical pumping saturation parameter

$$I_0 = \hbar\omega/\sigma\tau = (2\tau/T) \, I_S \ , \tag{2}$$

with optical transition energy $\hbar\omega$, absorption cross-section $\sigma \simeq 3\lambda^2/2\pi$, excited state radiative lifetime τ, and light-atom interaction time (i.e. effective ground state lifetime) T. Since $\tau \sim 10$ nsec. and $T \sim \mu$sec. the saturation threshold can be sizably reduced.

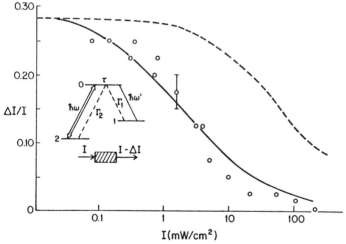

Fig. 2 Saturation curve for sodium D_1 line. The solid line is a least squares fit of Eq. (3a) to the data, giving $I_1 = 0.7$ mW/cm^2. The dashed line is the calculated saturation behavior assuming no optical pumping.

197

Consider the simplified energy level scheme of Fig. 2, a Doppler-
broadened three level system in which two degenerate or near-degenerate
ground state M-levels, 1 and 2, are optically coupled to a common excited
state M-level, 0, by electric dipole transitions. Level 0 can decay via
spontaneous emission to levels 1 and 2 with rates Γ_1 and Γ_2, respectively,
resulting in a radiative lifetime $\tau = 1/(\Gamma_1 + \Gamma_2)$. Thus, after absorbing
a pump photon at the 2-0 transition the atom can either return to state 2,
with probability $\Gamma_2 \tau$ or decay to state 1, with probability $\Gamma_1 \tau$. Accordingly, the applied field will cause atoms initially in level 2 to accumulate
in level 1. In particular, in the case of a short-lived excited state with
large branching ($\Gamma_1 >> \Gamma_2$) the probability of transferring an atom from
level 2 to 1 in a time t is given by $(\sigma I/\hbar\omega)t$. However, this process
cannot continue indefinitely, since the atom can only interact with the
beam for duration T. Thus, an atom initially in state 2 will have a large
probability of being transferred to state 1 when $(\sigma I/\hbar\omega)T = 1$, i.e. when
$I = I_0$ [Eq. (2)]. Hence, a laser beam of intensity $I = I_0$ [<< I_s, Eq. (1)]
will substantially deplete the population of level 2.

A straightforward calculation of the Doppler broadened absorption
coefficient of the 2-0 transition along these lines yields

$$\alpha(I) = \alpha_2/\sqrt{1 + I/I_1} , \qquad (3a)$$

with α_2 the small signal 2 - 0 absorption coefficient and

$$I_1 = \frac{\hbar\omega}{\sigma\tau} \frac{(1 + \tau/T)}{(2 + \Gamma_1 T)} \simeq \frac{\hbar\omega}{\sigma\tau} \cdot \frac{1}{2 + \Gamma_1 T} . \qquad (3b)$$

For no optical pumping ($\Gamma_1\tau \to 0$), $I_1 \to I_s$ and Eq. (3a) reduces to the
standard result. [12] However, for large branching ($\Gamma_1\tau \to 1$), $I_1 \to I_0$,
as anticipated above.

To demonstrate this effect the fractional absorption of a cw single-
mode linearly polarized dye laser beam by an optically thin ($\alpha_2 L = 28\%$)
sample cell containing sodium vapor was measured as a function of I [Fig.2].
The laser was tuned to the center of the 5896Å sodium D_1 Doppler profile.
To insure uniform intensity over the beam cross-section, a 1 mm diameter
pinhole was placed in front of the detector to eliminate the outer portion
of the 3 mm laser spot. As can be seen, saturation occurs at $I_1 \simeq 0.7$
mW/cm^2, $\sim 3\%$ of the value of $I_s \simeq 25$ mW/cm^2, calculated assuming no optical
pumping from the weighted average of Eq. (3a) with $I_1 = I_s$ for each (M_F -
degenerate) hfs component of the D_1 transition (broken line of Fig. 2).

4. Ultra-narrow spectral features

Optical pumping effects must be considered in all saturation experiments
where hyperfine structure exists. In special situations ultra-narrow
features can be observed within a normal saturation resonance. In an
atomic vapor having two closely-spaced resonance transitions (e.g. hfs)
sharing a common upper level (inset, Fig. 2, with $\omega \neq \omega'$) interacting with
two co-propagating laser fields, a fixed frequency saturating beam (I,Ω)
resonating with the 2-0 transition and a weak tunable beam (I', Ω') which
probes the 1-0 transition, a resonant change in the probe transmission
called laser-induced line narrowing,[13] occurs when the tuning condition
$\Omega' = \Omega_0 \equiv \omega' + (\Omega - \omega)(\omega'/\omega)$ is satisfied. This signal can be narrower
than that predicted by population considerations due to the occurrence of
Raman-type (e.g. double-quantum) processes.[6] In the absence of optical

pumping for $\omega'/\omega \simeq 1$ and no dephasing collisions, the change signal is a Lorentzian of width $\gamma_1 + \gamma_2$, γ_j = level j decay rate, and thus independent of the upper level lifetime. (This narrowing also forms the basis of two-photon Doppler free spectroscopy.)[14] In the present case, however, optical pumping modifies the form of the change signal. A density matrix calculation for the probe absorption coefficient, α_p, is similar to those of Refs. 13 except that excited state branching into the ground states must be included. In the case of interest, $\omega'/\omega \simeq 1$, $\tau/T \ll 1$, $\Gamma_1 T \gg 1$, one obtains for weak saturation ($I \ll I_0$):

$$\alpha_p = \alpha_1 [1 + \frac{I}{2I_0} \frac{\Gamma_1 \tau}{(1 + (\Omega - \Omega_0)^2 \tau^2)} - \frac{1}{1 + (\Omega - \Omega_0)^2 T^2)}] \qquad (4)$$

which has the form of a narrow resonant decrease in absorption of width 1/T (Raman-type contribution) superimposed on a broader resonant increase of width $1/\tau$ (population saturation contribution). Also note that this change signal should become sizable at a reduced intensity $I \simeq I_0$, as explained above. Similar conclusions have been reached in a calculation by ARIMONDO and ORRIOLS.[15]

In an experiment designed to test these ideas, a sodium cell was illuminated by two copropagating single mode dye laser beams. One laser, a linearly polarized saturating beam (6 mW/cm²) was locked to the center of one of the hyperfine transitions of the D_1 line and a weaker (3 mW/cm²) circularly polarized probe beam was electronically frequency scanned. Figure 3 shows the results of this experiment which appears <u>qualitatively</u> like a normal sodium Lamb dip experiment. The nine major resonances observed however, are due to several different physical phenomena associated with 3 and 4 level systems.

The resonances labelled 4 and 6 are due to laser interaction with the coupled three level systems formed by the F = 2 and F = 1 hyperfine levels of the $^3P_{1/2}$ excited state and a common lower level, either F = 1 or F = 2,

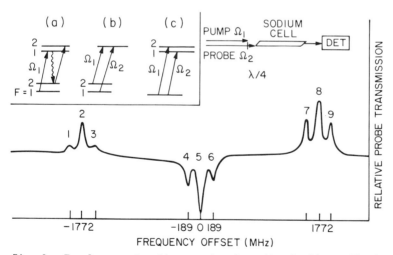

<u>Fig. 3</u>   Two laser saturation spectra in sodium D_1 line. The inset shows the 3 and 4 level systems which contribute to the various resonances.

of the ground state (inset c). These resonances occur at probe frequency offsets of ± 189 MHz, the excited state hyperfine splitting. Their observed widthes, 18 MHz are determined by excited state spontaneous emission (10 MHz) plus power broadening and relative laser jitter.

Features 1, 3, 7 and 9 are 4 level resonances in which the two lasers interact with distinct hyperfine transitions and coupling is due to "branching" radiative decay between the upper level of one laser transition and the lower level of the other (inset a). Hence they are displaced by the ground state hyperfine splitting plus or minus the excited state splitting. These resonances are inverted as are resonance 2 and 8 because of optical pumping.

Features 2 and 8 are due to three level systems of the F = 2 or F = 1 $^2S_{1/2}$ ground state and a common upper level (inset b). They occur at a frequency offset equal to the ground state hyperfine splitting and should exhibit the narrow feature, associated with the ground state levels, described by the second term of Eq. (4). A higher resolution scan (Fig. 4) yields a result with HWHM of 6 MHz when fit to Eq. (4) vs 18 MHz for the broader resonance. The 6 MHz width is primarily due to jitter between the lasers and power broadening. At low power levels the broader resonance was observed without the narrow dip. This result, which disagrees with Eq. (4), is possibly due to relative laser jitter which can obscure a very narrow feature.

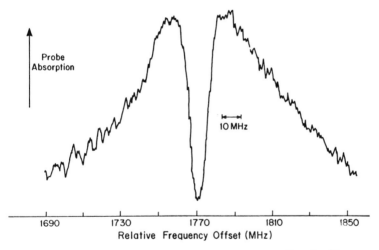

Fig. 4   Laser induced line narrowing in the sodium D$_1$ line.

Feature 5, when the two laser frequencies coincide, is composed of contributions from energy degenerate three and four level systems when the laser fields interact with sets of M$_F$ levels of particular hyperfine transitions. A narrow feature should be observed in this case also, but only one laser is really necessary and no hyperfine spectroscopy information is obtained.

In a closely related experiment we observed this effect as the transmission of a single saturating beam in the same sample cell was studied as

a function of an applied transverse static magnetic field. In this "stimulated level crossing" experiment [16] the width of the narrow component,
determined by the short term laser linewidth, was found to be 0.2 gauss,
corresponding to 350 kHz. Such narrow Hanle effect resonances have been
observed previously in other contexts, for example in an atomic beam
optical pumping experiment by SCHIEDER and WALTHER.[17] The present two
laser experiment, however, allows hyperfine structure spectroscopy with rf
spectral linewidths in the optical saturation spectrum.

5. Conclusion

Laser saturation spectroscopy with optical pumping leads to several novel
effects. Saturation signals of anomalous sign and amplitude dependent on
laser polarization, reduced saturation thresholds, and narrow spectral
features have been described and experimentally demonstrated.These effects,
which must be considered in the analysis of any saturation measurement
where optical pumping is possible, provide a new means for studying
M-changing collisions, transit time effects and other properties of laser
optical pumping. They also make possible ultra-high resolution studies in
special situations and open the possibility of doing saturation spectroscopy in the UV and elsewhere where only weak laser sources are available.
Finally, these effects should be applicable to narrow gamma-ray anisotropy
resonances produced by laser-induced nuclear orientation.[18]

Stimulating discussions with Martin Deutch, Ali Javan, and Hyatt Gibbs
are gratefully acknowledged.

1. G. W. Series, Laser Spectroscopy IV, H. Walther, K. W. Rothe (eds.)
   Springer-Verlag, Heidelberg 1979.
2. A. Kastler, J. de Physique II, 255 (1950).
3. M. S. Feld and D. E. Murnick, Laser Spectroscopy IV, H. Walther, K. W.
   Rothe (eds.) Springer-Verlag, Heidelberg 1979.
4. G. W. Series and W. Gawlick, Laser Spectroscopy IV, H. Walther, K. W.
   Rothe (eds.) Springer-Verlag, Heidelberg 1979.
5. L. Pinard, C. G. Aminoff and F. Laloe, Phys. Rev. A, 1979, in press.
6. V. S. Letokhov and V. P. Chebotaev, Nonlinear Laser Spectroscopy,
   Vol. 4, edited by D. L. McAdam (Springer, Berlin, 1977); M. S. Feld,
   Fundamental and Applied Laser Physics, edited by M. S. Feld, N. A.
   Kurnit and A. Javan (Wiley, New York, 1973), p. 369; and references
   therein.
7. W. Happer, Revs. Mod. Phys. 44, 169 (1972) and references therein.
8. P. G. Pappas, M. M. Burns, D. D. Hinshelwood, M. S. Feld and D. E.
   Murnick, Phys. Rev. A, to be published.
9. W. E. Lamb, Jr., Phys. Rev. A, 134, 1492 (1964).
10. H. R. Schlossberg and A. Javan, Phys. Rev. 150, 267 (1966).
11. T. Hänsch, I. S. Shahin, and A. Schawlow, Phys. Rev. Lett. 27, 707
    (1971); Wieman et al., Phys. Rev. Lett. 36, 1170 (1976); Cahuzac et
    al., Opt. Commun. 20, 111 (1977).
12. See, for example, M. S. Feld in Frontiers in Laser Spectroscopy,
    Vol. 1, edited by R. Balian, S. Haroche and S. Liberman (North Holland,
    Amsterdam, 1977), p. 203.
13. G. E. Notkin, S. G. Rautian, and A. A. Feoktistov, Sov. Phys. JETP
    25, 1112 (1967); M. S. Feld and A. Javan, Phys. Rev. 177, 540 (1969);
    T. Hänsch and P. Toschek, Z. Physik 236, 213 (1970).
14. L. S. Vasilenko, V. P. Chebotaev and A. V. Shishaev, JETP Lett. 12,
    161 (1970); B. Cagnac, G. Grynberg and F. Biraben, J. de Phys. 34,

845 (1973).
15. E. Arimondo and G. Orriols, Lettere al Nuova Cimento 17, 333 (1976).
16. M. S. Feld, A. Sanchez, A. Javan and B. J. Feldman, Proceedings of the Aussois Conferences on High-Resolution Molecular Spectroscopy, Colloques Internationaux de C.N.R.S. 1973, p. 87; and footnote 25 of B. J. Feldman and M. S. Feld, Phys. Rev. A12, 1013 (1975).
17. R. Schieder and H. Walther, Z. Physik 270, 55-58 (1974).
18. M. Burns, P. Pappas, M. S. Feld and D. E. Murnick, Nuc. Inst. and Meth. 141, 429 (1977).

Doppler-Free Intracavity Polarization Spectroscopy

W. Radloff and H.-H. Ritze

Central Institute of Optics and Spectroscopy
Academy of Sciences of GDR
DDR-1199 Berlin, German Democratic Rep.

The high resolution polarization spectroscopy by means of
quasitravelling waves allows to enlarge the signal-to-background ratio of spectroscopic measurements and furthermore
to obtain information about the rotational quantum numbers
of the resonant levels. If it would be possible to apply polarization methods also to standing waves, one could place
the absorption cell inside the laser resonator and combine
on this way the advantages of polarization and intracavity
spectroscopy. That means, however, equal intensities and
equal polarization states for the two counterpropagating waves in contrast to the methods used up to now /1, 2, 3/.

We have found, that this goal can be achieved by means of
elliptically polarized light. Therefore we have treated the
saturation behaviour of an inhomogeneously broadened dipole
transition interacting with two counterpropagating monochromatic waves of the same frequency and intensity polarized elliptically in the same way. By means of a perturbation theory
up to the third order we have calculated the saturated absorption coefficients of the right (+) and left (-) hand circularly polarized components, in which the elliptically polarized wave can be decomposed,

$$\alpha_{\pm} = \alpha_0 \left[1 - \frac{1}{2I_s} \left(I_{\pm} + 2d I_{\mp} \right) \left(1 + \frac{\Gamma^2}{\Gamma^2 + \Omega^2} \right) \right] . \qquad (1)$$

Here are : α_0 – the unsaturated absorption coefficient,
$I_{\pm}$ – the intensities of the two circularly polarized components,
I_s – the saturation intensity for circularly
polarized light,
Γ – the dipole relaxation constant and
Ω – the laser frequency detuning from the molecular transition frequency.

The parameter d can be expressed by

$$
d = \begin{cases} \dfrac{(2J-1)(2J+3)}{2(2J^2+2J+1)} & \longrightarrow \quad 1 \qquad\qquad\qquad J \longleftrightarrow J \\[4mm] & \qquad\qquad\text{(for } J \to \infty) \qquad\qquad (2) \\[2mm] \dfrac{2J^2+3}{2(6J^2-1)} & \longrightarrow \quad \dfrac{1}{6} \qquad\qquad\quad J \longleftrightarrow J-1, \end{cases}
$$

where J is the rotational quantum number of one of the resonant levels. This parameter d is the same as obtained by
Wieman and Hänsch /1/ if we take there equal population relaxation rates (see also /4/).

According to eqs. (1), (2) for elliptically polarized
light ($I_+ \neq I_-$) and if $d \neq 1/2$ we obtain $\alpha_+ \neq \alpha_-$,
i.e., the saturation leads to an ellipse deformation. This
self-induced dichroism effect increases in the centre of the
absorption line because here the resonant molecules interact
with both counterpropagating waves.

Near the centre of the Doppler profile we can also observe
a rotation of the polarization ellipse due to the related
difference of the refractive indexes (self-induced circular
birefringence)

$$
n_+ - n_- = \frac{\alpha_0 c}{4\omega I_s}(I_+ - I_-)(1-2d)\,\frac{\Gamma\Omega}{\Gamma^2+\Omega^2} \; . \qquad (3)
$$

In the following only the induced circular dichroism is
discussed. In order to detect this effect with a high signal-
to-background ratio we use a combination of a quarter wave
plate and an analyzer. Adjusting the optical axis of the
$\lambda/4$ plate parallel to the large axis of the ellipse, we obtain behind the plate linearly polarized light, the direction
of which is determined by the excentricity of the ellipse.
Placing the analyzer perpendicularly to this direction, we define the angle $\theta = 0$. If we tune now the laser frequency
across the line centre, the small deformation of the ellipse
($\Delta I_+/I_+ \ll 1$, $\Delta I_-/I_- \ll 1$) leads (at $\Omega = 0$) to the relative intensity variation behind the analyzer depending on
its position

$$
\frac{I(\theta)}{I_0} \approx \left[1+\frac{\Delta I_+ + \Delta I_-}{I_+ + I_-}\right]\sin^2\theta
$$
$$
+\left[\frac{1}{2}\left(\frac{\Delta I_-}{I_-} - \frac{\Delta I_+}{I_+}\right)\frac{\sqrt{I_+ I_-}}{I_+ + I_-}\right]\sin 2\theta + \frac{I_+ I_-}{(I_+ + I_-)^2}\left[\frac{1}{4}\left(\frac{\Delta I_+}{I_+} - \frac{\Delta I_-}{I_-}\right)^2\right] . \qquad (4)
$$

The first term - dominating at $\theta = \pi/2$ - describes the
usual inverted Lamb dip with the contrast $K = (\Delta I_+ + \Delta I_-)/I_+ + I_-$
on the Doppler background. The second term, which determines
the signal at small angles θ , is the dichroism term; the

quantity $\Delta\varphi = \frac{1}{2}\left(\frac{\Delta I_-}{I_-} - \frac{\Delta I_+}{I_+}\right)\frac{\sqrt{I_-I_+}}{I_-+I_+}$ represents the small angle, about which the linear polarization is rotated behind the $\lambda/4$ plate due to the ellipse deformation. The small quadratic term gives the signal for $\theta = 0$, no background appears in this case.

In order to have a comparison in the first step we have calculated the quantities K and $\Delta\varphi$ for a standing light field in an external absorption cell

$$K = \frac{1+4ad+a^2}{(1+a)^2}\left(\frac{\alpha_0 I_0}{2I_s}\right) = \frac{10}{12}\left(\frac{\alpha_0 I_0}{2I_s}\right)$$

$$\Delta\varphi = \frac{1}{2}\cdot\frac{\sqrt{a}\,(1-a)}{(1+a)^2}(2d-1)\left(\frac{\alpha_0 I_0}{2I_s}\right) = -\frac{1}{12}\left(\frac{\alpha_0 I_0}{2I_s}\right). \tag{5}$$

Here $a \equiv I_-^{(0)}/I_+^{(0)}$ characterizes the excentricity of the ellipse outside the Lamb dip, and $I_0 = I_+ + I_-$ is the total intensity. The results at the right hand of eq. (5) are obtained for the optimum value $a = 3 - 2\sqrt{2}$ and for $d = 1/6$, and may be compared with the results for a pure circularly polarized field ($a = 0$). In the case of high degeneracy ($J \gtrsim 5$) the dichroism term is $1/10$ of the Lamb dip term

$$\Delta I(\theta) \propto \left(\sin^2\theta \mp \frac{1}{10}\sin 2\theta\right). \tag{6}$$

The upper sign holds for P, R branch lines, the lower for Q lines. This fact allows a significant discrimination between lines of the different branches.

Turning now to the intracavity absorption we have to take into account the influence of the active laser medium on the absorption signals. For the corresponding experiments we have used a He-Ne laser at 3.39 μm tunable by Zeeman effect. The power density inside the resonator of this laser is high enough for efficient saturation of carbon-hydrogen molecules /5/.

We have derived the following rate equations for the two circularly polarized components of the elliptically polarized laser field

$$g_+ = \varkappa_0 + \alpha_+ + \Delta\varkappa\left(1 - \sqrt{\frac{I_-}{I_+}}\right) - 2\delta\sin 2\gamma\cdot\sqrt{\frac{I_-}{I_+}}, \tag{7}$$

$$g_- = \varkappa_0 + \alpha_- + \Delta\varkappa\left(1 - \sqrt{\frac{I_+}{I_-}}\right) + 2\delta\sin 2\gamma\cdot\sqrt{\frac{I_+}{I_-}}. \tag{8}$$

The quantities $g_\pm$ represent the saturated gain of the amplifying laser line for the two polarization states, $\varkappa_0$ the isotropic resonator losses and $\alpha_\pm$ the losses due to saturated absorption. The two terms on the right hand of each equation are due to additional optical elements, by means of which the elliptical polarization was performed.

We have to distinguish two cases of laser operation - with and without applied magnetic field.

The first case is characterized by $\delta = 0$ (no phase retardation, see below) and by $\vartheta_- \approx 0$ (if we treat, e.g., the σ_+ Zeeman component, well separated from the σ_- one). The originally circular polarization of the laser field was changed into elliptical polarization with the help of inserted Brewster plates, which mean additional losses $2\Delta\varkappa$ for the field component perpendicularly to the entrance plane of these plates. Due to the fact that the smaller component I_- experiences no amplification, the excentricity of the polarization ellipse is determined alone by eq. (8)

$$a = \frac{I_-^{(0)}}{I_+^{(0)}} = \left(\frac{\Delta\varkappa}{\varkappa_0 + \Delta\varkappa + \alpha_0}\right)^2 . \qquad (9)$$

The shape of the ellipse depends only on the losses and is independent especially on the laser gain and its fluctuations.

By means of eqs. (7), (8) we have calculated for a strongly saturated and homogeneously broadened active medium

$$K = \frac{1 + a(1 + 8ad + 3a - a^2)}{(1+a)^2(1+\sqrt{a})}\tilde{K} . \qquad (10)$$

$$\Delta\varphi = \frac{\sqrt{a}(1-a)(a+2d)}{(1+a)^2(1+\sqrt{a})}\tilde{K} ,$$

where $\tilde{K} = \frac{\alpha_0}{\varkappa_0 + \alpha_0} \cdot \frac{I_0}{2I_S}$ is the well-known expression for the peak contrast, obtained for circularly polarized light ($a = 0$) in usual intracavity Lamb dip spectroscopy. For our experimentally adjusted value $a \approx 0.2$ the θ - dependence of the Lorentzian peak magnitude follows from eqs. (4) and (10)

$$\Delta I(\theta) \propto \begin{cases} \sin^2\theta + 0.12\sin 2\theta & d = \frac{1}{6} \\ \sin^2\theta + 0.27\sin 2\theta & \text{for} \quad d = 1 \end{cases} \qquad (11)$$

In this case we have a positive dichroism term for P, R as well as for Q branch lines in contrast to the result for an external cell. If we look at the relative peak size as a function of the analyzer position θ (Fig. 1), we find nevertheless typical differences between the lines of these different branches and furthermore see the essential enlargement of the relative signals for small θ - values, which is typical for polarization spectroscopy.

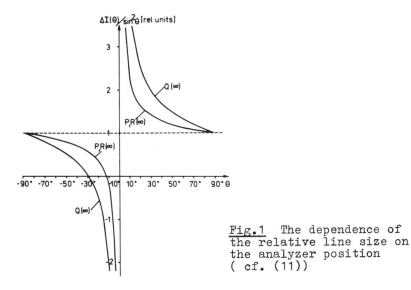

Fig.1 The dependence of the relative line size on the analyzer position (cf. (11))

These features we have clearly detected in the corresponding experiments. In Fig. 2 the scheme of our equipment is represented. Two Brewster quartz plates were inserted in the resonator to achieve the elliptical polarization. The combination of a λ/4 plate and an analyzer allows the detection of the dichroism. By means of the interferometer in front of the detector the unwanted Zeeman component of the laser was suppressed. The absorbing gas was methane enriched by ^{13}C. We use for our investigations the P(6) line of ^{13}CH$_4$ at -940 MHz with respect to the well-known ^{12}CH$_4$ line at zero magnetic field. The relative strength of the inverted Lamb dip at a pressure of 8 mTorr was 8%, measured at θ = 90°.

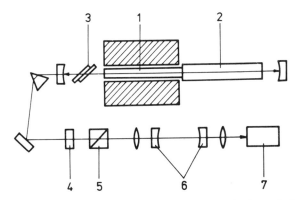

Fig.2 Experimental arrangement·for detection of intracavity polarization effects (1 - laser tube, 2 - absorber cell, 3 - polarization plates, 4 - phase retarder, 5 - analyzer, 6 - interferometer, 7 - InSb detector)

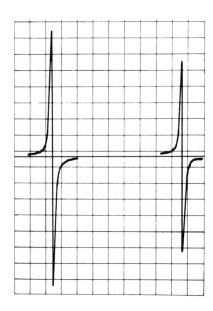

Fig.3 Derivative nonlinear absorption signals of a $^{13}CH_4$ line (P(6) transition of the ν_3 mode) for symmetric analyzer positions

θ=+50°   θ=-50°

When we detect this peak at symmetrical analyzer positions $\theta = \pm 50°$, we found exactly the same background intensity for both positions, but strongly different Lorentzian signals. In Fig. 3 the derivative signals obtained by phase sensitive detection are represented for both angles. The measured ratio of both signal sizes is nearly 1.4, which is somewhat smaller than the theoretical value 1.5 .

At smaller angles θ we have observed a further improvement of the signal-to-background ratio in comparison to conventional Lamb dip spectroscopy.

By means of this example we have clearly demonstrated the dichroism effect for a Zeeman-tuned laser. But generally tunable lasers are not circularly but mostly linearly polarized. Therefore we have studied also a corresponding example at zero magnetic field, where the laser field in our experimental arrangement (see Fig. 2) is linearly polarized. Inserting now additionally a birefringent plate with a phase retardation δ ($\delta \ll 1$) and an angle γ between its optical axis and the polarization direction (see eqs. (7), (8)) we achieve elliptical polarization also for this case. Assuming $g_+ \approx g_-$ we have calculated the relevant quantities

$$K = \left(\mathscr{H}_0 + \alpha_0 + \Delta \mathscr{H} \frac{(1-\sqrt{a})^2}{1+a} \right)^{-1} \cdot \frac{(1+2d)}{2} \cdot \frac{\alpha_0 I_0}{2 I_s}$$

$$\Delta \varphi = \frac{(1-a)}{4 \Delta \mathscr{H}(1+a)} (2d-1) \cdot \frac{\alpha_0 I_0}{2 I_s} .$$

$$(12)$$

We see, that the dichroism term changes the sign for P,R
and Q branch lines, respectively, like in the case of an external absorption cell. Our first preliminary measurements
on the well-known P(7) line of $^{12}CH_4$ confirm this result.
We have detected at $\theta = -50°$ a larger Lorentzian peak
than at $\theta = +50°$ in contrast to the case with an applied
magnetic field, which we have discussed before.

On this way we have clearly recognized that also for laser
types, which have originally linearly polarized light, intracavity polarization spectroscopy with elliptically polarized
light can be carried out.

These results have encouraged us to introduce a tunable
colour centre laser in this field of experiments. We have
built up recently such a laser with relatively well properties on the basis of crystals obtained from the group in
Hannover. Now we investigate the intracavity absorption of
suitable molecules in the tuning range of the laser between
2.5 and 2.9 μm in order to study the conditions for application of our new technique, the intracavity polarization
spectroscopy. with elliptically polarized light.

1. C. Wieman and T.W. Hänsch, Phys.Rev.Lett. 36, 1170 (1976)

2. V. Stert, R. Fischer, E. Meisel and H.-H. Ritze, Kvantovaya Elektronika (Sov. Journal) 4, 2620 (1977)

3. J.C. Keller and C. Delsart, Opt.Commun. 20, 147 (1977)

4. S. Saikan, J.Opt.Soc.Am. 68, 1184 (1978)

5. W. Radloff, Kvantovaya Elektronika 5, 2358 (1978)

Forward Scattering and Polarization Spectroscopy

W. Gawlik[1] and G.W. Series

J.J. Thomson Physical Laboratory, The University of Reading, Whiteknights
Reading RG6 2AF, U.K.

We recall some of the essential points concerning the forward scattering of
radiation through a gas and show how the treatment that proved convenient
for the analysis of optical pumping experiments can be used also for polarization spectroscopy with lasers. We distinguish between polarization spectroscopy as a saturation phenomenon and as a pumping phenomenon – the latter
being realisable when the lower state is not single – and demonstrate how
the long relaxation time of ground states allows polarization spectroscopy
to be carried out with very weak light beams. We investigate some of the
consequences of using strong beams: the decoupling of the hyperfine structure by the light and the manifestation of the Autler-Townes effect.

1. Forward Scattering

Two important points to notice about the forward scattering of monochromatic
light through a gas are (i) the frequency of the forward scattered light is
not Doppler-shifted with respect to the incident light, and (ii) the forward-
scattered optical paths, for scattering from atoms in random positions, are
all equal (see Fig.1).

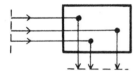
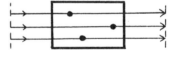

Fig.1 Lateral scattering: random   Forward scattering: equal paths.
      paths between reference planes.

These points, together with the principle of indistinguishability applied
to atoms which return to their original state after the scattering process,
secure that the forward-scattered light from different atoms is mutually
coherent, and is coherent with the exciting light.

The properties of the atoms – which is our concern as spectroscopists –
are reflected in the amplitude of the scattered light. To separate the forward-scattered light from the primary beam two methods have been used:
(i) interferometric method based on beam splitting, which cancels the primary
beam by introducing a controlled phase difference in that beam which does

---
[1] permanent address: Instytut Fizyki, Uniwersytetu Jagiellonskiego, ul.
Reymonta 4, 30-059 Krakow, Poland.

not traverse the sample [1, 2, 3], (ii) polarization method (see figure 2) where the primary beam is rejected by a crossed polarizer [4, 5, 6, 7, 8].

With either technique it is sometimes useful to allow a small fraction of the primary light to be transmitted to heterodyne the scattered light. If all the primary light is transmitted the antiphase component of the scattered light appears as absorption, the quadrature component as dispersion.

Method (ii) will form the basis of our analysis. Although the scattered light is coherent with the primary light, it must be of different polarization in order to pass the analyser. It is, therefore, of the essence of this work that the scattering atoms introduce some asymmetry into the scattering process: the symmetry of the perturbation acting on the atoms must be different from the symmetry of the probe beam. While many different symmetry-breaking perturbations may be envisaged, we shall be concerned here with two: (i) external, longitudinal magnetic fields, and (ii) a counter-propagating 'pumping' beam, circularly polarized. When this is the same frequency as the probe beam there exists, of course, a velocity-selective interaction leading to narrow resonances. This is the basis of Doppler-free polarization spectroscopy [9]. We shall consider the two perturbations acting separately, and together.

1.1 Calculation of the Scattered Intensity

The object of the analysis is to relate the intensity of light which passes the crossed analyser (the signal) to the polarizability tensor of the medium, which is obtained by summing the polarizabilities of the atoms in a macroscopic volume. It is in the calculation of these polarizabilities that one finds the differences between optical pumping and laser saturation spectroscopy. The beam illustrated in figure 2 constitutes a probe beam. In simple cases it is so weak that its interaction with the atoms is linear, but we have found that non-linear interactions must also be reckoned with.

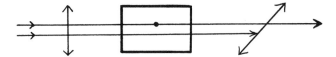

Fig.2 The primary beam is rejected but the scattered beam is transmitted if the polarization has been changed in the process of scattering.

Let the bulk medium be characterized by a polarizability tensor $\hat{\chi}$ or refractive indices $\eta_\pm = \eta_\pm^r + i\eta_\pm^i$. We describe the tensor in a spherical basis with the z-axis in the direction of propagation of the light. Suffixes $\pm$ on η refer to circularly polarized light. For small optical depth the intensity transmitted by the analyzer in Fig.2 is

$$I = \tfrac{1}{4}I_0\ (\omega L/c)^2\ \left[(\eta_+^r - \eta_-^r)^2 + (\eta_+^i - \eta_-^i)^2\right],$$

where I_0 is the intensity of the incident light and L is the length of the sample. In terms of the components of χ, this becomes

$$I = \pi^2 I_0\ (\omega L/c)^2\ \{\left[(\chi_{++}^r - \chi_{--}^r) - (\chi_{+-}^r - \chi_{-+}^r)\right]^2$$
$$+ \left[(\chi_{++}^i - \chi_{--}^i) - (\chi_{+-}^i - \chi_{-+}^i)\right]^2\}. \tag{1}$$

211

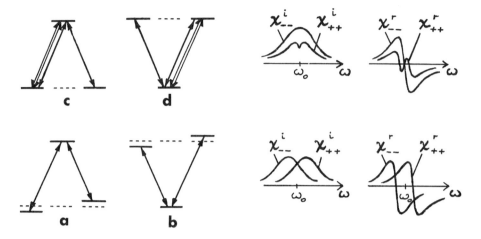

Fig.3 Three-level system interacting with a probe beam. In cases (a) and (b) there is a displacement of levels induced by an external electric or magnetic field. In cases (c) and (d) the levels remain degenerate but are unequally populated through the operation of a pumping beam (double lines). On the right of the transition diagrams are sketched the real and imaginary parts of the corresponding polarizability tensor, as a function of frequency.

$\hat{\chi}$ will depend on the intensity and polarization of the pumping beam and, in the case of non-linear interactions with the probe beam, may be of rank higher than two, in which case the necessary summations must be carried out to contract it to a rank-2 tensor.

In contrast to saturated absorption spectroscopy, where a 2-level system provides an adequate model, the simplest atomic structure which will allow a discussion of the necessary asymmetries in polarization spectroscopy is a three-level structure of either of the types shown in Figs. 3(a) and (b) or (c) and (d).

2. Forward-Scattering Perturbed by a Magnetic Field

In this section we treat the propagation of the probe beam in the presence of a longitudinal magnetic field (Fig.4).

If the probe beam is weak the $\chi_{\pm\mp}$ are zero and $I \propto |\chi_{++}-\chi_{--}|^2$. These functions are Doppler-broadened, and $I(B)$ is a Doppler-broadened Hanle curve [4]. If, however, the probe beam is strong enough to induce stimulated transitions between upper and lower levels (cases (a) and (b) of Fig.3), the

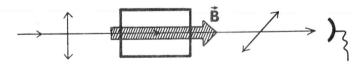

Fig.4 The forward-scattered signal attributable to the magnetic field $\underset{\sim}{B}$ is recorded by the detector.

induced coherence between the lower levels or upper levels, as the case may be ('Zeeman coherence') generates non-vanishing values of $\chi_{\pm\mp}$ [6, 10, 11]. (Coherence of this kind may also be induced by transverse r.f. fields [4].) This Zeeman coherence gives rise to narrow resonances around B = 0 superimposed on the broad background (see inset to Fig.10). The width of these resonances is determined by the relaxation constants of the levels involved.

3. Forward-Scattering Perturbed by a Counter-Propagating Pumping Beam

To induce the anisotropy in $\hat{\chi}$ the polarization of the pumping beam needs to be different from that of the probe beam (Fig.5).

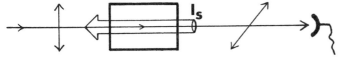

Fig.5 The forward-scattered signal attributable to the pumping beam I_s is recorded by the detector.

We consider the case when the pumping beam is circularly polarized, although other polarizations have been investigated [9, 12, 13]. The frequency of the pumping beam is taken to be the same as that of the probe beam.

The characteristics of the observed signals depend on the intensities of the two beams in relation to the relaxation rates of the levels involved. We use Γ, γ for the relaxation rates of the upper and lower levels, respectively, in the absence of light, and $\Gamma_+ = \Gamma/2 + aI_s$ for the power-broadened width of the optical transition. We use also the saturation parameters $G_\gamma = (E_s \cdot D/\hbar\gamma)^2$ and $G_\Gamma = (E_s \cdot D/\hbar\Gamma)^2$. E_s, I_s relate to the pumping beam.

It is not difficult to appreciate that, when the pumping beam is weak in the sense that stimulated emission is negligible ($G_\Gamma \ll 1$), the induced anisotropy depends solely on the difference between the populations, $N_\pm$, of the sub-levels illustrated in Fig.3(c) or (d). The mechanisms responsible for maintaining these population differences are, however, quite different in the two cases and will be discussed further below. The signal detected by a weak probe beam is proportional to $(N_+-N_-)^2$ in both cases. Detailed calculation yields

$$I \propto G_\gamma^2 \frac{\Gamma_+^2}{4\delta^2 + \Gamma_+^2} \qquad \text{for the } \Lambda\text{-system (Fig.3c),} \qquad (2)$$

$$\text{and } I \propto G_\Gamma^2 \frac{\Gamma_+^2}{4\delta^2 + \Gamma_+^2} \qquad \text{for the } V\text{-system (Fig.3d).} \qquad (3)$$

Here, δ is the frequency-offset from resonance. Notice that the equations predict a line-width of Γ_+, which tends to $\frac{1}{2}\Gamma$ in the limit of weak pumping fields. This contrasts favourably with the value 2Γ, the limiting line-width in saturation spectroscopy.

It is more especially important to notice the factors G_γ^2, G_Γ^2, different in the two cases. The significance is that, in the frequently-encountered situation $\gamma \ll \Gamma$, the Doppler-free Lorentzian signals represented by (2) and

(3) are achieved with much weaker pumping beam intensities for the Λ-system than for the V-system. In the Λ-system the main effect of the pumping beam is to re-distribute the population between the sub-levels of the ground state - optical pumping. The dynamics of the re-distribution is determined by G_γ, the ratio between the perturbation due to the pumping beam and the relaxation of the lower states.

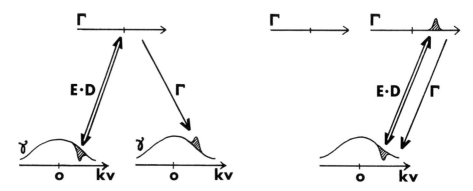

Fig.6 Population difference to be sustained against relaxation from the ground states.

Fig.7 Population difference to be sustained against relaxation from the excited state.

In the V-system, on the other hand, the population difference on which the signal depends is between the upper state sub-levels. To sustain this population difference the pump beam must work against the relaxation of the upper states. If this is faster than that of the lower state, the intensity of the pumping beam needs to be correspondingly greater.

We distinguish the two mechanisms by the terms 'paramagnetic' and 'dia-magnetic'. 'Paramagnetic polarization spectroscopy' (PPS) is represented by the Λ-system, 'diamagnetic polarization spectroscopy' (DPS) by the V-system. In PPS the signal depends on the existence of lower state struc-tures, in DPS, on excited state structures.

In general, we shall find structure both in the upper and in the lower states, so that, in principle, we might expect both mechanisms to contribute to the signal. However, since PPS generally requires far weaker pump inten-sities, this will generally be the dominant mechanism. DPS will be the operative mechanism when J = 0 in the lower state, or, with J ≠ 0, when the population difference is destroyed by additional perturbations (r.f. fields, field gradients, depolarizing collisions). Some early analyses of polariza-tion spectroscopy [9, 14] were carried through under the approximation $G_\Gamma \gg 1$. This represents, effectively, the pure DPS case. Our conclusion is that analyses of this type are not representative of polarization spectroscopy. It will generally be the case that the appropriate mechanism for the inter-pretation of the signal will be PPS. Our analysis has been concerned with the dispersive as well as with the absorptive components of $\hat{\chi}$. Saturated absorption spectroscopy concerns itself with the absorptive components only. The possibility that optical pumping effects might be important in this kind of work has recently been pointed out also by other investigators [15, 16].

4. Experimental Studies

We have studied forward-scattering and polarization spectroscopy in sodium vapour in the neighbourhood of the D-lines, which offer themselves as examples of the general case, having Zeeman structure in both upper and lower levels. The vapour, without buffer gas, was contained in a closed cell with strain-free windows at a vapour density of 10^{11} cm^{-3}. Stray magnetic fields were cancelled to within a few mG. Both probe and pumping beams were taken from a single-mode, tunable dye laser, line-width about 10 MHz. The probe beam followed a path through the vapour in exact opposition to the pumping beam, constrained by diaphragms to travel along the inner region of the pumping beam where the intensity was uniform. Crystal polarizers of high quality were used: a Babinet-Soleil compensator was employed to secure exact circular polarization of the pumping beam. For most experiments the polarizer and analyzer in the probe beam were exactly crossed to avoid heterodyning the forward-scattered light with the probe beam itself. No modulation was employed. The signal was recorded with a photo-multiplier.

4.1 Polarization Spectrum with Weak Beams

A high-resolution polarization spectrum of the NaD$_1$ line is shown in Fig.8. Of the nine resonances shown, four are the allowed transitions between the hyperfine levels F = 1, 2 in the lower and upper states; the remaining resonances are 'crossovers'. The pump intensity indicated in the legend corresponds to $G_\Gamma \approx 0.1$, $G_\gamma \approx 100$, so that the DPS contribution to the signals is negligible. The probe intensity was very much weaker.

4.2 Intensity of the Probe Beam

If the purpose of the probe beam is to monitor the asymmetry introduced by the pumping beam, its intensity must be much less. In Fig.9a it is 50 times weaker, and the resonances appear as simple near-Lorentzians. In Fig.9b the intensity of the probe beam has been increased by a factor 50 (though it is still very weak by the standards of saturation spectroscopy). Structure appears in the spectrum, and the relation between signal strength and beam intensity is strongly non-linear. These effects are due to the pumping action

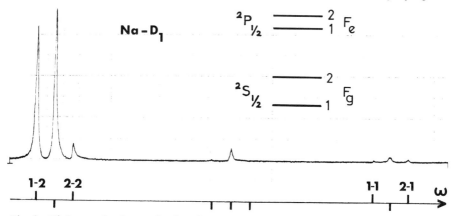

Fig.8 High resolution polarization spectrum of Na D$_1$ line. The intensity of the pumping beam was 25 μW mm^{-2}. On the frequency scale below, hyperfine components are marked F_e-F_g. Crossovers are marked below the line.

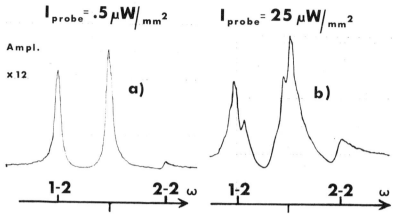

Fig.9 The effect of increasing the intensity of the probe beam.

of the probe beam: it tends to equalise the populations of levels of the same $|m|$, and in this way acts in opposition to the circularly-polarized pumping beam.

The strong probe beam, however, also induces coherence between Zeeman levels [6, 11], and hence generates non-vanishing components χ_{+-}. The resulting contributions to the signal are zero if the levels are degenerate, but not if the degeneracy is removed by a magnetic field. This will be the subject of the next section.

An analysis of the signal to be expected for the Λ-system of Fig.3(c) yields the result

$$I \propto G'_\gamma \; G^2_\gamma \left[\frac{1}{4\delta^2 + \Gamma^2_+}\right] \left[1 - CG'_\gamma \; \frac{1}{4\delta^2 + \Gamma^2_+}\right]$$

where $G'_\gamma = (E_{probe} \cdot D/\hbar\gamma)^2$ and C is a constant. This reproduces the important features of Fig.9b, the dip in the line, and the non-linear dependence on the intensity of the probe beam.

4.3 Interference Between Polarization and Alignment in the Presence of a Longitudinal Magnetic Field

A strong probe beam ($G_\gamma \approx 1$: transverse linear polarization) propagating alone in the vapour generates (through stimulated emission) coherences between Zeeman levels of the ground state. Though the intensity transmitted through a crossed analyzer is zero if no other perturbations are applied to the vapour, a signal, illustrated in the inset to Fig.10, does appear when a longitudinal magnetic field is applied [6]. It should be noticed that this signal is narrow in relation to the Doppler width (about 500 G). In Fig.10c we see the signal, at fixed magnetic field, as a function of frequency. There is no velocity-selection: the signal is Doppler-broadened. The narrow polarization spectroscopy signal induced by a counter-propagating pumping beam with zero magnetic field is seen in Fig.10b: in this experiment the probe beam was not negligibly weak, though not as strong as in Fig.9b.

216

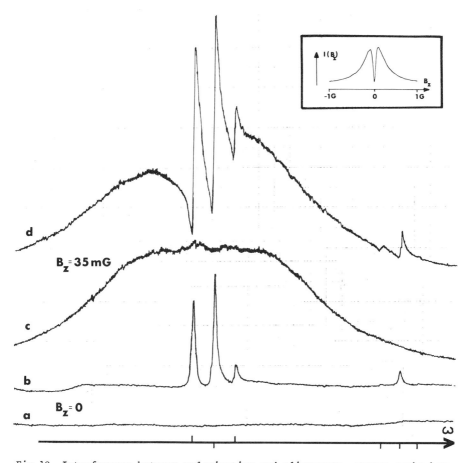

Fig.10 Interference between polarization and alignment: strong probe beam.
a) No magnetic field. No pumping beam. Null signal recorded on probe beam.
b) No magnetic field. Pumping beam. Polarization signal.
c) Magnetic field. No pumping beam. Alignment signal.
d) Magnetic field and pumping beam simultaneously applied.
In the inset is shown the dependence of probe signal on magnetic field for
fixed laser frequency near resonance [6].

   In Fig.10d we see the interference between the polarization induced by
the pumping beam and the alignment induced by the strong probe beam in the
presence of the very weak longitudinal magnetic field. It is important to
appreciate that effects of this kind can manifest themselves in polarization
spectroscopy.

4.4 Signal Attributable to Ground-State Population Differences: Direct Demonstration

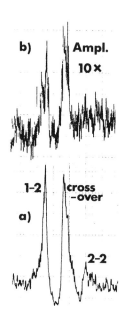

In Fig.11a is to be seen a polarization spectrum taken with weak pumping and probe beams. The signal persists, as shown in Fig.11b, when the beams are spatially separated. Since the time which the atoms would have taken, a few μs, to have travelled the distance between the beams is very much longer than the lifetime of the excited state, it is clear that the signals cannot be attributed to any population of the excited state resulting from the operation of the pumping beam. They must result from a net polarization of the ground state.

Fig.11 a) Pumping and probe beams coincident.
b) Probe beam displaced laterally by 2 mm.

4.5 Coherent Background: Transition from Dispersive to Absorptive Signal

In Fig.12 is to be seen the effect of adding a coherent background to the polarization signal, a device used to increase the signal strength by heterodyning [10]. It can be achieved by rotation of the analyzer with respect to the polarizer in the probe beam. When polarizer and analyzer are crossed (0° in the Fig.) the signal is due to the modulus of $\hat{\chi}$, differential dispersion and absorption. With rotation of the analyzer by a few degress one sees the effect of interference between the signal and a transmitted component of the primary beam. The result is a series of dispersion-shaped resonances. Differential absorption now plays a minor role. At a rotation angle of 90° (analyzer parallel to polarizer), only that component of the signal is seen, the absorptive component of $\hat{\chi}$, which is in phase or antiphase with the strong, primary beam. The variations of the dispersive component, being in quadrature, result in this case in variations of the net scattered field which contribute negligibly to the detected signal.

The observation of Doppler-free, absorptive signals, in circumstances (as above) when the pumping beam is too weak to saturate the transition indicates again that it is inequalities of population in the ground, rather than in the excited state which are responsible for the observed effects (see also [16]).

218

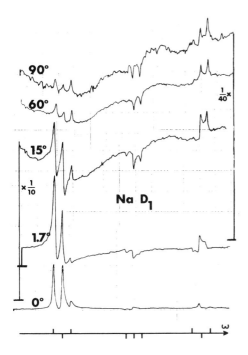

Fig.12 Polarization spectrum
of Na D_1 line with coherent
background introduced by rotation of the analyzer in the
probe beam. The angle 0° represents analyzer crossed with
polarizer. The intensities of
the pumping and probe beams
were unchanged throughout.

4.6 Decoupling of Hyperfine Structure by the Pumping Beam, and Autler-Townes Effect

Finally, we report the effect of increasing the intensity of the pumping
beam. In Fig.13 B is to be seen a resolved hyperfine structure under a
moderately weak pumping beam ($G_\Gamma = (E_s.D/\hbar\Gamma)^2 = 0.6$). In part A of the figure the spectra are shown for increasing values of G_Γ up to 130. It is to
be noticed that the first effect of increasing the intensity is that the
lines are broadened until the resolution is lost: the interaction with the
pumping beam has de-coupled the hyperfine structure.

At the higher intensities a new structure develops; a doubling of the
line which becomes more symmetrical. The peaks move apart as G_Γ is increased.
We believe this is a manifestation of Autler-Townes effect. In the inset,
C, we have plotted the observed separation of the peaks as a function of
$(I_s)^{\frac{1}{2}}$. The broken line on the same graph is a plot of the Rabi frequency of
the strongest Zeeman hyperfine component, $F_e = 3$, $m = 3 \to F_g = 2$, $m = 2$.
The observed signals are a blend of all the Zeeman hyperfine components of
the D_2 line which originate from the $F = 2$ ground level, which have a variety
of Rabi frequencies. Moreover, averaging over velocities leads to a more
complicated situation than in the case of simple resonances. For these
reasons we do not expect the experimental points in Fig.13C to lie on the
broken line, but we interpret the fact that they lie near the line as evidence
that the observed splitting is a manifestation of the Autler-Townes effect.

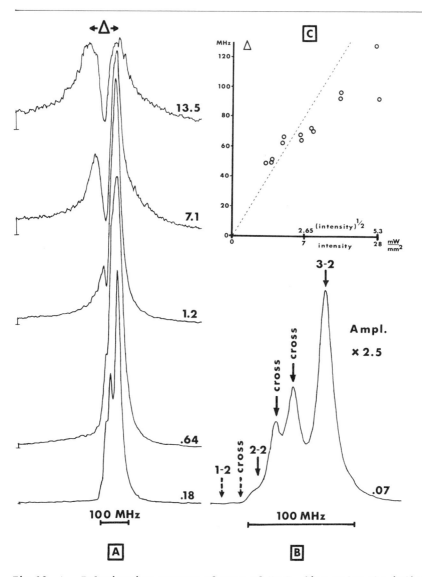

Fig.13 A. Polarization spectra of part of Na D₂ line under the indicated pump beam intensities (mW mm⁻²).

B. The hyperfine structure (F_e-F_g) shown resolved at the lowest intensity.

C. Separation of peaks observed at high intensity plotted against $I_s^{\frac{1}{2}}$ (see text).

5. Conclusions

We have investigated polarization spectroscopy from the point of view adopted in previous studies of the forward-scattering of resonance radiation. We are convinced of the importance, in the interpretation of the signals, of lower-state population changes arising from optical pumping. We regard polarization spectroscopy as a velocity-selective optical pumping technique, rather than as a variant of saturated absorption spectroscopy. We have demonstrated the complications that may arise from too strong a probe beam, or pumping beam, or from a longitudinal magnetic field. (We have observed, but not reported here, the destruction of the signals by static and r.f. transverse magnetic fields.) We believe polarization spectroscopy to be a most sensitive and delicate technique, but it is our experience that the signals need to be interpreted with especially great care and circumspection.

References

1. R.L. Fork and L.C. Bradley, Appl. Opt. $\underline{3}$, 137 (1964).
2. C. Bordé, G. Camy and B. Decomps, "Laser Spectroscopy II", Springer, Berlin, Heidelberg, N. York (1975).
3. R. Schieder, Opt. Commun. $\underline{26}$, 113 (1978).
4. A. Corney, B.P. Kibble and G.W. Series, Proc. Roy. Soc. $\underline{293A}$, 70 (1966).
5. A.V. Durrant and B. Landheer, J. Phys. B $\underline{4}$, 1200 (1971).
6. W. Gawlik, J. Kowalski, R. Neumann and F. Träger, Opt. Commun. $\underline{12}$, 400 (1974).
7. W. Winiarczyk, Acta Phys. Polon. $\underline{A52}$, 157 (1977).
8. L.N. Novikov, Opt. Spektrosk. $\underline{24}$, 866 (1968); Opt. Spectrosc. $\underline{24}$, 465 (1968).
9. C. Wieman, and T.W. Hänsch, Phys. Rev. Lett. $\underline{36}$, 1170 (1976).
10. G.W. Series, Proc. Phys. Soc. $\underline{88}$, 995 (1966).
11. W. Gawlik, J. Phys. B $\underline{10}$, 2561 (1977).
12. V. Stert and R. Fischer, Appl. Phys. $\underline{17}$ 151 (1978).
13. J.C. Keller and C. Delsart, Opt. Commun. $\underline{20}$, 147 (1977).
14. M. Sargent, Phys. Rev. $\underline{A14}$, 524 (1976).
15. M. Pinard, C.G. Aminoff and F. Laloë, to be published in Phys. Rev. A. (Feb. 1979).
16. M.S. Feld and D.E. Murnick, Proceedings of this Conference.

Spectroscopic Studies into Elastic Scattering of Excited Particles

S.N. Bagayev

Institute of Thermophysics,
Siberian Branch of the USSR Academy of Sciences
Novosibirsk-90, USSR

Up to now much information on potentials of the interaction of colliding particles and on elastic scattering cross-sections was gained from studies of atomic and molecular beams [1]. The use of narrow optical resonances that arise in the nonlinear interaction of optical fields with a gas enabled one to study mechanisms of collisions of excited particles in atomic and molecular low-pressure gases [2]. It has become possible to study scattering of short-lived molecules. Information on scattering cross-sections and potentials of interaction is gained from studies into the shape of nonlinear resonances, from measurements of their broadening and shift due to collisions.

Recent progress in obtaining narrow resonances of the width of 10^3-10^5 on vibrational-rotational transitions (VRT) of molecules permits precision measurements of the influence of elastic collisions in scattering at small angles and gain of information on scattering cross-sections.

A long lifetime of excited molecules on VRT gives rise to a specific character of the influence of collisions on the nature of saturation and the narrow resonance shape [3, 4]. Since a radiative width on the VRT of molecules is small, the width of an optical resonance is determined by collisions and transit effects even with very low pressures. Ref.[4] paid attention to peculiarities in broadening of narrow resonances due to the influence of spatial inhomogeneity of the field in a cavity caused by sphericity of a light wave and a Gaussian profile of the field along the beam cross-section. In [5, 6] we reported on new qualitative peculiarities of the behaviour of collisional broadening and shift of narrow resonances in a low-pressure gas. The collisional broadening and shift nonlinearly depended on the gas density, which was explained by the influence of elastic particle scattering at small angles. Recently, the nonlinear dependences of collisional resonance broadening in a low-pressure gas were observed in NH_3 [7], J_2 [8], CO_2 [9, 10], and Xe [11].

Theoretical studies into the influence of collisions on the resonance shape carried out earlier were based on the use of phenomenological constants that characterized the collision frequency. These studies could not explain the observed nonlinear dependence of resonance width and shift on pressure. The first strict theoretical studies using kinetic equations for a density matrix where the terms of arrival and departure are expressed through scattering amplitudes on the upper and lower levels have been made in [12, 13]. The solution of these equations in the standing-wave field enables us to find the expression for a nonlinear resonance and, consequently, to associate the resonance characteristics with scattering cross-sections. The results of [12-14] have explained the nonlinear dependence of broadening and shift on pressure.

The report presents the results of experiments carried out at the Institute of Thermophysics of the Siberian Branch of the USSR Academy of Sciences on studying the influence of elastic particle collisions on the shape and position of narrow resonances in molecular low-pressure gases. We discuss the experiments carried out earlier and the new results on observation of the nonlinear dependence of broadening and shift of narrow resonances at the VRT of CH_4 and CO_2 molecules. The results of investigations of collisional relaxation in SF_6 by the method of unidirectional and oppositely traveling waves are presented. The recent results on direct observation of the influence of elastic scattering of excited molecules at small angles of about 1° on the shape of resonances in methane are stated.

1. Peculiarities of Broadening of Narrow Resonances on Molecular VRT

Let us consider the physical picture of the influence of collisions on a Lamb dip shape in molecular low-pressure gases. The formation of the Lamb dip is mostly determined by the atoms whose Doppler frequency shift kv_z (v_z is the velocity projection onto the light wave propagation) is of an order of the homogeneous width 2Γ. The dip broadening arises from collisions that either destroy an excited atom (phase randomization, quenching) or change its velocity by a value larger than $v_z = \Gamma/k$. Since at collisions an electron state does not change for the molecular VRT, the scattering amplitudes on the upper and lower levels may be considered to be little different and there is practically no phase randomization. If the scattering amplitudes on the upper and lower levels are equal, the particles deflect along the same trajectory, coherence between levels is conserved (no phase randomization of the oscillator) and only the particle velocity is varied. In this case the resonance is broadened due to the Doppler frequency shift at collisions $kv\bar{\theta}$ (v is an average thermal particle velocity) at a typical angle $\bar{\theta}$. The character of the influence of these collisions depends on the ratio between the homogeneous resonance width 2Γ and the Doppler frequency shift $kv\theta$. When the Doppler frequency shift $kv\theta$ at the elastic scattering is more than Γ, after collision the molecules go out of the region of interaction with the field and contribute to the resonance broadening. In this range of pressures a linewidth is determined by the total scattering cross-section

$$2\Gamma = N\upsilon\,(\sigma_e + \sigma_i) + \gamma_o, \tag{1}$$

γ_o is the resonance width with no collisions, N is the density of scattering particles, σ_e and σ_i are the total cross-sections of elastic and inelastic scattering, respectively. With an increase of gas pressure the homogeneous linewidth is increased and the situation has set in where the homogeneous width may be more than the Doppler shift ($\Gamma \gg kv\bar{\theta}$). In this case after collision the particle continues interacting with the field, and the elastic collisions result in no resonance broadening. The resonance broadening is determined only by phase-randomizing and quenching collisions:

$$2\Gamma = N\upsilon\sigma_i + \gamma_o. \tag{2}$$

With $\Gamma \sim kv\bar{\theta}$ only a part of particles scattered at the angles $\theta > \Gamma/kv$ contributes to the resonance width.

Thus, with the elastic collisions and with no phase randomization there is a considerable difference in the resonance broadenings at low ($\Gamma \ll kv\bar{\theta}$) and high ($\Gamma \gg kv\bar{\theta}$) pressures.

2. Measurement of Broadening of Narrow Resonances

In 1972 we observed for the first time the nonlinear dependence of broadening of the Lamb dip on the $F_2^{(2)}$ line in methane ($f_2^{(2)} = 3.39$ µm) by using a He-Ne laser with an internal absorption methane cell [5].

Figure 1 shows the experimental installation. We used two similar He-Ne/CH$_4$ lasers. The cavity length of each laser was 200 cm, the absorption cell length 50 cm. The range of methane pressures under investigation was 1-25 mTorr. The radiation frequency of laser 1 was stabilized to the resonance in methane.

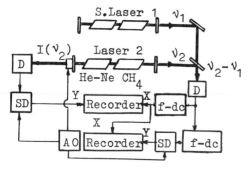

Fig.1 Experimental installation for measuring resonance broadening in methane: SD - synchronous detector, AO - audio oscillator

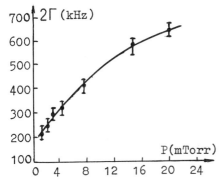

Fig.2 Dependence of the resonance width in methane at
$\lambda = 3.39$ μm on pressure

The resonance was recorded in a recorder with the dispersion
curve of the generation power peak by modulating the radiation
frequency of laser 2. The system enabled us to eliminate the
influence of nonlinear frequency pulling on the form of the
recorded curve. The frequency pulling to the absorption line
center results in nonlinear dependence of the lasing frequency
on the frequency of the cavity when changing its length. So
the dispersion curve was recorded as a function of variation
of the difference frequency of the lasers. At the same time
the change of frequency deviation amplitude of laser 2 near
the absorption line center was registered. The values of the
resonance widths for each pressure were obtained by extrapolating to zero values of the field in the cavity for the purpose
of eliminating the effects of field broadening.

Figure 2 depicts the dependence of the resonance width in
methane on pressure. In the pressure range of 1-5 mTorr the
curve slope is 30 ± 2 MHz/Torr. With an increase of pressure the
slope is decreased and with a pressure of about 20 mTorr is
about 10 ± 5 MHz/Torr. The difference in the curve slopes at low
and high pressures evidences a great role of elastic scattering
without phase randomization. According to the data of [15] 90%
of methane molecules are scattered at the angle θ exceeding
10^{-2} rad. The corresponding Doppler shift $kv\theta$ is about 1 MHz.
So in the pressure range of 1-5 mTorr the resonance width is
$\Gamma \ll kv\theta$, and almost all atoms contribute to the resonance
broadening. With the pressure increase, a part of atoms begins
scattering at the angles $\theta \sim 10^{-2}$ rad with no loss of coherence,
and the curve slope is decreased. The curve $\Gamma(p)$ bends at the
width of about 600 kHz, which corresponds to the typical scat-
tering angle $\bar{\theta} \sim 2 \cdot 10^{-3}$ rad. The difference in the slopes of the
curve $\Gamma(p)$ permits determination of the elastic scattering
cross-section in methane that is $\sigma_e = (4.5 \pm 0.5) \cdot 10^{-14}$ cm^2.

The similar results were obtained when studying the behaviour of the resonance width in CO_2 at the P(20) transition of
the 00°1-10°0 band ($\lambda = 10.6$ μm) against an absorbing gas pressure [10]. The investigations were carried out with a CO_2 laser

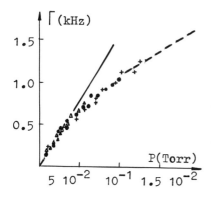

Fig.3 Dependence of the resonance halfwidth in CO_2 on pressure
at the P(20) transition of the
$00°1-10°0$ band. Experimental
points have been obtained at
temperatures Δ - 690°K,
● - 590°K, + - 470°K

with an external CO_2 absorption cell. The experimental scheme
is similar to that described above.

Figure 3 shows the dependence of the resonance halfwidth on
pressure. The nonlinear character of the dependence is well defined. The curve slope in the range of low pressures is 2.7
times as large as that in the range of high pressures ($13.0{\pm}2$
MHz/Torr and $4.8{\pm}0.9$ MHz/Torr). The total elastic scattering
cross-section calculated from the obtained dependence amounts
to $4.5 \cdot 10^{-14}$ cm^2, the typical scattering angle is $\bar{\theta} \sim 10^{-2}$rad.
An average variation of the particle velocity in scattering is
$\upsilon\bar{\theta} \simeq 4$ m/s.

3. Investigation of Collisional Relaxation of Molecular VRT

The role of quenching and phase-randomizing collisions in the
broadening of the lines of the molecular VRT may be determined
in observing the lineshape of absorption of a weak wave in the
presence of a strong one for unidirectional and oppositely
traveling waves. As it can be seen from theoretical studies
made in [16, 17], there is a substantial difference in the
lineshape of absorption of a weak signal for unidirectional
and oppositely traveling waves. For oppositely traveling waves
there is one resonance associated with population difference
saturation in the absorption coefficient of a weak wave. For
unidirectional waves the absorption line of a weak wave has
several resonances with widths determined by the time of relaxation of an upper and lower levels and by the homogeneous
linewidth. Since the lifetimes at the molecular VRT are long,
the level relaxation is mostly determined by inelastic scattering.

The expression for the absorption coefficient of a weak
wave in the presence of a strong one for unidirectional waves
derived in [17] with allowing for level degeneration is of the
form

$$\alpha \sim 1-x \frac{G^2}{2\Gamma - \Gamma_m - \nu_m} \frac{\Gamma_m + \nu_m}{(\Gamma_m + \nu_m)^2 + \varepsilon^2} - x \frac{G^2}{2\Gamma - \Gamma_n - \nu_n}.$$

$$\frac{\Gamma_n + \nu_n}{(\Gamma_n + \nu_n)^2 + \varepsilon^2} - (2\Gamma - \Gamma_m - \nu_m - \Gamma_n - \nu_n) \frac{2\Gamma G^2}{(2\Gamma)^2 + \varepsilon^2}$$

$$\left[\frac{1}{(\Gamma_n + \nu_n)(2\Gamma - \Gamma_m - \nu_m)} + \frac{1}{(\Gamma_m + \nu_m)(2\Gamma - \Gamma_n - \nu_n)} \right] \quad , \quad (3)$$

where x is the factor allowing for level degeneration, ε is detuning between the frequencies of the strong and weak fields, Γ_m and Γ_n are the level halfwidths, ν_m and ν_n are the quenching collision frequencies, Γ is a homogeneous line halfwidth. The second and third terms describe the resonances with halfwidths inverse to the relaxation times of levels m and n. The fourth term describes the resonance associated with the population difference saturation with the halfwidth 2Γ. Studying the behaviour of the resonance width on gas pressure one can gain information on the nature of collisions.

In [18] there have been observed the resonance shape and measured the resonance broadening in SF_6 at $\lambda = 10.6$ µm in unidirectional and oppositely traveling waves.

Figure 4 gives the record of resonances in SF_6 for unidirectional waves. It is seen that there is the only resonance that is described by a dispersion curve within the limits of a measurement error.

Figure 5 depicts the dependence of the resonance broadening for oppositely traveling and unidirectional waves on the absorbing gas pressure. The magnitude of the resonance broadening for unidirectional waves is twice as little as that for oppositely traveling waves and amounts to 3.8 ± 0.3 MHz/Torr and 8 ± 0.4 MHz/Torr, respectively.

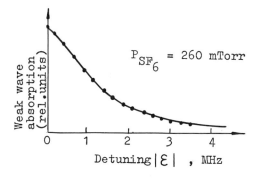

Fig.4 Record of the lineshape of absorption in SF_6 of a weak wave in the presence of a strong one for unidirectional waves

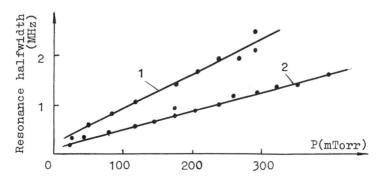

Fig.5 Dependence of broadening of nonlinear resonances in SF_6 for oppositely traveling (curve 1) and unidirectional (curve 2) waves

The analysis of experimental results made in accordance with (3) indicates that the relaxation constants of the upper and lower levels are equal and correspond to the homogeneous linewidth Γ, i.e., $\Gamma_m + \nu_m \simeq \Gamma_n + \nu_n \simeq \Gamma$. This evidences the absence of phase-randomizing collisions and of the equality of scattering amplitudes on the upper and lower levels ($\nu_m \sim \nu_n$). The main contribution to the resonance width in the pressure range under investigation is provided by quenching collisions.

4.   Investigation of Shifts of Nonlinear Resonances in Methane

Now let us consider the shift of nonlinear resonances. The first experiments on observation of the nonlinear dependence of resonance shift on pressure were carried out by us in the $F_2^{(2)}$ line in methane [6]. However due to the influence of a magnetic hyperfine structure (MHS) at the operating methane transition investigated in [19-21], an unambiguous interpretation of the experimental results in [6] presents difficulties. Recently we carried out additional investigations of the maximum resonance shifts on the $F_2^{(2)}$ line in methane on pressure.

The schematic of the experiment is given in Fig.6. Unlike the previous works, the shift of frequency ν_1 of laser 1 was measured with respect to the frequency ν_3 of laser 3 with a telescopic beqm expander that was stabilized to the central MHS component.

Figure 7 shows the experimental dependence of the maximum resonance shift in methane on pressure. The frequency for each pressure was determined by extrapolating to zero intensities of the field in the cavity. The nonlinear shift against pressure is observed. In the pressure range of about 1 mTorr the shift is small and amounts to 10-20 Hz/mTorr. The shift is increased with an increase of pressure. In the pressure range of 3-4 mTorr the shift slope is about 400 Hz/mTorr. The non-

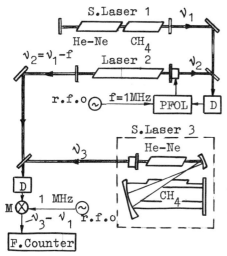

<u>Fig.6</u> Scheme of the apparatus for measuring the maximum resonance shift: PFOL - phase-frequency offset lock, M - mixer, r.f.o. - radio-frequency oscillator

linear shift of the maximum resonance is conditioned by the influence of the MHS and of the collisional shift in methane. The dotted line in Fig.7 indicates the calculated dependence of the resonance shift on the MHS. The difference between the experimental and calculated curves indicates the influence of the collisional shift in methane. The analysis has indicated that the collisional shift in methane nonlinearly depends on pressure. In the range of high pressures of 3-4 mTorr the shift slope is about 10 times as large as that in the range of low pressures ($\sim$1 mTorr). With the methane pressure of about 1 mTorr the collisional shift cancels the stabilized frequency shift due to the MHS and the resulting shift is little dependent on pressure.

The qualitative explanation of this phenomenon is as follows. In the range of low pressures where $\Gamma \ll k v \bar{\theta}$, after collision the particles do not interact with the field, and these collisions do not contribute to the nonlinear resonance shift. The resonance shift is associated with the molecules that are scattered at the angles less than a typical scattering angle. The magnitude of shift is increased with an increase of pressure. With $\Gamma \gtrsim k v \bar{\theta}$ almost all particles are scattered at angles $\theta \lesssim \bar{\theta}$ and participate in the resonance shift. In this pressure range the magnitude of shift is equal to the line shift in an ordinary theory of collisional broadening. These results dramatize a great role of elastic collisions with velocity change in broadening and shift of nonlinear resonances at the molecular VRT in low-pressure gases.

5. Investigation of the Resonance Shape in Methane

Let us discuss the results of direct observation of the influence of elastic scattering of excited particles on the shape

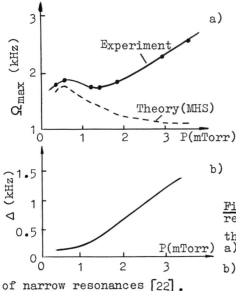

Fig.7 Dependence of the maximum resonance shift in methane at the $F_2^{(2)}$ line on pressure a) $\Omega_{max} = \omega_{max} - \omega_{7 \to 6}$; b) pressure shift in methane of narrow resonances [22].

Figure 8 shows the experimental installation. Investigations were carried out in methane on the $F_2^{(2)}$ line by using a He-Ne/CH$_4$ laser. The frequency of stable laser 1 is shifted with respect to the line center of methane by about 8 MHz. The laser 2 under investigation with an internal methane cell was locked to laser 1 by using the system of frequency offset-lock with the shift by a value Δf. The frequency of the laser under investigation was tuned with respect to the laser 1 in the range of Δf =1-16 MHz by supplying a frequency discriminator of the system of frequency offset-lock with sawtooth voltage. This enabled us to scan the frequency of the laser 2 relative to the methane absorption line center. A usable measuring system was the 512-channel computer analyzer. The signal of the difference frequency of the lasers ($\nu_2 - \nu_1$) was fed to one input of the analyzer, the signal that is proportional to the radiation power of the laser 2 to the other. Unlike the conventional methods of investigation of the resonance shape based on registration of a variable component of radiation power in modulating a laser frequency, we recorded directly a power resonance. This enabled us to increase the sensitivity of the resonance wings. However the influence of power fluctuations and drift increased sharply. The fluctuations were removed in processing the signal with a computer. After each passage of frequency through the resonance the power level was corrected. The experimental installation permitted us to examine the resonance shape with a contrast $\simeq 0.001$. An increase in the sensitivity is caused by a decrease in the generation power drift.

Figure 9 depicts the resonance shape at different helium pressures in the methane cell. With a methane pressure of about

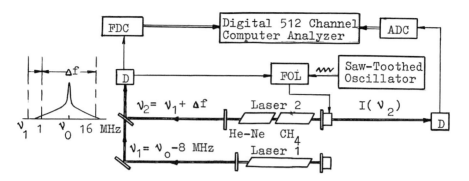

Fig.8 The experimental setup: FOL - frequency offset lock, FDC - frequency-digital converter, ADC - amplitude-digital converter

10^{-3} Torr the resonance has a Lorentzian form and a width of about 70 kHz. When helium is added, the resonance is widened and the pedestal of 2-3 MHz wide appears. A slight displacement of the pedestal relative to the resonance center was observed, which may be conditioned by their collisional separated shifts. At low pressures (about 10 mTorr) where the collisional frequency is small, the pedestal form is associated with the characteristics of a differential scattering cross-section. From the data of Fig.9 one can obtain the value of the typical angle θ that is about 1°. With an increase of pressure the width and amplitude of the pedestal are increased. At $P_{He}=0.1$

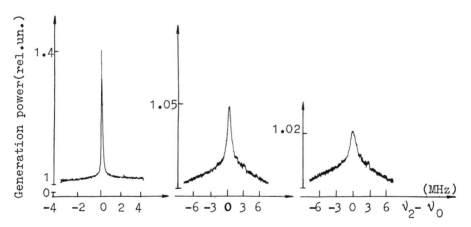

Fig.9 Record of the resonance shape in methane at different helium pressures: a) P_{CH_4} = 1 mTorr, P_{He}= 0; b) P_{CH_4} = 1 mTorr, P_{He}= 20 mTorr; c) P_{CH_4} = 1 mTorr, P_{He}= 43 mTorr

Torr the pedestal amplitude is compared with the amplitude of a sharp part of the resonance. With a further increase in pressure the resonance shape is determined by the pedestal whose width may be much more than a collisional one. In this region a molecule suffers many collisions for the lifetime on a rotational level. The pedestal width is determined by the change of a particle velocity at many events of collisions. This change is of diffusion character. The process of finding the resonance shape in this region is adequate to the process of finding the velocity distribution of particles at multiple collisions.

Figure 10 indicates that the width of the narrow part of the resonance nonlinearly depends on the helium pressure. The results were obtained in processing the resonance shape at various helium pressures. In the pressure range of 1-10 mTorr the resonant broadening amounts to about 20 MHz/Torr and is conditioned by the total elastic scattering cross-section.

Thus, it has been experimentally shown that at low pressures the resonance width in molecular gases is determined by the total scattering cross-section, the pedestal shape by the characteristics of the differential elastic scattering cross-section of excited particles. With a pressure of about 0.1 Torr the resonance shape is determined by particle diffusion at multiple events of scattering. These results indicate that spectroscopic methods of superhigh resolution are of prime importance in studying elastic scattering of excited particles at collisions.

New possibilities to study particle scattering at collisions in low-pressure gases are associated with the method of separated optical fields [23,24]. The use of atomic and molecular beams will permit one to employ this method in measuring differential scattering cross-sections of excited particles.

The author is indebted to V.P.Chebotayev, E.V.Baklanov, and L.S.Vasilenko for valuable discussions and help in the work.

References

1. Molecular Beams, Advances in Chemical Physics, vol. X, ed. by J.Ross (New York, London, Sydney, 1966)
2. V.S.Letokhov, V.P.Chebotayev: Nonlinear Laser Spectroscopy (Springer-Verlag, Berlin, Heidelberg, New York, 1977)
3. Yu.V.Brzhazovsky, L.S.Vasilenko, V.P.Chebotayev: ZhETF 54, 2095 (1968)
4. S.N.Bagayev, L.S.Vasilenko, V.P.Chebotayev: Preprint No.15 (Institute of Thermophysics, Novosibirsk, 1970)
5. S.N.Bagayev, E.V.Baklanov, V.P.Chebotayev: Pisma ZhETF 16, 15 (1972)
6. S.N.Bagayev, E.V.Baklanov, V.P.Chebotayev: Pisma ZhETF 16, 344 (1972)
7. A.T.Mattick, N.A.Kurnit, A.Javan: Chem.Phys.Lett. 38, 176 (1976)

8. C.Bordé: In Laser Spectroscopy III, ed. by J.L.Hall and J.L.Carlstein (Springer-Verlag, Berlin, Heidelberg, New York, 1977)
9. T.W.Meyer, C.K.Rhodes, H.A.Haus: Phys.Rev.A 12, 1993 (1975)
10. L.S.Vasilenko, V.P.Kochanov, V.P.Chebotayev: Opt.Comm. 20, 409 (1977)
11. P.Cahuraé, E.Marié, O.Robaux, R.Vetter, P.R.Berman: J.Phys.B: Atom.Molec.Phys. 11, 645 (1978)
12. V.A.Alexeyev, T.A.Andreyeva, I.I.Sobelman: ZhETF 62, 614 (1972); ZhETF 64, 813 (1973)
13. E.V.Baklanov: Optika i Spektroskopiya 38, 24 (1975)
14. V.P.Kochanov, S.G.Rautian, A.G.Shalagin: ZhETF 72, 1358 (1977)
15. H.S.W.Massey, E.H.S.Burhop: Electronic and Ionic Impact Phenomena (Oxford, 1952)
16. E.V.Baklanov, V.P.Chebotayev: ZhETF 61, 922 (1971)
17. S.G.Rautian, L.I.Smirnov, A.A.Shalagin: ZhETF 62, 2097 (1972)
18. L.S.Vasilenko, N.N.Rubtsova, M.N.Skvortsov, V.P.Chebotayev: Report at 4th Vavilov Conference on Nonlinear Optics (Novosibirsk, 1975)
19. S.N.Bagayev, L.S.Vasilenko, V.G.Goldort, A.K.Dmitriyev, V.P.Chebotayev: Pisma ZhTF 13, 291 (1977)
20. J.L.Hall, C.Bordé :Phys.Rev.Lett. 30, 1101 (1973)
21. J.L.Hall, C.Bordé, K.Uehara: Phys.Rev.Lett. 37, 1339 (1976)
22. S.N.Bagayev, A.S.Dychkov, V.P.Chebotayev: Pisma ZhETF 29, 570 (1979)
23. V.P.Chebotayev: Proceedings of the Sixth International Conference on Atomic Physics (Riga, USSR, 1978). Zinatne Riga, Plenum Press New York-London, p. 585
24. V.P.Chebotayev, this volume.

Part III

Rydberg States

Stark Ionization of Rydberg States

P. Jacquinot

Laboratoire Aimé Cotton, Centre National de la Recherche Scientifique
F-91405 Orsay, France

The behaviour of atoms in high lying excited states submitted to a static electric field during the excitation or immediately after has been studied by different groups during the last years. Interesting, and sometimes rather unexpected phenomena have been observed and the situation is now becoming rather clear thanks to numerous and refined experiments and to successful interpretations.

The aim of this paper is to report on the work recently done at Laboratoire Aimé Cotton, Orsay * . An effort will be made to insert this work in the general context, although this paper is not a review article and does not pretend to be complete. Most of the work has been done on alkali metals and this paper will be restricted to this case.

1. The meaning of the $F_c = E^2/4$ law.

In the case of an electron in a Coulomb field (purely hydrogenic model) + a static field F , the well known potential energy surface has a saddle point (SP) at $E = -2F^{1/2}$ (in a.u.). This relationship is sometimes written $F = (2n^*)^{-4}$. The existence of this SP means only that something can happen for energies above it : some states can ionize while others remain stable. In fact if the problem is treated in parabolic coordinates [1] it appears that the SP limit plays no role but one can still define a critical field F_c^H for each Stark level : in the simplest case ($|m| = 1$ states) this critical field is given by $F_c^H = E^2/4Z_2$.

For a Stark manifold n , the usual separation constant Z_2 can take n values from practically 1 for the lowest state ($n_1 = 0$, $n_2 = n - 1 - |m|$ in the weak field limit) to $1/n$ for the highest one $(n_2 = 0)^2$. The result is shown, in a very simplified form on Fig. 1 for $n = 10$ and 11 . For the lowest states of the different manifolds ionization begins on the "saddle point line" (SPL) $F_c = E^2/4$: F_c has thus really the meaning of a critical field since a continuum exists above the SP line.

---

* by A. Bachelier, S. Feneuille, S. Liberman, E. Luc-Koenig, J. Pinard and A. Taleb.

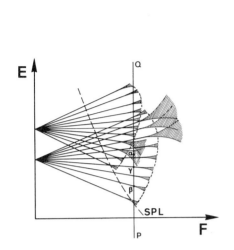

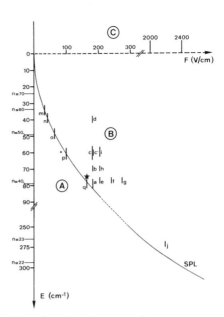

Fig. 1 : Stark manifolds for n = 10
and 11. For the sake of clarity
the broadening of only a few levels
has been represented.

Fig. 2 : The three regions,
and the type of experiments
made at LAC.

So far it is all that we can say. But the SP line can at least serve
as a guide to discuss and classify the phenomena by considering three regions A , B and C (Fig. 2) , the last one being above the field free
ionization limit E = 0 .

- Region A is usually studied at constant F by excitation with a tunable laser. It corresponds to pure spectroscopy of Stark levels.

- Going from region A to region B , or vice versa, corresponds to threshold studies. This may be done either at constant E by varying F or at
constant F by frequency scanning.

- Region B and region C have recently shown evidence of interesting
structures. In this talk the emphasis will be put on experiments made at
Laboratoire Aimé Cotton (LAC) in the vicinity of the SP line, essentially
in region B .

2. Experimental.

The experiments are usually made with different alkalis, sometimes
with rare gases , in a collisionless atomic beam. The Rydberg states are
excited in a one step process or by step excitation with one or several
pulsed lasers . The line width is usually .2 to .5 cm^{-1} . At LAC a
higher resolution has been used thanks to a special monomode pulsed laser
already described [2] : practically the final resolution is better than

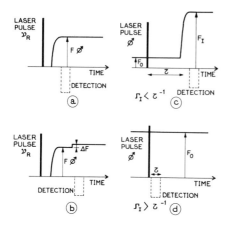

Fig. 3 : Different modes of detection of the ions.

150 MHz (.005 cm^{-1}). Essentially Rb has been studied by direct excitation from the ground state $(\lambda \sim 300 \text{ nm})$.

D.C. electric fields can be applied and ions are counted through a counting gate. Different modes for the timing of the laser pulse, the D.C. fields and the counting gates are shown on Fig. 3 and correspond to the following measurements :

a) threshold determination by varying F . This arrangement is also used for the detection of unperturbed states ;

b) differential threshold determination by varying F [3] ;

c) spectroscopy of "stable" Stark levels in the field F_0 (region A). Only the states with an ionization rate $\Gamma_I < \tau^{-1}$ give a signal ;

d) photoionization in the presence of a D.C. electric field F_0 . Only the states with an ionization rate $\Gamma_I > \tau^{-1}$ give a signal.

(the delay of the gate is corrected from the transit time of the ions ; the same in d) .

3. Studies accross the SP line.

3.1 Since several years different determinations of critical ionization fields have been made by the rather crude method "a" of Fig. 3 . Rather generally a well defined threshold has been found obeying the simple law $F_c = K(2n*)^{-4}$, where n* is the effective quantum number of the unperturbed Rydberg level and K a coefficient very close to 1 (within 10%).

Refined measurements [3][4] have shown several thresholds corresponding to different m_ℓ values. Cooke and Gallagher [5] have proposed a satisfactory explanation of this fact and given numerical values for K(m) . Since then the simple law for the critical field F_c is well understood.

3.2 The experiments made at constant F and variable E give more information. Littman et al. [6] have examined the Stark structures of Li n = 18, 19, 20 both in region A (mode "c" of Fig. 3) and in region B

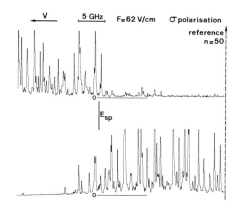

V   5 GHz   F=62 V/cm   σ polarisation

reference
n = 50

E_{sp}

Fig. 4 : Spectra obtained
in mode "d" (above) ,
and in mode "c" (below)

(mode "d") corresponding in both cases to $\Gamma_I = 10^5$ s^{-1} . They have
clearly shown that the two types of resonances are separated by the SP line.
The same type of study has been made recently at LAC at higher resolution
on Rb in the regions labelled m, n, o, p, q on Fig. 2 . An example of
the results is shown on Fig. 4 . The resonances in region A (mode "c")
cease a little further than the limit $E_c = -2F^{1/2}$ whereas the resonances
in region B (mode "d") start a little earlier than this limit. The overlapping region extends over about 15 GHz for n around 40 , to 3 GHz for
n around 60 , which corresponds to a variation of K of about .3% .

3.3 The role played by the SP line in the field ionization of alkalis
is thus clearly shown. The decisive step for its explanation has been done
by Littman et al. [6] who gave the following argument : in hydrogen the
ionization of any state of a manifold should take place only at F_c^H since,
because of symmetry reasons, these states are not coupled to the continuum
arising from the $(n , n_2 = 0)$ states. But in alkalis the hydrogenic super-
symmetry is broken since the central field is no longer coulombis so that
the coupling becomes possible and ionization can begin at $F = F_c$. We
shall see later an experimental fact in favour of this explanation.

4. Studies in region B .

The first evidence of resolved structures in region B (mode "d") has
been given at LAC by Feneuille et al. [7] who observed sharp photoionization
resonances in Rb around n = 50 in the vicinity of the SP line : the width
of the observed resonances, limited only by the residual Doppler effect in
the beam, was such that the hyperfine and the isotopic structures of the
ground state were completely resolved. Such resonances were observed mostly
in σ polarization $(\vec{\varepsilon} \perp \vec{F})$ for an excitation from the ground state S$_{1/2}$.
Then Littman et al. [6] showed the same phenomenon on Li (σ excitation)
around n = 19 at a lower resolution but in a wider range. Freeman et al.
[8] have also reported such resonances in Rb around n = 20 with a limit
of resolution of .5 cm^{-1} in a paper principally devoted to broad structures in region C . Quite recently Feneuille et al. [9] have shown, in π
excitation that some resonances show a typical Fano profile (Fig. 5). This
is a direct proof of the coupling between a discrete state and a continuum :

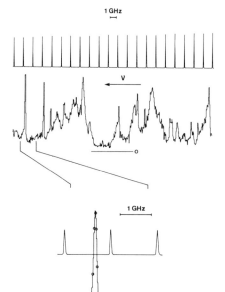

Fig. 5 : A typical Fano profile due to the coupling of a stable level with the continuum.

the π excitation had been chosen because it excites $m = 0$ states which are the most non hydrogenic ones. In the same paper, a much broader wavy structure in π excitation in the same region was also reported. Finally Freeman et al. [10] have confirmed and extended their previous observations in π and σ in the vicinity of the SP line.

By considering Fig. 1, relative to hydrogen, we se that, with an alkali, we can expect three types of photoionization resonances in region B . For a given F if E varies (vertical line PQ) we can cross the lines representing Stark states of the different manifolds (for the sake of clarity only two manifolds have been represented, but all should be there !) in three different situations :

α - The hydrogenic state has become extremely broad : this should give rise to broad resonances observable both in hydrogen and in alkalis . For a reason explained later these structures are visible only in π excitation.

β - The hydrogenic state is beginning to get broadened so that $\Gamma_I \geqslant \tau^{-1}$: one can then observe sharp resonances since τ is of the order of $10^{-5} - 10^{-6}$ s . Here also the resonances could be observed in hydrogen and in alkalis.

γ - The hydrogenic state is not yet broadened so that $\Gamma_I < \tau^{-1}$ and no photoionization could take place in hydrogen . But in alkalis, because of the coupling with the underlying continuum an additional ionization rate Γ'_I is introduced and one can have $\Gamma'_I \geqslant \tau^{-1}$. These resonances, exist only with alkalis and can be observed in both π and σ polarizations but the coupling is stronger for $m_1 = 0$ states than for others. They should give rise to more or less pronounced Fano profiles : this is what has been observed [9]. Of course in the case of alkalis the positions of the Stark levels given are not the same as in Fig. 1 . In Fig. 1 only one example of each of the three types α , β , γ has been represented ; of course there are many of them all along the PQ line . But as the excitation energy E increases, for a given field F , the sharp resonances γ and β disappear and only the broad α structures remain in π excitation. The essential difference between these three types of resonances is their widths and shapes ; but for all of them their positions are those of the Stark levels and can be calculated on the same manner. A new series of experiments has been made at LAC in region B in different regions labelled "a" to "j" in Fig. 2 . The main results are the following :

i – One series of recordings have been obtained at a fixed electric field strength, and is labelled a , b , c (c') and d . In the "a" region near the SPL , narrow resonances (line width less than 300 MHz) have been observed in σ polarization as well as in π polarization (a , c'). As the energy is increased the π resonances disappear within a few cm^{-1} (less then 5 cm^{-1}), whereas the σ resonances do not. In fact the σ resonances are still observed in the "c" region located at 25 cm^{-1} above the SPL (and disappear only at about 30 cm^{-1}). It is to be noticed that in the π spectrum the narrow resonances are superimposed to broad undulations which can be observed everywhere in the B region.

ii – This has been confirmed by the results obtained on a second series of π recordings (a , e , f , g) done in the same energy interval but for various D.C. field strength values.

iii – The three types of resonances have thus been observed on the same recording. It is in general impossible to distinguish between the β (hydrogenic) and the γ (non hydrogenic) resonances except for some lines showing an obvious Fano profile (e.g. observations marked by a star in Fig. 2).

More complete details will be published in a forthcoming article.

5. Studies in region C .

Freeman et al. [8][10] have observed on Rb broad undulations extending from the critical energy $E_c = 2F^{1/2}$ to well above the zero field ionization limit. These structures exist in π excitation but not in σ excitation. This phenomenon is of course of a same nature as the structures observed in a strong magnetic field by different authors . Classical and semi-classical theories have been proposed (see [8] or [10] and references therein) with a hydrogenic model and give a satisfactory agreement for the period of the structures in the vicinity of or above the zero field ionization limit. No more can be said in the frame of this paper. Only a point should be emphasized : according th these theories the upper states of the transitions must have m = 0 for the structures to exist, which explains that they are observed only in π excitation. We shall see a little further that this may be questioned.

6. An interpretation of the undulations.

These undulations are nothing but the α type resonances mentioned in the preceding paragraph : but in this region they take a more or less sinusoïdal character and the sharp α and β resonances have disappeared. Quite recently E. Luc-Koenig and A. Bachelier [11] have given a complete interpretation based on a quantum mechanical hydrogenic model, and it seems worthwhile to summarize here their main results. They first compute without approximations, a "partial density of states" $c_{n_1}^{|m|}(E)$ for each value of the generalized parabolic quantum number n_1 as a function of E ; then the total density $c^{|m|}(E)$ is obtained by summing over n_1 . Figure 6 gives the result for $m = 0$ and $|m| = 1$ in the vicinity of the zero field ionization limit (labelled E_0 on this figure). One sees indeed undulations in $c^0(E)$ and $c^1(E)$ but the depths of modulation are very small, resp. 3% and 1% for this field. Although this is valid only for a hydrogenic atom, it probably does not suffice to explain the observations in alkalis. But what is actually observed in the experiments is a photoionization cross section and not a density of states . It is then necessary to compute the density of oscillator strength df/dE $(|m|,E)$ obtained by summing the partial densities relative to n_1 . Now df/dE $(n_1,|m|,E)$ is proportional to $|\langle\Psi_i|\vec{r}|\Psi_{n_1}^{|m|}(E)\rangle|^2$ ($\vec{r}$ being the excitation operator, z for π excita-

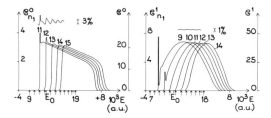

Fig. 6 : Partial and total densities of states.

tion, $x \pm iy$ for σ_+ excitation) so that the dependence on E can be very different for $c_{n_1}^{|m|}$ and for df/dE $(n_1,|m|)$. In fact Fig. 7 shows for $m = 0$ a deep valley between two maxima for each value of n_1 so that the summation over n_1 gives strong undulations with a depth of $\sim 15\%$; for $|m| = 1$ there is only a broad maximum for each n_1 so that the summation gives practically no modulation. The minimum in the df/dE $(n_1,0)$ curves is due to the following fact. For each $c_{n_1}(E)$ curve the starting point corresponds to $z_1 \simeq 1$, the ending point to $z_1 \simeq 0$ and the middle to $z_1 = z_2 = 1/2$. For this value of the energy, the charge distribution near the nucleus is symmetrical with respect to the $z = 0$ plane : since Ψ_i has the same symmetry (S state), then $\langle\Psi_i|z|\Psi_{n_1}\rangle \simeq 0$. This would not happen if Ψ_i were asymmetrical with respect to $z = 0$: the depth of modulation is then actually related not to the m value of the upper state but to the symmetry properties of both the initial and the final states . For the same reason the curves df/dE $(n_1,1)$ show only a maximum since now $\Delta m = 1$ so that the transition operator is symmetrical with respect to $z = 0$ (σ excitation). But even with $|m| = 1$ for the upper state, if the initial

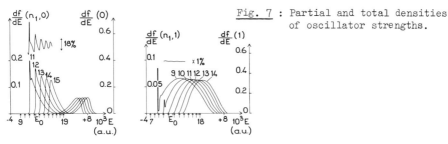

Fig. 7 : Partial and total densities
of oscillator strengths.

state had $|m| = 1$ (π polarization) and an even value of $(\ell + m)$ one would now observe a valley in each df/dE $(n_1, 1)$ and a deep modulation in the summation. Of course these results are valid only for hydrogenic atoms ; with an alkali the results are certainly different for the depths of modulation, but the same symmetry considerations still hold. These results concerning the polarizations are then contradictory to those of the classical models ; but no experiment has yet been made to check them.

References.

[1] H. A. Bethe and E. E. Salpeter, Quantum Mechanics of One- and Two-Electrons Atoms (Academic Press, New York 1957).

[2] J. Pinard and S. Liberman, Opt. Commun. 20, 344 (1977).

[3] J.-L. Vialle and H. T. Duong, J. Phys. B : Atom. Molec. Phys. 12, 1407 (1979).

[4] T. F. Gallagher, L. M. Humphrey, W. E. Cooke, R. M. Hill and S. A. Edelstein, Phys. Rev. A 16, 1098 (1977).

[5] W. E. Cooke and T. F. Gallagher, Phys. Rev. A 17, 1226 (1978).

[6] M. G. Littman, M. M. Kash and D. Kleppner, Phys. Rev. Lett. 41, 103 (1978).

[7] S. Feneuille, S. Liberman, J. Pinard and P. Jacquinot, C. R. Acad. Sci. Paris 284, B 291 (1977).

[8] R. R. Freeman, N. P. Economu, G. G. Bjorklund and K. T. Lu, Phys. Rev. Lett. 41, 1463 (1978).

[9] S. Feneuille, S. Liberman, J. Pinard and A. Taleb, Phys. Rev. Lett. 42, 1404 (1979).

[10] R. R. Freeman and N. P. Economu, Private communication, to be published.

[11] E. Luc-Koenig and A. Bachelier, to be published.

Rydberg States and Microwaves: High Resolution Spectroscopy, Masers and Superradiance

S. Haroche, C. Fabre, P. Goy, M. Gross, and J.M. Raimond

Laboratoire de Physique de l'Ecole Normale Supérieure, 24, rue Lhomond
F-75231 Paris Cedex 05, France

1. Introduction

Radioemission of microwaves by Rydberg atoms in the interstellar medium have been first observed in the early sixties [1] and have been widely studied since [2], providing a lot of informations about the physical conditions existing in the emitting regions. In the laboratory, microwave-Rydberg atoms interaction experiments, though more recent, have also undergone a big development in the last few years in the areas of spectroscopy [3] [4] [5] and multiphoton ionization studies [6]. Among the very attractive features of the interaction of Rydberg atoms with microwave radiation, certainly the most striking ones are the very narrow ultimate resonance widths of the transitions due to the very long radiative lifetime of the levels and the extreme sensitivity to microwave processes due to the huge size of the Rydberg electric dipoles for microwave transitions. The first feature makes Rydberg atoms very attractive for high resolution spectroscopy and metrological applications. The second one can be taken advantage of in order to develop new types of far infrared or microwave amplifiers, masers or superradiant systems operating with extremely low thresholds and microscopic energy outputs. These devices can have a big potential interest for microwave detection technology.

The high resolution and high sensitivity features of Rydberg state microwave interaction processes are now being investigated at Ecole Normale Supérieure in a series of experiments whose latest results are presented in this paper.

2. General description of experiments

The general set-up used in all these experiments is shown in Fig.1-a. An atomic beam of alkali (Na or Cs) is excited into a Rydberg nS or nD level ($20 \lesssim n \lesssim 50$) using a stepwise excitation process with two colinear-pulsed-Nitrogen-laser-pumped-dye-laser beams. (In the following, we call i for sake of simplicity the initially prepared Rydberg level). Evolution of the system after its excitation is monitored by a Rydberg state detector (R.S.D.) made of a set of condenser plates allowing to apply an ionizing electric field on the atoms and of a high gain electron multiplier collecting the ionization electrons (E.M.). The R.S.D. makes use of the well known selective field ionization method [5], whose principle is recalled on Fig.1-b : at a given time t_0 after the laser excitation, a ramp of electric field $F(t)$ is applied to the atoms via the condenser plates. The field reaches at time t_i the threshold value   corresponding to ionization of level i. The atoms in this level are then ionized and the electrons accelerated and collected by the E.M. which delivers an ion-peak signal around time t_i on a fast scope. If

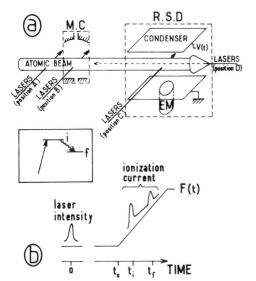

a. Sketch of set-up for
Rydberg-microwave experiments. b.
Time analysis of field ionization
detection procedure

some atoms have been transferred to another level f during the time interval
between the laser excitation and t_0, an ion peak is also produced at the time
t_f when the electric field reaches the threshold   for ionization of this
latter level. The time resolved ionization-signals thus obtained allow to
monitor the population of various Rydberg states of interest at time t_0. By
varying t_0, one can sample the evolution of the system and make dynamical
studies. The method is extremely sensitive and selective. Single Rydberg
atoms can be detected. Absolute calibration of the number of atoms in each
level is possible. The ion current is indeed proportional to the number of
ionized atoms and the proportionality constant can be estimated by reducing
the pumping light intensity with calibrated filters until individual atom
counts are registered. Since each time-resolved peak can be assigned to a
well defined electric field threshold, the R.S.D. acts indeed as a very sensitive Rydberg state spectrometer.

Before reaching the R.S.D., the atomic beam crosses a semi-cofocal millimeter
Fabry-Perot cavity (M.C.) (quality factor $Q \sim 10.000$, finesse $\mathscr{F} \sim 200$ at
$\lambda = 2$ mm), whose axis is perpendicular to the atomic beam. The cavity length
ℓ (22 or 75 mm) is adjustable around focus by a micrometric screw-drive. Depending on the experiment, the laser excitation is applied either in front
of the M.C. (Lasers in position A), or in the cavity (Lasers in position B)
or in the R.S.D. after the M.C. (position C) or else along the atomic beam
(position D). In the three cases A, B and C, the excited Rydberg state volume
is a small cylinder whose axis is perpendicular to the atomic beam direction
(its length L is typically 5 mm and its diameter a ~ 1 mm). After the laser
excitation, this volume spreads out along the atomic beam due to atomic velocity dispersion. Well defined Rydberg state velocities can be selected by
choosing the delay time t_0 between the excitation and the detection occuring
at different locations. When the lasers are in position D, the excited region
is a long pencil shaped volume whose shape remains essentially unchanged during the atomic flight time. There is no possible velocity selection in this
case.

3. Lifetime studies. Effects of blackbody radiation

This set-up, with separated preparation and detection zones, looks very similar to a ground state molecular or atomic beam apparatus. It can be used with excited atoms only if their lifetime is of the order of or larger than the flight time along the beam. Let us recall that the radiative lifetime of Rydberg atom increases as n^3. It is of the order of 100 µs for n ∿ 45. In this lapse of time, a Na atom at mean thermal velocity (v = 600 m/s) should fly in our set-up the 6 cm distance between position A and the R.S.D. Spontaneous emission however is not the only relaxation mechanism able to affect Rydberg atoms : collisions with other atoms or with impinging blackbody radiation photons could also cause transitions and reduce the lifetime below its purely radiative value. In order to check that these processes are not severely limiting our set-up performances, we have carried a set of preliminary lifetime measurements in levels nS (25 < n < 40) of Na. The excitation was performed with lasers in position D and the ion-signal corresponding to the excited level was detected using a boxcar whose gate was set around the ionization time t_i of this level. The boxcar output was recorded as a function of the delay time t_0. Figure 2 shows recordings of the signals thus

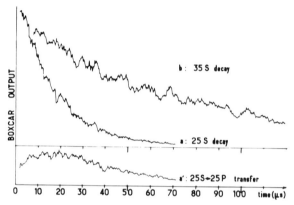

Fig.2 : Boxcar-averaged ionization signal recorded as a function of delay t_0 showing a) 25S level decay (following laser excitation at t=0); a') 25S→25P transfer following 25S level excitation at t=0; b) 35S level decay.

obtained for n = 25 (Fig.2 , trace a) and n = 35 (Fig.2-b). Note that the total observation time (100 µs) corresponds to an average flight distance of 6 cm, much larger than the 1 cm aperture of the E.M. window. The finite escape time from the detector does not however affect the measurements : the long pencil-shaped volume of excited atoms just flies pass the detector, the atoms leaving the detection region being merely replaced by other ones excited at the same time nearer to the oven. This excitation geometry with lasers head-on in position D is thus particularly well suited for the measurement of very long lifetimes. The results are in fair agreement with theoretical Coulomb approximation values [7] . (One finds for example 18 ± 6 and 65 ± 20 µs respectively for n = 25 and n = 35, whereas the calculated figures are 18.3 and 52.6 µs). Experimental uncertainties though are too large to make sure that spontaneous emission is the only or even the dominant decay cause of these states. A much more sensitive way of detecting other relaxation mechanisms consists in recording the ion-signal corresponding to a final state f different from the initially pumped one : trace a' of Fig.2 shows the signal appearing at the onset of ionization of level 25P, lying above the 25S excited state. This level should of course not be at all popu- lated by spontaneous emission. One clearly observes however a transfer of about 25% of the atoms to this level. The measured rate for the 25S → 25P transition is $3.10^3 s^{-1}$, in good agreement with the estimated transition rate induced by blackbody radiation at room temperature [8]. From similar mea- surements performed on lower nS states [9], COOKE and GALLAGHER have re- cently estimated that the blackbody effect should account for about 20% of the measured lifetime in the nS levels of Na. (The effect is much more im- portant on the nP lifetimes [8]). Our observations on nS states are con- sistent with their findings. As for possible collisional effects, we have verified that they should be negligible by carrying measurements under va- rious pressures of background gas (10^{-7} to 10^{-5} torr) in the atomic beam apparatus. These preliminary lifetime studies have thus shown that it is indeed possible to produce long lived nS Rydberg atoms, with lifetimes essentially limited by the spontaneous emission radiative decay processes.

4. High resolution microwave spectroscopy : towards a new Rydberg constant measurement ?

Using the set-up of Fig.1, we have then performed various microwave double resonance studies. The lasers were set in position A and the R.S.D. was de- tecting the Rydberg state level populations after a delay t_0, while micro- wave radiation applied to the atoms in the M.C. was frequency tuned accross the atomic resonance lines. The sketch of the microwave source used for these experiments is shown in Fig.3. Its main element is a Thomson CSF carcinotron delivering a power of about 500 mWatts. Several carcinotrons with frequencies ν_c centered around various values (55 GHz, 76 GHz, 96 GHz, 112 GHz) have been successively used in the course of the experiments. The carcinotron is powered by a stabilized high voltage power supply. The frequency jitter of the free-running carcinotron is of the order of 1 MHz, enabling to perform without stabilization double resonance experiments with moderate resolution[5].

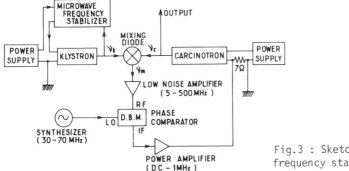

Fig.3 : Sketch of microwave frequency stabilized source

For high resolution, the carcinotron is locked on a very stable reference signal provided by a klystron whose frequency $\nu_K \sim 10$ GHz is stabilized on a 5 MHz quartz oscillator. A mixing silicon diode generates the pth harmonic (p = 5 to 12) of the klystron and beats it against the carcinotron microwave radiation. The amplified $\nu_m = \nu_c - p\nu_K$ beat note signal is phase-compared to a local oscillator (L.O) by a double-balanced mixer (D.B.M.) and an error voltage proportional to the phase mismatch is, after amplification, used to correct the high voltage driving the carcinotron. Scanning of the microwave frequency ν_c is achieved by sweeping the L.O. frequency. Absolute frequency determinations are obtained by measuring ν_K and ν_m with frequency counters previously calibrated on a signal from a Cs atomic clock. Short term frequency stabilities better than 10^{-8} are achieved with this source ($\Delta\nu_{source} < 10^3$ Hz).

The microwave radiation is sent on the resonant cavity through a 4 mm waveguide. The cavity is tuned at resonance by adjusting its length around focus while monitoring the transmitted or reflected microwave radiation. The power level at atomic beam location is controlled by using calibrated attenuators on the waveguide.

High resolution spectroscopic investigations of various kinds involving single- and double-photon microwave transitions are presently under way with this apparatus (quantum defects, fine and hyperfine structure studies in S, P and D states of Na and Cs). An analysis of these experiments being beyond the scope of this paper, we restrict ourselves here to the description of the ultimate resolution achieved so far with this set-up and to the discussion of a possible Rydberg constant measurement.

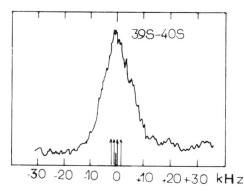

39S–40S

-30 -20 -10 0 +10 +20 +30 kHz

Fig.4: 39S-40S two-photon resonance induced in the M.C. ($\nu_c \sim 59$ GHz). Vertical arrows show positions of 4 unresolved h.f. components.

The narrowest lines have been obtained on nS $\leftrightarrow$ (n+1)S two-photon transitions. Fig.4 shows as an example the 39S-40S resonance of Na with microwave at $\nu_c \sim 59$ GHz. The signal is the ratio of the boxcar-averaged ionization currents corresponding to levels 40S and 39S. The delay t_0 is 100 μs, the effective interaction time in the M.C. being ~ 80 μs. The linewidth is 11 kHz only. Let us note that these resonances are Doppler broadening-free, for the same reason as the optical two-photon resonance experiments : the Doppler effects of the two-photon propagating in the M.C. in opposite directions cancel each other and each atom , whatever its velocity is, contributes to the same resonant frequency. The Doppler effect can be restored if, instead of interacting with a standing field, the atoms are placed in a running wave.

Fig.5 shows as an example the 36S $\rightarrow$ 37S two photon transition around $\nu_c = 76$ GHz when the lasers are set near position C with a delay $t_0 = 3$ μs. The atomic velocity selection is poor in this case and the resonance is induced on a large part of the maxwellian velocity distribution by microwave radiation escaping from the M.C., with wave vectors essentially along the atomic beam direction. The Doppler effect is then clearly apparent : it broadens ($\Delta\nu \sim 350$ kHz) and shifts the resonance lines towards higher frequencies (the position of the Doppler-free line is shown by the arrow on Fig.5; the

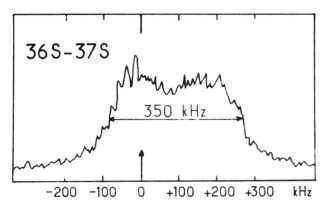

Fig.5 : 36S-37S two-photon resonance induced in a running wave. Arrow shows position of the Doppler-free two-photon line center.

complicated shape of the signal depends on the non-controlled k wave-vector distribution of the microwave).

The nS → (n+S)S Doppler-free two photon resonances we have observed are -with the 5S-nS optical two photon line recently observed in the Rb spectrum [10]- the narrowest electronic transition ever recorded in the alkali spectra. The nS-(n+1)S line centers can be presently determined with a precision of 3.10^{-8}. At this resolution level, it is important to analyze in details the effects on the resonance shape and position of small perturbation terms in the atomic Hamiltonian such as hyperfine structure (h.f.s.), Zeeman and Stark interactions with stray magnetic and electric fields. One can estimate Stark shifts $\Delta\nu_C$ on the 39S - 40S transition to be of the order of 1 kHz only for electric fields as big as 50 mV/cm. Although it has not yet been precisely measured, it is reasonable to assume that the stray electric field in the well shielded M.C. does not exceed the above value and thus does not affect the line center position [11]. (This is obviously true also for two photon transitions with lower n values that we have observed with similar resolution). The zero-field h.f.s. splitting $2A_n$ of the nS levels is well estimated by the Fermi-Segré formula which yields for n = 39 and 40 the values 130 kHz and 120 kHz respectively. These structures are decoupled in the non-compensated earth magnetic field H_0 ($\sim$ 600 mG), the sublevel energies being $m_J g_J \mu_B H_0 + A_n m_I m_J$ where m_I and m_J are the nuclear and electronic magnetic quantum numbers ($-1/2 \leqslant m_J \leqslant 1/2$; $-3/2 \leqslant m_I \leqslant 3/2$), g_J = 2 the electronic Landé factor and μ_B the Bohr magneton. Due to the selection rule $\Delta m_I = \Delta m_J = 0$, the two photon spectrum in fact consists of four lines [12] whose distances from the line center are $\pm 3/8$ $(A_{n+1} - A_n)$ and $\pm 1/8$ $(A_{n+1} - A_n)$, i.e. ± 2 kHz and ± 0.7 kHz (vertical arrows on Fig.4). These lines are not resolved and contribute for about 4 kHz in the experimental linewidth. The intrinsic width of each unresolved hyperfine component is thus 7 kHz only and can be entirely attributed to the finite 80 μs transit-time across the gaussian profile of the cavity mode.

Note that the line center does not depend on the hyperfine structure and magnetic field magnitudes. Its determination is thus a good measurement of the nS - (n+1)S electronic transition frequency. On the 39S - 40S line, we have found $2\nu_C$ = 118, 520, 443, 100 $\mp$ 4000 Hz. Similar measurements have been carried on the 40S - 41S, 36S - 37S, 34S - 35S, 33S - 34S and 32S - 33S

249

transitions. They yield the quantum defects ε_S of the nS states ($30 < n < 40$) with an absolute 4.10^{-7} precision. Systematic relative variations of the order of 2.10^{-6} between the quantum defects of successive Rydberg levels are clearly put in evidence. The following formula for $\varepsilon_S(n)$ is obtained :

$$\varepsilon_S(n) = \varepsilon_S(\infty) + \beta \ [\ n - \varepsilon_S(\infty)\ ]^{-2}$$

with $\varepsilon_S(\infty) = 1.3479692 \pm 4.10^{-7}$

$$\beta = 0.06137 \mp 0.00010$$

An increase by a factor 10 of the present resolution would allow to perform a new determination of the Rydberg constant. Such an improvement in resolution could be obtained by lengthening the transit time with very large single cavities or Ramsey-type-multiple cavity designs. The widths could ultimately be reduced to the natural width value, of the order of 1 kHz only. The major difficulty for such an experiment is certainly that it would have to be performed on Hydrogen, whose Rydberg states are much more sensitive to electric field perturbations than Na nS ones (to get Stark shifts smaller than 1 kHz on nS $\rightarrow$ (n+1)S transition with $n \sim 40$, the electric field should be controlled to better than 1 µV/cm in Hydrogen, whereas any field smaller than 50 mV/cm is good enough in Na). The advantage of such a microwave-Rydberg constant measurement might at first sight not be obvious, since the optical two photon lines recently observed in the Rb spectrum [10] are almost as narrow as the one reported here and have thus obviously a much larger intrinsic quality factor $\nu/\Delta\nu$. However, it is not yet possible to measure directly frequencies in the optical domain and hence the line positions would have in these experiments to be determined by comparison with the length etalon, thus limiting practically the absolute precision. The interest of a microwave Rydberg constant measure is that it can be performed directly in frequency units. Comparing the value thus obtained with wavelength optical measurements of the Rydberg [13] would -among other informations- yield a new precise value for the velocity of light.

5. Rydberg state "microscopic" masers [14]

A very spectacular feature of these microwave resonance experiments is certainly that they can be performed with extremely low microwave powers. Two photon transitions ($n \sim 35$) need an effective power flux as little as 10^{-6} Watts/cm^2 and single photon nS - (n-1)P lines have been detected with effective powers flux as low as 10^{-13} Watts/cm^2. It is in fact possible to observe them directly with only the harmonic of the klystron irradiating the atoms, when the carcinotron is switched off.

It is even possible to switch off completely the microwave source and to observe a resonant transfer from an nS to an (n-1)P Rydberg state by merely tuning the cavity across the atomic frequency (the laser excitation in that case is performed in position B inside the cavity, see Fig.1-a). Fig.6-a shows the atomic signal observed when the 27S Na level is excited and the cavity tuned on the 27S-26P$\frac{1}{2}$ transition (the delay t_o is here 30 µs). A strong signal corresponding to the 26P$\frac{1}{2}$ level is observed. When the cavity is tuned 40 MHz off this frequency (by a 5 µm change of the cavity length) the transfer signal drops down as shown on Fig.6-b. (A much smaller residual transfer, probably due to blackbody radiation transition [8] induced during the time interval t_o, is still observed in this latter case). This resonant effect, obtained without applying any external microwave signal, shows that

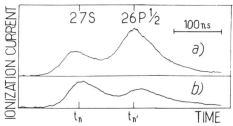

Fig.6 : Time-resolved Na ion-
signal recordings averaged
over 200 pulses exhibiting
maser effect :
a) cavity is tuned on
$27S \rightarrow 26P\frac{1}{2}$ transition
b) cavity is 40 MHz off reso
nance

the Rydberg atoms crossing the cavity have themselves provided the microwave
necessary to induce the transition : this is a genuine maser effect occuring
at λ = 1.49 mm. The difference with other regular masers operating at similar
wavelengths lies in the orders of magnitude of the threshold and energies
involved. Whereas the NH_3-beam maser operates with about 10^9 atoms at a time
in the cavity, the Rydberg Na atom maser with a cavity finesse of about 200
has a threshold corresponding to only 500 excited atoms ! This six orders
of magnitude difference is merely due to the three orders of magnitude increase in the electric dipole when going from NH_3 to a n $\sim$ 30 Rydberg state
(the atomic electric dipoles are proportional to n^2). As a direct consequence of their low thresholds, these Rydberg masers have extremely low output
energies, of only a few electron-volts per pulse. It would be very difficult
to detect directly these very small radiation bursts and the indirect electric field detection is certainly a great asset in these experiments. These
systems are, to our knowledge, the smallest electromagnetic coherent sources
of the maser or the laser kind existing to this date. Maser effects similar
to the one observed in Fig.6 have been seen for several nS $\rightarrow$ (n-1)P transitions in Na (23 < n < 35). When sweeping the cavity length ℓ, the maser signal appears around positions corresponding to different modes, separated
by $\Delta\ell = \lambda/2$. For each mode, one observes two close resonant positions (separation $\delta\ell \ll \lambda/2$) corresponding to the P state fine structure splitting.
From the $\delta\ell$ measurement, we have deduced for the splittings of the 24P,27P,28P
and 33P levels respectively 507±30, 314±30, 295±30 MHz and 159 ± 30 MHz.
These results are in good agreement with the values 488 ± 5 MHz, 300 ± 5 MHz,
266 ± 4 MHz and 161.5 ± 1 MHz measured in the course of moderate resolution
microwave experiments. The resonant maser transfer to lower states can thus
be used as a very convenient (and cheap) way of microwave spectroscopy, performed without applying any external microwave at all on the atoms. The moderate resolution (± 30 MHz) of the present measurement is due to the relatively small finesse of our cavity and could certainly be improved by several
orders of magnitude with a higher finesse. Improving the cavity quality
could have other interesting applications. The threshold of the maser is
reached when the multiple pass gain of the microwave radiation in the cavity
becomes equal to one. Increasing the finesse beyond the present value 200
would allow to increase the number of effective passes and to get much smaller maser systems. The ultimate limit would be the case of a single Rydberg
atom inside a very high finesse cavity ($\mathcal{F} \sim 10^5$). Although it is hardly
possible to view such a system as a maser, it is clear that it should have
very interesting features and would allow to test the emission properties
of a single atom interacting with perfectly conducting walls, a problem which
has been given a lot of theoretical interest.

6. Rydberg State Superradiance [14]

If instead of increasing the cavity finesse, one reduces it to the value 1
-i.e. if the cavity is simply removed- the Rydberg state maser becomes a

251

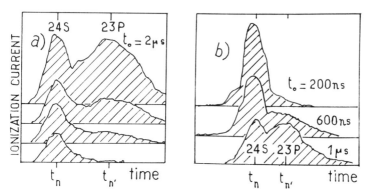

Fig.7 : Time resolved ion signal recordings showing superradiance on the
24S → 23P transition in Cs (λ = 0.68 mm). a) The delay t_0 is 2 μs. The four
traces correspond from bottom to top to increasing initial inversion
(2.10^5, 4.10^5, 6.10^5 and 12.10^5 excited atoms at t=0). b) The number of
excited atoms is fixed ($\sim 2.10^6$) and the delay t_0 is increased from the
upper to the lower trace.

superradiant system. In that case, its threshold is reached when the single
pass gain along the amplifying medium becomes larger than one, which implies
atomic inversion number of the order of 10^5 instead of 500. In order to observe Rydberg state superradiance, we have excited the atoms (in that case
Cs) directly in the R.S.D. with lasers in position C. The excited medium is
still in this case a small pencil shaped volume of length L $\sim$ 5 mm and diameter a $\sim$ 1 mm. Fig.7-a shows the time-resolved ion-signal observed 2 μs after
excitation of the 24S Cs level. The traces correspond, from bottom to top
to increasing initially excited atom numbers. One clearly observes on the
upper trace a large transfer of atoms towards the 23P level.(The superradiant
transition has a wavelength λ = 0.68 mm). The threshold effect is clear, the
transfer occuring only at high pumping intensities. These recordings have
shown that the threshold on this transition occurs for an inversion of the
order of 5.10^5, a figure in fair agreement with theory. This emission phenomenon exhibits all the distinctive characteristics of superradiance. In particular, it is delayed in time as shown in Fig.7-b where we have displayed
time-resolved ion-signals corresponding to the same number ($\sim 2.10^6$) of initially excited 24S Cs levels, with delays t_0 increasing from top to bottom.
The peak of the emission clearly occurs after a delay of about 1 μs. Here
also, the orders of magnitude are quite different from the ones of superradiant experiments performed on less excited atomic or molecular species.
The far-infrared molecular HF system [15] emits with about 10^{12} inverted
molecules, again a six order of magnitude difference with the Rydberg atom
case. Even near-infrared alkali atom superradiant systems [16] , which
operate in a wavelength domain where the radiative coupling is intrinsically
much stronger than for microwaves (due to the λ^{-3} term in the radiation
transition rate), have threshold levels at least a factor 1000 higher than
the superradiant Rydberg state systems.

We have observed Rydberg superradiant emission similar to the ones shown on
Fig.7 on many transitions starting from various nS and nD states in Cs
(20 < n < 50). The dynamics of these Rydberg superradiant systems is slightly
more complicated than the one of the Rydberg masers, since there is obviously
a competition of several possible emission frequencies towards a set of lower

states with quantum numbers n' < n, whereas in the maser case, cavity tuning always allows to unambiguously select a single emission frequency. In the absence of cavity selection, superradiance occurs on the transitions with the largest gain. For initially excited nS levels with n in the range 20 to 30 all the nS → n'P possible transitions have a millimeter or submillimeter wavelength smaller than the sample dimension (L = 5 mm, a = 1 mm). Simple arguments involving diffraction theory and Coulomb approximation calculations of oscillator strengths then show that the maximum gain occurs for the longest possible wavelength transition towards the (n-1)P closest state. This is consistent with the signals observed on Fig.7 for n = 24. If n is in the range 40-50, our sample dimensions become of the order of or smaller than the centimeter-wavelength of the transitions linking the excited level to its closest neighbours. One can then show that these transitions have a lower gain than shorter wavelengths ones towards more bound states. These latter ones are then favored and the superradiant emission from an nS level (n ∼ 40 - 50) should occur towards n'P states with n - n' ≫ 1. This seems to be confirmed by preliminary observations showing that in this range of n values, the superradiant does indeed transfer the atoms towards a bunch of n' levels with n - n' ∼ 5 to 10. It thus appears that the dynamics of superradiance strongly depends on the shape of the emitting sample and on its size relative to the emission wavelength. It would certainly be very interesting to make a more detailed quantitative study of this phenomenon in order to test various theories dealing with small sample superradiance.

The transient Rydberg maser and superradiant systems described here are unstable devices which are triggerred by the blackbody radiation background at room temperature. (At these long wavelengths, this is a much more efficient noise source than spontaneous emission fluctuations.). These systems can thus be considered as single or multiple pass amplifiers of the blackbody radiation field. By cooling the apparatus, it should be possible to greatly reduce this background and to use the same systems as amplifiers of very small far-infrared or microwave input signals. Tunability of these highly frequency selective amplifiers could be achieved over a very wide range by Stark shifting the highly polarizable Rydberg levels. The possibility of using such devices -or their c.w. counterparts- for very sensitive microwave detectors certainly deserves further investigations. It thus appears that Rydberg-microwave interaction experiments might have a very promising future for fundamental studies in radiation-matter interaction problems and for technological applications as well.

References

1. B. Hoglund, P.B. Mezger, Science, 150, 339 (1965)
2. A. Duprée and L. Goldberg, Annual Rev. Astronom. Astrophysics, 8, 231 (1970)
3. K.B. Mc Adam, W.H. Wing, Phys. Rev. A, 15, 678 (1977)
4. T.F. Gallagher, R.M. Hill, S.A. Edelstein, Phys. Rev. A, 13, 1348 (1976)
5. C. Fabre, P. Goy and S. Haroche, J. Phys. B, 10, L-183 (1977)
   C. Fabre, S. Haroche and P. Goy, Phys. Rev. A, 18, 229 (1978)
6. J.E. Bayfield, Proceedings of the C.N.R.S. International Conference on Atomic and Molecular States Coupled to a Continuum, Highly Excited Atoms and Molecules, Aussois, France (1977)
7. J.F. Gounand, J. Phys. Paris, 40, 457 (1979)
8. T.F. Gallagher and W.E. Cooke, Phys. Rev. Letters, 42, 835 (1979)
9. W.E. Cooke and T.F. Gallagher, Appl. Phys. Letters, 34, 369 (1979)
10. J.L. Hall, S.A. Lee, J. Helmcke, these proceedings

11. Any stray field in the M.C. would certainly be quite inhomogeneous and would give rise to line broadening of the same order as line shifts. Furthermore, this broadening should increase with n. As there is no evidence of any such n-dependent line broadening, we can assume that systematic Stark shifts are certainly negligible.

12. For an analysis of hyperfine and Zeeman effects on Na S-S transitions, see N. Bloembergen, M.D. Levenson and M.M. Salour, Phys. Rev. Letters, 32, 867 (1974)

13. T.W. Hänsch, in these proceedings and refs in.

14. M. Gross, P. Goy, C. Fabre, J.M. Raimond and S. Haroche, to be published

15. N. Skribanowitz, I.P. Herman, J.C. Mc Gillivray and M.S. Feld, Phys. Rev. Letters, 30, 309 (1973)

16. M. Gross, C. Fabre, P. Pillet and S. Haroche, Phys. Rev. Letters, 36, 1035 (1976)
H.H. Gibbs, Q.M.F. Vrehen and H.M.J. Hikspoors, Phys. Rev. Letters, 39, 547 (1977)

High Resolution Spectroscopy of Rydberg States

G. Leuchs, and S.J. Smith[1]

Sektion Physik, Universität München, D-8000 München, Fed. Rep. of Germany

H. Walther

Sektion Physik, Universität München and
Projektgruppe für Laserforschung der Max-Planck-Gesellschaft
zur Förderung der Wissenschaften e.V.
D-8046 Garching, Fed. Rep. of Germany

Introduction

High resolution spectroscopy of Rydberg states with sub-Doppler resolution has so far been performed using two-photon spectroscopy, the quantum beat and the double resonance methods. For the very highly excited states detection cannot be performed by observation of fluorescence since the lifetimes of these states scale as the principal quantum number cubed. Field or collisional ionization is used instead. As a result of the relatively long lifetimes Rydberg states have small natural widths which allow one to resolve splittings of less than 1 MHz. Therefore a very narrow laser linewidth is necessary if the two-photon absorption technique is used. In addition the Ramsey method of separated laser fields has to be applied in order to avoid transit time broadening [1]. In the case of the quantum beat and the double resonance methods there is no such requirement on the laser.

The use of the quantum beat method previously had been restricted to lower lying levels. However, it has recently been demonstrated that field ionization from Rydberg levels can also be used for the detection of quantum beats [2,3]. In the double resonance experiment the Rydberg states excited by a laser pulse interact with a microwave field for a certain time. Then a pulsed electric field, rising linearly in time, ionizes the Rydberg atoms. If the threshold for field ionization is different for the two states coupled by the microwave field the field ionization signals of these two states can be discriminated in time. In this way the microwave resonance can be detected via the field ionization signal [4,5].

In this paper quantum beat measurements of the fine structure of sodium ^2D states for n=21 to n=31 are reported. Furthermore it will be shown that the angular distribution of photoelectrons could be used to detect microwave resonances between Rydberg states.

Detection of Quantum Beats by Field Ionization

In the standard quantum beat experiment a short light pulse is used to excite a coherent superposition of the two states $|1\rangle$ and $|2\rangle$. This is possible if the duration of the pulse is short com-

[1]Permanent address: JILA, National Bureau of Standards, Boulder, Co 80302, USA

pared to $\hbar/\Delta E$ where ΔE is the energy difference between the levels $|1\rangle$ and $|2\rangle$. The detection is performed by observing the temporal behaviour of the fluorescence which is emitted due to a transition into a lower lying final state $|f\rangle$. Quantum beats can also be observed in absorption of a second short light pulse, as a function of the time delay between the two light pulses [6].

In the case of high Rydberg states the quantum beats can neither be observed in fluorescence nor in absorption to higher bound states. However, when the second step is photoionization the quantum beats remain observable. The photoelectrons can be used to detect the transition [7,8]. This has the disadvantage that the photoionization cross sections are rather small. In addition a variable delay of up to $\sim\mu$s between the two laser pulses has to be accomplished.

For quantum beats observed in field ionization, the second step $|1\rangle,|2\rangle \rightarrow |f\rangle$ is also a bound-free transition. Its nature, however, is quite different [9]. When interacting with photons the atoms jump from the initial to the final state whereas a slowly rising electric field pulse takes the atoms from zero to high field through atomic states which are stationary solutions of the corresponding Hamiltonian. The quantum interference can only be observed if the two channels $|g\rangle \rightarrow |1\rangle \rightarrow |f\rangle$ and $|g\rangle \rightarrow |2\rangle \rightarrow |f\rangle$ are indistinguishable, where $|f\rangle$ now is a continuum state of the atom in a high electric field and $|g\rangle$ the ground state. Consequently, for the observation of quantum beats the rapid risetime of the electric field pulse is critical.

GALLAGHER et al.[10] and VIALLE et al.[11] have investigated the field ionization behaviour of sodium ^2D states for slowly rising electric field pulses. They recorded the field ionization signal as a function of the electric field amplitude and found individual thresholds for the $|m_l|$ Stark levels (m_l is the quantum number for the projection of the orbital angular momentum on the direction of the electric field). The threshold for $|m_l|=0$ was found to be smaller than the one for $|m_l|=1$ and $|m_l|=2$. These $|m_l|$-states can be related to the initially excited $|m_j|$-states of the $^2D_{3/2}$ and $^2D_{5/2}$ fine structure levels which are excited by the laser in zero electric field.

With respect to the quantum beat experiment it is clear that for the observation of the coherences between the $^2D_{3/2}$ and $^2D_{5/2}$ states, both must contribute to the final states $|m_l|$ from which the ionization is performed. This can be achieved by using fast rising electric field pulses so that diabatic mixing between the Stark levels becomes possible.

In the quantum beat experiment the total number of electrons emitted by field ionization is recorded as a function of the delay between the exciting laser pulse and the field pulse. In order to avoid saturation the amplitude of the electric field pulse has to be kept low enough so that only a fraction (about 30% - 50%) of the atoms are ionized. With such an amplitude mainly the $|m_l|=0$ channel is observed, which originates only from the $^2D_{5/2}$, $|m_j|=1/2$ state when the electric field is changed so that the atoms go through an adiabatic passage. Thus the required condition for the detection of the quantum interference is that the electric field rises so fast that the $|m_l|=0$ channel is effectively mixed with the neighbouring channels ($|m_l| =1$ and/or $|m_l|=2$). GALLAGHER et al. [10] have shown that for an electric field slope of 2×10^{10} V/(cm s) no mixing between $|m_l|$ channels is observed for n<18 whereas for n>18 the $|m_l|=2$ channel mixes with other channels but still no mixing of the $|m_l|=0$ and $|m_l|=1$ channels occurs. Therefore for the quantum beat experiment a risetime faster than 2×10^{10} V/(cm s) has to be applied. An estimate of the required risetime can be obtained in the following way.

The Stark manifold of a level with principal quantum number n consists of n states. Due to a number of avoided crossings these states level out at high electric field and are about equally spaced in energy [12]. The separation between neighbouring states can be roughly estimated by dividing the energy difference of neighbouring n-states at zero field $2R/n^3$ by n, which is the number of states in the manifold. The resulting separation is $2R/n^4$ which corresponds to about 8 GHz for n=30 and 40 GHz for n=20. This estimate implies that if the risetime of the electric field is faster than $n^4 \hbar/(2R)$, which is about 0.01 ns for n=25, the mixing between neighbouring $|m_l|$ channels is guaranteed. But since during the passage from low to high field the $|m_l|$ states come even closer than $2R/n^4$ the required minimum risetime is in general less than $n^4 \hbar/(2R)$ which was actually the case in the experiment described in this paper.

Quantum Beat Experiments

The sodium Rydberg states were populated by a two step excitation using dye lasers pumped simultaneously by a nitrogen laser. In order to avoid collisions of the Rydberg atoms an atomic beam was used. The sodium atoms were excited via the $3\,^2P_{3/2}$ intermediate state to the levels of the $n\,^2S$ and $n\,^2D$ Rydberg series. The optical excitation was performed between two parallel field plates.

Fig.1 Pulse sequence and timing used for excitation and detection (a) and schematics of the experimental set up (b)

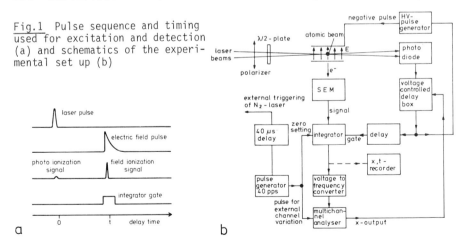

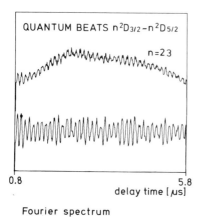

QUANTUM BEATS n²D₃/₂ - n²D₅/₂

n = 23

0.8   5.8
                    delay time [μs]

Fourier spectrum

0   10
        frequency [MHz]

Fig.2 Quantum beat signal of the fine structure splitting of the 23²D state of sodium. The upper trace shows the experimental recording. The lower trace shows the same recording at another scale with the slowly varying background subtracted. The splitting frequency is obtained by a Fourier analysis of the quantum beat signal

The Rydberg atoms were detected by applying a voltage pulse to the field plates immediately after the excitation and detecting the emitted field electrons (Fig.1a).

For the quantum beat measurement the field ionization signal had to be measured as a function of the delay between the laser excitation and the electric field pulse used for field ionization. The scheme of the experimental set up is shown in Fig.1b. For details see [3].

Fig.2 shows the field ionization signal of the 23 ²D Rydberg state as a function of the delay between laser and electric field pulse. The delay time was varied between 0.8 and 5.8 μs. The observed periodic variation is the quantum beat signal of the two fine structure levels 23²D₃/₂ and 23²D₅/₂ . The periodic variation due to the quantum beat is 5 - 10 % of the total field ionzation signal. The frequency of the modulation giving the splitting of the two fine structure levels was determined by a Fourier analysis. The linewidth in the Fourier spectrum is 250 kHz which is close to the theoretical limit of 200 kHz corresponding to the finite observation time of 5 μs.

For this experiment the earth's magnetic field in the excitation region was compensated by three pairs of Helmholtz-coils not shown in Fig.1b. The signal in Fig.2 was obtained by averaging over 1000 laser pulses per channel. This corresponds to a total measuring time of 1.5 hours. The slowly varying background of the field ionization signal shown in Fig.2 is caused by the change of the detection sensitivity when the Rydberg atoms pass through the field of view of the electron multiplier. The dependence of the quantum beat signal on the polarization of the laser beams was investigated. For details see [3].

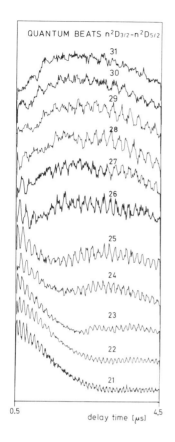

QUANTUM BEATS n²D₃/₂ - n²D₅/₂

0.5   4.5
           delay time [μs]

Fig.3 Quantum beat signals of high lying ²D states of sodium. The variation of the beat frequency with principal quantum number is shown. Several quantum beat frequencies appear due to a Zeeman splitting of the fine structure levels in the earth's magnetic field. The frequencies were determined by a Fourier analysis. The results are

n	Δ_{FS} [MHz] =
21	10.23(25)
22	9.13(10)
23	8.00(10)
24	7.03(10)
25	6.22(10)
26	5.50(30)
27	4.75(30)
28	4.63(30)
29	4.10(30)
30	3.60(50)
31	3.10(50)

Fig.3 shows a series of quantum beat measurements on sodium n^2D states for n=21 to n=31. The earth's magnetic field was not perfectly compensated so that in addition to the fine structure a Zeeman splitting also is present. Therefore several quantum beat frequencies are observed due to coherences between Zeeman levels of the same and of different fine structure levels. Since the first type is very slow only half a period can be seen. The latter type results in a signal which is determined by several higher frequencies which differ by the amount of the Zeeman splitting; the superposition of these frequencies leads to the periodically varying amplitude of the quantum beat signal. The variation of the quantum beat frequency with the principal quantum number can easily be recognized.

Results on the Fine Structure Splitting of the n^2D States of Na

Using the quantum beat method described in this paper the fine structure splittings have been measured for the n ²D levels from n=21 to n=31. The results are given in the caption of Fig.3. The hyperfine interaction in the n^2D states is expected to be smaller than the natural width and can therefore be neglected.

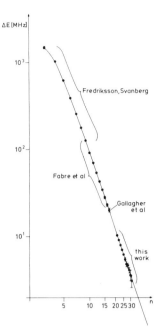

Fig.4 The fine structure splitting of sodium ^2D states is plotted versus principal quantum number. The solid line resulting from the extrapolation of the results obtained from measurements on lower ^2D states fits well with the fine structure splitting measured in this work (FREDRIKSON et al. [22], FABRE et al. [13], GALLAGHER et al. [5]).

In the following our results are compared with the n^2D fine structures measured previously for lower lying levels.

FABRE et al. [13] have investigated the splitting for the levels n=9 to n=16 by observing quantum beats in fluorescence. Their results could be summarized by the formula

$$\Delta_{FS} = -A/n^{*3} + B/n^{*5}.$$

They are compared with our measurements in Fig.4. Our new values agree quite well with a $1/n^{*3}$ dependence since the straight line obtained in the double logarithmic plot (Fig.4) has a slope of -2.95(10). The $1/n^{*5}$ term which is important for the lower levels can be neglected for n larger than 20.

Collisions of Rydberg Atoms

The investigation of the collisional properties of Rydberg states is important because the large mean square radii of the charge distribution of electrons in states with high principal quantum numbers lead to large cross sections for quasi elastic collisions [14]. Experiments have been performed investigating collisions of alkali Rydberg atoms with inert gas or ground state atoms [15, 16,17]. In general it is found that the collisional cross sections increase in proportion to the geometric cross section up to a certain principal quantum number $\check{n}$, and then slightly decrease for $n > \check{n}$.

In the case of quantum beat measurements the amplitude of the modulation can be attenuated by phase changing collisions and

260

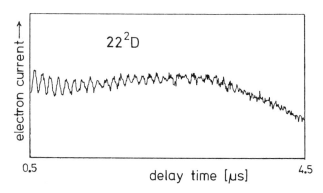

22^2D

electron current →

0.5   delay time [μs]   4.5

Fig.5 Quantum beat signal of the 22^2D state of sodium under the influence of collisions

quasi elastic collisional energy transfer to neighbouring states. For quantum beats in fluorescence a collisionally induced decrease has been observed by PENDRILL [18].

In our experiment we have observed a decrease of the quantum beat amplitude under atomic beam conditions where collisions between the atoms could not be excluded (sodium vapour cloud). A measurement on the 22^2D state of sodium is shown in Fig.5. This decrease has also been seen for ^2D states with n=21 to n=30. The experiments show that there is no significant difference in the decrease of the quantum beat signal as a function of n. This indicates that for Na**-Na collisions the range of n=21 to n=30 is already beyond $\tilde{n}$.

A crude estimate for $\tilde{n}$ can be obtained in the following way. The perturbing particle interacts with the Rydberg atom mainly via the smeared out charge distribution of the highly excited electron. The orbiting time of the electron in a state n is $\tau_{orb}= h a_0 n^3/e^2$ [19] and the interaction time of the perturbing particle with the electron is given by the low energy electron scattering length of the perturber L divided by the thermal velocity $\tau_{int}= |L|/v$. If $\tau_{orb} < \tau_{int}$ the perturbing particle interacts with the total charge distribution smeared out over the atom. Therefore the collisional cross section is given by the geometric cross section. In the other case $\tau_{orb} > \tau_{int}$ this is not guaranteed. $\tilde{n}$ is determined by $\tau_{orb} = \tau_{int}$, $\tilde{n} = (e^2|L|/(ha_0v))^{1/3} = 131|L|^{1/3}$, with L in Å. For sodium ground state atoms L =5 Å [20] giving n=22 which is in reasonable agreement with the experimental observation.

Angular Distribution of Photoelectrons in Three Photon Ionization of Sodium

The angular distribution of photoelectrons produced by multiphoton ionization of atoms is quite sensitive to the intermediate states. We report measurements of the angular distribution which have been obtained for resonant three-photon ionization. The results demonstrate that the angular distribution would be changed by mixing the intermediate states by microwave transitions. This can be used as mentioned above for optical microwave experiments to determine the energy splitting of Rydberg states differing in their angular momentum quantum number by one.

As for the quantum beat experiment, two nitrogen laser pumped dye lasers were used to excite the atoms of the atomic beam. The sodium atoms were ionized in the resonant three-photon process $3\ ^2S_{1/2} \rightarrow 3^2P_{1/2}, 3^2P_{3/2} \rightarrow n^2D \rightarrow |1,k\rangle$. The electrons emitted in the plane perpendicular to the direction of the laser beams were detected with an angular resolution of 0.35 rad. The angular distribution in that plane was probed by rotating the direction of the linear polarization of the laser light.

In the case of single photon ionization of atoms with equal populations of the m-sublevels the photoelectron angular distribution can be described by the general formula

$$I(\theta) = 1 + \beta P_2(\cos\theta).$$

Here θ is the angle between the direction of emission of the photoelectron and the laser polarization. β is the anisotropy parameter and P_2 is the second Legendre polynomial. In the case where m-sublevels are selectively populated in an n-photon ionization process, higher powers of $\cos^2\theta$ have to be included:

$$I_n(\theta) = \sum_{\nu=0}^{n} a_\nu \cos^{2\nu}\theta .$$

Thus for a resonant multiphoton ionization process sharp maxima appear in the angular distribution of photoelectrons which critically depend on the quantum numbers of the intermediate states.

The angular distribution of photoelectrons produced by resonant three-photon ionization $3^2S_{1/2} \rightarrow 3\ ^2P_{1/2} \rightarrow 6^2D \rightarrow |1,k\rangle$ is shown in a polar diagram (Fig.6a). The solid line was obtained in a least squares fit of the analytical function $I_3(\theta)$ to the experimental data giving $I_3(\theta)=0.23-0.48\cos^2\theta+\cos^4\theta$. The experimental errors for the coefficients are smaller than ±0.01. The coefficient of $\cos^6\theta$ is zero within this error. The arrow indicates the direction of laser polarization. Theoretical calculations by LAMBROPOULOS [21] yield $I_3(\theta)=0.19-0.36\cos^2\theta+\cos^4\theta+0.03\cos^6\theta$. A similar experiment where instead of $3^2P_{1/2}$, the $3^2P_{3/2}$ state was tuned into resonance is shown in Fig.6b. The least squares

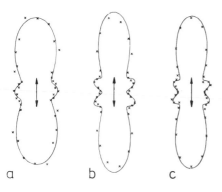

a   b   c

Fig.6 Angular distribution of photoelectrons in resonant three-photon ionization of sodium via the $3^2P_{1/2}$ and 6^2D states (a), via the $3^2P_{3/2}$ and 6^2D states (b), and via the $3^2P_{3/2}$ and 10^2D states (c)

fit yields $I_3(\theta)=0.03+0.32\cos^2\theta-\cos^4\theta+0.87\cos^6\theta$ and the theory gives $I_3(\theta)=0.01+0.29\cos^2\theta-\cos^4\theta+0.90\cos^6\theta$. We also measured the angular distribution in three-photon ionization via the $3^2P_{3/2}$ and the 10^2D state. The result is shown in Fig.6c and the least squares fit yields $I_3(\theta)=0.03+0.30\cos^2\theta-\cos^4\theta+0.88\cos^6\theta$. Thus the shape of the photoelectron angular distribution does not depend strongly on the principal quantum number of the intermediate ^2D state.

Acknowledgement

The support of the Deutsche Forschungsgemeinschaft is gratefully acknowledged. One of us (SJS) would like to thank the Alexander von Humboldt Foundation for a Senior U.S. Scientist Award.

References

1. S.A.Lee, J.Helmcke, J.L.Hall: this volume
2. G.Leuchs, H.Walther: in Laser Spectroscopy III, eds. J.L.Hall, J.L.Carlsten (Springer, Berlin, Heidelberg, New York 1977) pp.299-305
3. G.Leuchs, H.Walther: submitted to Z.Physik A
4. C.Fabre, P.Goy, S.Haroche: J.Phys.B10,L183(1977)
5. T.F.Gallagher, L.M.Humphrey, R.M.Hill, W.E.Cooke, S.A.Edelstein: Phys.Rev.A15,1937(1977)
6. T.W.Ducas, M.G.Littman, M.L.Zimmerman: Phys.Rev.Lett. 35,1752 (1975)
7. R.Zygan-Maus, H.H.Wolter: Phys.Lett.64A,351(1978)
8. A.T.Georges, P.Lambropoulos: Phys.Rev.A18,1072(1978)
9. P.Jacquinot: this volume
10. T.F.Gallagher, L.M.Humphrey, W.E.Cooke, R.M.Hill, S.A.Edelstein,: Phys.Rev.A16,1098(1977)
11. J.L.Vialle, H.T.Duong: J.Phys.B12,1407(1979)
12. M.G.Littman, M.M.Kash, D.Kleppner: Phys.Rev.A17,1226(1978)
13. C.Fabre, M.Gross, S.Haroche: Opt.Comm.13,393(1975)
14. A.Omont: J.Physique 38,1343(1977)
15. T.F.Gallagher, S.A.Edelstein, R.M.Hill: Phys.Rev.A15,1945(1977)
16. M.Hugon, F.Gounand, P.R.Fournier, J.Berlande: subm.to J.Phys.B
17. B.P.Stoicheff, E.Weinberger: this volume
18. L.R.Pendrill: J.Phys.B10,L469(1977)
19. A.Sommerfeld: Atombau und Spektrallinien, 8th ed. (Vieweg, Braunschweig 1969) p.93
20. H.S.W.Massey, E.H.S.Burhop, H.B.Gilbody: Electronic and Ionic Impact Phenomena Vol.I (Clarendon Press, Oxford 1969) p.564
21. P.Lambropoulos, private communication
22. K.Fredrikson, S.Svanberg: J.Phys.B9,1237(1976)

Term Values, Pressure Broadenings and Shifts in High Rydberg Levels of Rubidium

B.P. Stoicheff and E. Weinberger

Department of Physics, University of Toronto
Toronto, Ontario M5S 1A7, Canada

1. Introduction

The Doppler-free linewidth and dramatic increase in sensitivity
characteristic of two-photon absorption using counter-propagating laser beams, has made this an important technique for
studies of highly-excited Rydberg states. In this paper, we
present the results of recent investigations of two-photon
spectra in atomic rubidium as examples of the type of information obtainable from such spectra. Wavelengths of the transi-
tions $n^2S \leftarrow 5^2S$ (with n=9 to 116), and $n^2D \leftarrow 5^2S$ (with n=7 to 124)
have been measured with a precision of 1 part in 10^7. Series
formulae have been computed for the term values, leading to
accurate values for the ionization limit and quantum defects.
Fine and hyperfine structure splittings have been measured, and
the isotope shift for ^{85}Rb and ^{87}Rb evaluated. Measurements of
self-broadening and frequency shifts have been made for Rb
pressures in the range 1 to 100 mTorr. For both series of
lines, frequency shifts are to the red; the rate increases
rapidly with n, and remains constant for n>40. For the ^2S
series, the linewidths exhibit a highly oscillatory behaviour
with n, the peaks being approximately equally spaced in n.

2. Experiments and Observed Spectra

The experimental arrangement shown in Fig.1 is essentially that
described by HARVEY and STOICHEFF [1]. The beam from a CW
tunable dye laser was focused at the centre of a thermionic
detector enclosed in the Rb sample cell, and was reflected back
on itself by a concave mirror, to provide the two counterpropagating beams. As the frequency of the dye laser was
scanned over two-photon resonances, Rb atoms were excited to
high Rydberg states and ionized on collision with neighbouring
atoms, resulting in increased current in the detector. A part
of the laser beam was diverted to a wavemeter (KOWALSKI,
HAWKINS and SCHAWLOW [2], for wavelength measurement by

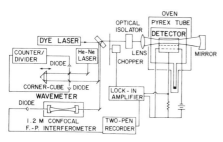

Fig.1 Schematic diagram of the experimental arrangement.

Fig.2 Energy level diagrams of the $5s^2S_{1/2}$ ground state, and $ns^2S_{1/2}$ and $nd^2D_{3/2,5/2}$ excited states of ^{85}Rb and ^{87}Rb, showing allowed two-photon transitions and expected spectra.
▼

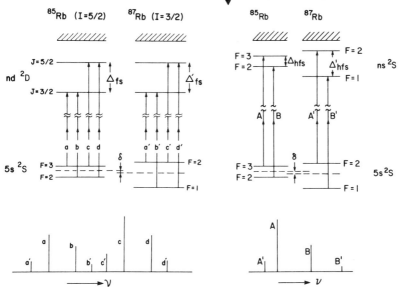

comparison with the known wavelength of a stabilized He-Ne laser, and apart to a 1.2 m confocal interferomenter (spectral free range =62.5 MHz) to provide a frequency scale.

In Fig.2 are shown the energy levels $5^2S_{1/2}$, $n^2S_{1/2}$, and $n^2D_{3/2,5/2}$ of ^{85}Rb and ^{87}Rb, together with the allowed two-photon transitions and expected spectra. Some examples of observed spectra are reproduced in Fig.3. For each spectrum, the most intense component was identified (F=3←F=3 for n^2S←5^2S transitions, and J=5/2←F=3 for n^2D←5^2S transitions) and its wavelength measured. All of the measured wavelengths were corrected for the small dispersion in air relative to the He-Ne laser wavelength. The wavenumbers corresponding to the two-photon transitions, or term values $T_{n,l}$ were determined taking into account the weighted average of the fine structure of the ^2D excited levels, and of the weighted average of the hyperfine structure for the ^2S levels. Each of the term values was also corrected for frequency shifts due to Rb pressure.

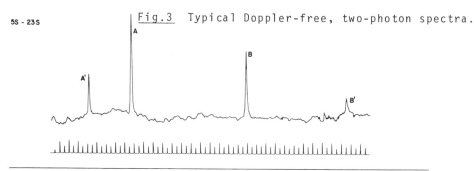

Fig.3 Typical Doppler-free, two-photon spectra.

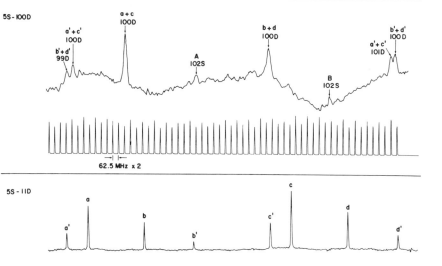

Earlier measurements of term values for ^{85}Rb, including up to 53^2S and 54^2D are listed by MOORE [3]. Most of these values were obtained from absorption spectra, some at relatively high Rb pressure, and are quoted to be reproducible to $\sim$0.03 cm^{-1}. Comparison of the present and earlier term values for both series, indicates that many of the values differ by more than $\pm$0.03 cm^{-1}, some by as much as $\pm$0.2 cm^{-1}, and 17^2D and 51^2S by $\sim$1 cm^{-1}. The present values are more extensive, and are considered to have an absolute accuracy of $\pm$0.003 cm^{-1}, or $\pm$100 MHz.

The wavenumber separations of various components (Fig.2) were measured: those of c-a gave the fine structure spacings ^{2}D$_{5/2}$-^{2}D$_{3/2}$; those of B-A and B'-A' were used to determine the hyperfine structure spacings of ^{85}Rb and ^{87}Rb, respectively; and the spacings c-c', a-a' were used to obtain the isotope shift of the ground states.

For measurement of frequency shifts and linewidths with pressure of Rb, two thermionic detectors were used in series. One served as a refernce, with Rb pressure maintained at 2 mTorr, while the Rb pressure in the second cell was carried between 2 to 100 mTorr by control of the cell temperature. A three-pen chart recorder simultaneously monitored the signals from the two detectors and the fringes from the confocal inter- ferometer. The tunable CW dye laser (Coherent CR-599-21) used in this work had a linewidth and stability of $\sim$2 MHz. Measure- ments of frequency shifts and linewidths to within an accuracy of 10% were easily possible.

3. Analysis of Spectra: Term Values, Quantum Defects, and Spectroscopic Parameters

For ^{85}Rb with its valence electron outside a spherically sym- metric core, the term values $T_{n,1}$ are given by the Rydberg formula

$$T_{n,1} = T_\infty - R/n^2_{eff} \tag{1}$$

Here, T_∞ is the ionization limit, R is the appropriate Rydberg constant (R_{85}=109736.605 cm^{-1}), and n_{eff} the effective quantum number, given by

$$n_{eff} = n-\delta_{n,1} \tag{2}$$

where n is the principal quantum number, and $\delta_{n,1}$ is the quantum defect.

In the theory of quantum defects [4] $\delta_{n,1}$ is written as a polynomial:

$$\delta_{n,1} = \sum_{k=0}^{N} P_{2k}(1/n_{eff})^{2k} \tag{3}$$

The coefficients P_{2k} can be calculated if the form of the potential is known, or they can be determined experimentally; $-R/n^2_{eff}$ is the energy of the nth level measured from the series limit, and k is an interger with values 0,1, ...N.

We have analyzed the data by least-squares, making use of Eqs. (1,2,3) to determine values for the ^{2}S and ^{2}D series limits, the coefficients P_{2k}, and the corresponding values of $\delta_{n,1}$ and n_{eff}. The "best-fit" values of T_∞, P_0, P_2, P_4, P_6 (ie N=3) are given in Table 1. Graphs of the wave number

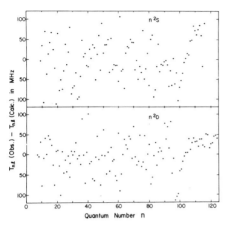

Fig.4 Wavenumber differences
$T_{n,1}(obs)-T_{n,1}(calc)$ for the
^2S ^2D states

differences $T_{n,1}(obs)-T_{n,1}(calc)$ are shown for both series, in
Fig.4. These differences are seen to be reasonably random, and
scattered within ±80 MHz, well within the stated errors of
measurement. Considerably larger differences, from ±200 to
±800 MHz, were found between the calculated and Moore's term
values for the low levels 6^2S to 8^2S and 4^2D to 6^2D. These
results suggest that an effort should be made to measure the
term values for these levels by two-photon, doppler-free
spectroscopy, even though it is presently more difficult to
achieve tunable, monochromatic radiation in the 700 to 1000 nm
region required for these measurements.

Table 1 Term value parameters T_∞, P_0,...P_{2N}, for n^2S and n^2D
series of ^{85}Rb

	n^2S Energy levels (n=9 to 116)	n^2D Energy levels (n=7 to 124)
$T_\infty cm^{-1}$	33690.856(3)	33690.857(3)
P_0	3.13114(2)	1.34717(2)
P_2	0.1891(9)	-0.5987(8)
P_4	-0.578(32)	-1.39(2)
P_6	13.5(12)	-7.4(9)

The values for the ionization limit obtained from ^2S and ^2D
series were found to be the same, namely T_∞=33690.8565(15) cm^{-1}.
This is in excellent agreement with the value T_∞=33690.7989(2)
cm^{-1} recently determined by LEE, HELMCKE, HALL, and STOICHEFF
[5]. The difference of 0.058 cm^{-1} results from the fact that
in the present work, measurements are quoted relative to the
F=2 sublevel of the ground state, whereas in LEE et al,
measurements are relative to the centre of gravity of the

ground state (a difference of 1770.8 MHz or 0.059 cm^{-1}). The value $T_\infty = 33690.8565(15)$ cm^{-1} determined here is considerably more accurate than the earlier value $T_\infty = 33691.02(3)$ cm^{-1} [3]. The value for the ^{87}Rb ionization limit (relative to the F=1 sublevel of the 5s^{2}S level) becomes $T_\infty = 33690.8565(15)+0.0890 = 33690.9455(15)$ cm^{-1}, in comparison with $33691.10(1)$ cm^{-1} found earlier [3].

The dependence of quantum defects δ_{ns} and δ_{nd} on n is shown in Fig. 5. For the higher Rydberg states, the relative accuracy of the small difference $T_\infty - T_{n,1}$ decreases markedly with n, and this is reflected in the increasing errors of δ_{n1} with n. Nevertheless, it is evident that the value of δ_{ns} is essentially constant at 3.135 for n≳20, after decreasing rapidly at lower n; and that of δ_{nd} increases rapidly at low n to reach a constant value of 1.3465 for n≳30. These asymptotic values correspond to the constants P_0 in Table 1. Their nonzero values are an indication that the interaction potential of these highly excited electrons with the core is non-Coulombic. The + and - values of P_2 for δ_{ns} and δ_{nd} (Table 1) are representative of penetrating ns, and non-penetrating nd, Rydberg "orbits", respectively. The nonzero values for P_2, P_4, P_6 are most important for term values of low n, and are a measure of core-polarization.

The fine structure splittings for the levels 7^2D to 11^2D were measured and found to be in good agreement with recent measurements of NILSSON and RYDBERG [6], and those for 32^2D and 50^2D

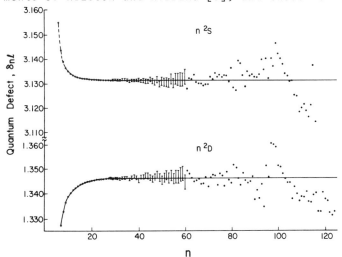

Fig.5 Graphs of the quantum defect $\delta_{n,1}$ vs quantum number n, for ^{2}S and ^{2}D states

agreed with our earlier values [1]. All of these values
(including that for 70²D) also agree with the relation Δ_{fs}
(GHz)=10,800 n_{eff}^{-3}-84,870 n_{eff}^{-5} obtained for the dependence of
Δ_{fs} on n[1].

All of the known values for the hyperfine structure splittings in ⁸⁵Rb and ⁸⁷Rb, from 5²S [7], 7²S and 8²S [8], and 9²S
to 13²S measured here, plotted on log Δ_{hfs} vs log n_{eff} graphs
show linear relations with slopes of -3.0. This is in agree-
ment with the dependence $\Delta_{hfs} \alpha n_{eff}^{-3}$, expected from theory.
The value for the isotope shift of the ⁸⁵Rb and ⁸⁷Rb ground
states was determined from 13 spacings c-c' and a-a' over the
range n=7 to 90, and found to be δ=167±10 MHz.

4. Frequency Shifts and Self-Broadening

Frequency shifts were measured directly from the shift of line
centres of spectra obtained with the high pressure cell (up to
100 mTorr) against the cell with 2 mTorr pressure. Each line
was found to be shifted to lower frequencies with increasing
pressure of Rb. This red shift increased linearly with pressure.
The dependence of the shift rate on quantum number is shown in
Fig.6 and is seen to be similar for the S and D series of lines.
The shift rate increases almost linearly, from ∿0.1 MHz/mTorr
at n=10 to ∿2.0 MHz/mTorr at n∿40, and remains at this value
for higher levels. This is similar to the behaviour predicted
[9] for shifts of high Rydberg levels by foreign gases at
pressures of ∿1 atmosphere, yet observed here at pressures 4 to
6 orders of magnitude lower.

The measured linewidths (FWHM) were found to increase linearly
with increasing pressure of Rb. Extrapolation to zero-pressure
resulted in linewidths of ∿10 MHz, this being the contribution

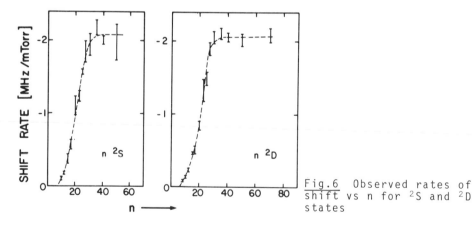

Fig.6 Observed rates of
shift vs n for ²S and ²D
states

of natural linewidth (<1 MHz), transit-time broadening (∿5 MHz),
and perhaps residual foreign-gas broadening (due to a small
impurity of Cs). There was no evidence of Zeeman, Stark, or
power broadening even in level 124D, and since these effects
scale as n to the sixth or higher power, their contribution to
lower levels is negligible. Linewidths as narrow as ∿10 MHz
were measured at 2 mTorr, and as large as ∿300 MHz at 100 mTorr.
Again, as for the frequency shift rate, both the S and D series
showed the same general behaviour for the broadening rate with
quantum number. The broadening rate was 0.3 MHz/mTorr at n=10,
increased to a maximum of 3.0 at n∿23, and then decreased to a
constant rate of ∿0.5 MHz/mTorr at n=70 to 90. This general
behaviour is expected for foreign-gas broadening at high
pressures [9], but has not been reported at the low pressures
used here.

When the linewidths in the region of the maximum broadening
rate (for n∿15 to n∿35) were investigated more thoroughly by
measuring the widths for each n value, a surprising dependence
on n was observed for the ^{2}S series of lines. This is shown in
Fig.7 for a pressure of 60 mTorr. A highly oscillatory
dependence on n was found in the region n=20 to 30, with the
peaks having an almost constant spacing. At this time, the
the origin of this dependence is not known. It should be
pointed out that for n∿30, the diameter of the classical electron
orbit is ∿1000 Å, which is also the spacing of the Rb atoms at
this pressure. Thus, at least one Rb atom would be expected
within the large volume of the Rydberg atom, possibly forming
a quasi-molecule, and leading to an increase in the lifetime
of the excited electron (at n∿21, 25, 28, 32). Further studies,
including the broadening due to krypton are being considered.

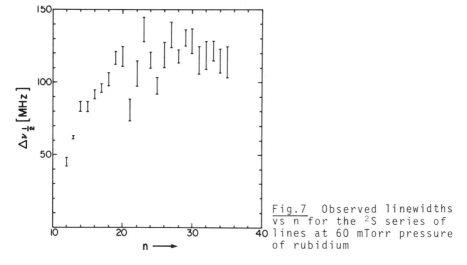

Fig.7 Observed linewidths
vs n for the ^{2}S series of
lines at 60 mTorr pressure
of rubidium

271

In summary, the technique of Doppler-free, two-photon spectroscopy, introduced by Vasilenko, Chebotaev and Shishaev [10], and by Cagnac, Grynberg and Biraben [11], is an important one for spectroscopy and for investigating the interaction of highly-excited Rydberg atoms.

References

1 K.C. Harvey and B.P. Stoicheff, Phys. Rev. Lett. 38, 537 (1977)

2 F.V. Kowalski, R.T. Hawkins, and A.L. Schawlow, J. Opt. Soc. Am. 66, 965 (1976)

3 C.E. Moore, "Atomic Energy Levels", U.S. Gov't. Printing Office, Washington, Vol. II, 180 (1971)

4 M.J. Seaton, Proc. Phys. Soc. 88, 815 (1966)

5 S.A. Lee, J. Helmcke, J.L. Hall, and B.P. Stoicheff, Opt. Lett. 3, 141 (1978)

6 L. Nilsson and S. Rydberg, Physica Scripta 17, 53 (1978)

7 E. Arimondo, M. Inguscio, and P. Violino, Rev. Mod. Phys. 49, 31 (1977)

8 R. Gupta, S. Chang, and W. Happer, Phys. Rev. A6, 529 (1972)

9 R.G. Breene, Jr. in Handb. der Phys. XXVII, Spectroskopie I, 1 (1964)

10 L.S. Vasilenko, V.P. Chebotaev, and A.V. Shishaev, Zh. Eksp. Teor. Fiz. Pis'ma Red. 12, 161 (1970) [JETP Lett. 12, 113 (1970)]

11 B. Cagnac, G. Grynberg, and F. Biraben, J. Phys. (Paris) 34, 845 (1973)

Doubly Excited Alkaline Earth Atoms

T.F. Gallagher, W.E. Cooke, and K.A. Safinya

Molecular Physics Laboratory, SRI International
Menlo Park, California, USA

I. Introduction

The spectra of two electron, He-like, alkaline earth atoms are much richer
than the corresponding H-like alkali spectra. The added complexity comes both
from the far greater number of possible states due to the larger number of
degrees of freedom and from the fact that most of the doubly excited states
lie above the first ionization limit and therefore autoionize. The doubly
excited states are quite fascinating in their own right, and an understanding
of their properties is important for several applications. For example, both
RHODES [1] and HARRIS [2] have suggested the use of autoionizing states for
the generation of far uv radiation. Another example, far removed from laser
spectroscopy, is the inverse of autoionization, dielectronic recombination,
which is an important energy loss mechanism in Tokamaks [3].

Here we describe experiments to probe both intrinsic features of doubly
excited autoionizing alkaline earth atoms as well as one of the familiar
features of the study of autoionizing states, the BEUTLER-FANO interference
profiles [4].

II. Experimental Approach

In our experiments we have used multi-step laser excitation of atoms in an
atomic Beam. In Fig.1 we show the energy levels [5] used in the laser exci-
to the Ba$6p_{\frac{3}{2}}15d$ state. The first two lasers are simultaneous and excite atoms
from the ground state to the 6s6p state and then to the bound 6snd state.
That is, we excite only the "outer" electron. In the absence of an electric
field we can excite either the 6s15s or 6s15d states. We can extend the technique to an arbitrary 15ℓ state by a Stark switching technique [5]. If we
apply an electric field we may excite the s and d component of any of the
6s15k Stark states. If we then turn off the field slowly (500 ns), the atoms
relax back adiabatically to one zero field ℓ state. The no crossing rule
produces a 1:1 correspondence between the Stark states and zero field ℓ states
[7].

The third laser excites only the inner electron, driving the transition
6s15d → $6p_{\frac{3}{2}}15d$. By sweeping the wavelength of the third laser we are able
to determine the position and width of the $6p_{\frac{3}{2}}15d$ level.

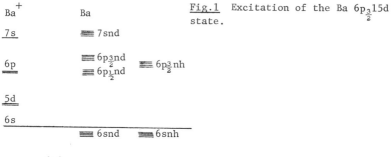

Fig.1 Excitation of the Ba $6p_{\frac{3}{2}}15d$ state.

We detect the fact that we have induced the transition to the autoionizing state by collecting the ions or electrons produced by autoionization. Ion detection is suitable for experiments where all we wish to determine is the total ionization produced. Electron detection, though, offers more interesting possibilities. Since the ejected electron carries off all the kinetic energy from the autoionization, analyzing the energies of the ejected electrons enables us to determine the final state of the Ba[+] ion and to observe transitions between autoionizing states. The physical arrangement for energy analysis is shown in Fig.2. The atom beam is crossed by the laser beams midway between a plate and a grid 1.12 cm apart. The grid is grounded and up to $\pm$ 15 V is applied to the plate to accelerate or retard the ejected electrons. Scanning the voltage on the plate yields an energy spectrum of the ejected electrons.

One aspect of this method which we would particularly like to stress is that all the transitions are strong "one electron" transitions. It is particularly important that the third transition $6sn\ell \rightarrow 6pn\ell$, to the autoionizing state, is basically the resonance line of the Ba[+] ion, while the transition

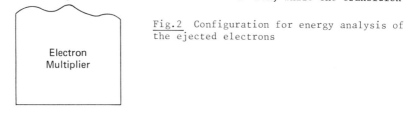

Fig.2 Configuration for energy analysis of the ejected electrons

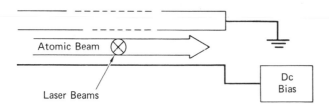

to the continuum $6sn\ell \rightarrow 6st\,\ell'$ is very weak. Thus, we <u>never</u> observe Beutler-Fano profiles from interference with the continuum, a very real advantage over conventional methods.

III. Experiments

A. Variation in Autoionization widths with n and ℓ

In Fig. 3 we show a picture of the Ba atom with the inner electron in the 6p state and the outer electron in several n,ℓ states. This simple picture helps to convey in a qualitative way the dependence of the autoionization rate of the $6pn\ell$ states on n and ℓ of the outer electron. Autoionization normally proceeds via the electrostatic interaction, $1/r_{12}$, between the two electrons. Since the 6p electron is localized near the Ba^{++} core, it is clear that the autoionization rate will directly reflect the amplitude of the outer electron's wave function at the core. Thus, we expect lower n states with smaller orbital radii to have higher autoionization rates. Qualitatively, the autoionization rate should vary as $(n^*)^{-3}$ where n^* is the effective quantum number of the outer electron. Fig. 3 also leads us to expect that the low ℓ states with highly eccentric orbits would have higher autoionization rates than the circular high ℓ orbits which never come close to the ion core.

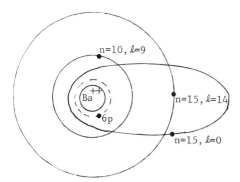

Fig.3 Ba $6pn\ell$ atom with the outer electron in several $n\ell$ states

In our initial studies [6] of Sr, we observed the variation of the autoionization rates of the Sr $5pn\ell$ levels (no difference was ever observed between $5p_{\frac{1}{2}}n\ell$ and $5p_{\frac{3}{2}}n\ell$). These autoionization rates were obtained from the widths of the third transition ($5sn\ell \rightarrow 5pn\ell$). In Fig.4 we show the n^* dependence of the widths of the 5pns states which exhibits the expected $(n^*)^{-3}$ behavior, and in Fig.5 we show the ℓ dependence of the $5p15\ell$ states which was obtained using the Stark switching technique to populate the $5sn\ell$ states.

B. Branching ratios for autoionization

If we look carefully at Fig.1, we see that for the $6pn\ell$ states, autoionization can lead to several states of the Ba^+ ion. For example, for the $6p_{\frac{1}{2}}nd$ states, it is possible to form Ba^+ in the $5d_j$, or $6s_{\frac{1}{2}}$ states, but for the $6p_{\frac{3}{2}}nd$ states, formation of Ba^+ in the $6p_{\frac{1}{2}}$ state is also possible.

275

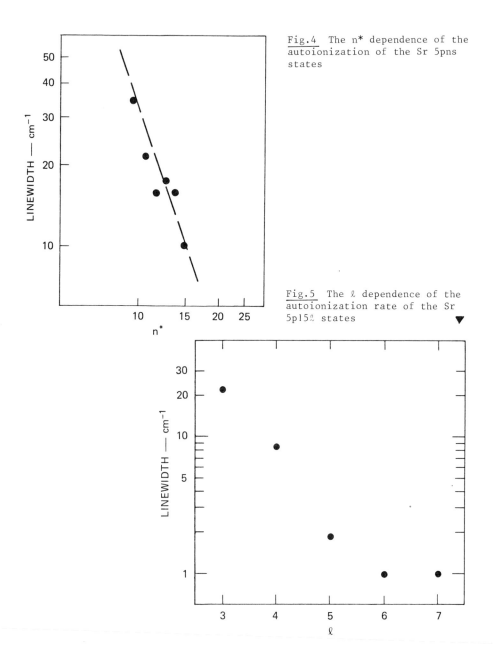

Fig.4 The n* dependence of the autoionization of the Sr 5pns states

Fig.5 The ℓ dependence of the autoionization rate of the Sr 5p15ℓ states ▼

Although it was of no consequence whether the Sr^+ core was in the $5p_{\frac{1}{2}}$ or $5p_{\frac{3}{2}}$ state, in Ba we have found pronounced differences between the autoionization rates of the $6p_{\frac{1}{2}}n\ell$ and $6p_{\frac{3}{2}}n\ell$ states for several series of ℓ states. One of the best such examples is the $Ba(6p_j20s_{\frac{1}{2}})_J$ states [8]. Both the J = 0

276

and $1(6p_{\frac{1}{2}}20s_{\frac{1}{2}})_J$ states have widths, due to autoionization, of 3 cm^{-1}, as does the $(6p_{\frac{3}{2}}20s_{\frac{1}{2}})_2$ state. However, the $(6p_{\frac{3}{2}}20s_{\frac{1}{2}})_1$ state has a width of 12 cm^{-1}. In the $p_{\frac{3}{2}}$ doublet, the J = 2 state lies above the J = 1 state, so the energy ordering of the states is anomalous. This led us to suggest that the excess width of the $(6p_{\frac{3}{2}}20s_{\frac{1}{2}})_1$ state was due to a spin-other orbit coupling which allowed the J = 1 but not the J = 2 state to autoionize to the Ba$^+$6p$_{\frac{1}{2}}$ state. However, in these experiments we only observed the ions produced and were thus only sensitive to the total ionization, so there was no concrete evidence for the suggestion. Recently, though, we have analyzed the energies of the ejected electrons to test the hypothesis [9].

Atoms in the 6p$_{\frac{3}{2}}$20s state autoionizing to the Ba$^+$6p$_{\frac{1}{2}}$ state yield 0.2 eV electrons whereas those which autoionize to the Ba$^+$6s state yield 2.5 eV electrons, so it is not particularly difficult to see a difference if it exists. Since our retarding voltage energy analyzer was configured with precisely this not very taxing experiment in mind, it does not have high

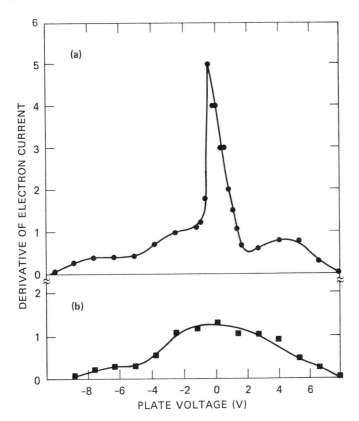

<u>Fig.6</u> Energy analysis of electrons ejected from (a) the Ba$(6p_{\frac{3}{2}}20s_{\frac{1}{2}})_1$ state and (b) the Ba$(6p_{\frac{3}{2}}20s_{\frac{1}{2}})_2$ state showing the presence of 0.25 eV electrons in the former

resolution, in fact electrons of a given energy produce a feature which extends twice the energy on either side of zero retarding voltage. Nonetheless, it is quite adequate for its intended purpose.

In Fig.6a we show the derivative of the scan of the $(6p_{\frac{3}{2}}20s_{\frac{1}{2}})_2$ state showing one feature ~ 10 V wide which is indicative of only 2.5 eV electrons. Scans for the $(6p_{\frac{1}{2}}20s_{\frac{1}{2}})_J$ states look essentially the same. In Fig.6b we show the analagous trace for the $(6p_{\frac{3}{2}}20s_{\frac{1}{2}})_1$ state which shows two features, one ~ 10 V wide indicating 2.5 eV electrons and one 1 V wide indicating 0.2 eV electrons, which indicates autoionization to the $Ba^+6p_{\frac{1}{2}}$ state. The signal from the slow electrons is 2/3 of the total signal, in reasonable agreement with the width measurements which imply that 3/4 of the signal should be due to autoionization to the $Ba^+6p_{\frac{1}{2}}$ state.

C. Excitation profiles

Frequently autoionizing states are studied by single photon absorption from the ground state. Thus the transition to the doubly excited autoionizing state is necessarily a weak "two electron" transition of a strength comparable to "one electron" photoionization at the same frequency. Interference between these two possible transitions leads to the familiar asymmetric Beutler-Fano lineshapes. Ironically, although an important feature of our multistep laser approach is the avoidance of such interference effects, we can also use the method to study such effects in a systematic calculable way [10].

Let us consider as an example the excitation of the Ba $6p_{\frac{1}{2}}$nd states near n = 15. We can measure the width and position of each of these states by exciting the bound 6snd levels and driving the 6snd → 6pnd transition as described above with no interference effects.

If we now excite the bound 6s15d state and use a low laser intensity for the third laser we see only the strong 6s15d → 6p15d transition. If, however, we increase the laser power we are able to see not only the power broadened 6s15d → 6p15d transition, but the weaker two electron 6sn'd transitions in its wings which interfere with the strong transition to give the familiar Beutler-Fano lineshapes. Since we have independently measured the position and width of each state we can fit the observed spectrum to derive the ratio of the matrix elements for the 6s15d → 6pnd to 6s15d → 6p15d transitions. In Fig.7 we show the observed spectrum from the 6s15d(^{1}D) state to the $6p_{\frac{1}{2}}$nd series. The fit is indistinguishable from the data and the dotted line shows the Lorentzian curve of the 6s15d → 6p15d transition ignoring the satellite peaks.

Since the transition is basically a transition of the inner electron from the Ba^+6s state to the Ba^+6p state with the outer electron only making a small change in its orbit it is reasonable to write the matrix element as a product

$$\langle 6s15d/\mu/6pnd \rangle = \langle 6s/\mu/6p \rangle \; \langle 15d/nd \rangle \tag{1}$$

where the second term indicates the overlap of the outer electron in the two

278

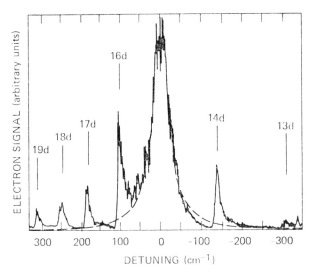

DETUNING (cm^{-1})

<u>Fig.7</u> Excitation spectrum from the Ba 6s15d state to the 6p$_{\frac{1}{2}}$nd state showing the interferences between the discrete states. The fit is indistinquishable from the data and the dashed line indicates the Lorentzian fit to the central line ignoring the satellites.

states. Knowing the change in the quantum defect, we can calculate this overlap using the BATES-DAMGAARD method [11], and we find excellent agreement with our measured values. To our knowledge, this is the first study of a Beutler-Fano profile where it has been possible to independently measure all the relevant parameters.

D. Spectroscopy of higher lying levels

Now let us consider the effect of increasing the size of the inner electron's orbit. In the 6snℓ and 6pnℓ states the inner electron's orbital radius r_1, is much smaller than the outer electron's, r_2, and the inner electron sees a charge of 2 and the outer electron is fully screened by the inner electron and only sees a charge of 1. However, as we increase the size of the inner electron's orbit this is no longer true, and in addition the motions of the electrons become highly correlated. Finally, as r_1 approaches r_2, the atoms become what PERCIVAL [12] has termed planetary atoms.

We have now begun the study of the higher lying levels with the study of the Ba7snd states [13]. In raising the inner electron from the 6s to the 6p and then to the 7s state, we increase its orbital radius from 4 au to 8 au. The Ba$^+$7s state has a much larger orbital radius because it has a node in its radial wave function outside the Ba^{++} core, whereas the 6s and 6p states do not. In this experiment we are able to differentiate between energy of the state in the coulomb (large r) part of outer electron orbit and the energy in the perturbed inner region containing the Ba$^+$ core.

The experimental technique is a slight extension of that described earlier. A fourth laser simultaneous with the third excites the Ba atoms from the 6pnd states to the 7snd states. Since this is a transition between autoionizing states there is no change in the total ionization produced, so we detect preferentially electrons which have energies > 5 eV.

An interesting feature of this experiment is that the spatial constraint of the outer electron's orbit plays a key role. The form of the matrix element for the third and fourth transitions is given in (1). Eq. (1) makes it clear that the outer electron state is simply projected from the initial nd state onto the overlapping final n'd states. For example, if the change in the Ba$^+$ core state produces no change in the spatial wave functions of the nd electron in the large r, coulomb region, then we expect to see one tran-

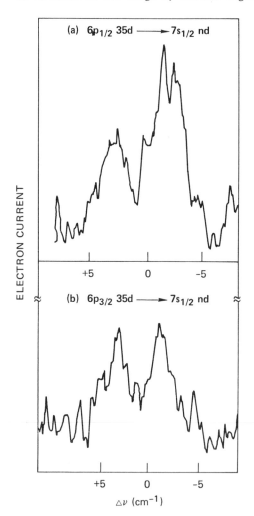

Fig.8 Excitation spectrum from the autoionizing Ba 6p$_J$35d states to the 7s nd states. The frequencies are measured relative to the ion 6p$_J$ → 7s frequencies. Note that the observed transitions are not symmetrically located about the ion line.

sition at the frequency of the Ba^+ ion transition. If, on the other extreme, the nd wave functions are spatially shifted so that the initial nd wave function overlaps equally with two final n^1d wave functions, we would expect to see two transitions equally spaced on either side of the ion line. If they are not equally spaced, the frequency discrepancy must be due to the interactions of the two electrons and the Ba^{++} core at small radius.

When we excite the $Ba6s_{\frac{1}{2}}nd \rightarrow 6p_Jnd$ we see one strong transition red shifted from the ion line by a negligible amount. However, when we excite the Ba $6p_{\frac{3}{2}} \rightarrow 7s_{\frac{1}{2}}nd$ states we see two equally strong transitions which are not equally spaced about the ion $6p_{\frac{3}{2}} \rightarrow 7s$ line, as shown in Fig.8. The energy of the coulomb part of the outer electron orbit would require that the transitions be located symmetrically about the ion $6p_{\frac{3}{2}} \rightarrow 7s$ transitions, but they are not. The discrepancy Δ implies that an energy $0.14\ n^{*-3}$ is invested in the short range interactions. In fact, this may be estimated using HEISENBERG'S choice for the potential [12]

$$V = \frac{-2}{r_1} - \frac{-1}{r_2} - \left\{ \frac{1}{r_2} - \frac{1}{r_{12}} \right\} \tag{2}$$

and treating the part outside the brackets as the unperturbed potential and the term in the brackets as a perturbation. This amounts to assuming that the inner electron sees an ion charge of 2 and the outer electron sees a ion charge of 1. Using estimated Ba^+ 6s, 6p and 7s wave functions and coulomb nd wave functions, we found that each of the terms in the brackets were $\sim 0.3(n^*)^{-3}$ in qualitative agreement with our observations.

IV. Conclusion

Here we have described experimental investigations of the intrinsic properties of doubly excited alkaline earth atoms, as well as techniques for their excitation. In our opinion the studies to date have only served to indicate potentially interesting areas for further investigation, and we are confident that the refinement of the techniques presented here will lead to many new insights. We are grateful to D. C. Lorents, D. L. Huestis, and R. Morgenstern for helpful discussions and encouragement. This work has been supported by the Air Force Office of Scientific Research, under Contract F44620-74-C-0069, the National Science Foundation, under Grant PHY 76-24541, and the Office of Naval Research, under Contract N00014-79-C-0202.

REFERENCES

1. C. K. Rhodes, private communication.

2. S. E. Harris, private communication.

3. A. Burgess, Astrophys J. 139, 776 (1964).

4. U. Fano, Phys. Rev. <u>124</u>, 1866 (1961).

5. C. E. Moore, Atomic Energy Levels, NBS Circular No. 487 (U. S. GPO, Washington, D. C., 1949).

6. W. E. Cooke, T. F. Gallagher, S. A. Edelstein, and R. M. Hill, Phys. Rev. Lett. <u>40</u>, 178 (1978).

7. V. W. Hughes and L. Grabner, Phys. Rev. <u>79</u>, 829 (1950).

8. W. E. Cooke and T. F. Gallagher, Phys. Rev. Lett. <u>41</u>, 1648 (1978).

9. T. F. Gallagher, K. A. Safinya, and W. E. Cooke (to be published).

10. K. A. Safinya and T. F. Gallagher (to be published).

11. D. R. Bates and A. Damgaard, Phil. Trans. Roy. Soc. London <u>A242</u>, 101 (1949).

12. I. C. Percival, Proc. Roy. Soc. London <u>A353</u>, 289 (1977).

13. T. F. Gallagher, K. A. Safinya, and W. E. Cooke (to be published).

14. H. A. Bethe and E. A. Salpeter, <u>Quantum Mechanics of One and Two Electron Atoms</u> (Academic Press, New York, 1957).

Selective Detection of Single Atoms by Multistep Excitation and Ionization Through Rydberg States

G.I. Bekov, E.P. Vidolova-Angelova, V.S. Letokhov, and V.I. Mishin

Institute of Spectroscopy of the Academy of Sciences USSR
SU-142092, Troitzk, Moscow, USSR

1. Introduction

Considerable attention has been paid lately to the development of laser detection technique of ultrasmall substance amount in order to achieve ultimate sensitivity of single atoms and molecules level. Among all the methods used for detection one of the most perspective at atomic approach is the method of selective stepwise ionization of atoms by laser radiation [1]. This method at saturation regime makes it possible to convert every atom into ion [1,2]. Under such conditions the method of stepwise ionization can be efficiently used for atom detection [3].

According to this method the atoms are excited by laser radiation into the intermediate state by one or several steps and then ionization of excited atoms is accomplished. Conventionally two approaches can be outlined depending on the way of the atom ionization from the intermediate state. They are: 1) non-resonant ionization (NRI), 2) resonant ionization (RI) (Fig.1). In the first case the excited atom is ionized

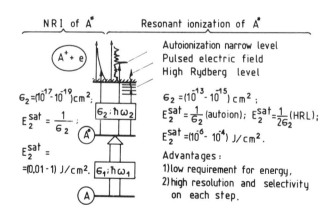

Fig.1 Methods of selective stepwise ionization of atoms

by the additional or by the same laser radiation. Low cross-sections ($\sigma_2 = 10^{-17} \div 10^{-19}$ cm^2) are characteristic for such non-resonant photoionization and corresponding saturation energy densities are $E_2=(0.01-1)$ J/cm^{-2}. In the second case the atom is excited either into the autoionization state having little lifetime or it is excited into the highlying Rydberg state with subsequent ionization by the electric field pulse. Ionization of Rydberg atoms by the electric field was investigated in [4]. As it was shown in [5] at electric field strength larger than the critical one, the ionization cross-section is determined by the resonant excitation cross-section of the atom into the Rydberg state. This cross-section exceeds the non-resonant ionization cross-section into continuum by several orders.

The first experiment on single atom detection by stepwise photoionization was successfully carried out in [6]. Cesium atoms were excited in a buffer gas into the state $7^2P_{3/2}$ by the pulsed dye laser radiation. The excited atoms were ionized by the same laser pulse. Electron - ion pairs formed were detected by the proportional counter. In such experiment the flight time of detected atoms through the laser beam is long enough since it is determined by the atom diffusion in a buffer gas (Fig.2). And in spite of the large energy density value of the ionizing pulse the moderate laser radiation power

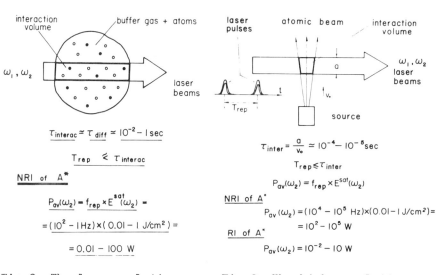

Fig.2 The low resolution buffer-gas experiment

Fig.3 The high resolution atomic beam experiment

is required for efficient ionization. However it is impossible to achieve high spectral resolution in such experiment since the buffer gas leads to considerable collisional broadening of absorption lines. In the cases when maximum spectral resolution is to be realized, detection should be carried out in a vacuum in the atomic beam (Fig.3). The application of the method of non-resonant ionization here makes high demands to the energetics of ionizing laser radiation. The average power should be about several kilowatt since in order to trap larger part of atoms the pulse repetition rate of 20÷50 kHz is required.

At average laser radiation power available in laboratory conditions ($P \sim 1W$) most of atoms can be ionized efficiently if the atoms are ionized through Rydberg states. In this case the non-resonant atom photoionization from the intermediate state is replaced by the process of resonant excitation of the atoms from the same state into the highlying state with subsequent ionization by the electric field pulse. The efficiency of ionization in this process is close to one. Since the excitation on all steps is resonant, moderate laser pulse energy density, which can be obtained by the existing dye lasers, is required for the saturation absorption on these transitions. The detection of single sodium and ytterbium atoms was carried out by such method of ionization in [7,8]. The results of those works allow to make the following conclusion. Such method provides registration of the element in the amount $10^5 \div 10^6$ atoms ($10^8 \div 10^{10}$ atoms are required to record the hyperfine or isotope structure) if dye lasers with low pulse repetition rate are used. Besides the method proposed has high selectivity since, when the atom is excited into several steps, the selectivity is achieved on every step at correct choice of transitions. And the total selectivity is determined by the product of selectivities on every step [9]. Thus the method will make it possible to detect isotopes of the element with relative content $10^{-10} \div 10^{-15}$.

This report informs about the results of the investigation of isotope and hyperfine structure of the 6^1S_0-$6^3P_1{}^o$ ytterbium transition carried out by the above described method. The samples containing only 10^{10} atoms of the certain ytterbium isotope were used in this experiment.

2. Experimental

Figure 4 shows the scheme of ytterbium atom energy levels and transitions used for excitation. Ytterbium atoms were excited into the Rydberg state by radiation of three tunable pulsed dye lasers (the laser spectrum width $\Delta \mathcal{V}$las $\simeq 1$ cm^{-1}, the pulse duration on the half maximum $\mathcal{T}_\rho$ =7 nsec). Laser pumping was carried out simultaneously by one nitrogen laser with the pulse repetition rate 12 Hz. The laser of the first step of excitation with the wavelength λ_1=5556.5 Å transferred ytterbium atoms from the ground $6_0^1S_0$ state into the $6^3P_1{}^o$ state. The second laser with λ_2=6799.6 Å excited atoms further into the state 7^3S_1. The wavelength of the third laser was tuned to the transition $7^3S_1 \rightarrow 17^3P_2{}^o$ (λ_3=5843 Å). The 17 P-state was chosen because the cross-section of its excitation is large enough and the critical electric field strength for this state is easily obtained in the experiment ($\mathcal{E}_{crit}^{17P}$ =11.5 kV/cm).

Laser beams were directed into the vacuum chamber and intersected the atomic beams in the region between the electrodes (Fig.5). The atomic beam of the isotope to be investigated intersected the laser beams at an angle of 90°. The reference atomic beam formed at the evaporation of the metal ytterbium of the natural isotope abundance from the additional oven intersected the laser beams at an angle of 45°. This led to the absorption line shift into the longwave edge by $\delta\nu \approx 0.015$ cm^{-1}.

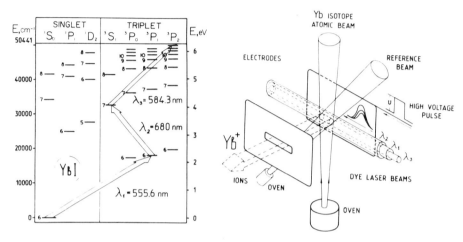

Fig.4 The scheme of energy levels of the ytterbium atom and transitions used for excitation

Fig.5 The mutual arrangement of electrodes, the atomic beams and laser beams

After excitation of atoms into the highlying state by dye laser pulses the electric field pulse of a rectangular shape was sent to the electrodes with a delay τ =20 nsec. The pulse generator was a concordant spark-gap with the cable line triggered by the nitrogen laser radiation pulse. Ions formed as a result of ionization of Rydberg atoms by the pulsed electric field acquired in this field a lateral velocity component which exceeded the atomic velocity in the beam by two orders. After passing through the slit in the electrode with the zero potential the ions were registered by a secondary emission multiplier (SEM). The region of excitation, i.e. the region where all the laser beams and investigated atomic beam intersected was a cyllinder of d=0.8 mm and l=4 mm. The size of the slit through which the ions where extracted was 3x10 mm². Thus the geometry of the excitation region made it possible to extract practically all the ions from the space between the electrodes. Under our experiment conditions the ion registration efficiency by the secondary emission multiplier was close to one. The signal from SEM was amplified and registered simultaneously by the boxcar averager and an oscilloscope. The experimental set-up is described in detail in [10].

The atomic beam under investigation was formed at heating up the tantalum foil in which about 10^{10} ions of investigated ytterbium isotope were implanted preliminarily. The implanta-

tion was carried out in the following way (Fig.6). The tantalum target heated up to the temperature 2000-3000°C was irradiated by an intense beam of protons. As a result of interac-

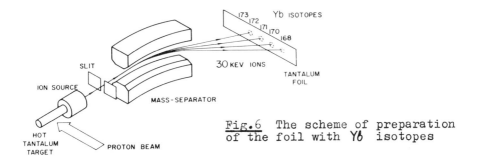

Fig.6 The scheme of preparation
of the foil with Yb isotopes

tion of the target nuclei with protons the whole series of
ytterbium isotopes including unstable ones was formed. The reaction products in the ion source were ionized and after collimation got into the mass-separator. The spatial separation of
isotopes on masses was carried out in the mass-separator and
ions of various isotopes with the energy of about 30 keV got
into different regions of 0.05 mm thick tantalum foil. The
ions penetrated into the foil into the depth about 100 Å and
were neutralized. Thus the atoms turn out to be isolated from
the surrounding medium and under normal conditions are preserved in the foil. Hence the techniques of ion implantation into
foil is the convenient way of transportation and storage of the
ultrasmall amount of the element. As it has been mentioned it
is possible to implant the unstable isotopes of the element as
well.

The atomic beam out of the foil was formed in the following
way. The above described foil with ytterbium was placed into
the crucible with narrow cylindrical channel. The crucible was
inside the tantalum thin-walled tube heated up by the electric
current. The design of the oven is shown in Fig.7. Such system
made it possible to increase the temperature of the crucible
up to 1800°C. At temperature more than 1200°C noticeable diffusion of ytterbium atoms from the tantalum foil takes place [11].
Free atoms after collimation by the narrow channel creat a well
directed beam. One of the aims of the work was to define the
atom yield coefficient from the foil and crucible. Two diaphragms under potentials of different signs (U_2=100 V, U_3=15 V)
were used to eliminate thermal ions and electrons.

3. Results and Discussion

Using the reference ytterbium beam, laser radiation was tuned to the corresponding transitions using ionization signal
from the Rydberg state under the electric field action. Three
simultaneous laser pulses distribute atoms among the four unperturbed atomic levels. Such expression for the level popula-

tion is obtained in the steady state rate equation approximation: $n_i=g_i/\sum g_k$, where n_i and g_i are relative population and degeneracy factor of the i-level. The 5/12 of all atoms which are in the region of registration are excited at saturation absorption on all three transitions into the Rydberg state $17^3P_2^0$. At the electric field strength which is more than the critical one each highly excited atom is ionized (it was investigated in detail in [10]). Energy density measurements carried out in our experiment showed that according to the saturation curves shown in [8,10] about 1/3 of all atoms interacting with laser radiated is excited into the Rydberg state 17 p. Thus the conditions, when practically every third atom is ionized reaching the excitation region at the moment when laser pulses arrive there, are achieved.

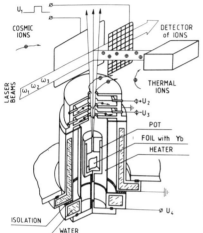

Fig.7 The design of the high-temperature oven for the element evaporation from the foil

After this tuning the reference beam was shut. And the background level caused by both non-selective ions and selective ions from ytterbium atoms that are drifting in the vacuum chamber was investigated. The measurements showed that the total background was about 2-3 pulses of SEM for 3 minutes. After this the crucible with tantalum foil containing about 10^{10} atoms of one of the stable ytterbium isotopes was heated up. The spectrum width of lasers of all steps was larger than the isotope and hyperfine structure width of the corresponding transitions. Therefore the laser line tuning by the reference beam provided laser lines resonance with any isotope absorption lines.

At the crucible temperature t $\simeq$ 1200°C the selective signal already exceeded the background signal two-three times. At temperature t=1350-1400°C the selective ion was registered almost per every laser pulse during 10 minutes. Then the signal gradually decreased. The background level at such temperature did not exceed 3 pulses per minute. When the temperature grew up to 1500°C the background noticeably increased due to the ther-

mal ions and only slight enhancement of useful signal is observed. As the experiment showed the ratio signal/noise was not
less than 20 within the temperatures 1300°÷1400°C during 10
minutes. Sich time is enough to record the structure of one of
the transitions by a narrowband laser. For this purpose pressu-
re tunable dye laser with line width $\Delta \mathcal{V} < 0.04$ cm^{-1} was used
for excitation on the first step. The lasers on the other steps
remained the same. To control the tuning of the first step laser wavelength laser radiation after the vacuum chamber was directed through the Fabry-Perrot etalon to the photodiod. The
signal from the photodiod was sent to one of the two recorders
synchronized with each other. The signal from the SEM averaged
by the boxcar was sent to the second recorder. The Fabry-
Perrot etalon transparancy maxima were used as the references

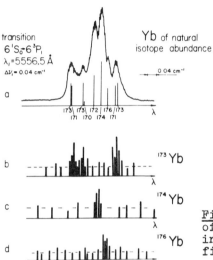

transition
$6\,^1S_0\!\div\!6\,^3P_1$
$\lambda_1 = 5556.5$ Å
$\Delta\mathcal{V}_1 = 0.04$ cm^{-1}

a

Yb of natural
isotope abundance

0.04 cm^{-1}

173\ 173\ 172 | 176 /173
171 170 174 171

b ^{173}Yb

c ^{174}Yb

d ^{176}Yb

Fig.8 The ion current dependence
of the reference (a) and to be
investigated (b,c,d) beams on the
first step laser wavelength

of the relative frequency coupling of the tunable laser when
the retuning cycles were repeated. Figure 8 a shows the dependence of the ion current on the first step laser wavelength
for the reference beam. Vertical lines denote the position of
the ytterbium monoisotope absorption lines. The height of those lines corresponds to the relative natural abundance of the
Yb isotopes. The total width of the whole structure is 0.2 cm^{-1}.
Then the reference beam was shut. After that crucible with the
foil containing an ytterbium monoisotope was heated up to
1350°C and a new cycle of the first step laser tuning was
carried out.
    Since the density of the atomic beam under investigation is
very low, the registration of the ion signal was accomplished
as detection of separate ions during laser frequency scanning.
The signal from SEM was sent to the boxcar averager with a
small time constant and then was registered by the recorder
in the form of separate peaks with the amplitude proportional

to the amount of detected ions. Typical pulse spectra obtained
for various isotopes are shown in Fig.8 b,c,d. Dotted lines
correspond to the total background level. The mutual coupling
of the reference and investigated spectra was fulfilled by the
coincidence of the etalon transparancy maxima obtained for
these two tuning cycles. The spectra shape (Fig.8 b,c,d)allows
to identify them only by the ytterbium isotopes $^{173}Yb, ^{174}Yb, ^{176}Yb$
respectively. The experiment showed that it was enough to re-
gister about 3 ions per resolved spectral interval for the
stable recording of the isotope absorption lines. Hence knowing the geometry of the beam system it is possible to estimate the minimum amount of atoms necessary for the spectrum registration and the coefficient of the atom yield from the
crucible. To obtain 3 ions 10 atoms are required at ionization
probability 0.3. At the atom velocity in the beam of $5 \cdot 10^4$ cm/
sec and laser beam diameter 1 mm the duty ratio between laser
pulses for interception of all atoms should be $5 \cdot 10^4$ times less
than the duty ratio used in the experiment. This increases the
amount of atoms up to $5 \cdot 10^5$. The registration region occupies
only 1/7 part of the atomic beam cross-section, hence the total number of atoms flying out of the crucible should be $3.5 \cdot$
$\cdot 10^6$. Such amount of atoms is required to obtain one spectral
interval. About 50 spectral intervals with the resolution
0.02 cm^{-1}, i.e. $2 \cdot 10^8$ atoms, are necessary to obtain the total spectrum with the width of about 1 cm^{-1}. The amount of
ions implanted into the foil was about 10^{10}. There were obtained 3-4 spectra in the experiment for each of the isotopes,
i.e. the number of atoms in one cycle was $3 \cdot 10^9$. Thus the coefficient of the atom yield from the crucible is 0.07 which is
quite a real value.

The experiment showed that the technique of the atom detection in the beam from the foil containing the atoms of the
element could be successfully applied even with the dye lasers
with a small pulse repetition rate to obtain the spectra of
elements available in the amount $10^9 \div 10^{10}$ atoms. The sensitivity of the technique can be essentially increased if the copper
vapor laser having the pulse repetition rate 10÷20 kHz is used

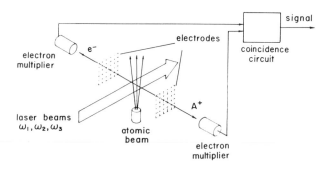

Fig.9 The application of the scheme of coincidences for the
thermal ion background discrimination at simultaneous registration of the ion and the electron

for pumping the dye lasers. The improvement of the atomic beam collimation or the increase of laser beam diameter will make it possible to decrease almost by an order the required amount of atoms. Thus $10^5 \div 10^6$ atoms implanted into the foil will be enough to record the spectrum.

One of the ways of further improvement of the possibilities of the method is the elimination of the background level by applying the scheme of coincidences at simultaneous registration of the ion and the electron formed during ionization process (Fig.9). This will considerably expand the range of application of the method to the group of elements for evaporation of which temperatures 2000°C-2500°C are required.

Another means of increase the atom ionization cross-section is the excitation of atom on the last step into the autoionization state. For multielectron atoms such states can be rather narrow and such transition cross-section exceeds the nonresonant ionization cross-section more than by an order [12]. We have investigated autoionization states of the gadolinium atom and registered the state with the lifetime about 10^{-9}sec [13].

Gadolinium atoms in a beam were excited into the states lying higher than the ionization limit by 300 cm^{-1} in three steps by pulsed dye laser radiation. Figure 10 shows the simplified scheme of the Gd atom energy levels and transitions used for excitation. The registration system and the configuration of the laser and atomic beams were analogous to the ones described in the experiment on ytterbium. Figure 11 presents spectra of Rydberg and autoionization states of gadoli-

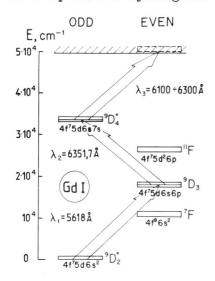

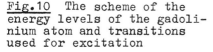

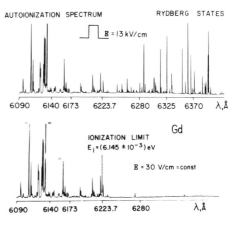

Fig.10 The scheme of the energy levels of the gadolinium atom and transitions used for excitation

Fig.11 The Rydberg and autoionization states spectra of the gadolinium atom in pulsed (upper) and constant (low) electric fields

291

nium obtained in a pulsed electric field (upper) and in the constant field(low).As the figure shows the amplitudes of Rydberg and autoionization resonances are close. Autoionization resonance with $\lambda \sim 6133$ Å is distinguished particularly.

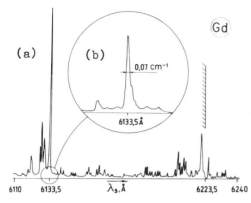

Fig.12 (a) - The ion signal dependance on the third step laser wavelength, $\Delta \nu_3 = 1$ cm^{-1}; (b) - the same dependance in the vicinity $\lambda = 6133.5$ Å, $\Delta \nu_3 = 0.03$ cm^{-1}

The record of the this spectrum region with high resolution is shown in Fig.12 b. The excitation on the third step was carried out by the pressure tunable dye laser with the radiation line width $\Delta \nu_3 = 0.03$ cm^{-1}. The autoionization resonance width on the half-maximum is about 0,07 cm^{-1}, that gives the estimation of the state lifetime as $\tau \sim 0.5$ nsec. The cross-section of this autoionization transition was measured by the saturation method proposed in [2]. The following value was obtained $\sigma_{aut} = = 8 \cdot 10^{-16}$ cm^2.

This experiment showed that autoionization states with the lifetime of about 10^{-9} sec are possible in the spectrum of multielectron atoms. The excitation cross-section of such states is comparable with the excitation cross-section of Rydberg states and these long-life autoionization states may be efficiently used in the method of stepwise ionization of atoms by laser radiation. There is no need of the electric field and the continuous decay channel allows during a laser pulse to ionize practically all the atoms which are in the ground state.

4. Conclusion

Thus the suggested technique of selective detection of single atoms provides comprehensive spectral information about the element when there is an ultrasmall amount of its atoms at the disposel. Perspectives of application of this method in nuclear physics are worth of special attention. One of the examples is the investigation of shortliving isotopes. Modern mass-separators make it possible to obtain ion flows of 10^6 ions per sec. Using the method in "on-line" it is possible to obtain spectral information from isotopes with the life time of several seconds and less.

High sensitivity and accuracy of measuring of spectral line

shifts in the method of laser detection of single atoms allow
to organize experiments on determination of the mean-square
sizes of various nuclei of isotopes and isomers. It is possible to trace how the nucleus form changes for the isotopes with
neutron excess or shortage. Experiments on the search of atoms
with superdense nuclei are also real [14]. In such systems the
line shifts to be expected are tens cm^{-1} and the amount of
atoms is about 10^3. Probably such method of detection may be
used to solve the problem of the existence of spontaneously
fissible nuclear isomers of the form, for instance, in ^{242}Am.

High selectivity and sensitivity of the method allow to
differentiate a literally calculable number of atoms of one
isotope in large mass of the substance. It is especially important, for instance, for detection of isotopes ^7Be, ^{35}Ar,
^{55}Fe and ^{71}Ge which are final products in the reactions of solar neutrino trapping in radiochemical detectors. And finally
the method can be used as the basis for the development of new
highly selective analytical methods.

5. Acknowledgments

The authors express their deep gratitude to professor
E.E.Berlovich for placing of the foil with ytterbium isotopes
at their disposal.

References

1. V.S.Letokhov, USSR. Patent No.65743 (Appl.30.03.1970);
   R.V.Ambartzumian, V.P.Kalinin, V.S.Letokhov. Pis'ma ZhETF
   (Russian) 13, 305 (1971) /JETP Lett., 13, 217 (1971)/.
2. G.S.Hurst, M.G.Payne, M.H.Nayfeh, J.P.Judish and E.B.Wag-
   ner: Phys.Rev.Lett. 35, 82 (1975)
   R.V.Ambartzumian, V.M.Apatin, V.S.Letokhov, A.A.Makarov,
   V.I.Mishin, A.A.Puretzky, N.P.Furzikov: Sov.Phys.JETP 43,
   866 (1976)
3. V.S.Letokhov, in "Frontiers in Laser Spectroscopy", vol.2,
   R.Balian, S.Haroshe, and S.Liberman, eds.(North-Holland,
   Amsterdam, 1977) p.771 and in "Tunable Lasers and Applications," vol.3, A.Mooradian, T.Jaeger and P.Stokseth, eds.
   (Springer-Verlag, Heidelberg, 1976), p.122.
4. R.V.Ambartzumian, G.I.Bekov, V.S.Letokhov, V.I.Mishin:
   JETP Lett. 21, 279 (1975); R.F.Stebbings, C.J.Lattimer,
   W.P.West, F.B.Dunning, and T.B.Cook: Phys.Rev. A12, 1453
   (1975); T.W.Ducas, M.G.Littman, R.R.Freeman, and D.Klepp-
   ner: Phys.Rev.Lett. 35, 366 (1975).
5. G.I.Bekov, V.S.Letokhov, V.I.Mishin: Sov.Phys.JETP 46, 81
   (1977)
6. G.S.Hurst, M.H.Nayfeh, and J.P.Young: Appl.Phys.Lett. 30,
   299 (1977); Phys.Rev. A15, 2283 (1977); in "Laser Spectroscopy III", J.L.Hall and J.L.Carlsten, eds.(Springer-Verlag, Heidelberg, 1977), p.44
7. G.I.Bekov, V.S.Letokhov, V.I.Mishin: JETP Lett. 27, 47
   (1978)
8. G.I.Bekov, V.S.Letokhov, O.I.Matveev, V.I.Mishin: Opt.Lett.
   3, 159 (1978)

9. V.S.Letokhov, V.I.Mishin, Opt.Comm. (to be published)
10. G.I.Bekov, V.S.Letokhov, O.I.Matveev, V.I.Mishin: Zh.Eksp.
    Teor.Fiz. <u>75</u>, 2092 (1978) (in Russian)
11. A.Lyatishinski, K.Zuber, Ya.Zuber, A.Potempa, V.Zhuk:
    Preprint OIYaI 6-7469, Dubna (1974) (in Russian); G.-Yu.
    Baier, A.F.Novgorodov, V.A.Khalkin: Preprint OIYaI R6-9917,
    Dubna (in Russian)
12. J.A.Paisner, R.W.Solarz, L.R.Carlson, C.A.May, and S.A.
    Johnson: Lawrence Livermore Laboratory, Preprint UCRL-
    77590 (1975)
13. G.I.Bekov, V.S.Letokhov, O.I.Matveev, V.I.Mishin: Pis'ma
    ZhETF <u>28</u>, 308 (1978)(in Russian)
14. V.A.Karnaykhov, S.M.Polikanov: Pis'ma ZhETF <u>25</u>, 328 (1977)
    V.A.Karnaykhov: Preprint OIYaI R13-12000, Dubna (1978)
    (in Russian)

Part IV

**Multiphoton Dissociation,
Multiphoton Excitation**

Classical Model of SF$_6$ Multiphoton Dissociation

W.E. Lamb, Jr.

University of Arizona, Tucson, AZ 85721, USA

The interaction of a polyatomic molecule with intense laser radiation represents a very interesting but difficult problem. The study has obvious applications to isotope separation, to the production of highly excited vibrational and rotational states of the system, and the possible production of frequency selective chemical reactions. There have been many experimental observations of such phenomena. However, in general, the theory has not kept up well. It would be very useful to have an adequate theory with good predictive features.

Despite the title, there is no real need to treat the strong electromagnetic field of the laser in terms of photons, as in quantum electrodynamics, for such fields can be described with the very good approximation of classical Maxwellian electro-magnetic theory. The word "multiphoton" means only that if perturbation theory were used for the interaction between the electromagnetic field and the molecule, it would have to be of very high order. The radiation may or may not be coherent, although in the present work it is taken as monochromatic and plane polarized. (We will neglect collisions throughout.)

A typical molecule of interest is sulfur hexafluoride. It has seven nuclei and seventy electrons. Experiments on laser induced isotope separation indicate that it should suffice to treat only states of the molecule which correspond to the ground electronic state for any configuration of the nuclei. According to the Born-Oppenheimer approximation, the seven heavy particles can be regarded as moving in an interparticle potential given by the electronic ground state energy. In principle this could be calculated by methods of ab initio quantum chemistry. We cannot rely very much on these calculations because of the amount of computer time involved in order to describe the small amplitude vibrational motion with the accuracy necessary to deal with the resonant phenomena in the early stages of the break-up process. These are vital for the isotopically selective dissociation. Hence, we use a semi-empirical potential, depending on a number of adjustable parameters, which will be consistent with a large amount of experimental data obtained from spectroscopic and thermochemical observations. It would be easy to find the potential function in the neighborhood of equilibrium, but it is nontrivial to determine a suitable extrapolation of the poten-tial for an arbitrarily deformed configuration of the seven nuclei. Murray Sargent, Lionel Menegozzi and I have developed a powerful computer program for finding a suitable potential. This potential has a number of adjustable parameters which are fitted so as to reproduce various observed properties of the molecule such as: equilibrium configuration, normal mode frequencies and their anharmonicities, dissociation energies, etc. One of the major complications is proper allowance for ionic and multi-particle forces in a way which is simple enough to use for the dynamical calculations of the dissociation process. This part of the work is currently active, and the results discussed below are based on a potential derived from two body inter-actions.

It is also necessary to have a realistic description of the way in which the laser radiation perturbs the molecule. This can be obtained from a time dependent potential energy function designed to give the observed infrared transition probabilities, and behaving in a reasonable manner as the atoms move far apart.

With these potential functions, it would be possible to try to solve the time dependent quantum mechanical Schrödinger wave equation for the seven heavy particles. This is still too diffi-cult a problem. Fortunately, a strong case can be made that a treatment of the corresponding classical problem should be a very good approximation. It is possible to improve on this approximation if desired. However, even the solution of the problem of classical dynamics has considerable interest and value.

Following from the theory of small vibrations, the mole-cule has fifteen normal modes of small amplitude vibration, as well as three translational and three rotational normal modes of zero frequencies. The vibrational modes will be shown in a computer generated motion picture. Since interatomic potential functions depart greatly from Hooke's law restoring force form, the normal modes have validity only in the limit of small amplitudes. In the real problem, several new effects have vital importance: (1) the frequencies of motions are amplitude dependent, and (2) there is strong coupling of the normal modes. All kinds of harmonics and combination tones are present in the motion. Molecular theory is simple only when these complications are ignored.

According to the prevailing view (from Bloembergen, Yablonovitch and others), the photodissociation process proceeds through three regions of molecular energy levels: (1) a resonance region with approximately simple harmonic small amplitude vibration, (2) moderately higher rungs on the vibrational ladders where there is strong combination tone coupling of many modes (the so-called quasi-continuum), and (3) the real continuum, where the molecule has more than enough energy to dissociate (but may or may not do so before the end of the laser pulse, or a collision of some kind.)

This picture undoubtedly has a great deal of truth. However, The proposed methods for dealing with the quasi-continuum have been far too simple. The processes are characterized by many rate constants, with no reliable way to determine them. The processes in the real continuum are also rather remote from real quantification.

The classical dynamical approach makes it possible to describe in detail what is happening in both quasi-continuum and real continuum. Once we have decided on suitable potentials for interactions of the atoms in the molecule, and their coupling to the laser, the equations of motion can be integrated numerically by use of a powerful minicomputer, starting from some suitable initial state. A three dimensional representation of the molecular configuration can be displayed on a cathode ray screen. A short motion picture of the results will be shown, admittedly for the case of an oversimplified interatomic potential. The motions are very complex.

The dissociation of such a molelcule has certain similarities with the problem of the evaporation of a speck of carbon or a drop of liquid by application of heat or laser radiation. These latter problems could be described in thermodynamic terms, giving the system a time dependent temperature, and a temperature dependent evaporation coefficient. Our molecular problem is described in mechanical terms. If the number of degrees of freedom were large, one would expect the two treatments to lead to very similar conclusions. It is not apriori obvious how large a system must be before one can use the macroscopic point of view. Calculations of the kind outlined above make it possible to examine the interface between mechanics and statistical thermodynamics.

The dynamical calculations can be analyzed in a number of other ways. A plot of energy transfer from the laser to the molecule as a function of time shows that enormous fluctuations are superimposed on a constant rate. It is possible to test for the applicability of a statistical description in the following manner: The whole of the dynamical record can be subdivided into (say) eight time segments. Each of these might have four thousand integration steps. One can then form an ensemble from the kinetic energies of any number or all of the seven particles for each segment. By counting the number of kinetic energies which fall in various ranges it is possible to form a histogram of the kinetic energy distribution. Under suitable circumstances, this histogram will have the canonical Maxwell-Boltzmann-Gibbs form expected for an ensemble in thermodynamic equilibrium at a certain temperature T.

On the basis of calculations to date, it seems that the use of a temperature affords a good description, even for our seven particle system. However, this temperature does not increase monotonically throughout the motion, but seems to be rather constant in all but the initial stages of any particular integration. It will be interesting to learn how general this phenomenon may be, as more computer runs are made with better interaction potentials.

Another feature of the work so far is that, even when
the molecule has absorbed several times the energy required for
dissociation, the actual dissociation is a relatively rare event,
and can probably be adequately described by techniques used in
the theory of evaporation (for the temperature found above.)

This research was supported in part by the Office of Advanced
Isotope Separation of the U. S. Department of Energy.

Collisionless Multiple Photon Excitations in SF$_6$: Thermal or not?[1]

J.R. Ackerhalt and H.W. Galbraith

Theoretical Division - MS569, Los Alamos Scientific Laboratory
University of California
Los Alamos, NM 87545, USA

1. Introduction

Recently, BLACK, KOLODNER, SCHULTZ, YABLONOVITCH, and BLOEMBERGEN (BKSYB) [1] made the claim that IR laser excitation in polyatomic molecules leads to a thermal population distribution. The far reaching implications of this result for laser induced chemistry stimulated several doubting researchers to perform experiments to either prove or disprove this result [2]. The experiments so far are not conclusive. One might then hope that existing theory could resolve this question, but to date theory remains incomplete, and divided into two regimes: detailed low level coherent excitation including first order rotation-vibration structure [3.a] and high level quasi-continuum (qc) absorption typically described by incoherent rate equations [3.b]. Joining the ν_3 low level excitation with the higher lying qc in a smooth continuous manner has been the major stumbling block in forming a complete theoretical model of multiple photon excitation (MPE) and dissociation (MPD). In fact the lack of a complete model has forced researchers to speculate on the effects of the so called ν_3 ladder "bottleneck" [1]. Therefore, the problem of whether laser excitation in polyatomic molecules leads to a thermal population distribution has remained unresolved.

In the talk we will answer this question theoretically by describing a complete model of MPE and MPD applied to SF$_6$ [4]. We have combined our earlier work [3.a] on low level ν_3 excitation in a smooth continuous manner with a fundamentally consistent model of the qc [5]. Spectroscopy of the $3\nu_3$ overtone by PINE [6] uniquely determines the v-v anharmonic coupling parameter in our model. Spectroscopy of the fundamental as interpreted by MCDOWELL, ALDRIDGE and HOLLAND [7] for the anharmonicity constants determines the average "hotband" anharmonicity. Finally a knowledge of the size of the ordering parameter in the Hamiltonian [8] independently specifies the strength of the self mixing leaving us <u>no</u> remaining parameters to fit data.

We can now examine the conclusions of BKSYB by looking at the laser induced population distributions predicted by our model. Our understanding of the bottleneck allows us to study their model coupled to our bottleneck, which when fit to experiment, makes predictions in relatively close agreement with our model. (This is surprising since our qc cross sections at low levels are very different from theirs.) We find that the bottleneck inherently makes the population distribution nonthermal even for very high laser intensities. In addition, we rederive their model showing it to be funda-

[1]Work performed under the auspices of the U.S.D.O.E.

mentally inconsistent at low levels in the qc, but at very high levels of excitation our model asymptotically approaches their model. However, if we are forced to base our conclusions solely on experimental data (<n> vs. fluence and absolute dissociation vs. fluence), then the models (theirs coupled to our bottleneck and ours) are indistinguishable. We will comment on new experiments to test some basic differences between these models.

2. A Complete Model of MPE and MPD

During the past year we made a detailed study of MPE in the ν_3 mode of SF_6 based on spectroscopy of the fundamental and third overtone [3.a]. We found that at laser powers $\lesssim 100$ MW/cm^2 less than 3% of the ground state population reached $4\nu_3$ in the frequency range 938-950 cm^{-1} where experiment shows the dominant multiple photon absorption. Since the molecule can absorb ~ 40 photons and dissociate, we concluded that substantial anharmonic coupling to the qc must occur at or below $3\nu_3$ in SF_6.

The gradual decay of the normal mode picture due to anharmonic coupling is explicitly taken into account in the model with the assumption that those vibrational levels having the most similar ν_3 character are the most strongly interacting, i.e. levels with $n\nu_3$ character are coupled to levels with $(n\pm 1)$ ν_3 character. Terms in the Hamiltonian which change ν_3 quanta by more than one are of higher order and are neglected [8]. If the density of states is sufficiently large, then the coupling can be described by a Fermi Golden Rule (FGR) rate. For states with $n\nu_3$ character at an energy equal to $m\nu_3$ quanta we have a leakage into states of $(n\pm 1)\nu_3$ character at the rate

$$\Gamma_{n,n\pm 1}(m) = 2\pi g^2 \chi_{n\pm 1}(m) \tag{1}$$

where $\chi_n(m)$ is the density of states at energy $m\nu_3$ with $n\nu_3$ character. The parameter g can represent either an effective coupling constant or the square root of the dispersion of a random coupling to the background density [9]. As we will see this coupling leads to a consistent description of both ν_3 ladder coupling to the qc and of intra-mode coupling within the qc.

From our previous discussion we now recognize that the ν_3 ladder is coupled to a background density of states at each level of ν_3. At $3\nu_3$ the third overtone spectra of PINE [6] shows a residual width of .007 cm^{-1} after subtracting off both the doppler and pressure broadened linewidths implying a FGR leakage rate of .76 ns. Using (1) and the density of states function from WHITTEN and RABINOVITCH [10] as illustrated by STONE, GOODMAN, and DOWS (SGD) [3.b] we find g = .035 cm^{-1}. In our computer calculations at $1\nu_3$ we have used a FGR rate consistent with the density $\chi_0(1)$. While this is not rigorously justified at this level in the molecule, a more detailed level to level coupling is beyond the scope of our general model. However, the error we make is not significant, since the leakage into states $\chi_0(1)$ and the laser excitation out of states $\chi_0(1)$ is small.

Once population is in the qc it can flow between densities of states at the same energy, but with different ν_3 character, as described by (1). Each state with $n\nu_3$ character at energy $m\nu_3$ carries dipole strength, $D^2(m) = \frac{n+3}{2} d_{01}^2$ where d_{01} is the bare Rabi frequency, and can be pumped by the laser into states with $(n+1)\nu_3$ character at $(m+1)\nu_3$. A description of the molecule showing this complexity in the qc is illustrated in Fig. 1 where the qc leakage begins at $1\nu_3$.

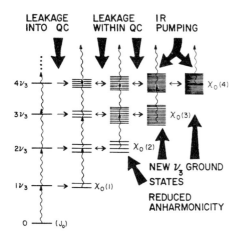

LEAKAGE INTO QC   LEAKAGE WITHIN QC   IR PUMPING

$4\nu_3$

$3\nu_3$

$2\nu_3$

$1\nu_3$

$X_0(4)$

$X_0(3)$

$X_0(2)$

NEW ν_3 GROUND STATES

$X_0(1)$

REDUCED ANHARMONICITY

0 — (J_0)

Fig. 1 Energy level diagram for SF$_6$.
Regions where primary characteristics of the molecule play a role are indicated by large arrows. Vertical (horizontal) arrows indicate laser excitation (intramolecular V-V relaxation).

The dipole strength carried by the density of states $\chi_n(m)$ depends on the amount of self coupling between the levels in the effective states sense of SGD. Since self coupling requires more non-ν_3 quanta interchanges than that implied by g where we changed mode character only the amount necessary to roughly equal $1\nu_3$ quanta, we estimate the self coupling parameter $\bar{g}$, to be smaller than g by roughly two orders in the molecular Hamiltonian [8],

$$\bar{g} = (\chi_{33}/\omega_3)\, g \quad , \qquad (2)$$

where X_{33} is the anharmonicity constant and ω_3 is the frequency of the pumped mode. The number of levels strongly coupled at $m\nu_3$ with character $n\nu_3$ is simply

$$N_n(m) = \bar{g}\chi_n(m) \quad . \qquad (3)$$

If $N_n(m)$ is less than unity, it simply means that the coupling $\bar{g}$ is not strong enough to mix states as they are too far apart, and $N_n(m) = 1$. The total effective squared dipole strength is

$$D_n^2(m) = N_n(m)\, \frac{n+3}{2}\, d_{01}^2 \quad . \qquad (4)$$

If we substitute values for the density of states $\chi_n(m)$ into (4) and (1), and calculate the Rabi frequency d_{01} for typical laser intensities used in SF$_6$ experiments, we find that in SF$_6$ the intramolecular v-v rates (1) dominate the Rabi frequencies at energy step in the qc. The dynamics in the qc is, therefore, not coherent, and the perturbation theory rate, $R_n(m)$, for a laser induced transition from the states $\chi_n(m)$ to the states $\chi_{n+1}(m+1)$ [11] is

$$R_n(m) = \frac{D_n^2(m)\, \gamma_n(m)/4}{\Delta_n^2(m) + \gamma_n^2(m)/4} \quad , \qquad (5)$$

where

$$\gamma_n(m) = \Gamma_{n,n+1}(m) + \Gamma_{n,n-1}(m) + \Gamma_{n+1,n+2}(m+1) + \Gamma_{n+1,n}(m+1) \quad , \qquad (6)$$

and $\Delta_n^2(m)$ is the laser transition detuning in the molecule. We calculate $\Delta_n(m)$ using the anharmonicity parameters of MCDOWELL, ALDRIDGE, and HOLLAND [7] for the "hotband" anharmonicity which depends on m, and using the ν_3 ladder anharmonicity as calculated in our previous work which depends on n^3 [3.a].

Since the intramolecular v-v relaxation is very fast in SF_6, the dynamics of the complex qc, shown in Fig. 1, can be simplified to a single rate ladder. Each state in this ladder for example, at level $m\nu_3$ consists of all the states with $n\nu_3$ character where $n \leq m$; i.e.

$$P(m) = \sum_n p_n(m) \quad , \tag{7}$$

where $P(m)$ is the total population at $m\nu_3$ energy and $p_n(m)$ is the population with $n\nu_3$ character at $m\nu_3$. The dynamics of the sub-populations are completely described by rate equations,

$$\dot{p}_n(m) = - (R_n(m) + R_{n-1}(m-1) + \Gamma_{n,n+1}(m) + \Gamma_{n,n-1}(m)) \, p_n(m)$$

$$+ \Gamma_{n+1,n}(m)p_{n+1}(m) + \Gamma_{n-1,n}(m)p_{n-1}(m) \tag{8}$$

$$+ R_n(m) \, p_{n+1}(m+1) + R_{n-1}(m-1)p_{n-1}(m-1) \quad .$$

Rapid intramolecular v-v relaxation means that the sub-populations in states $p_n(m)$ are in statistical equilibrium [5],

$$p_n(m) = \frac{\chi_n(m)P(m)}{\sum_n \chi_n(m)} \equiv \frac{\chi_n(m)}{\chi(m)} P(m) \quad . \tag{9}$$

Substituting (9) into (8) and summing over n gives rate equations governing the laser dynamics of the populations $P(m)$,

$$\dot{P}(m) = - \left(\sum_{n=0}^{m} \frac{R_n(m)\chi_n(m)}{\chi(m)} + \sum_{n\neq 0}^{m} \frac{R_{n-1}(m-1)\chi_n(m)}{\chi(m)} \right) P(m)$$

$$+ \sum_{n=0}^{m-1} \frac{R_n(m-1)\chi_n(m-1)}{\chi(m-1)} \, P(m-1) \tag{10}$$

$$+ \sum_{n\neq 0}^{m+1} \frac{R_{n-1}(m)\chi_n(m+1)}{\chi(m+1)} \, P(m+1) \quad ,$$

where the rates in (10) have been derived in a manner consistent with spectroscopy and standard molecular concepts. We should emphasize here that this simplification of our complex model (Fig. 1) to a single ladder rate description in the qc results from specializing the problem to SF_6 and may be very different for other molecules. Eq. (10) makes it easy for us to compare our qc model with other rate models, in particular, the model of BKSYB.

We complete our model by including dissociation in the form of standard RRKM rates [12] which for SF_6 begin at $33\nu_3$ [13]. Since our model of the qc reduced naturally to a statistical one for SF_6, we find a statistical theory of dissociation completely consistent with our picture of the SF_6 molecule.

3. Derivation of the BKSYB Model

The qc rate model recently proposed by BKSYB is based entirely on Fermi's Golden Rule at every excitation step. The validity of FGR rates rests on the assumption that at each energy mv_3 all the levels are strongly mixed; i.e. the single pair strength (sps) concept of SGD is valid.

$$(sps)^2 = \frac{(3/2)\chi_4(m)}{G\chi(m)\chi(m+1)} d_{01}^2 \quad , \tag{11}$$

where $\chi(m)$ is defined in (9), G is energy spread of the qc levels which are strongly mixed, and $\chi_4(m)$ is the density of states assuming v_3 is quadruply degenerate as defined by SGD. The FGR rates in the qc from mv_3 up and $(m+1)v_3$ down are simply

$$R^{up}(m) = 2\pi(sps)^2\chi(m+1) \quad \propto \quad \frac{\chi_4(m)d_{01}^2}{G\chi(m)} \tag{12a}$$

$$R^{down}(m+1) = 2\pi(sps)^2\chi(m) \quad \propto \quad \frac{\chi_4(m)d_{01}^2}{G\chi(m+1)} \tag{12b}$$

which reduce exactly to the thermal rates of BKSYB when $\chi(m)(\chi_4(m))$ is approximated as a 15 (16) fold degenerate oscillator as done by BKSYB. The significance of this simple derivation is that their assumption of constant cross section and a value for that cross section chosen to fit the $<n>$ vs. fluence data requires $G \gtrsim 500$ cm^{-1} at every step in the qc. While a G of this size very high in the qc is feasible, it is inconsistent with spectroscopy [6,7] and standard notions governing molecular interactions [8] very low in the molecule. While we reject this qc model as inconsistent on theoretical grounds, we will still show that coupling the BKSYB qc with our v_3 ladder can be consistent with experiment.

Let us return to our model momentarily and look at our rates (5) very high in the qc. In this region $\gamma(m) \gg \Delta_n(m)$ and the widths $\gamma_n(m)$ are proportional to $\chi_{n+1}(m+1)$ making the rates

$$R_n(m) \quad \propto \quad \frac{\chi_n(m)\left(\frac{n+3}{2}\right)}{\chi_{n+1}(m+1)} d_{01}^2 \quad . \tag{13}$$

Substituting (13) into (10) we find for our model

$$R^{up}(m) \quad \propto \quad \left(\frac{3/2\chi_4(m)}{\chi(m)} -1\right) d_{01}^2 \tag{14a}$$

$$R^{down}(m) \quad \propto \quad \left(\frac{\chi_4(m)}{\chi(m+1)} - \frac{3/2\chi_0(m)}{\chi(m+1)}\right) d_{01}^2 \tag{14b}$$

which reduce to (12) asymptotically in the high qc. This calculation shows that only the high qc rates in our model are consistent with strong mixing and the FGR rates of BKSYB.

4. Comparison, Results and Conclusions

Even though we have shown that the thermal model suffers from unphysically large (500 cm^{-1}) level widths very low in the molecule, it is still useful to compare the population distributions predicted by both models.

In Fig. 2 we show our theory compared with the optoacoustic excitation data of DEUTSCH [14] which has been calibrated by O. JUDD (LASL) to agree with the absorption data of LYMAN et al. [15]. The agreement is extraordinary for a zero parameter theory. However, since the thermal qc model coupled to our ν_3 excitation ladder can also fit this excitation data, we can only conclude that data of this type is not sufficient to uniquely distinguish between different theoretical models. The data on dissociation is even less precise due to the measurement process and SF$_5$ absorption. Our agreement with experiment is shown on the right side of the figure. The upper curve is all the population in levels above the dissociation limit (33ν_3 in SF$_6$) plus the RRKM dissociation during the pulse. The lower curve is the RRKM dissociation during the pulse. We show both the beam data of BRUNNER and PROCH [16] and the cell data of BKSYB [1].

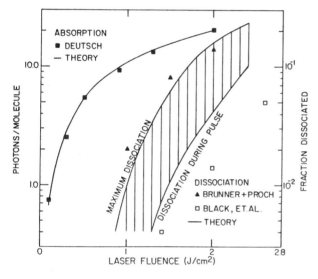

Fig. 2

In Fig. 3 we show the qc transition cross sections (both for absorption and emission) at each level $n\nu_3$ both for our model and for the thermal model including our ν_3 ladder. Considering the very different shapes and the differences between up and down cross sections it is surprising both models fit the same <n> vs. fluence data. As we previously stated this type of data cannot by itself uniquely distinguish between different theoretical qc models. The figure clearly shows that our qc cross sections traverse three regions in SF$_6$ (5, 6, 10): (1) ($\Delta_n^2(m) > \gamma_n^2(m)/4$, $N_n(m) = 1$) the qc cross sections increase proportional to the density of states (2) ($\gamma_n^2(m)/4 > \Delta_n^2(m)$, $N_n(m) = 1$) the qc cross sections decrease proportional to the density of states, and

305

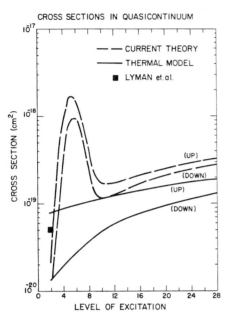

CROSS SECTIONS IN QUASICONTINUUM

— — CURRENT THEORY
—— THERMAL MODEL
■ LYMAN et.al.

CROSS SECTION (cm²)

(UP)
(DOWN)
(UP)
(DOWN)

LEVEL OF EXCITATION

Fig. 3

(3) $(\gamma_n^2(m)/_4 > \Delta_n^2(m),\ N_n(m) > 1)$
the qc cross sections level off
due to the presence of strong
self mixing. D. HAM (Rochester)
has suggested that a test of the
peak in our cross sections could
be made using short pulses. If
the experimental conditions are
right, then population should collect at levels $8-10\nu_3$ allowing
some type of excited state chemistry to be performed. J. GOLDSTEIN
(LASL) has also suggested that a
pulse shape measurement might test
the form of our qc cross sections.
We would expect to see an exceptionally large depletion of the pulse
initially.

In Fig. 4 we show our very non-
thermal population distributions in
the qc at four different fluences.
Note the very long tails on the distributions due to the ν_3 ladder
bottleneck [17]. The large peak
at $1\nu_3$ is due to the very small qc
cross section for going from $1\nu_3$
to $2\nu_3$.

FLUENCE DEPENDENCE OF LEVEL POPULATIONS

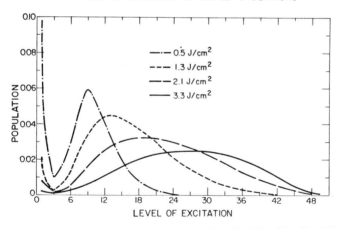

POPULATION

—·— 0.5 J/cm²
— — — 1.3 J/cm²
— — 2.1 J/cm²
—— 3.3 J/cm²

LEVEL OF EXCITATION

Fig. 4

In Fig. 5 we show the effect of the ν_3 ladder bottleneck on the thermal
model's Boltzmann distribution. By adding our ν_3 ladder excitation (as
discussed in Sec. 2) we find a very nonthermal distribution which is very
close to that predicted by our model for the same conditions (100 ns pulse.
$\langle n \rangle \sim 14$, collisionless).

306

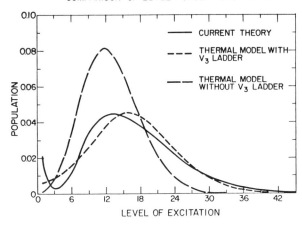

COMPARISON OF LEVEL POPULATIONS <n> ~14

CURRENT THEORY

THERMAL MODEL WITH V₃ LADDER

THERMAL MODEL WITHOUT V₃ LADDER

POPULATION

LEVEL OF EXCITATION

Fig. 5

BKSYB have made the claim that at high intensities the ν_3 ladder bottleneck is not important, and the population distribution is "close" to thermal. Taking their advice we calibrated their thermal qc model with their 500 ps data finding a qc cross section. Using this calibrated qc model with our ν_3 ladder bottleneck we found that this model no longer fit the <n> vs. fluence data of DEUTSCH [14] for 100 ns pulse lengths. The qc cross section was too large by a factor of three. We concluded that the ν_3 ladder bottleneck was fundamental for MPE and MPD of polyatomic molecules at all laser intensities. Therefore, the population distribution (collisionless) is never thermal.

References

1. J. G. Black, P. Kolodner, M. J. Schultz, E. Yablonovitch, and N. Bloembergen, Phys. Rev. A 19, 704 (1979) and references therein.
2. E. R. Grant, P. A. Schulz, Aa. S. Sudbo, and Y. T. Lee, Phys. Rev. Lett. 40, 115 (1978); M. Rothschild, W. S. Tsay, and D. O. Ham, Opt. Commun. 24, 327 (1978); M. M. Arvedson and D. A. Kohl, preprint; R. Dupperrex and H. van den Bergh, preprint.
3. (a) J. R. Ackerhalt and H. W. Galbraith, J. Chem. Phys. 69, 1200 (1978); H. W. Galbraith and J. R. Ackerhalt, Optics Lett. 3, 109 (1978); D. M. Larsen, Optics Commun. 19, 404 (1976); C. D. Cantrell and K. Fox, Optics Lett. 2, 151 (1978); H. W. Galbraith and J. R. Ackerhalt, Optics Lett. 3, 152 (1978); H. W. Galbraith, Optics Lett. 3, 154 (1978).
(b) Ref. 1; J. L. lyman, J. Chem. Phys. 67, 1868 (1977); E. R. Grant, P. A. Schulz, Aa. S. Sudbo and Y. T. Lee [2]; M. Tamir and R. D. Levine, Chem. Phys. Lett. 46, 208 (1977); S. Mukamel and J. Jortner, Chem. Phys. Lett. 40, 150 (1976); D. P. Hodgkinson and J. S. Briggs, J. Phys. B: Atom. Molec. Phys. 10, 2583 (1977); J. Stone, M. F. Goodman, and D. A. Dows, J. Chem. Phys. 65 5062 (1976).
4. J. R. Ackerhalt and H. W. Galbraith, submitted to Optics Lett.; H. W. Galbraith and J. R. Ackerhalt, in preparation.
5. M. Goodman, J. Stone, J. Horsley, D. Dows and E. Thiele are simultaneously working on a complete model of MPE and MPD. For a description of this work see J. Stone and M. F. Goodman to be published J. Chem. Phys. and J. C. Stephenson, D. King, M. F. Goodman and J. Stone, J. Chem. Phys. in press.

6. A. Pine, private communication.
7. R. S. McDowell, J. P. Aldridge and R. Holland, J. Phys. Chem. <u>80</u>, 1203 (1976).
8. <u>Rotation-Vibration of Polyatomic Molecules</u>, G. Amat, H. H. Nielsen, G. Tarrago (Marcel Dekker Inc., New York, 1971).
9. J. Jortner, Adv. Laser Spect. <u>113</u>, 88 (1977); B. Carmeli and A. Nitzan, Chem. Phys. Lett. (to be published).
10. G. Z. Whitten and B. S. Rabinovitch, J. Chem. Phys. <u>38</u>, 2466 (1963).
11. J. R. Ackerhalt and J. H. Eberly, Phys. Rev. A <u>14</u>, 1705 (1976); J. R. Ackerhalt and B. W. Shore, Phys. Rev. A. <u>16</u>, 277 (1977).
12. P. J. Robinson and K. A. Holbrook, <u>Unimolecular Reactions</u> (Wiley, New York, 1972); M. J. Shultz and E. Yablonovitch, J. Chem. Phys. <u>68</u>, 3007 (1978) compare QRRK with RRKM finding only small differences.
13. T. Kiang, R. C. Estler and R. N. Zare (submitted to J. Chem. Phys.); J. F. Bott and T. A. Jacobs, J. Chem. Phys. <u>50</u>, 3850 (1969); S. W. Benson, Chem. Revs. <u>78</u>, 23 (1978); J. L. Lyman, J. Chem. Phys. <u>67</u>, 1868 (1977).
14. T. F. Deutsch, Optics Lett. <u>1</u>, 25 (1977).
15. J. Lyman, R. Anderson, R. Fisher and B. Feldman, Optics Lett. <u>3</u>, 238 (1978).
16. F. Brunner and D. Proch, J. Chem. Phys. <u>68</u>, 4936 (1978).
17. This is consistent with one of the interpretations of BKSYB [1] of their data.

Resonant Multiphoton Dissociation of Small Molecules

A.H. Kung, H.-L. Dai, M.R. Berman, and C.B. Moore

Department of Chemistry, University of California and
Materials and Molecular Research Division of the Lawrence Berkeley Laboratory
Berkeley, CA 94720, USA

Introduction

Studies of multiphoton dissociation (MPD) over the past few
years have led to a good semi-quantitative understanding of the
process for SF_6 and similar molecules. Several reviews have
appeared recently.[1] Applications have progressed to the point
that a pilot plant is operating for the production of ^{13}C by
MPD of CF_3I.[2] The model developed for understanding MPD divides
the vibrational energy levels of a molecule into three regions.[3]
Region I includes the lowest, "discrete" levels of the vibrational modes that are nearly resonant with the laser frequency.
Vibrational anharmonicity prevents exact resonances for the sequential absorption of several photons. However, the combined
effects of rotational energy compensation, Coriolis shifts, and
other perturbations, along with power broadening, permit the absorption of several photons. In region II, strong coupling
among vibrational modes drains energy from the pumped mode and
broadens vibration-rotation levels so that absorption occurs
over a broad frequency range. Sequential absorption of photons
becomes possible. In region III, where levels are above the
dissociation threshold, this absorption continues, but with the
competing process of dissociation occurring simultaneously.
The boundary between regions I and II is poorly defined. From
an operational point of view, region II begins when absorption
becomes non-frequency selective for any molecule at any laser
frequency of interest to the experimenter. In region II and
especially in region III there is good evidence that energy is
transferred rapidly among all vibrations of the excited molecule
and that the rate of dissociation as a function of energy may
be estimated using the RRKM theory.[4] The main features of the
dynamics of molecules in regions II and III appear to apply
rather generally to all types of molecules,[5] with the possible
exception of very large ones.[6]

The excitation mechanism in region I has been studied extensively only for SF_6. The dynamics in this region and in
the transition from I to II depend strongly on detailed features
of molecular structure and spectra. The great diversity of the
latter implies a great variety in the corresponding MPD phenomena. Many questions remain to be answered. How small a molecule will exhibit MPD? How does MPD change when photons outside
the 9 - 11 μm range are used? Can narrow resonances in MPD
yield vs laser frequency occur, and for which molecules and fre-

quencies? What is the nature of the transition between regions I and II? How does it depend on energy level density, molecular structure, and vibrational mode pumped by the laser? In this paper two molecules C_2H_5Cl and D_2CO are studied. The spectroscopic and structural features of these two molecules are intermediate between those of SF_6 and of diatomic molecules. For EtCl, hydrogen stretching vibrations are excited. Qualitatively new phenomena are observed. One purpose of the work on EtCl was to search for sharp resonances by using higher frequencies for excitation. The sharp resonances found may be useful for applications in selective photochemistry and laser isotope separation. D_2CO is studied to test the molecular size limit of collisionless MPD. The data also provide new and essential information for understanding the predissociation of formaldehyde.

Ethyl Chloride MPD

Let us first consider the thermochemistry and linear spectroscopy of EtCl. The lowest reaction channel is molecular elimination to hydrogen chloride and ethylene. Only seven photons with $\nu \geq 2900$ cm^{-1} are needed to overcome the 58 ± 2 kcal/mole activation barrier.[7] The molecule is nearly a prolate symmetric top ($\kappa = -0.959$) with 18 non-degenerate vibrational modes.[8] There are 5 C — H stretch modes and they provide the dominant coupling between the molecule and the radiation field in this experiment. The fundamental absorption spectrum shows sharp rotational structures with measured widths limited only by instrument resolution of 0.06 cm^{-1}. At $\nu_{CH} = 2$ there are 15 overtone and combination levels. The 2.5 cm^{-1} wide Q-branch of the first overtone indicates a fair amount of mixing of these levels with the low frequency vibrations. At $\nu_{CH} = 3$ the total vibrational level density reaches 1300 per cm^{-1}.[9] The second overtone spectrum shows level widths that are wider than the P, R branch widths. Level mixing by anharmonic coupling and Coriolis coupling is strong.[10] Since the anharmonic defect of each individual C — H stretch mode is ~ 50 cm^{-1}, the probability of populating high quantum states by exciting a single mode becomes miniscule.[11] However, the presence of a multitude of nearly degenerate vibrational modes makes it possible to find a path of excitation that involves combination levels. This path is resonant for the $v = 0$ to 1, 1 to 2, and higher transitions. Figure 1 is a pictorial illustration of such a path to $v = 2$ and the quasi-continuum. Only molecules with the proper rotational quantum number will be excited at each laser frequency. Substantially more molecules will come into resonance if one transition in the path falls on a Q-branch. With the help of these resonances, the probability of absorbing n photons becomes large at modest intensities. Due to the large quantum of excitation at the stretch frequency, the quasi-continuous levels may be reached at n = 3 or 4. Sequential absorption of more photons will drive the molecule to the dissociative states. Since the quasi-continuous levels are expected to have broad spectral characteristics, MPD yield would reflect the resonances of discrete level excitation and show enhancements that coincide with peaks in the fundamental and overtone spectra.

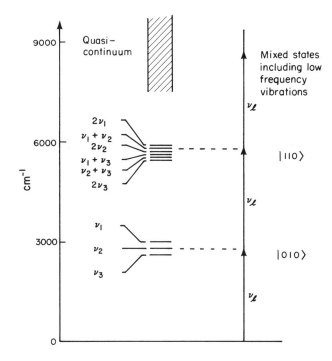

Fig. 1 Vibrational level diagram for molecule with 3 nearly
degenerate modes. Arrows indicate resonant transition pathway
to v = 3 via: |000> → |010> → |110> → quasi-continuum. Low
frequency vibrations are not included.

The experimental arrangement has been described elsewhere.[12,13]
Briefly, the tunable 3.3 μm beam from a Nd:YAG laser pumped
$LiNbO_3$ parametric oscillator was focused to the center of a
Brewster angle cell where photolysis took place. From 3000 to
7200 pulses were used for each photolysis. Photolysis products
were then analyzed with a flame ionization gas chromatograph
(GC). The photolysis yield $W_d(\nu)$ is given by the total area
beneath the GC ethylene peak. An absolute calibration of the
GC, combined with the effective focal volume estimate, gives an
approximate absolute yield scale.

Results of the experiment have been reported in Refs. 12 and
13. We find that dissociation occurs throughout the entire
spectral range studied. The shape and structure of the yield
spectrum match those of the fundamental absorption spectrum
(Fig. 2). If we divide the dissociation yield $W_d(\nu)$ by the
fundamental absorbance $A_{01}(\nu)$ $(A_{01}(\nu) = \sigma(\nu)N\ell)$, the resulting
curve correlates well with those of the first and second over-
tone spectra (Fig. 3) Yields measured at the Q-branch at 2944
cm^{-1} show that the frequency resolution of this MPD process is
as good as 0.4 cm^{-1} and may be limited only by the combined
laser bandwidth (0.15 ± 0.03 cm^{-1}) and the power-broadened

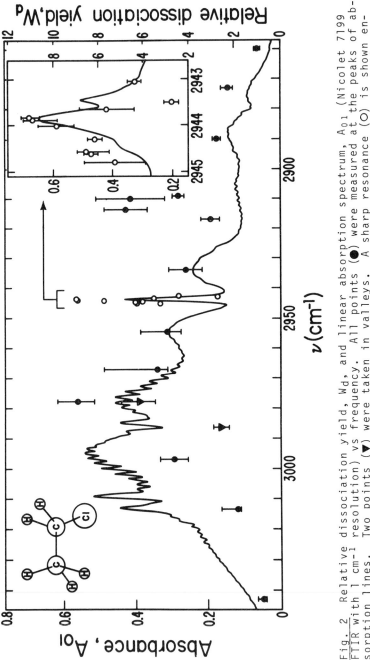

Fig. 2 Relative dissociation yield, W_d, and linear absorption spectrum, A_{01} (Nicolet 7199 FTIR with 1 cm^{-1} resolution) vs frequency. All points (●) were measured at the peaks of absorption lines. Two points (▼) were taken in valleys. A sharp resonance (○) is shown enlarged with an absorption spectrum at 0.24 cm^{-1} resolution. Two points near 2977 cm^{-1} show sharp structure in a P-branch.

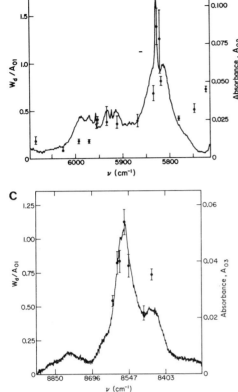

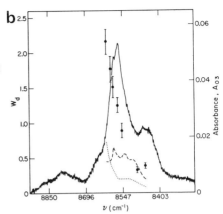

Fig.3 (a) Relative dissociation yield, $W_d(\nu/2)$, divided by the absorbance, $A_{01}(\nu/2)$ at 0.24 cm^{-1} resolution, from ($\bullet$) in Fig.2. The first overtone absorption spectrum, $A_{02}(\nu)$, is also plotted at 1 cm^{-1} resolution (FTIR). (b) Relative dissociation yield $W_d(\nu/3)$ and second overtone spectrum $A_{03}(\nu)$ (2.5 cm^{-1} resolution, Cary 17). $A_{01}(\nu/3)$ ..... and $A_{02}(2\nu/3)$ --- are also shown. (c) $W_d/A_{01}(\nu/3)$ and the second overtone spectrum. The monotonic decrease of W_d in (b) becomes a broad resonance matching the second overtone spectrum when normalized.

spectral width. Although the overtone spectra do not provide information on the strength of excited-state absorption, they do give the frequencies and approximate bandwidths of the more anharmonic transitions. The frequency match of the yield spectra (W_d and W_d/A_{01}) with these linear absorption spectra establish the resonance excitation provided by 0 to 1, 1 to 2, and 2 to 3 transitions as discussed earlier. The excitations from 1 to 2 and then 2 to 3 represent a gradual transition from the discrete levels (v = 1) to the quasi-continuum (v = 3).

After establishing the dynamics of excitation through region I, it is possible to calculate the MPD yield following the formalism used by Grant et al.[4] Absorption in regions II and III is treated by stepwise, incoherent, single photon transitions in a driven harmonic oscillator. The process can be described by a series of rate equations. In region III, the competing process of dissociation is included. The set of rate equations can be solved for a given molecule when the absorption cross section $\sigma(\nu)$ and laser intensity I(t) are specified. For this calculation, absorbed energy is assumed to be randomized before dissociation occurs so that rate constants obtained from the RRKM theory[7] can be used. The excited state absorption

313

cross section is not known. However, several recent studies on SF_6 and similar molecules have shown that the cross section of highly excited molecules is very slowly decreasing or even remains constant until dissociation is reached.[14,15,16,17] This is reasonable because when intramolecular energy transfer is fast compared to the excitation rate, the absorbing mode would remain in a state with a vibrational quantum number that puts it in region II. At the same time, energy is funneled from the laser field through this mode to other modes of the molecule. By adopting a cross section that is constant and equal to the average cross section for transition $v = 1$ to 2, dissociation yields are calculated and compared with measured yields at several frequencies (Table 1). The agreement is quite satisfactory. The discrepancy at 2913 cm^{-1} arises most probably because the absorbance at 2913 cm^{-1} for the 1 to 2 transition is a few times higher than the average due to the presence of a Q-branch. The phenomenological model also successfully explains the fluence dependence (Fig. 4) and the pressure dependence of the dissociation.[13]

<u>Table 1</u> Calculated and measured C_2H_4 yield at selected frequencies.

	P Branch		Q Branch	Q	Overtone
$\nu(cm^{-1})$	2976.6	2977	2943.8	2913	967.7
σ^a_{01} (10^{-19} cm^2)	30	2	2	7	3
σ^a_{12}	3	3	3	3	-
$\sigma^a_{\ell,\ell+1}$ ($\ell \geq 2$)	3	3	3	3	-
Yield (%) calculated	1.5	1.3	1.7	0.48	-
Yield (%) measured	1.8±0.2	0.9±0.3	1.8±0.2	1.3±0.2	0.09±.01

One of the new features of EtCl MPD is the sharp resonances in the yield spectrum. Based on this study it is possible to obtain some general ideas as to when structures should occur in MPD. The molecule must have a structured fundamental spectrum. It must have a path of at or near resonance excitation through the discrete levels to the quasi-continuum so that power broadening does not destroy the structure. Hence, small and medium size molecules with a single, isolated vibrational mode will usually not fall into this category. Molecules with three or more nearly degenerate modes have a high probability of exhibiting dissociation with sharp resonances. A careful examination of fundamental and overtone spectra should often reveal possible resonant excitation paths. Level widths can show the

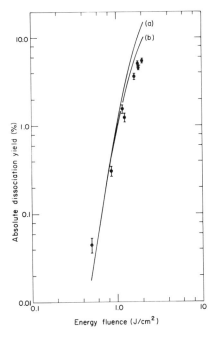

Fig. 4 Comparison of calculated ethylene yield with experimental yield vs laser fluence. (a) calculated, 2976.8 cm^{-1}, $\Delta \nu_\ell = 0.15$ cm^{-1}; (b) calculated, 2977 cm^{-1}, $\Delta \nu_\ell = 2.5$ cm^{-1}; (⦁) experimental data, $\Delta \nu_\ell = 2.5$ cm^{-1}. $\Delta \nu_\ell$ is laser bandwidth.

onset of the quasi-continuum. More accurate assessments can be obtained if absolute intensities of transitions among excited states are available.

Formaldehyde MPD

Several attempts to induce collisionless MPD in diatomic molecules have met with failure. What is then the limitation on molecular size and structure for collisionless MPD? The simple four-atom molecule, formaldehyde, is a good test for this limit. Only the deuterated isotope D_2CO has absorption that straddles the CO_2 laser wavelengths. The vibrational density of states computed by direct count is only about 50 per cm^{-1} at $30,000$ cm^{-1} of internal energy.[18] Experiments have been performed to study total ir fluorescence, fluorescence from the C — D stretches, and the yield of CO from dissociation following TEA CO_2 laser excitation at 975.9 cm^{-1}. Intensity of emission from the C — D and C ═ O stretches is found to increase as the square of D_2CO pressure (Fig. 5). Although there is strong Coriolis coupling between the C — D stretch and C — D wagging modes,[19] no emission is observed at the stretch frequency around 2000 cm^{-1} at pressures where gas kinetic collision times are longer than the laser pulse duration. These observations indicate that excitation to high vibrational levels does not occur under collisionless conditions in D_2CO. The lack of close compensation mechanisms to overcome anharmonicity has prevented absorption of several photons per molecule. Dissociation yield measurements support these observations. CO production occurs only at high pressure.[20] The yield increases as a high order of

315

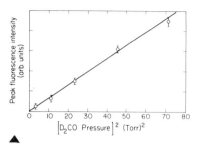

Fig.5 5 μm fluorescence from multiphoton excited D_2CO as a function
of the square of D_2CO pressure.

Fig.6 Total CO yield vs D_2CO pres- ▶
sure. ⊕ pure D_2CO; △ D_2CO / 15 Torr
NO. NO replaces D_2CO as the collisional partner for rotational relaxation and also acts as a radical
scavenger.

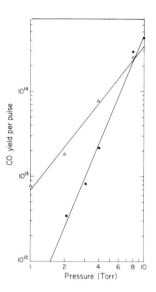

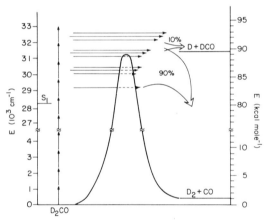

Fig.7 Product distribution
from D_2CO MPD at 75 J/cm^2.
Molecular products result
from both tunneling through
and passing over the bar-
rier.

total reactant pressure (Fig. 6). This indicates that collisional V — V energy transfer and up-pumping play a crucial role in
facilitating excitation. The study of D_2CO MPD is particularly
interesting because its uv dissociation is well studied. Formation of radical products (D + DCO) has a threshold at 323 ± 3
nm.[21] Theory and experiment have established a barrier for
molecular product formation.[22] During infrared excitation, the
molecule remains on the ground potential surface. It has been
observed that 10% of the products appear through the radical
channel (Fig. 7) when pumped by a CO_2 laser at 75 J/cm^2. Since
the barrier height is close to the radical threshold, study of
ratio of radical yield to molecular yield as a function of
energy fluence may allow the estimation of absolute molecular
dissociation rates and the barrier height for molecular dissociation.

Summary

The experiments on EtCl and D_2CO have provided answers to some of the questions that were raised earlier in this paper. MPD dynamics are probably quite similar for all molecules in regions II and III but the dynamics in region I depend greatly on molecular structure and spectra. MPD behavior will vary greatly with structure. For D_2CO, intensities of 1 GW/cm^2 were not sufficient to get the molecule through region I. Collisions are required. For C_2H_4, collision-free excitation is possible but dissociation still occurs only at such high intensities that spectral structure is almost completely smoothed by power broadening.[23] In EtCl, the presence of a sharp fundamental spectrum and anharmonic compensation by resonant combination band transitions have enabled the observation of sharp features in the dissociation spectrum. For molecules such as SF_6, dissociation occurs at low fluences but structure is lost due to a more complex spectrum.[24] For very large molecules like $UO_2(hfacac)_2 \cdot THF$, resonances only a few cm^{-1} wide may occur in quite a different way.[6]

Acknowledgments

We thank the Division of Advanced Systems Materials Production, Office of Advanced Isotope Separation, U.S. Department of Energy for research support under contract No. W-7405-Eng.48.

References

1. P.A. Schulz, Aa.S. Sudbo, D.J. Krajnovich, H.S. Kwok, Y.R. Shen, and Y.T. Lee, Annual Review of Physical Chemistry 30 1979, in press, and references therein.
2. V.N. Bagratashvili, V.S. Dolzhikov, V.S. Letokhov, and E.A. Ryabov, Proc. 2nd Intl. Symp. on Gas-Flow and Chemical Lasers, Brussels, Belgium, Sept. 1978.
3. J.G. Black, E. Yablonovich, N. Bloembergen, and S. Mukamel, Phys. Rev. Lett. 38, 1131 (1977).
4. E.R. Grant, P.A. Schulz, Aa.S. Sudbo, Y.R. Shen, and Y.T.Lee, Phys. Rev. Lett. 40, 115 (1978).
5. Aa.S. Sudbo, P.A. Schulz, Y.R. Shen, and Y.T. Lee, J. Chem. Phys. 69, 2312 (1978).
6. A. Kaldor, R.B. Hall, D.M. Cox, J.A. Horsley, P. Rabinowitz, and G.M. Kramer, J. Am. Chem. Soc., in press.
7. P.J. Robinson and K.A. Holbrook, Unimolecular Reactions (Wiley, 1972).
8. F.A. Miller and F.E. Kiviat, Spect. Chim. Acta 25A, 1363 (1969).
9. Whitten-Rabinovitch approximation, ref. 7, Chap. 5.
10. S. Mukamel and J. Jortner, J. Chem. Phys. 65, 5204 (1976).
11. N. Bloembergen, C.D. Cantrell, and D.M. Larson in "Tunable Lasers and Applications", eds. A. Mooradian et al. Springer-Verlag 1976, p. 162.
12. Hai-Lung Dai, A.H. Kung, and C. Bradley Moore, submitted to Phys. Rev. Lett. (to be published).
13. Hai-Lung Dai, A.H. Kung, and C. Bradley Moore, in preparation.
14. A.V. Nowak and J.L. Lyman, J. of Quant. Spectrosc. Radiat. Transfer 15, 945 (1975).
15. W. Fuss, J. Hartmann, and W.E. Schmid, Appl. Phys. 15, 297 (1978).
16. J.G. Black, P. Kolodner, M.J. Shultz, E. Yablonovich and N. Bloembergen, Phys. Rev. A, Feb. (1979), in press.
17. J.L. Lyman, W.C. Danen, A.C. Nilsson, and A.V. Nowak, to be published.

18. Calculated by direct counting of vibrational levels.
19. T. Nakagawa and Y. Morino, J. Mol. Spectrosc. 38, 84 (1971) and T. Nakagawa, private communication.
20. See also G. Koren, M. Okon, and U.P. Oppenheim, Opt. Commun. 22, 351 (1977).
21. J.H. Clark, C.B. Moore, and N.S. Nogar, J. Chem. Phys. 68, 1264 (1978).
22. J. Goddard and H.F. Schaefer III, J. Chem. Phys., in press, and J.C. Weisshaar and C.B. Moore, J. Chem. Phys., in press.
23. V.N. Bagratashvili, I.N. Knyazev, V.S. Letokhov, and V.V. Lobko, Opt. Commun. 14, 426 (1975).
24. R.V. Ambartzumian and V.S. Letokhov in Chemical and Biochemical Applications of Lasers, Vol. III, ed. C.B. Moore, Academic Press, New York 1977, p. 167.

Double Resonance Spectroscopy of Multiple-Photon Excited Molecules

J.I. Steinfeld, C.C. Jensen[1], T.G. Anderson, and Ch. Reiser

Department of Chemistry, Massachusetts Institute of Technology
Cambridge, MA 02139, USA

1. Introduction

A crucial problem in the interpretation of multiple infrared photon absorption and dissociation effects in polyatomic molecules is that of the distribution of vibrational energy among the excited molecules. At issue is not only the distribution over vibrational modes within and individual molecule at total energy E $\{f(n_1,...,n_s|E)\}$, but also of E given an incident infrared fluence Φ $\{f(E|\Phi)\}$ and, indeed, of <E> in a bulk ensemble given a spatially varying $\Phi(r,z)$ resulting from focussing of the laser beam into the sample. This problem has been attacked from a number of points of view, including those of master-equation models [1,2], information theory [3], quantum-mechanical equations of motion [4], and geometrical optics [5,6]. These models are typically compared with measurements of either total energy deposition in the irradiated gas, <n|Φ>, or dissociation yield, <Y|Φ>, at an over-all fluence Φ [7]. It is not possible, however, to determine a distribution function uniquely from its first moments alone, such as <n> or <Y> [8]. Additional information is required, and the best source of this information is some sort of spectroscopic probe of the multiple-photon excited molecule. Such methods include:

(1) Optical/Infrared Double Resonance (Laser-Induced Fluorescence Excitation): a tunable ultraviolet or visible laser is used to excite fluorescence from the vibrationally excited molecules prior to relaxation or dissociation. This is a very sensitive probe for species processing suitable electronic transitions, such as biacetyl [9] or propynal [10].

(2) Infrared/Infrared Double Resonance: systems in which a laser-induced fluorescence technique cannot be applied may be probed by a low-power c.w. (or pulsed) infrared source. This technique has been extensively applied SF_6 [11-18] and a number of other systems since its initial demonstration by STEINFELD and co-workers in 1969.

(3) Other methods, such as laser intracavity absorption or infrared fluorescence due to multiple-photon excited molecules, are just now being developed.

In this paper, we describe a diode-infrared laser-double resonance investigation of the initial photon absorption step ($\nu_3 \leftarrow 0$) in SF_6.

---

[1]Present address: Joint Institute for Laboratory Astrophysics,
                 Boulder, Colorado, USA

A preliminary account of this work was presented at the Conference on Tunable Lasers and Applications [19].

2. Experimental Arrangement

The apparatus used is similar to that employed in the first IRDR experiments on SF_6 [11], except that the cw CO_2 line-tunable probe laser is replaced by a tunable semiconductor diode laser. The pulsed CO_2 laser is a continuous discharge, Q-switched at 150 Hz, providing 2 mJ, 400 nsec (FWHM) pulses on the 10.6 μm P(12)-P(22) lines. The diode laser is a Laser Analytics Model SDL-30 PbSnSe element mounted in a Model TCR closed-cycle cooler, which is temperature stabilized to ± 0.001°c over the operating range of 14°K to 100°K. Current to the diode is provided by an Arthur D. Little TDLS-II power supply, which is current stabilized to ± 0.0001 amp over the range 0.5-0.2 amps. This combination provides tunable infrared output over the range 930-960 cm^{-1} in a series of modes from 0.2 to 1.0 cm^{-1} in width, with 1-2 cm^{-1} gaps between modes, and a wavenumber stability ≤ 0.001 cm^{-1}. Multimode power output is 0.1 to 1.0 mW.

The two beams are made to overlap in a 30 cm gas sample cell containing 0.05-0.1 Torr SF_6 with added buffer gases. The beams are counterpropagating and overlap each other a considerable distance in the cell, unlike the crossed-beam geometry of Ref. 17. This produces larger amplitudes and narrower linewidths for the double resonance signals, but it is then necessary to take considerable precautions in order to prevent scattered light from the high-intensity CO_2 pump laser from reaching either the infrared detector or the diode laser itself. The diode laser beam is detected with an Arthur D. Little Hg: Cd/Te detector the signal amplified and sent to either a PAR Model 160 Boxcar Integrator or a Biomation Model 820 Transient Recorder. In the former case, the integration gate (0.1 or 0.5 μsec) is set at a fixed time relative to the CO_2 laser pulse, and the diode laser is scanned in frequency. This produces a spectrum of the change in transmitted infrared intensity correlated with the pump pulse, which is displayed on an X-Y recorder.

Wavenumber calibration of the diode is accomplished by using a 1" Ge etalon (FSR = 0.0499 cm^{-1}), in combination with reference lines of CO_2, N_2O [20], and NH_3 [21]. The assignments for SF_6 absorption lines in the regions studied were provided by MCDOWELL and co-workers at Los Alamos [22-25].

3. Results

Two diode modes were used in this work. One was centered at 950.4 cm^{-1} and encompassed the R(45) through R(56) lines of the SF_6 ν_3 fundamental. The other extended from 945.7 to 946.4 cm^{-1}, including the SF_6 P(28) - P(38) lines and the CO_2 00°1 - 10°0 P(18) pump line. A variety of double resonance effects were observed.

3.1 Two-Level Double Resonance

The "hole-burning" resulting from pumping a single SF_6 fine-structure transition is shown in Fig. 1. As expected, there is a strong saturation, or induced transparency, at the frequency of the line pumped by the laser. There is also satellite structure appearing at frequencies above and below that of the pump laser. These signals correspond to relaxation out of the directly pumped level; these effects are discussed further in Section 3.3.

The effect of varying CO_2 laser pump intensity is shown in Fig. 2. As the laser power is increased, the strong central feature broadens, and the satellite structure becomes more pronounced. A rough estimate of the power-broadened widths (FWHM) gives a slope of 74 MHz/$[KW/cm^2]^{1/2}$ proportional to the square root of the intensity (I). This slope should be of the order of magnitude of the Rabi precession frequency,

$$\Delta\nu \approx \mu_{01}|E|/h$$

where μ_{01} is the transition dipole moment and the electric field amplitude $|E|$ is proportional to $I^{1/2}$. Using the value $\mu_{01} = 0.388 \times 10^{-18}$ esu-cm [26], we calculate an expected slope $\Delta\nu/I^{1/2} \approx 117$ MHz/$[KW/cm^2]^{1/2}$. Considering the highly approximate nature of both the experimental and the theoretical determinations of this quantity, this is very satisfactory agreement. More exact calculations of the power-broadened line shapes will be carried out following the methods of MOLLOW [27] and WU et al. [28].

The time dependence of the double resonance signals is shown in Fig. 3. The initial signal is just the "hole-burning" at the laser frequency. The satellite structure is fully developed after only 100 nsec, and persists for the duration of the CO_2 laser pulse (~ 600 nsec). The "hole" has a relaxation time of 470 nsec; four separate determinations, at 0.05 and 0.08 Torr SF_6, give $p\tau = (24 \pm 4)$ nsec-Torr. Moulton et al. [17] have obtained a value of $p\tau = 32$-36 nsec-Torr, from measurements on three-level double resonance signals. The difference may reflect a shorter relaxation time when the pump and probe fields have both an upper and a lower level in common.

3.2 Three-Level Double Resonance

3.2.1 Ground-State Depletion

Double resonance signals were also observed in which the pump and probe beams interactions involved only a single SF_6 level in common. An example is shown in Fig.4, in which the pump beam is the CO_2 P(14) line at 949.48 cm^{-1}, which is near the $(E^1+F_2^2)$ and $(A_1^0+A_2^0+F_1^1+F_2^2)$ components of the SF_6 R(28) manifold. Signals are observed in the P(28) manifold near 946.4 cm^{-1}, with the strongest signals corresponding to the same symmetry components as are pumped by the CO_2 laser line. Weaker signals are also observed in the P(30) manifold, resulting from overpopulation of a common level in the upper vibrational state (J'=29). As in the two-level system, satellite structure, in addition to the main double line, is observed on all the components of a given J-manifold.

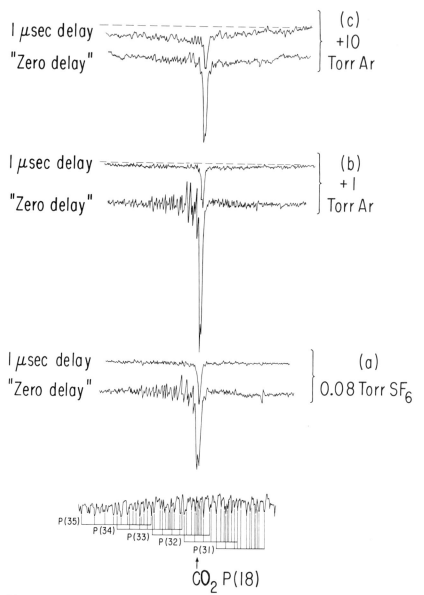

Fig.1 Two-level infrared double resonance in SF_6. The bottom trace is the normal absorption spectrum of SF_6 in the vicinity of 945.98 cm^{-1} [CO_2 P(18) line], with the P(31)-P(35) transitions identified. Successive scans show the double resonance effect in 0.08 Torr SF_6 with 0, 1.0, and 10.0 Torr Ar added. In the "zero delay" trace, the 0.5 μsec gate is set coincident with the CO_2 laser pulse; the delayed trace corresponds to ~ 1 μsec following the pulse.

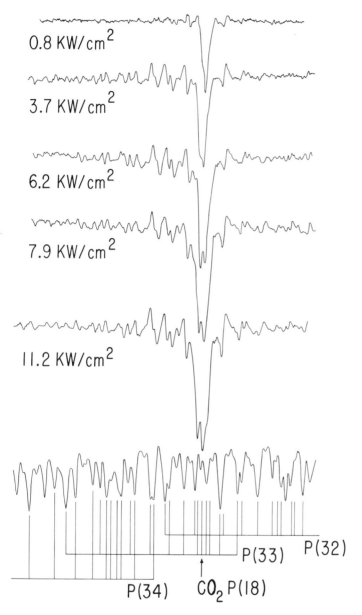

Fig.2 Power broadening of the two-level infrared double resonance in 0.08 Torr SF_6. The bottom trace is an expanded version of the absorption spectrum shown in Fig. 1, near the P(31) manifold. Successive scans shows the double resonance effect at peak laser intensities of 0.8 to 11 KW/cm^2.

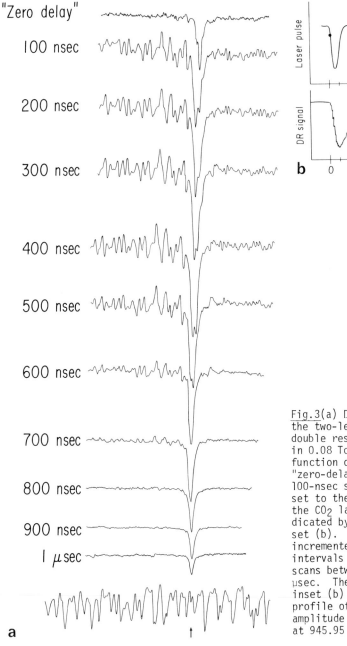

"Zero delay"

100 nsec

200 nsec

300 nsec

400 nsec

500 nsec

600 nsec

700 nsec

800 nsec

900 nsec

1 μsec

a

Laser pulse

DR signal

t, μsec

b   0 1 2 3

Fig.3(a) Development of the two-level infrared double resonance signal in 0.08 Torr SF_6 as a function of time. In the "zero-delay" scan, the 100-nsec sampling gate is set to the leading edge of the CO_2 laser pulse as indicated by the dot in inset (b). The gate is then incremented by 100-nsec intervals in the successive scans between 0.1 and 1.0 µsec. The lower trace in inset (b) is the time profile of the maximum-amplitude double resonance at 945.95 cm^{-1}.

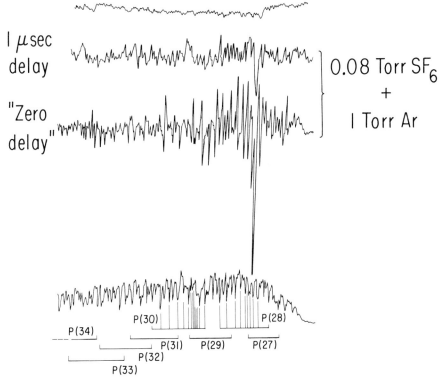

Fig.4 Three-level infrared double resonance in 0.08 Torr SF_6 + 1.0 Torr Ar. The diode scans across the P(27)-P(33) manifolds in the 946.0-946.4 cm^{-1} region; the laser pump line is the CO_2 P(14), at 949.48 cm^{-1}, which coincides with the A_1 + A_2 + F_1 + F_2 component of the SF_6 R(28) manifold [22].

3.2.2 Excited-State Absorption

In a separate series of experiments [8], a transient *increase* in absorption was observed for certain combinations of pump and probe frequencies. Specifically, pumping in the SF_6 P(32) region with the P(18) line led to two prominent features in the R-branch region, at 950.300 and 950.508 cm^{-1}. If we assign these to R-branch lines of the $2\nu_3 \leftarrow \nu_3$ transition involving rotational states J = 32 and J = 31, we can obtain a qualitative picture of the appearance of this band. Note that the R(32) or R(31) $2\nu_3 \leftarrow \nu_3$ transitions are occurring in the same frequency region as the R(50) to R(60) transitions of the ν_3 fundamental band. Given that there is a red shift of the centers of the $2\nu_3 \leftarrow \nu_3$ bands from the ν_3 fundamental, the $2\nu_3 \leftarrow \nu_3$ band envelope must be considerably broader than that for $\nu_3 \leftarrow 0$. A spectroscopic model of just this type was employed in the recent calculations of multiple photon excitation probabilities by ACKERHALT and GALBRAITH [29,30], with considerable success.

3.3. Four-Level Double Resonance: Collision and Relaxation Effects

Relaxation effects on the double resonance spectrum manifest themselves
in two ways: appearance of signals corresponding to transitions not
possessing a level in common with the laser-pumped transition, and
decay of the two- and three-level double resonance signals in time. From
the latter [cf. Fig. 3], we obtain a relaxation time for SF_6-SF_6 collisions, p_τ = 24 nsec-Torr, corresponding to an effective cross section
$\sigma \simeq 170 Å^2$. This is in reasonable agreement with a predicted value of
$230 Å^2$, based on a dipole-dipole interaction model [31].

   The spectra shown in Figs. 1 and 4 indicate that the initial relaxation
is to other symmetry ("hyperfine") components within a given J-manifold,
with $\Delta J = \pm 1$ occurring with much smaller probability. The rise time of
these satellite components [cf. Fig. 3] is a factor of 3 or 4 faster
than the measured relaxation time of the hole-burning signal, however.
This situation is reminiscent of that in Ref. 12, in which a collisionless mechanism had to be invoked to explain the short relaxation times
observed at low pressures. In our case, the level of excitation is too
low to allow for purely intramolecular coupling. A possible explanation
of these results is that the power broadening caused by the pump laser
itself [cf. Sec. 3.1] mixes the nearly-degenerate symmetry components
of a given rotational level in the ground vibrational state (v_3 = 0).

4. Discussion

Diode-Infrared Laser Double Resonance is a powerful technique for elucidating the dynamics of multiple photon excitation in polyatomic molecules.
The strength of the off-resonant coupling to the pump laser field can
be observed as a power-broadening of the saturated transition. Analysis
of congested molecular spectra is greatly simplified by locating lines
in different branches possessing common rotational levels; in the case
of SF_6, we merely confirm the existing analysis [22-25], but there are
many other molecules whose infrared vibration-rotation spectra have not
been assigned, and which are very suitable for this technique. Relaxation processes can be followed in detail; indeed, the quantity of information available in SF_6 is greater than can be sensibly dealt with on a
"state-to-state" levels and we should seek instead some grouping procedure
which will permit the significant features of the measurements to be recognized. Such a procedure may be available in the "cluster analysis"
developed by Harter et al. [32] to model the SF_6 vibration-rotation spectrum. In this analysis, groups of levels are labelled by total angular
momentum J and by symmetry properties A (corresponding to O_h irreducible
representations A_1, A_2, E, F_1, F_2) and P(=3 or 4), and by a "cluster
momentum" n = N, N-1, ..., -N, which is the "body-fixed" component of the
rotational angular momentum N = J-1. The conclusion which is drawn
from the four-level double-resonance experiments in Sec. 3.3 is that the
propensity rule in collisions is $\Delta J \simeq 0$, but there are no apparent
restrictions on ΔA or Δn. By contrast, the dipole-like selection rules
for photon absorption or emission lead to $\Delta J = \pm 1$, $\Delta A = 0$, $\Delta n = 0$. More
exhaustive measurements, at a series of pump and probe frequencies, will
be required to determine whether more subtle, symmetry-related restrictions
exist for the outcomes of rotationally inelastic collisions.

   The Double Resonance technique can also provide information on vibrational levels above the first excited state, and on into the quasi-

continuum. From the small number of excited-state absorptions observed thus far (Sec. 3.2.2, Refs. 8 and 33) one can conclude only that the $2\nu_3 \leftarrow \nu_3$ transition has a band shape significantly different from the fundamental, but more data will be needed to complete the analysis. Double Resonance experiments using a high-power TEA CO_2 pump laser, currently in progress [33], should yield direct information concerning the distribution over vibrational energy states at high levels of excitation.

Acknowledgments

This work was supported by the Office of Advanced Isotope Separation Technology, U.S. Department of Energy, under Contract EY-76-S.02-2793. We thank Drs. R. McDowell and H. Galbraith (Los Alamos Scientific Laboratory) and P. Moulton and A. Mooradian for helpful discussions and for making available data in advance of publication.

References

1. E.R. Grant, P.A. Schulz, Aa. S. Sudbø, Y.R. Shen, and Y.T. Lee, Phys. Rev. Letts. 40, 115 (1978).
2. J.L. Lyman, J. Chem. Phys. 67, 1868 (1977).
3. C.C. Jensen, J.I. Steinfeld, and R.D. Levine, J. Chem. Phys. 69, 1432 (1978).
4. I. Scheck and J. Jortner, J. Chem. Phys. 70, 3016 (1979).
5. S. Speiser and J. Jortner, Chem. Phys. Letts. 44, 399 (1976).
6. C. Reiser and J.I. Steinfeld, Opt. Eng. (to be published).
7. J.G. Black, P. Kolodner, M.J. Shultz, E. Yablonovitch, and N. Bloembergen, Phys. Rev. A19, 704 (1979).
8. C.C. Jensen, Ph.D. Thesis, Massachusetts Institute of Technology, Cambridge, Mass. (1979).
9. I. Burak, T.J. Quelly, and J.I. Steinfeld, J. Chem. Phys. 70, 334 (1979).
10. M.L. Lesiecki, J.A. Stewart, and W.A. Guillory, Physical Chemistry Abstract No. 138, ACS/CSJ Chemical Congress (Honolulu, April 1979).
11. I. Burak, A.V. Nowak, J.I. Steinfeld, and D.G. Sutton, J. Chem. Phys. 51, 2275 (1969); J.I. Steinfeld, I. Burak, D.G. Sutton, and A.V. Nowak, J. Chem. Phys. 52, 5421 (1970).
12. D.S. Frankel, J. Chem. Phys. 65, 1696 (1976); D.S. Frankel and T.J. Manuccia, Chem. Phys. Letts. 54, 451 (1978).
13. A.B. Peterson, J. Tiee, and C. Wittig, Opt. Commun. 17, 259 (1976).
14. W. Fuss, J. Hartmann, and W.E. Schmid, Appl. Phys. 15, 297 (1978).
15. K.S. Rutkovski and K.G. Tokhadze, Zhur. Eksp. Teor. Fiz. 75, 409 (1978).
16. V.M. Gordienko, A.V. Mikheenko, and V. Ya. Panchenko, Soviet J. Quantum Electronics 8, 1013 (1978).
17. P.F. Moulton, D.M. Larsen, J.N. Walpole, and A. Mooradian, Opt. Letts. 1, 51 (1977).
18. T.F. Deutsch and S.R.J. Brueck, J. Chem. Phys. 70, 2063 (1979).
19. C.C. Jensen and J.I. Steinfeld, in "Tunable Lasers and Applications: Proc. Loen Conf., Norway, 1976 (A. Mooradian, T. Jaeger, and P. Stokseth, eds.)" pp. 190-192, Ser. Opt. Sci. 3, Springer-Verlag, Berlin, 1976.
20. B.G. Whitford, K.J. Siemsen, H.D. Riccius, and G.R. Hanes, Opt. Commun. 14, 70 (1975).
21. N.G. Nereson, J. Mol. Spectroscopy 69, 489 (1978).

22. R.S. McDowell, H.W. Galbraith, B.J. Krohn, C.D. Cantrell, and E.D. Hinkley, Opt. Commun. 17, 178 (1976).
23. R.S. McDowell, H.W. Galbraith, C.D. Cantrell, N.G. Nereson, P.F. Moulton, and E.D. Hinkley, Opt. Letts. 2, 97 (1978).
24. R.S. McDowell, H.W. Galbraith, C.D. Cantrell, N.G. Nereson, and E.D. Hinkley, J. Mol. Spectroscopy 68, 288 (1977).
25. R.S. McDowell, private communication.
26. K. Fox, Opt. Commun. 19, 397 (1976).
27. B.R. Mollow, Phys. Rev. A5, 2217 (1972).
28. F.Y. Wu, S. Ezekiel, M. Ducloy, and B.R. Mollow, Phys. Rev. Letts. 38, 1077 (1977).
29. H.W. Galbraith and J.R. Ackerhalt, Opt. Letts. 3, 152 (1978).
30. J.R. Ackerhalt and H.W. Galbraith, J. Chem. Phys. 69, 1200 (1978).
31. M.M. Mkrtchyan and V.T. Platonenko, Sov. J. Quantum Electronics 8, 1187 (1978).
32. W.G. Harter and C.W. Patterson, Intl. J. Quantum Chem. 11, 479 (1977); W.G. Harter, H.W. Galbraith, and C.W. Patterson, J. Chem. Phys. 69, 4888, 4896 (1978).
33. P. Moulton and A. Mooradian, private communication.

Spectroscopy of Photodissociation Products

H. Zacharias, R. Schmiedl, R. Böttner, M. Geilhaupt, U. Meier, and K.H. Welge

Fakultät für Physik, Universität Bielefeld
D-4800 Bielefeld 1, Fed. Rep. of Germany

Introduction

The photodissociation is a most important elementary process in the interaction of light with molecules, of basic and practical interest from various points of view. A detailed investigation of the dynamics of the dissociation of a given molecule would involve the preparation of the molecule
before absorption in a defined initial state and measurement of the fragment energy and momentum distribution in the excit channels. For polyatomic molecules of three and more atoms this would include the rotational and
vibrational fragment motion, in addition to translation. Experiments in
such detail have not yet been performed with polyatomic molecules, though
considerable progress in the investigation of the dynamics of photodissociation processes has been made through the application of photofragment
spectroscopy with mass spectroscopic detection combined with the time-of-
flight technique. The potential of this method has recently been used for
instance by LEE and coworkers [1] in studies of the multiple-photon dissociation of polyatomic molecules by infrared light. The method is generally
applicable and yields inter-fragment recoil energy and momentum distributions. Quantum state specific measurements are normally not possible because of limitation of the velocity resolution.

The intra-fragment energy content can be obtained state specifically
by optical fragment spectroscopy [2,3]. Such experiments have recently
been performed on multiple-photon dissociation processes using laser excited fluorescence for detection [4]. In this paper we report first studies on the one-photon dissociation of the NO_2 molecule into $NO + O$, where
we have measured the complete vibrational-rotational energy distribution
of the NO fragment. This is an interesting case since the NO_2 dissociation
dynamics has been previously investigated by other methods, and the question arises whether the fragmentation occurs statistically or not.

Although the applicability of the optical photofragment spectroscopy
is determined by appropriate spectroscopic properties of the fragments
and thus limited largely to small polyatomics it allows investigations in
great detail. In addition to intra-fragment energy measurements also inter-fragment recoil experiments can be performed state-specifically. This
is possible under beam condition using the time-of-flight technique or
also Doppler spectroscopy for the measurement of recoil energy and momentum. In this paper we report Doppler experiments under crossed beam
conditions applied to the multiple-photon dissociation of a molecule,
CH_3NH_2. First experiments of this kind have been reported previously [5].

The main purpose of the work was to demonstrate and study the feasibility of the method.

Doppler Spectroscopy of MPD Fragments

Method

The application of Doppler line width measurement by laser induced fluorescence for the investigation of the dynamics of collision processes has been previously discussed by KINSEY [6]. He has shown that the three-dimensional velocity distribution $F(v;\theta,\varphi)$ of a particle can be unambiguously determined from the measurement of the line profile $D(v;\theta,\varphi)$ as a function of the excitation frequency v and the direction (θ,φ) of the exciting light beam. Each measurement in a given direction (θ,φ) yields a line profile determined by the projection of the three-dimensional velocity distribution onto this direction. From the set of measurement as function of (θ,φ) the full velocity distribution $F(v)$ can be reconstructed by a Fourier transformation procedure.

If the spatial fragment distribution is known and has some simple symmetry one does not need the whole transformation procedure but F can be deduced from D by simpler relations. In the present case of the multiple-photon dissociation process one may assume that the fragmentation occurs with spatially isotropic fragment distribution. This is based on the previous experiments by LEE and coworkers [1]. In this case of spherically symmetric velocity distribution the line profile $D(v)$ is given by

$$\frac{dD(v)}{dv} = - 2 \pi v F(v),$$

where $F(v)$ is the speed distribution.

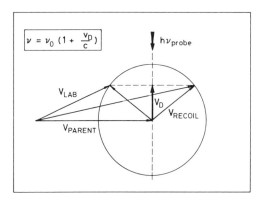

Fig. 1 Velocity diagramme for photofragmentation in a molecular beam and fragment probing by Doppler spectroskopy.

In the present experiments the dissociation was carried out under molecular beam condition. Fig. 1 shows the corresponding velocity diagramme; $\vec{v}_{par}$ indicates the parent molecule beam velocity, $\vec{v}_{rec}$ the fragment recoil velocity in the center-of-mass system, and $\vec{v}_D$ the velocity component in the direction of the probe laser beam. Since the probe laser and molecular beams crosses perpendicularly, $\vec{v}_{par}$ does not contribute to $\vec{v}_D$, which means that the Doppler linewidth is determined only by the distribution of the recoil velocity. With the perpendicular beam crossing the Doppler method probes the fragment recoil velocity directly in the center-of-mass system, an important difference to the time-of-flight technique which probes in the laboratory system.

Experimental

The experimental set up is shown schematically in Fig. 2. The molecular beam nozzle and skimmer had diameters of 0.08 mm and 0.6 mm, respectively. The distance between them was 11 mm. The beam divergence was 5° at FWHM. According to the vapour pressure of methylamine used in these experiments, the tube and nozzle were heated so that a stagnation pressure of up to 500 Torr was achievable.

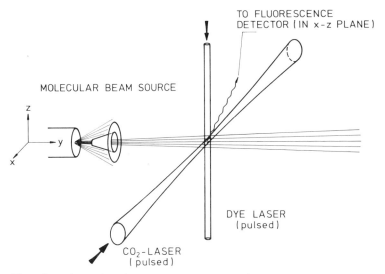

Fig. 2 Schematic diagramme of the experimental set up.

The CO_2 laser was a multimode type (Lumonics, TEA 801 A) operated in these experiments at the 9.586 μm P(24) line at an output of typically 1.0 J/ pulse. As usual, about 70% of the output was contained in a main pulse of about 0.2 μsec duration, and the rest in a 2 μsec tail. The CO_2-laser beam was focussed onto the molecular beam with 25 cm focal length lens. The probe laser was a pulsed tunable dye laser system pumped by a Nd : YAG laser with 10 Hz repetition rate. Without confocal resonator the bandwidth was about 1.5 GHz and with the resonator it was about 150 MHz. At 493 nm, for instance, the output was about 3 mJ/pulse and 1 mJ/pulse, respectively. The system has been described in detail elsewhere [7].

The intersection of the CO_2-laser and probe laser beams defined an effective detection volume of about 3 mm^3 on the axis of the molecular beam.

With this arrangement it thus was possible to avoid the mechanical skimmer, but nevertheless to retain crossed beam condition by this "optical skimming". Fluorescence from this region was imaged onto the photomultiplier (RCA C31034 A). The probe laser pulse was delayed with respect to the CO_2-laser by usually 1 μsec, measured from the CO_2 laser pulse peak, and the fluorescence was observed through a time gate of 10 μsec duration.

Results

The CH_3NH_2 molecule was chosen because previous experiments have shown that the multiple-photon dissociation by CO_2 laser light produces NH_2 (most likely by direct fragmentation into NH_2 + CH_3 [8,9]), and NH_2 has an open spectrum in the visible with many discrete lines. Excitation was performed in the $^PQ_{1N}$-branch around 493 nm from the $\tilde{X}(0,0,0)$ state to Σ-vibronic levels in the $\tilde{A}(0,13,0)$ electronically excited state. The red shifted fluorescence back to the $\tilde{X}(0,1,0)$ level around 533 nm was observed. Figure 3 shows a small section of the NH_2 excitation spectrum taken with a probe laser bandwidth of 1.5 GHz.

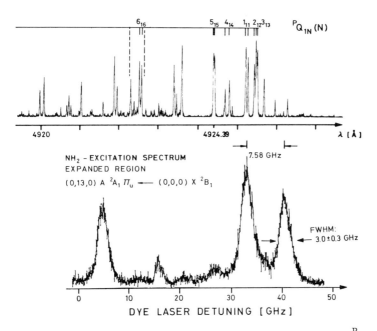

Fig. 3 Excitation spectrum of NH_2 in the vicinity of the $^PQ_{1N}$ branch of the $\tilde{A}\,^2A_1(0,13,0) \leftarrow \tilde{X}\,^2B_1(0,0,0)$ transition. Spectral resolution is 1.5 GHz, upper part, and 150 MHz for the expanded $^PQ_{16}(6)$-line, lower part.

High resolution ($\Delta\nu$ = 150 MHz) measurements yielded for the $^PQ_{16}$ (6)-line a Doppler width of (3.0 ± 0.2) GHz. Fitting the lines tentatively by Gaussian profiles shows that this representation is adequate well within the experimental precision limits. Thus the linewidth can be converted to a most probable speed of 890 ± 90 m/sec corresponding to a translational

NH$_2$-EXCITATION SPECTRUM

A $^2A_1\Pi_u$ (0,13,0) ← X $^2B_1$ (0,1,0); $^PQ_{16}$(6)-LINE (531.4 nm)

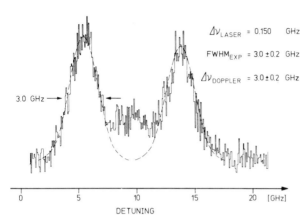

Fig. 4 High resolution ($\Delta\nu_L$ = 150 MHz) excitation spectrum of the
$^PQ_{16}$(6)-line originating from the first excited vibrational level of the
NH$_2$ bending mode: $\widetilde{A}$ $^2A_1$(0,13,0) ← $\widetilde{X}$ $^2B_1$(0,1,0) transition.

temperature of 760 ± 150 K, if one assumes a Maxwell-Boltzmann speed distribution. A more detailed description has been given previously [10].

Another measurement has been made for the N" = 6 level in the first
excited vibrational level, $\widetilde{X}$(0,1,0), of the NH$_2$ bending mode. For excitation we have chosen a line, $^PQ_{16}$(6), at 531.4 nm in the $\widetilde{A}$ $^2A_1$(0,13,0) −
$\widetilde{X}$ $^2B_1$(0,1,0) transition. Fig. 4 shows a scan of this $^PQ_{16}$(6) transition
from the $\widetilde{X}$ $^2B_1$(0,1,0) state taken with a probe laser bandwidth of 150 MHz.
The spin doublet splitting in the NH$_2$ lines are well resolved, but leads
to some overlapping. The curves drawn represent Gaussian profiles, which
evidently fit the measured lines quite well. For the lines taken with 150
MHz resolution we obtain a FWHM linewidth of (3.0 ± 0.2) GHz. Practically
the same linewidth is obtained from the lines taken with 1.5 GHz resolu-
tion, if one applies the relation $\Delta\nu_{exp}^2 = \Delta\nu_L^2 + \Delta\nu_o^2$ for deconvolution.
$\Delta\nu_L$ is the laser bandwidth, $\Delta\nu_{exp}$ the width of the measured line and $\Delta\nu_o$
that of the true transition. The FWHM width corresponds to a mean speed
of (960 ± 60) m/s or a translational temperature of (900 ± 110) K. The
distribution in the translational degree of freedom thus appears practically the same in the two different levels, althought the [6$_{16}$, (0,1,0)]
rotational level is higher than the [6$_{16}$, (0,0,0)] by 1496 cm^{-1}, or roughly 1.5 CO$_2$ photons. Further, preliminary measurements on other rotational
levels in (0,0,0) and (0,1,0) show roughly the same magnitude of translational energy. The explanation for this behaviour is not yet clear to us.

Laser Photofragment Spectroscopy of NO$_2$ One-Photon Dissociation

The dynamics of the NO$_2$ → NO + O dissociation by one-photon absorption has
been previously investigated at the doubled ruby laser wavelength
(347.1 nm = 28810 cm^{-1}) [11]. The results have been interpreted theoret-

ically by a dissociation mechanism of randomized energy distribution, in contrast to the finding in this work, where we have dissociated at 337 nm and measured the complete rotational and vibrational energy distribution of the NO fragment by laser excitation spectroscopy. NO_2 and other XNO molecules are particularly suitable for very detailed and systematic investigations since tunable laser can be used both for dissociation and probing. More over, using nozzle beam expansion, done in further studies, it will be possible to prepare the parent molecules state selectively.

Experimental

The dissociation was performed with a pulsed N_2-laser at 337.1 nm. Part of the N_2-laser light was used to pump a tunable dye probe laser consisting of an oscillator and two amplifier stages. To excite NO in its $A^2\Sigma+ -$ $X^2\Pi$ transition in the range from 224 to 250 nm the dye laser output was accordingly frequency doubled. The dissociating light beam passed through a reaction vessel and was crossed by the probe laser beam. Fluorescence was observed at right angle to the beams with a photomultiplier, and excitation spectra were recorded by conventional means. Typical intensities of the exciting and probe laser pulses of about 6 nsec duration were 1.8 mJ and 10 µJ, respectively. Linearity of the fluorescence intensity was checked accordingly. The probe laser pulse was delayed to the exciting pulse by 5 nsec. The pressure of NO_2 was kept below 500 mTorr, so that collision free conditions for testing the nascent NO were ensured.

Figure 5 shows the energetics of the dissociation and probe steps. As can be seen, at 337 nm NO can be excited up to $v'' = 2$. Excitation spectra of NO were taken from the $v'' = 0,1$ and 2 states always to the same upper state, $v' = 0$. In this way only the spectroscopic properties (line strength, Franck-Condon factors) for absorption had to be employed in the derivation of the relative level populations from the observed fluores-

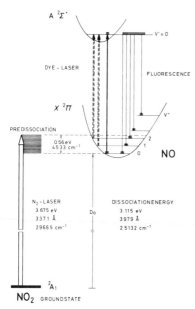

Fig. 5 Schematic diagramme of the energetics of the dissociation and probing steps for dissociation of NO_2 at 337.1 nm. The probe laser excites the nascent NO always to the same vibrational level $v' = 0$ of the upper electronic state $A\ ^2\Sigma+$.

cence intensities. The observed fluorescence signal, S, is thus proportional to the absorption flux, F_{abs}, from a given rotational-vibrational level (v", J") in the NO($X^2\Pi$) state. F_{abs} itself is essentially given by the probe laser intensity, the Franck-Condon factor, the line strength (i. e. the Hönl-London factors) and the number density in (v", J"). In the derivation of relative state populations Franck-Condon factors given by ANTROPOV et al [12] and Hönl-London factors given by BENNETT [13] have been used. Experimental parameters such as wavelength dependence of the probe laser intensity were calibrated and appropriately taken into account.

Results

Figure 6 shows for illustration excitation spectra of NO in the v' = 0 ← v" = 1 band region. The lower spectrum has been taken simply with NO in the fluorescence cell under thermal conditions at 296 K. The upper part is a section of the same transition of NO, however, produced in the photodissociation. Obviously the rotational distribution is non-thermal. Fig. 7 shows the rotational distributions in the three vibrational levels, separately for the two different electronic substates, $^2\Pi_{1/2}$ and $^2\Pi_{3/2}$. The dashed vertical lines correspond to the maximum J" values possible with 337 nm, if the NO_2 parent molecules were cold. The tails of the distributions beyond the maximum J" values correspond closely to the internal energy distribution of NO_2 at room temperature. The rotational distribu-

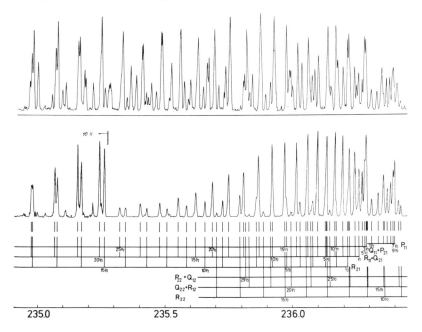

LASER WAVELENGTH [nm]

Fig. 6 Excitation spectrum of NO in the $\gamma(0,1)$ band. Lower trace: part of the spectrum taken under thermal conditions at 296 K; upper trace: section of the same transition, however, for NO produced in the photodissociation of NO_2.

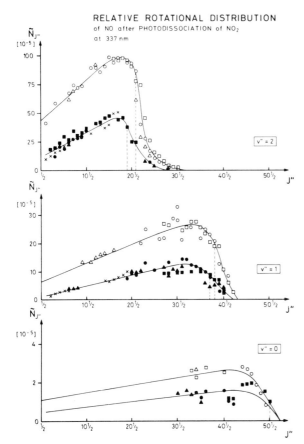

RELATIVE ROTATIONAL DISTRIBUTION
of NO after PHOTODISSOCIATION of NO_2
at 337 nm

Fig. 7 Population distribution for the rotational-vibrational states. Open symbols: $^2\Pi_{1/2}$ substate; full symbols: $^2\Pi_{3/2}$ substate. The different symbols refer to different rotational branches.

tion is evidently strongly non-thermal and inverted, peaking around the maximum possible J" value. This observation leads to conclude that the NO_2 molecule is decaying from a potential where the equilibrium configuration, i. e. the O – N – O angle, is very much different from that in the ground state. In such a situation the bending motion is transfered to high rotational momentum of the NO fragment which is compensated by a correspondingly large orbital momentum of the departing fragments O + NO.

In Fig. 8 the distribution is plotted for each state vs. the internal energy. The dashed line indicates the maximum excess energy. Population at higher energy is due to internal energy of NO_2 at room temperature. By far most of the excess energy evidently appears in rotation and vibration of NO. Summing over the rotational states, we find the vibrational population in v = 0,1 and 2 as 1 : 5 : 24, totally inverted. In summary, the energy partitioning is non-statistical at 337 nm, in contrast to the previous studies performed at a slightly longer wavelength.

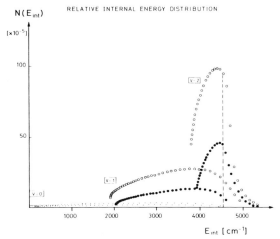

N(E$_{int}$)

[×10^{-5}]

RELATIVE INTERNAL ENERGY DISTRIBUTION

100

50

|v=0| [v=1] |v=2|

1000   2000   3000   4000   5000

E$_{int}$ [cm^{-1}]

Fig. 8 Population distribution per rotational state
plotted vs. internal NO energy.

References

1. M. J. Coggiola, P. A. Schulz, Y. T. Lee, Y. R. Shen, Phys. Rev. Lett., 38, 17 (1977)
   E. R. Grant, M. J. Coggiola, Y. T. Lee, P. A. Schulz, Aa. S. Sudbø, Y. R. Shen, Chem. Phys. Lett., 52, 595 (1977)
2. A. P. Baronavski and J. R. McDonald, Chem. Phys. Lett., 45, 172 (1977)
3. M. J. Sabety-Dzvonik and R. J. Cody, J. Chem. Phys., 66, 126 (1977)
4. D. S. King and J. C. Stephenson, Chem. Phys. Lett., 51, 48 (1977)
5. R. Schmiedl, R. Böttner, H. Zacharias, U. Meier, D. Feldmann, and K. H. Welge in "Laser-Induced Processes in Molecules" (eds K. L. Kompa and S. D. Smith), p. 186 ff, Springer Verlag, Berlin, Heidelberg, New York, 1979
6. J. L. Kinsey, J. Chem. Phys., 28, 349 (1977)
7. H. Zacharias, R. Schmiedl, and K. H. Welge, "Laser 79 Opto-Electronic" (ed. W. Waidelich) IPC Technologe Press, Guiltfort, in press
8. S. V. Filseth, J. Danon, D. Feldmann, J. D. Campbell, and K. H. Welge, Chem. Phys. Lett., 63, 615 (1979)
9. G. Hancock, R. J. Hennessy, and T. Villis, J. Photochem., 10, 305 (1979)
10. R. Schmiedl, R. Böttner, H. Zacharias, U. Meier, and K. H. Welge, Opt. Commun., in press
11. G. E. Busch and K. R. Wilson, J. Chem. Phys., 56, 3626 (1972)
    G. E. Busch and K. R. Wilson, J. Chem. Phys., 56, 3638 (1972)
12. E. T. Antropov, V. N. Kolesnikov, L. Ya. Ostrovskaya, and N. N. Sabolev, Opt. Spectrosc., 22, 109 (1967)
13. R. J. M. Bennett, Mon. Nat. R. Astr. Soc., 147, 35 (1970)

Part V

Nonlinear Processes, Laser Induced Collisions, Multiphoton Ionization

Recent Progress in Four-Wave Mixing Spectroscopy

N. Bloembergen

Division of Applied Sciences, Harvard University
Cambridge, MA 02138, USA

1. Introduction

The general framework for describing the large variety of nonlinear optical phenomena caused by an electric polarization cubic in the electric field amplitudes

$$P_i(\omega_4,\underset{\sim}{r}) = \frac{1}{2} \chi^{(3)}_{ijk\ell}(-\omega_4,\omega_1,-\omega_2,\omega_3) E_j(\omega_1) E_k^*(\omega_2) E_\ell(\omega_3)$$
$$\exp[i(\underset{\sim}{k}_1 - \underset{\sim}{k}_2 + \underset{\sim}{k}_3)\cdot r - i\omega_4 t] + \text{c.c.} \tag{1}$$

was introduced in 1962. In media with inversion symmetry this is the lowest order nonvanishing electromagnetic response. This nonlinearity describes a coupling between four electromagnetic waves [1]. In general, each of the four waves has its own frequency, wave vector and polarization direction. The polarization vectors are denoted by $\hat{e}_1$, $\hat{e}_2$, $\hat{e}_3$ and $\hat{e}_4$, respectively, and the nonlinear scalar coupling coefficients

$$\chi^{NL}_{1234} = \hat{e}_1 \hat{e}_2^*: \chi^{(3)}(-\omega_4,\omega_1,-\omega_2,\omega_3): \hat{e}_3 \hat{e}_4^*$$

$$\chi^{NL}_{ij} = \hat{e}_i \hat{e}_i^*: \chi(-\omega_i,\omega_i,\omega_j,-\omega_j): \hat{e}_j \hat{e}_j^*$$

are introduced. With the wave vector mismatch $\Delta kz = \underset{\sim}{k}_1 - \underset{\sim}{k}_2 + \underset{\sim}{k}_3 - \underset{\sim}{k}_4$, the four coupled complex amplitude equations take the form

$$\frac{\partial E_1}{\partial z} + \frac{1}{v_{g1}} \frac{\partial E_1}{\partial t} = 2\pi i (\omega_1/n_1 c)\left[\chi^{NL}_{1234} E_2 E_3^* E_4 \exp(-i\Delta kz)\right.$$
$$\left. + \sum_{j=1}^{4} \chi^{NL}_{1j} E_1 E_j E_j^*\right]$$

$$\frac{\partial E_2}{\partial z} + \frac{1}{v_{g2}} \frac{\partial E_2}{\partial t} = 2\pi i (\omega_2/n_2 c)\left[\chi^{NL}_{1234} E_1 E_3 E_4^* \exp(i\Delta kz)\right.$$
$$\left. + \sum_{j=1}^{4} \chi^{NL}_{2j} E_2 E_j E_j^*\right] \tag{2}$$

$$\frac{\partial E_3}{\partial z} + \frac{1}{v_{g3}} \frac{\partial E_3}{\partial t} = 2\pi i (\omega_3/n_3 c)\left[\chi^{NL}_{1234} E_1^* E_2 E_4 \exp(-i\Delta kz)\right.$$
$$\left. + \sum_{j=1}^{4} \chi^{NL}_{3j} E_3 E_j E_j^*\right]$$

$$\frac{\partial E_4}{\partial z} + \frac{1}{v_{g4}} \frac{\partial E_4}{\partial t} = 2\pi i (\omega_4/n_4 c) \left[\chi_{1234}^{NL} E_1 E_2^* E_3 \exp(+i\Delta kz) \right.$$
$$\left. + \sum_{j=1}^{4} \chi_{4j}^{NL} E_4 E_j E_j^* \right] \tag{2}$$

In many important cases two (or more) of the waves may be degenerate in frequency, and/or wave vector, and/or polarization. The coupling is especially strong if the conditions of energy and momentum conservation are satisfied,

$$\omega_4 = \omega_1 - \omega_2 + \omega_3 \quad \text{and} \quad \underset{\sim}{k}_4 = \underset{\sim}{k}_1 - \underset{\sim}{k}_2 + \underset{\sim}{k}_3 \tag{3}$$

In general, the nonlinear susceptibility $\chi^{(3)}$ is a fourth-rank tensor, and has 81 tensor elements $\chi_{ijk\ell}^{(3)}$. This number is, of course, drastically reduced by symmetry. For example, in the cubic symmetry $\bar{4}3m$, there are only four independent elements. In an isotropic fluid there are only three. These numbers may be further reduced by frequency degeneracies. Each element of $\chi^{(3)}$ consists of a sum of 48 terms. Explicit expressions for these have been published [2], and each term has a typical form, with three resonant factors in the denominator

$$\chi^{(3)} = \frac{1}{6} N \hbar^{-3} \sum_{gk,n,j} \frac{\mu_{gk} \, \mu_{kn} \, \mu_{nj} \, \mu_{jg} \, \rho_{gg}^{(o)}}{(\omega_{kg} - \omega_1 - i\Gamma_{kg})\{\omega_{ng} - i\Gamma_{ng} - (\omega_1 - \omega_2)\}\{\omega_{jg} - i\Gamma_{jg} - (\omega_1 - \omega_2 + \omega_3)\}}$$

$$+ \text{ 47 other terms} \tag{4}$$

where N is the number of particles per unit volume, μ_{gk} is the electric dipole matrix between states g and k, etc., $\hbar\omega_{kg}$ is the energy difference between this pair of states, and Γ_{kg} is the damping of the off-diagonal element of the density matrix, corresponding to the homogeneous width of the one-photon transition.

The different terms are distinguished by the time ordering of the photon creation and annihilation processes and the damping mechanism. The evolution of the density matrix operator can be obtained from standard higher order time-dependent perturbation theory. S. Y. YEE and coworkers [3,4] have applied a diagrammatic approach which facilitates a systematic accounting of the various terms. It is important to consider separately the evolution of <bra| and |ket> state functions, since the material system is also subjected to random interactions which lead to damping. Therefore each term of $\chi^{(3)}$ becomes a complex quantity. It is often possible to single out one or several resonant terms in $\chi^{(3)}$ and lump the remaining terms in a nonresonant contribution.

Various examples in Fig.1 include the nonresonant process, two-photon resonant processes, and combinations of one-photon and two-photon resonant processes. Terms resonant at the combination frequencies, $\omega_1 - \omega_2$, $\omega_3 - \omega_2$ and $\omega_1 + \omega_3$, correspond to Raman type processes and two-photon absorption, respectively. If there are no one-photon resonant terms, then $\chi^{(3)}$ may be written as

$$\chi^{(3)} = \chi^{NR} \left[1 + \frac{\alpha_R}{\omega_R - (\omega_1 - \omega_2) - i\Gamma_R} + \frac{\alpha_{R'}}{\omega_{R'} - (\omega_3 - \omega_2) - i\Gamma_{R'}} + \frac{\alpha_E}{\omega_E - (\omega_1 + \omega_3) - i\Gamma_E} \right] \tag{5}$$

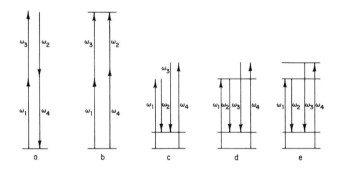

Fig.1 Various resonant situations of light generation at the combination frequency $\omega_4 = \omega_1 - \omega_2 + \omega_3$ by a material system in the ground state $|g>$ (a) nonresonant parametric mixing (b) two-photon absorption resonant mixing (c) Coherent Antistokes Resonant Scattering (CARS) (d) one-photon resonant CARS (e) all four one-photon transitions are resonant

The observed generated intensity at ω_4 is proportional to $\left|\chi^{(3)}\right|^2$. Here ω_R and $\omega_{R'}$ are resonant frequencies for a Raman transition and $\hbar\omega_E$ is the energy of an excitation reached by a two-photon absorption process. Far away from any resonance it is correct to describe this as a parametric process in which one quantum each at ω_1 and ω_2 is destroyed, and one quantum each is added to the beams at ω_3 and ω_4. Exactly at a Raman resonance $\omega_1 - \omega_2 = \omega_R$, a different language is more appropriate. The imaginary part of $\chi^{(3)}(-\omega_4, \omega_1, \omega_2, -\omega_3)$ should be considered as an interference term in the Raman transition probability between the two states with energy difference $\hbar\omega_R$. This transition can be accomplished by the absorption of $\hbar\omega_1$ and the emission of $\hbar\omega_2$ but also by the absorption of $\hbar\omega_4$ and the emission of $\hbar\omega_3$. This point has also been noted by TARAN [5,6], but was not taken into account in the analysis by ANDERSON [7].

Recently, experimental attention has also been devoted to one-photon resonant terms in $\chi^{(3)}$. In that case the medium becomes absorbing at one or more of the frequencies ω_1, ω_2, ω_3 and ω_4. One may distinguish situations in which two or even three factors in the denominator are simultaneously resonant. The simultaneous occurrence of a one-photon resonance and a Raman resonance leads to coherent resonant Raman scattering and resonant CARS.

Furthermore, a variety of distinct polarization geometries must be considered. The various nonresonant and resonant terms in $\chi^{(3)}$ will exhibit different tensorial properties. Consequently the polarization properties of the light generated at ω_4 will depend not only on the polarization directions of the incident beams at ω_1, ω_2 and ω_3, but also on the frequencies. This is exploited, for example, in nonlinear ellipsometry by AKHMANOV and coworkers [8] and also in the Raman Induced Kerr Effect Scattering [9] and polarization spectroscopy [10].

While the general framework sketched above is quite compact, it describes a rather bewildering array of phenomena, because of the many variations offered by different combinations of frequencies, wave vectors and polarization directions. Fortunately, a number of excellent reviews have been published. A concise general survey of nonlinear optics has been given by SHEN [11]. Nonlinear spectroscopy, in atoms and molecules and especially crystals, was the topic of an E. Fermi Summer School Proceeding [12]. An excellent review of coherent Raman spectroscopy was recently prepared by LEVENSON and SONG [10]. Coherent Antistokes Raman Scattering has been reviewed by TARAN, and others at recent conferences [5,6].

In the remainder of this paper a few examples will be taken from recent work on atomic vapors and molecular fluids. The choices are rather arbitrary, but serve to illustrate the close relationship between many different investigations. The reader will probably recognize how a large part of the material presented at this conference fits into this general framework. Since much attention has recently been devoted to phase conjugation, the case of four-wave mixing with degenerate frequencies is discussed in the next section. The last section will concentrate on interference and damping effects in coherent resonant Raman scattering and CARS.

2. Degenerate Four-Wave Mixing

Consider the nonlinear polarization in the case that all four waves have the same frequency [13]. The nonlinear susceptibility $\chi^{(3)}(-\omega,\omega,-\omega,\omega)$ becomes complex in the vicinity of a one-photon resonance at ω, or a two-photon resonance at 2ω.

Off resonance the real part of $\chi^{(3)}_{xxxx}(-\omega,\omega,-\omega,\omega)$ represents an intensity dependent index of refraction. Self-focusing and self-defocusing in alkali metal vapors has been studied extensively. The real part of $\chi^{(3)}_{xxyy}(-\omega,\omega,-\omega,\omega)$ leads to the self-induced Kerr effect [14], which has been used extensively in polarization spectroscopy during the past few years.

Near a one-photon resonance at ω, the (negative) imaginary part $\chi^{(3)''}_{xxxx}(-\omega,\omega,-\omega,\omega)$ describes "incipient" saturation. Near a two-photon resonance at 2ω, the (positive) imaginary part describes two-photon absorption.

A standing wave pattern occurs when two light waves at the same frequency have pair-wise equal and opposite wave vectors, as shown in Fig.2, the configuration used in phase conjugate generation results [16,17]. In the presence of the strong standing wave pattern produced by the pump waves with $\underset{\sim}{k}_1$ and $\underset{\sim}{k}_3 = -\underset{\sim}{k}_1$, an incident wave $\underset{\sim}{k}_2$ gives rise to a wave $\underset{\sim}{k}_4 = -\underset{\sim}{k}_2$. Furthermore, the field amplitude of this backward wave is proportional to

$$E_4 \propto \chi^{(3)}(-\omega,\omega,-\omega,\omega) \ E_1 \ E_2^* \ E_3$$

Thus the phase $\underset{\sim}{k}_4 \cdot \underset{\sim}{r} + \phi_4$ is equal and opposite to the phase $\underset{\sim}{k}_2 \cdot \underset{\sim}{r} + \phi_2$. The backward wave 4 equals the time reversed incident wave 2. Thus, all wave front aberrations that the wave 2 may have undergone will be corrected in wave 4. The effect may also be understood in terms of real-time holographic reproduction. The waves 1 and 2 create an interference pattern. Off resonance this is an index phase grating, but for ω on a sharp material resonance, it may be considered as a modulation of the absorption coefficient by saturation. Wave 3 diffracts from this grating to construct the "phase conjugate image" 4. The same wave is also created by diffraction of beam 1 from the grating produced by the interference patterns of beams 2 and 3. The two diffraction gratings have a different spacing, and they may decay with different time constants by diffusion. This may be verified experimentally by using time resolved geometry with a variable delay between the pulsed waves, and by giving beams 1 and 3 orthogonal polarizations [18].

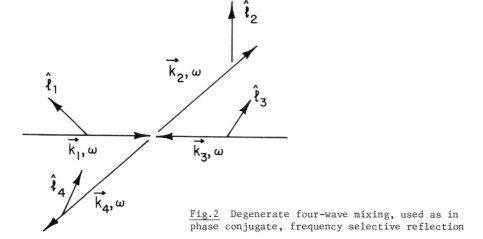

Fig.2 Degenerate four-wave mixing, used as in phase conjugate, frequency selective reflection

The effect has been observed at visible and infrared frequencies in solids and in fluids [19-24]. The nonlinear passive materials include $LiNbO_3$, Ge, liquid CS_2, Na vapor and SF_6 gas. It is particularly strong when ω corresponds to a resonant transition and has been demonstrated in laser media, including ruby [20], Nd YAG [21], as well as a CO_2 discharge [22,23]. The phase-conjugate mirrors thus obtained may display gain and also act as narrow frequency filters, since the $\chi^{(3)}$ response is determined by the width of the atomic resonance, with some further narrowing in the case of gain. BLOOM et al. have recently demonstrated [24] this last effect at the yellow resonance lines in Na vapor.

The use of single mode cw dye lasers with nearly equal frequencies ω, $\omega - \Delta\omega$ will create a wave at $\omega + \Delta\omega$. The addition of polarization discrimination may be advantageous. The slight phase mismatch leads to a coherence length which is in many cases still larger than the interaction length in the focal regions of the intersecting beams. This scheme has recently been used [25] to detect stimulated Brillouin scattering in CS_2, with $\Delta\omega \sim 0$ cm^{-1}. The method may also be used to study inelastic Rayleigh wing scattering and complements the well-established picosecond time resolved spectroscopy techniques.

3. Interference Effects in Raman and CARS Spectroscopy

High resolution coherent Raman gain measurements with cw dye lasers have been pioneered by OWYUNG [26]. The method can be used to probe the fine structure of Raman-active vibrations of polyatomic molecules such as H_2, CH_4 and SF_6. High resolution CARS data have also been obtained in many gases [5,6]. Interference effects with nonresonant background terms, as given by (3), are readily observed. ECKBRETH et al. [27,28] have recently used a broad-band pulsed dye laser, together with monochromatic pump beams at $\omega_1 = \omega_3$, to obtain a CARS spectrum over a frequency interval of several hundred cm^{-1} in a single pulse. The temperature in a sooty flame can be determined within about $50\,°K$ from the CARS spectrum of N_2 or flame reaction products in a volume element determined by the intersecting light beams.

BETHUNE et al. [29] have also used a broad band dye laser to obtain an intense broad band infrared pulse by means of the near-resonant electronic stimulated Raman effect in an alkali vapor. The initial and final states are $S^{\frac{1}{2}}$ states of alkali atoms K, Rb or Cs, as sketched in the diagram of Fig.3a. Although the effect is especially large in the vicinity of the mS-nP resonance lines, there is a destructive interference when $\hbar\omega$ corresponds to an energy one-third of the way between the P-doublet. The dominant term in the nonlinear polarization is proportional to

$$4\left(\omega_{nP^{3/2},mS} - \omega_1 - i\Gamma_{nP^{3/2},mS}\right)^{-1} + 2\left(\omega_{nP^{1/2},mS} - \omega_1 - i\Gamma_{nP^{1/2},mS}\right)^{-1}$$

The factors 4 and 2 represent the degeneracy of the P-levels. For the situation of Fig.3a the real parts have equal and opposite signs, and if the damping is small compared to the doublet splitting, the destructive interference is nearly complete. This interference effect was first demonstrated for two-photon absorption on the $3S^{\frac{1}{2}}-4D^{\frac{1}{2}}$ transition of Na by LIAO and BJORKHOLM [30], shown in Fig.3b.

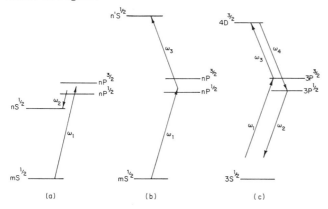

(a)   (b)   (c)

Fig.3 Near resonant two-photon processes and parametric four-wave mixing in alkali atoms (a) coherent resonant electronic Raman scattering (b) resonant two-photon absorption (c) four-wave resonant parametric mixing with Na atoms remaining in the ground state $3S^{\frac{1}{2}}$. For clarity the photon lines are drawn between the unpopulated excited states

Consider now the case that the damping is increased by increasing the partial pressure of a buffer gas. The destructive interference will disappear when the damping becomes comparable to or larger than the doublet splitting. Thus a coherent Raman signal may be increased by means of an incoherent damping process. The point is, of course, that the destructive interference between two coherent pathways is eliminated.

Similar effects undoubtedly play a role in CARS. Resonant CARS (compare Figs.1c-e) has been studied experimentally [31] and theoretically [32] in iodine vapor. DRUET et al. [32] have integrated all the terms in the CARS susceptibility over the inhomogeneous Doppler profile. The different denominators have, of course, different Doppler widths because of different energy spacings.

LYNCH et al. [2,33] have shown that half of the 48 terms in the general expression for $\chi^{(3)}$ are really small correction terms proportional to differences in damping constants. No experimental evidence for such terms has yet been found. They vanish when the system is in the ground state and all phase memory times are determined by the spontaneous lifetime of the levels.

The sign of the imaginary parts in the resonant denominators has been the subject of some controversy [7,10]. The experiments of LOTEM et al. [34] have shown conclusively that the sign of the imaginary part must be reversed with the sign of $\omega_1 - \omega_2$ in the Raman-type denominator of the CARS susceptibility, $\chi^{(3)}(-2\omega_1 + \omega_2, \omega_1, -\omega_2, +\omega_1)$. The imaginary part for two-photon absorption resonances is independent of the sign of $\omega_1 - \omega_2$. By interfering the imaginary part of $\chi^{(3)}$ for two-photon absorption in CS_2 with the Raman resonance in benzene, unambiguously different results were obtained in four-wave mixing for CARS with $\omega_1 - \omega_2 > 0$, and for CSRS with $\omega_1 - \omega_2 < 0$.

CARREIRA et al. have found [35] that the lineshapes for resonant CARS and resonant CSRS are the same for β-carotene. They conclude, therefore, that $\chi^{(3)}(-2\omega_1 + \omega_2, \omega_1, -\omega_2, \omega_1)$ for $\omega_1 - \omega_2 > 0$ and $\omega_1 - \omega_2 < 0$ must be the complex conjugates of each other. The formulae of LYNCH et al. [2] state, however, that two of the three resonant denominators change the sign of their imaginary part, but not the third. They predict, in general, different lineshapes [33] for resonant CARS and CSRS. CARREIRA treats the broad electronic resonance as a combination of homogeneously broadened structures. The line in such organic dyes has undoubtedly a large inhomogeneous distribution of electronic resonant frequencies. This makes the argument presented by CARREIRA less convincing, and it is clearly desirable to do similar experiments in an atomic vapor with well-defined energy levels, matrix elements and widths. Some preliminary experiments were carried out in our laboratory on four-wave mixing in sodium vapor with ω_3 and ω_2 in resonance with the $\omega_{3P^{3/2},4d}$ and $\omega_{3P^{1/2},4d}$ separations. The generated frequency ω_4 differs by 17 cm^{-1}, the 3P-doublet separation, from ω_1. Discrimination in polarization and wave vectors is also utilized. As indicated in Fig.3c, ω_1 and ω_4 were initially removed by 100 to 600 cm^{-1} from the 3S-3P resonance to keep all Na atoms in the ground state and to avoid any population in the excited states. The preliminary results could, nevertheless, be ascribed to Raman-type transitions between the 3P doublet states. About one part in a million of the Na atoms was excited from the ground state, presumably due to pumping of molecular transitions of Na_2 with subsequent dissociation. The intensity of this step-wise process varies as a higher power than the cubic function of the intensities of the incident beams, appropriate for four-wave mixing by ground state atoms. Further experiments are in progress to demonstrate coherent resonant four-wave mixing in alkali vapors. It is expected that the intensity of this coherent signal may increase with increasing pressure of the buffer gas.

In conclusion, fine details of the influence of incoherent homogeneous damping, as well as inhomogeneous broadening on the lineshapes of coherent four-wave mixing, will provide a rigorous test of the theory. Several crucial interferences between real or imaginary parts of different resonances, or with the real nonresonant part, have already been verified. While the main features of the $\chi^{(3)}$ formalism are well established and are used in an ever increasing number of experiments and applications, further tests on the influence of damping in nonlinear situations are required.

References

1. J. Armstrong, N. Bloembergen, J. Ducuing, P.S. Pershan: Phys. Rev. $\underline{127}$, 1918 (1962)
2. N. Bloembergen, H. Lotem, R.T. Lynch, Jr.: Indian J. Pure and Appl. Phys. $\underline{16}$, 151 (1978)
   The following misprint should be corrected in the expression for $K_1(\Omega_u,\Omega_v)$ on page 157: "ω_{kg}" on the right hand side should read "ω_{kj}".
3. S.Y. Yee, T.K. Gustafson, S.A.J. Druet, J.P. Taran: Opt. Commun. $\underline{23}$, 1 (1977)
4. S.Y. Yee, T.K. Gustafson: Phys. Rev. B $\underline{18}$, 1597 (1978)
5. J.P. Taran: in *Laser Spectroscopy III*, ed by J.L. Hall, J.L. Carlsten (Springer, Berlin 1977) p 315, and references quoted therein.
6. J.P. Taran: in *Tunable Lasers and Applications*, ed by A. Mooradian, T. Jaeger, P. Stokseth (Springer, Berlin 1976) p 378
7. H.C. Anderson: private communication, also quoted in L.A. Carreira, L.P. Goss, T.B. Malloy: J. Chem. Phys. $\underline{69}$, 855 (1978). R.S. Hudson, H.C. Anderson: *Molecular Spectroscopy*, a specialist periodic report (Burlington House, London 1978) Vol 5, p 142
8. S.A. Akhmanov, A.F. Bunkin, G.G. Ivanov, N.I. Koroteev: Zh. Eksp. Teor. Fiz. [Sov. Phys.-JETP] $\underline{74}$, 1272 (1978)
9. D. Heiman, R.W. Hellwarth, M.D. Levenson, G. Martin: Phys. Rev. Lett. $\underline{36}$, 189 (1976)
10. M.D. Levenson, J.J. Song: "Coherent Raman Spectroscopy" in *Advances in Coherent Nonlinear Optics*, ed by M.S. Feld, V.S. Letokhov (Springer, Berlin, to be published)
11. Y.R. Shen: Rev. Mod. Phys. $\underline{48}$, 1 (1976)
12. N. Bloembergen, ed: *Nonlinear Spectroscopy*, Course 64 of the E. Fermi International School of Physics (North-Holland Publishing Co., Amsterdam 1977)
13. See, for example, N. Bloembergen: Am. J. Phys. $\underline{35}$, 989 (1967) and references quoted therein.
14. P.D. Maker, R.W. Terhune, C.M. Savage: Phys. Rev. Lett. $\underline{12}$, 507 (1964)
15. See, for example, V.S. Letokhov, V.P. Chebotayev: *Nonlinear Laser Spectroscopy* (Series in Optical Sciences 4, Springer, Berlin 1976)
16. R.W. Hellwarth: J. Opt. Soc. Am. $\underline{67}$, 1 (1977)
17. A. Yariv: Opt. Commun. $\underline{25}$, 23 (1978)
18. D.S. Hamilton, D. Heiman, J. Feinberg, R.W. Hellwarth: Opt. Lett. $\underline{4}$, 124 (1979)
19. P.F. Liao, N.P. Economou, D.M. Bloom: Opt. Lett. $\underline{2}$, 58 (1978)
20. P.F. Liao, D.M. Bloom: Opt. Lett. $\underline{3}$, 4 (1978)
21. A. Tomita: Appl. Phys. Lett. $\underline{34}$, 463 (1979)
22. R.C. Lind, D.G. Steel, M.B. Klein, R.L. Abrams, C.R. Giuliano, R.K. Jain: Appl. Phys. Lett. $\underline{34}$, 457 (1979)
23. R.A. Fisher, B.J. Feldman: Opt. Lett. $\underline{4}$, 140 (1979)
24. D.M. Bloom, G.C. Bjorklund, P.F. Liao: J. Opt. Soc. Am. $\underline{68}$, 1367 (1978)
25. A.G. Jacobson, Y.R. Shen: Appl. Phys. Lett. $\underline{34}$, 464 (1979)
26. A. Owyung: Opt. Lett. $\underline{2}$, 91 (1978); see also his paper in these proceedings.
27. A.C. Eckbreth: Appl. Phys. Lett. $\underline{32}$, 421 (1978)
28. R.J. Hall, A.C. Eckbreth: *Proceedings Laser Spectroscopy* S.P.I.E. $\underline{158}$, 59 (1978)
29. D.S. Bethune, J.R. Lankard, P.P. Sorokin: Opt. Lett. $\underline{4}$, 103 (1979)
30. J.E. Bjorkholm, P.F. Liao: Phys. Rev. Lett. $\underline{33}$, 128 (1974)
31. B. Attal, O.O. Schnepp, J-P.E Taran: Opt. Commun. $\underline{24}$, 77 (1978)

32. S.A.J. Druet, J-P.E. Taran, Ch.J. Borde: J. Phys. (Paris) to be pub-
    lished (1979)
33. R.T. Lynch, Jr., H. Lotem, N. Bloembergen: J. Chem. Phys. 66, 4250
    (1977)
34. H. Lotem, R.T. Lynch, Jr.: Phys. Rev. Lett. 37, 334 (1976)
35. L.A. Carreira, L.P. Goss, T.B. Malloy: J. Chem. Phys. 69, 855 (1978)

Laser Induced Collisional and Radiative Energy Transfer[1]

S.E. Harris, J.F. Young, W.R. Green, R.W. Falcone, J. Lukasik[2], J.C. White,
J.R. Willison, M.D. Wright, and G.A. Zdasiuk

Edward L. Ginzton Laboratory, Stanford University
Stanford, CA 94305, USA

1. Introduction

This paper summarizes progress on two related methods of rapidly transferring energy which is stored in a metastable level of one species to a target level of a different species. The two methods are laser induced collision, and laser induced (two-photon) spontaneous emission and subsequent absorption. Both utilize high peak power, short pulse tunable lasers and both are capable of rapid and selective energy transfer.

In particular, the paper describes the recent experimental results of GREEN, ET AL., which demonstrate, we believe for the first time, laser induced charge transfer [1] and laser induced dipole-quadrupole collisions [2].

2. Laser Induced Collisions

The laser induced collision process was predicted by GUDZENKO and YAKOVLENKO in 1972 [3], and demonstrated experimentally by HARRIS, ET AL. [4] in 1976. An energy level diagram for the first experiment is shown in Fig. 1. A pump laser is first used to store energy in the radiatively trapped $Sr(5s5p\ ^1P_1^0)$ level. As an excited Sr and ground state Ca atom approach, a virtual dipole-dipole collision occurs. The process is completed by the absorption of a transfer laser photon at 4977 Å; producing an excited $Ca(4p^2\ ^1S_0)$ atom and ground state $Sr(5s^2\ ^1S_0)$ atom. Fluorescence at 5513 Å, and thereby the cross section for laser induced collision, is measured as a function of the transfer laser wavelength.

Theory predicts that for laser induced collisional processes where the interaction Hamiltonian of the colliding species is of the dipole-dipole or dipole-quadrupole type, that the cross section for laser induced collision should peak when the transfer laser is tuned to that wavelength $(\lambda_{R=\infty})$ which exactly satisfies energy conservation for the infinitely separated atoms. For quadrupole-quadrupole, charge exchange, and spin exchange processes, the peak should occur near to, but not at, the energy conserving wavelength.

---

[1]This work was jointly supported by the Office of Naval Research and by the Air Force under Contract F19628-77-C-0072.

[2]Laboratoire d'Optique Quantique du CNRS, Ecole Polytechnique, 91120 Palaiseau, France.

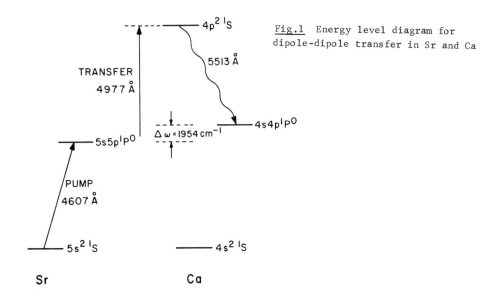

Fig.1 Energy level diagram for dipole-dipole transfer in Sr and Ca

The 1976 experiment [4] utilized pump and transfer lasers, each of which had an ~ 1 μsec pulsewidth. During this microsecond, collisional diffusion from the Sr(5s5p $^1P_1^0$) storage level populated other levels, particularly the lowest triplet levels of Sr and Ca. As the transfer laser was scanned, single photon transitions from these levels populated still higher levels, which in turn collided with and populated the Ca target level. Many peaks of amplitude much greater than that of the laser induced collision were observed.

During the last several years we have developed a picosecond time scale system, with time gated detection, which dramatically enhances the ability to observe laser induced collisional processes. The system (Fig. 2) consists of a mode-locked Nd:YAG oscillator-amplifier, the output of which is upconverted to 3547 Å and used to excite two synchronously pumped dye lasers. Each dye laser is cavity dumped to produce two independently tunable 40 ps pulses of several MW peak power with linewidths of about 10 cm^{-1}. The transfer laser pulse is delayed by ~ 5 ns from the pump pulse; and both are focused into a metal vapor cell to an area of about 10^{-3} cm^2. At each transfer laser wavelength the target state fluorescence is integrated over four consecutive 10 ns time intervals and recorded by a minicomputer.

The 5 ns spacing between the pump pulse and transfer laser pulse enormously reduces the collisional diffusion referred to above. The time gated detection allows the differentiation between the laser induced collision process, which occurs only during the 40 ps during which the transfer laser pulse is present, and other excitation-collisional processes which continue for much longer times.

A summary of the collisional processes which we have observed during the last several years is given in Table 1. Each process is discussed briefly below.

Process	Comments	Collision Cross Section (Laser Power Density)	Reference
Sr(5s5p $^1P_1^o$) + Ca(4s^2 $^1S_0$) + $\hbar\omega$(6409 Å) → Sr(5s^2 $^1S_0$) + Ca(4d ^1D)	Recall		6
Sr(5s5p $^1P_1^o$) + Ca(4s^2 $^1S_0$) + $\hbar\omega$(4977 Å) → Sr(5s^2 $^1S_0$) + Ca(4p^2 $^1S_0$)	Dipole-Dipole Weak Field	9×10^{-18} cm^2 (5×10^5 W/cm^2)	4
Sr(5s5p $^1P_1^o$) + Ca(4s^2 $^1S_0$) + $\hbar\omega$(4711 Å) → Sr(5s^2 $^1S_0$) + Ca(5d ^1D)	Dipole-Dipole Weak Field		4
Sr(5s5p $^1P_1^o$) + Ca(4s^2 $^1S_0$) + $\hbar\omega$(4977 Å) → Sr(5s^2 $^1S_0$) + Ca(4p^2 $^1S_0$)	Dipole-Dipole Strong Field	4×10^{-13} cm^2 (3×10^{10} W/cm^2)	5
Ca(4s4p $^1P_1^o$) + Sr(5s^2 $^1S_0$) + $\hbar\omega$(6217 Å) → Ca(4s^2 $^1S_0$) + Sr(5s6d $^1D_2$)	Dipole-Dipole Strong Field	10^{-14} cm^2 (3×10^9 W/cm^2)	5
Sr(5s5p $^1P_1^o$) + Ca(4s^2 $^1S_0$) + $\hbar\omega$(5307 Å) → Sr(5s^2 $^1S_0$) + Ca(3d4p $^1F_3^o$)	Dipole-Quadrupole	3×10^{-14} cm^2 (7×10^9 W/cm^2)	2
Ca$^+$(4s $^2S_{1/2}$) + Sr(5s^2 $^1S_0$) + $\hbar\omega$(4715 Å) → Ca(4s^2 $^1S_0$) + Sr$^+$(5p $^2P_{3/2}^o$)	Charge Transfer to an Excited Ionic State	5×10^{-15} cm^2 (1.5×10^9 W/cm^2)	1
Ba(6p $^1P_1^o$) + Ba(5d $^1D_2$) → 2Ba(6s^2 $^1S_0$) + $\hbar\omega$(3394 Å)	Radiative Fluorescence	$\sigma_{spontaneous} =$ 2.6×10^{-20} cm^2	8
Ba(6s^2 $^1S_0$) + Ba(6s^2 $^1S_0$) + $\hbar\omega$(3394 Å) → Ba(6p $^1P_1^o$) + Ba(5d $^1D_2$)	Pair Absorption		10
Ba(6s^2 $^1S_0$) + Tl(6p $^2P_{1/2}^o$) + $\hbar\omega$(3867 Å) → Ba(6p $^1P_1^o$) + Tl(6p $^2P_{3/2}^o$)	Pair Absorption		10

Table 1   Summary of experiments

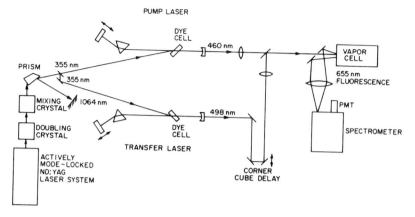

PUMP LASER

DYE CELL
460 nm

PRISM   355 nm

355 nm

MIXING CRYSTAL   1064 nm

DOUBLING CRYSTAL

ACTIVELY
MODE-LOCKED
ND:YAG
LASER SYSTEM

DYE CELL
498 nm
TRANSFER LASER

CORNER
CUBE DELAY

VAPOR
CELL

655 nm
FLUORESCENCE

PMT

SPECTROMETER

Fig.2 Experimental system for observation of laser induced collisions

Dipole-Dipole Processes

We have observed laser induced dipole-dipole collisional processes in three
systems (Table 1), two of which transfer from Sr to Ca [4][5] and one from
Ca to Sr [5]. Each peaks at the energy conserving $(\lambda_{R=\infty})$ wavelength.
Figure 3 shows Ca target state fluorescence, as a function of transfer laser
wavelength for the process of Fig. 1. The importance of time gated detection is clearly illustrated. Collision cross section as a function of laser
power density is shown in Fig. 4. Collision cross sections greater than
about 10^{-13} cm^2 were observed at a laser power density of 10^{10} W/cm^2.

We also note the work of CAHUZAK and TOSCHEK on the dipole-dipole process [7].

Dipole-Quadrupole Processes

The overall selection rule for a dipole-quadrupole process is that each of
the two participating atoms must <u>both</u> make parity allowed or parity non-
allowed transitions. This is to be contrasted with the dipole-dipole case
where if one species makes an allowed transition, the target species must
make a non-allowed transition, or vice versa.

Figure 5 shows the pertinent energy levels for the Sr-Ca dipole-quadrupole process which has recently been observed (Table 1) [2]. Figure 6 shows
collision cross section as a function of the applied laser power density.
The strong field regime begins when the laser power density is increased to
the point where the probability of collision becomes unity for an impact
parameter equal to the Weisskopf radius; and then increases as the cube
root of the applied power density.

Laser Induced Charge Transfer

Figure 7 shows an energy level diagram for the laser induced charge transfer process which has recently been studied by GREEN, ET AL. (Table 1) [1].
Energy is first stored in the form of <u>ground state</u> Ca ions. An intense
laser field at 4715 Å selectively transfers this energy to the excited

352

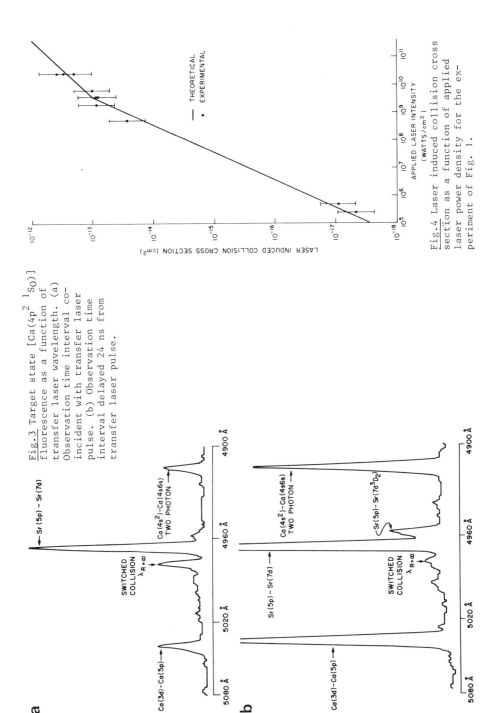

Fig.3 Target state [Ca(4p² ¹S₀)] fluorescence as a function of transfer laser wavelength. (a) Observation time interval coincident with transfer laser pulse. (b) Observation time interval delayed 24 ns from transfer laser pulse.

Fig.4 Laser induced collision cross section as a function of applied laser power density for the experiment of Fig. 1.

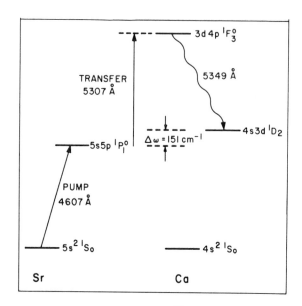

Fig.5 Energy level diagram for dipole-quadrupole transfer in Sr and Ca

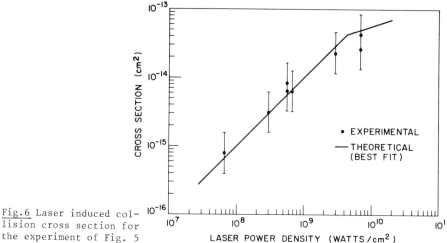

Fig.6 Laser induced col-
lision cross section for
the experiment of Fig. 5

LASER POWER DENSITY (WATTS/cm²)

$Sr^{+*}(5p\ ^2P^0_{3/2})$ level. Without the laser photon, the process is endothermic by 2.6 eV (~ 30 kT at 1000°K) and has a vanishingly small cross section. The laser photon supplies this energy and may be thought of as raising one of the two Sr valence electrons to a virtual level of approximately $Sr(5s5p\ ^1P^0_1)$ character; allowing the unexcited Sr electron to be captured by Ca^+; thereby producing the $Sr^+(5p\ ^2P^0_{3/2})$ excited state.

Experimental results [1] are shown in Fig. 8 which shows 4078 Å target fluorescence as a function of the transfer laser wavelength. The impor-tance of time gated detection is clearly seen. For example, the target

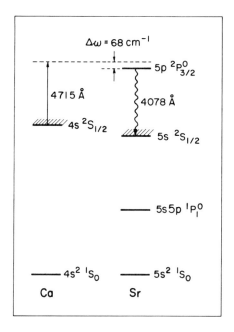

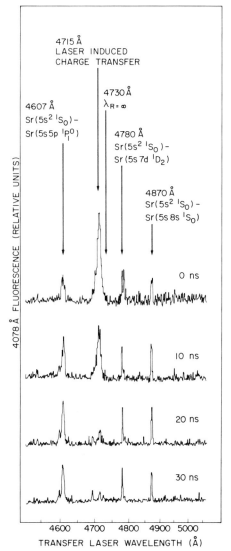

▲

Fig.7 Energy level diagram for laser induced charge transfer

Fig.8 Fluorescence from the Sr$^+$ (5p ^{2}P$^0_{3/2}$) target state (Fig.7) as a function of transfer laser wavelength

state excitation obtained when the transfer laser is tuned to the Sr resonance line transition at 4607 Å (Fig. 7) results from an exothermic (~ 500 cm^{-1}) charge transfer collision between excited Sr(5s5p ^{1}P$^0_1$) neutrals and ground state Ca$^+$ ions. Unlike the laser induced 4715 Å process, this two-step process persists for the lifetime of the excited Sr state and maintains nearly constant amplitude over the 40 ns observation period. This is also true for target state excitation which results from several other two-photon transitions which are identified in Fig. 8.

Note that unlike the case for the dipole-dipole and dipole-quadrupole processes, the cross section for the laser induced charge transfer collision peaks at a wavelength which is 15 Å shorter (68 cm^{-1}) than that which exactly satisfies energy conservation for the separated atoms. This is consistent with theory [4]. The lineshape of the laser induced process is roughly symmetrical and has a width of about 50 cm^{-1}.

A very rough estimate of collision cross section yields $\sigma_c \cong 5 \times 10^{-15}$ cm^2 at a laser power density of 1.5×10^9 W/cm^2.

Collisional Fluorescence and Pair Absorption

Two processes which are closely related to the laser induced collision process are that of radiative collisional fluorescence and pair absorption. In the dipole-dipole fluorescence process, two excited atoms collide and spontaneously emit a photon at the sum frequency of an allowed and non-allowed transition. In a Ba-Ba system (Table 1) with a virtual detuning of 901 cm^{-1}, WHITE, ET AL. [8] have measured a cross section for spontaneous de-excitation of 2.6×10^{-20} cm^2. For the measured linewidth of about 20 cm^{-1}, this implies a cross section for the stimulated (laser induced) process of $\sigma_c = 3.4 \times 10^{-24}$ (P/A) , where (P/A)(W/cm^2) is the incident laser power density.

In a pair absorption process, two atoms collide and absorb a single photon at a frequency corresponding to the sum energy of the two species. Pair excitation of molecular species at high pressure (10 to 100 atmospheres) has long ago been observed in the infrared [9]. Observation of pair absorption in Ba-Ba and Ba-Tℓ atomic systems is noted in Table 1. WHITE, ET AL. [10] have measured absorptions of several percent per cm at atom densities of about 10^{17} atoms/cm^3.

In a recent letter Falcone [11] has proposed the use of pair absorption processes to cause the inversion of atoms or molecules with respect to their ground state.

3. Radiative Energy Transfer

Energy may also be transferred from the storage species to the target species by radiative processes. If energy is stored in the upper level of a (parity) non-allowed transition, for example the He 2s^{1}S level, anti-Stokes scattering from an incident laser beam will produce tunable radiation at the sum and difference of the storage and laser frequency. This radiation may be tunable, quite narrowband, of picosecond time scale and linearly polarized. Its intensity is that of a two-photon blackbody radiator [12][13], i.e., as the intensity of the incident laser is increased so as to render the media two-photon opaque, the intensity of the source approaches that of a blackbody at the temperature of the non-allowed transition. A second species mixed with the storage species may absorb this radiation, thereby completing the energy transfer [14].

Alternately, if energy is stored in a radiatively trapped allowed transition, for example as in Fig. 1, an incident laser beam may excite a non-allowed level of a target species by two-photon absorption of the trapped (blackbody) radiation.

In a recent experiment, FALCONE, ET AL. [15] have shown how the first of the above radiative processes may be used for VUV spectroscopy, in a manner

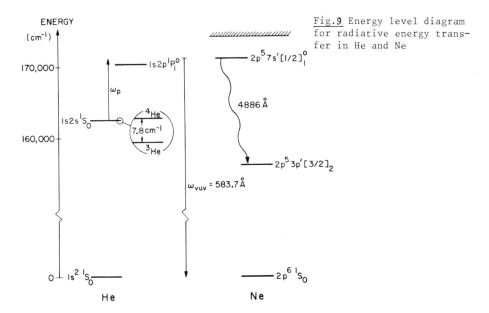

Fig.9 Energy level diagram for radiative energy transfer in He and Ne

that does not require any VUV optics or detectors. By using Ne as a target species, and substituting ^{3}He for ^{4}He, the 7.8 cm^{-1} isotopic shift of the He $2s^1$S level was clearly resolved (Figs. 9 and 10).

The second of the above radiative processes will allow spectroscopy of transitions which are non-allowed to ground.

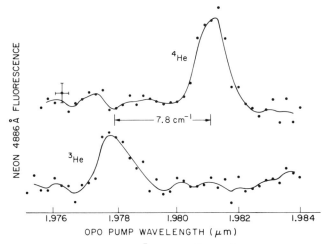

Fig.10 Observed Ne $2p^5 7s'[1/2]_1^0$ - $2p^5 3p'[3/2]_2$ fluorescence at 4886 Å as a function of OPO wavelength for ^{3}He and ^{4}He (Fig. 9). In each case the peak corresponds to the emission by He and the absorption by Ne of 583.69 Å anti-Stokes radiation.

4. Applications

Both the collisional and radiative transfer processes have implications
for laser-device technology. The collisional process may lead to a class
of active upconverters, where upconversion gain is attained from an elec-
trically excited media. Efficient use of the incident laser photons will
require product densities of $\sim 10^{34}$ atoms/cm^6; and thus devices of this
type will probably be best suited for the IR and visible regions of the
spectrum.

At much shorter wavelengths, where the atom density must be low to
achieve high electron energy, the radiative processes are probably pre-
ferable. Either anti-Stokes pumping of an allowed transition, or two-
photon pumping via trapped blackbody radiation of a non-allowed transition, should yield large excitation rates, with picosecond rise times.

5. References

1. W. R. Green, M. D. Wright, J. F. Young, and S. E. Harris, "Laser Induced Charge Transfer to an Excited Ionic State" (submitted for publication).

2. W. R. Green, M. D. Wright, J. Lukasik, J. F. Young, and S. E. Harris,
   "Observation of a Laser-Induced Dipole-Quadrupole Collision" (submitted for publication).

3. L. I. Gudzenko and S. I. Yakovlenko, Zh. Eksp. Teor. Fiz. 62, 1686
   (1972) [Sov. Phys. JETP 35, 877 (1972)]; Phys. Lett. 46A, 475 (1974).

4. S. E. Harris, R. W. Falcone, W. R. Green, D. B. Lidow, J. C. White,
   and J. F. Young, in Tunable Lasers and Applications, edited by A.
   Mooradian, T. Jaeger, and P. Stokseth (Springer-Verlag, New York,
   1976); and R. W. Falcone, W. R. Green, J. C. White, J. F. Young, and
   S. E. Harris, Phys. Rev. A 15, 1333 (1977).

5. W. R. Green, J. Lukasik, J. R. Willison, M. D. Wright, J. F. Young,
   and S. E. Harris, Phys. Rev. Lett. 42, 970 (1979).

6. D. B. Lidow, R. W. Falcone, J. F. Young, and S. E. Harris, Phys.
   Rev. Lett. 36, 462 (1976); Phys. Rev. Lett. 37E, 1590 (1976).

7. Ph. Cahuzak and P. E. Toschek, in Laser Spectroscopy, edited by
   J. L. Hall and J. L. Carlsten (Springer-Verlag, New York, 1977).

8. J. C. White, G. A. Zdasiuk, J. F. Young, and S. E. Harris, Phys.
   Rev. Lett. 41, 1709 (1978).

9. H. L. Welsh, M. F. Crawford, J. C. F. MacDonald, and O. A. Chisholm,
   Phys. Rev. 83, 1264 (1951).

10. J. C. White, G. A. Zdasiuk, J. F. Young, and S. E. Harris, Optics
    Lett. 4, 137 (1979).

11. R. W. Falcone, Appl. Phys. Lett. 34, 150 (1979).

12. S. E. Harris, Appl. Phys. Lett. 31, 498 (1977).

13. L. J. Zych, J. Lukasik, J. F. Young, and S. E. Harris, Phys. Rev. Lett. 40, 1493 (1978).

14. S. E. Harris, J. Lukasik, J. F. Young, and L. J. Zych, in Picosecond Phenomena, edited by C. V. Shank, E. P. Ippen, and S. L. Shapiro (Springer-Verlag, New York, 1978).

15. R. W. Falcone, J. R. Willison, J. F. Young, and S. E. Harris, Optics Lett. 3, 162 (1978).

Laser Bandwidth and Intensity Effects in Multiphoton Ionization of Sodium

S.J. Smith[1] and P.B. Hogan

Joint Institute for Laboratory Astrophysics,
University of Colorado and National Bureau of Standards
Boulder, CO 80309, USA

Recently there has been interest in the development of theoretical methods for incorporating the effects of finite laser bandwidth in the treatment of the ac Stark effect in the two-level atom approximation, and in the extension of this model to approximate descriptions of "weak probe" double resonance experiments [1]. In this paper we briefly describe some experimental observations which have played a role in motivating recent developments in the mathematical framework for treating finite laser bandwidth effects in phase diffusion models [2] and, more recently, in chaotic field models which include the effects of amplitude fluctuations [3].

We have studied resonant three-photon ionization in an atomic sodium beam, using two similar flashlamp pumped tunable dye lasers. The first of these is used to pump the $3S_{1/2}$-$3P_{1/2}$ transition (see Fig. 1) with enough intensity so that the on-resonance Rabi frequency $\Omega = \vec{\mu}_{ab} \cdot \vec{E}/\hbar$ corresponding to the interaction energy of the transition dipole moment $\vec{\mu}_{ab}$ in the electric field $\vec{E}$ the laser, is large compared to the approximately 2 GHz laser bandwidth, and to the substructure of each of these two states; and is comparable to and usually larger than the detunings, Δ, used in this study. However, Ω is much smaller than the $3P_{1/2}$-$3P_{3/2}$ fine structure interval. This laser is approximately 90% linearly polarized along the sodium beam axis (as is the second laser). Under these circumstances, the behavior of the $3S_{1/2}$-$3P_{1/2}$ system should approximate that of a two-level system [4], and the second laser, strongly attenuated and operated

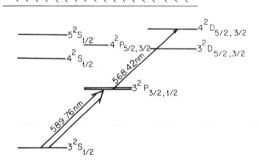

Fig. 1. Structure of the unperturbed lower levels of atomic sodium, showing the transition driven near resonance by the intense pump laser (broad arrow) and the transition to the 4D state scanned by the weak probe laser.

[1]Presently at Sektion Physik, Universität München, 8046 Garching, Federal Republic of Germany, under the Special Program for Senior U.S. Scientists of the Alexander von Humboldt Foundation.

synchronously with the first, can be used to probe the structure of the upper level in a double resonance configuration as indicated in Fig. 1. Ionization is accomplished by a second absorption from the intense pump laser.

The laser beams, in a coaxial, co-propagating configuration, intersect a well collimated sodium beam from a diffusive source, at right angles. The spread of transverse beam velocities corresponds to a Doppler width of approximately 10 MHz. We use a weak electric field, which may be gated off during the laser pulses, to collect all ions, and time-of-flight discrimination to eliminate molecular ions. The measurement is carried out in high vacuum. There is no detectable signal from backgound sodium, and there are no effects of collisions with background gas.

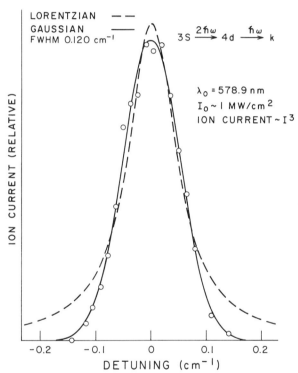

Fig. 2. In a single laser experiment, the pump laser is scanned over the $3S_{1/2}-4D_{3/2}$ two-photon resonance. The detuning equals $(2\nu_{Laser}-\nu_{3S-4D})$. Least-squares fitted Lorentzian and Gaussian line shapes are shown for comparison with the tuning curve, which is seen to be approximately Gaussian. The FWHM width is in satisfactory agreement with the measured laser line width of approximately 0.08 cm^{-1}, increased by $\sqrt{2}$ as would be expected for the convolution of two equal width Gaussian line shapes. The intensity dependence over the curve is approximately I_0^3 where I_0 is the peak axial intensity of the laser.

The pump laser spectral line shape has been studied with Fabry-Perot interferometry as well as by means of a single laser study of the $3S_{1/2}$-$4D_{3/2}$ two-photon resonance, with sequential ionization (Fig. 2). The laser line is seen to have a Gaussian spectral distribution to a good approximation. The figure also contains an impression about the shot-to-shot intensity variations. Each point in these and other data represents the average of ionization yields from ten laser shots. There are about 10^5 sodium atoms in the interaction volume. On the order of 50-100 ions are collected from each shot.

For the two laser double resonance observations, the second laser, strongly attenuated, is tuned across the $3P_{1/2}$-$4D_{3/2}$ resonance as a probe. Figure 3 shows the dependence of ionization current on probe laser intensity. All measurements were carried out within the linear regime.

Double resonance measurements carried out with monochromatic lasers would display an intermediate ($3P_{1/2}$) level structure consisting of a Rabi doublet with the splitting $(\Omega^2+\Delta^2)^{1/2}$, with an asymmetry which increases monotonically as the detuning, Δ, becomes large, until, in the limit of large pump laser detuning a single peak would be observed representing an energy-conserving two-photon resonance where the probe laser

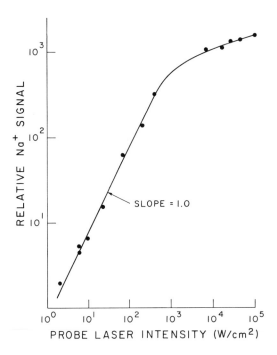

Fig. 3. Dependence of ionization current on probe laser intensity. Intensity of the pump laser ~1 MW/cm². Here and in other intensity measurements, neutral density filters are used to avoid altering spectral or geometrical properties of the laser beams.

detuning is $-\Delta$. In Fig. 4 are shown typical results of double resonance measurements with detection of the ionization current, using our broadband lasers. In each case the asymmetry is normal. The central minima are quite deep.

The broad outer wings on these curves are unlike those on the relatively narrow peaks with spreading bases predicted by phase diffusion models, and it is likely that they contain information about the role of amplitude fluctuations in the formation of such lines. Ideally, for the purpose of comparison with theories of the effects of finite laser bandwidth, one would like to present a pure spectrum of the Rabi frequencies which derive from the amplitude fluctuations of the pump laser beam. However, the Rabi frequency spectra such as shown in Fig. 4 are inevitably distorted in several ways. The most serious problem is the distribution of Rabi frequencies corresponding to the radial variation of intensities in the laser beam. This is greatly complicated if saturation of the ionization step occurs at the high axial intensities. A somewhat less troublesome effect is due to the time dependence of the probing and ionizing pulses. Doppler shifts and transit time effects are neglected here.

We illustrate the problem generated by the spatial intensity dependence with a calculation based on idealized laser beam geometry. We assume that the laser beam intensity is independent of position along the beam axis but has a Gaussian radial distribution. Then the ratio of intensity off-axis at distances r to the on-axis intensity I_0 is

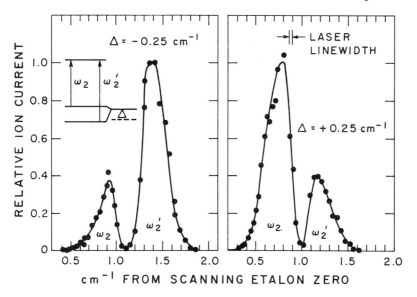

Fig. 4. Double resonance measurements in three-photon ionization, for opposite detunings of the probe laser. Relative ion current is plotted against increasing probe laser photon energy in cm^{-1}. In these experiments the pump laser bandwidth had been narrowed to ~0.05 cm^{-1}. The plotted points are the measured values. The solid curves in this and the next figure are visual averages.

$$\frac{I(r)}{I_0} = \left(\frac{\Omega(r)}{\Omega_0}\right)^2 = e^{-(r^2/r_2^2)\ln 2}$$

where Ω is the frequency corresponding to the on-resonance interaction energy of the dipole moment in the intense field of the pump laser, and r_2 is the radius at which $I(r)/I_0 = 1/2$. We assume the probe laser is unsaturating, has no functional dependence on frequency, and has a Gaussian radial distribution which, however, may have a different radial dependence, reaching half intensity at radius Cr_2, where C is a scaling constant.

If the ionization cross section for an atom in the 4D state is σ_i, and τ is the effective mean life of the 4D state against induced and spontaneous emission, the degree of saturation in the ionization step is given by the ratio of the D state ionization rate to the total D state loss rate

$$\frac{\sigma_i I/h\nu}{\sigma_i I/h\nu + (1/\tau)} \quad .$$

This factor would be $\approx 1/2$ at about 4 MW/cm^2, assuming the mean radiative lifetime of the 4D state is just the natural lifetime against spontaneous emission and taking $\sigma_i = 10^{-18}$ cm^2 [5]. A monochromatic laser with such a radial intensity distribution will thus produce a Rabi split doublet which is broadened and shifted, due to purely geometrical considerations, according to

$$\frac{dN(\Omega)}{d\Omega} \propto \left(\frac{\Omega}{\Omega_0}\right)^{[(2/C^2)-1]} \cdot \left(\frac{\sigma_i I_0 \tau}{h\nu}\right) \left(\frac{\Omega}{\Omega_0}\right)^2 \left/ \left[\left(\frac{\sigma_i I_0 \tau}{h\nu}\right) \left(\frac{\Omega}{\Omega_0}\right)^2 + 1\right]\right.$$

when $\Omega \gg \Delta$. For $C = 1$, roughly the case in the work described here, the predicted shape is proportional to $(\Omega/\Omega_0)^3$ at low intensity. However, at such high intensities that the on-axis ionization rate may be comparable to, or greater than, the spontaneous decay rate of the D state, $dN(\Omega)/d\Omega$ will tend to (Ω/Ω_0) near the beam axis, and the shape of the response to probe laser scanning will be significantly more complicated. Considerable uncertainty remains with respect to the role of saturation of the ionization step due to some ambiguity in the intensity scale, and in the effective ionization cross section. Some aspects of this problem were discussed by AGOSTINI et al. [6], who carried out an experiment with similar objectives. In any case, we conclude that spatial dependence of intensity (and to a lesser extent the time dependence) prevents detailed quantitative comparison with theory but that the qualitative aspects, particularly as they differ from the predictions of phase diffusion theory, are significant.

GEORGES, LAMBROPOULOS, and ZOLLER [3] have considered the role of amplitude fluctuations and have found potentially important contributions to the observed line shapes, which are also discussed in another paper in this volume [1]. Our experimental results are consistent with their findings but do not constitute confirmation because of the uncertainties discussed above.

A point of interest is the rather shallow central dip obtained for zero detuning (Fig. 5a). Since this central region corresponds to small Rabi splittings and therefore to low instantaneous intensities in the laser field (in the amplitude fluctuation model), one might have expected a deep central minimum corresponding to low ionization rates, since the pump laser also produces the ionizing field. However, on resonance, a fluctuation to low field intensity will leave a P state population which persists with the natural P state lifetime against spontaneous decay [7], which continues to be subject to excitation for probe laser tuning close to the 3P-4D resonance, leading to strong ionization when the pump laser field recovers. Therefore, the pumping and ionizing fields are effectively decorrelated. In the phase diffusion model, a sudden shift in phase or frequency can lead to a qualitatively similar result. Low pump fields deriving from radial intensity fall off should not contribute to filling in the center in a comparable way.

Off resonance, the behavior of the ionization signal for probe tuning near the P-D resonance is different. It can be noted that the P state population dynamics are complicated [7]. However, the essential point is that the pump laser detuning Δ enters the expression for the splitting $(\Omega^2+\Delta^2)^{1/2}$ of the Rabi doublet and contributes a central zone of width $\sim\Delta$ where, for perfectly monochromatic lasers the signal is zero. With our broadband lasers the extent to which zero minimum signal is approached depends on the spectral widths of the pump and probe lasers, somewhat less care having been lavished, typically, on the latter than on the former. The characteristic behavior of the central dip for the off-resonance case, deeper than for the on-resonance case, is evident in Fig. 5 (b, c, and d).

Figure 5 also illustrates the asymmetry reversal, explained by Georges and Lambropoulos [2] on the basis of a phase diffusion model. In this case symmetry is observed, not only at zero detuning, but also at positive and negative detunings $\Delta = \pm0.14$ cm^{-1}. For $|\Delta| > 0.14$ the two-photon $3S_{1/2}-4D_{3/2}$ process is dominant, but for $|\Delta| < 0.14$ the sequential $3S_{1/2}-3P_{1/2}-4D_{3/2}$ process dominates. These results are observed at axial intensities ~2 MW/cm^2. Results at higher and lower intensities are similar but the outer symmetry points are displaced to higher values $|\Delta|$ as the intensity is increased.

Finally, we show the results (Fig. 6) of asymmetry measurements at a fixed negative value of Δ (=-0.14 cm^{-1}) but with different pump laser field intensities. Note that we use as a definition of asymmetry (A) the ratio of the difference of peak heights to their sum. In the steady state adiabatic approximation $A \propto \Delta/(\Omega^2+\Delta^2)^{1/2}$ which is represented by the dashed line in the figure. In this approximation the asymmetry takes the sign of Δ, and for finite Δ goes to zero only as $\Omega \to \infty$. However, we find the observed asymmetry fails to saturate on the limit A = 0 but continues through the point of symmetry into the positive asymmetry regime. While the chaotic field model [3], which is discussed in the succeeding paper [1], has not been applied to the prediction of this behavior, it nevertheless indicates that saturation effects may develop quite differently from the predictions of phase diffusion models.

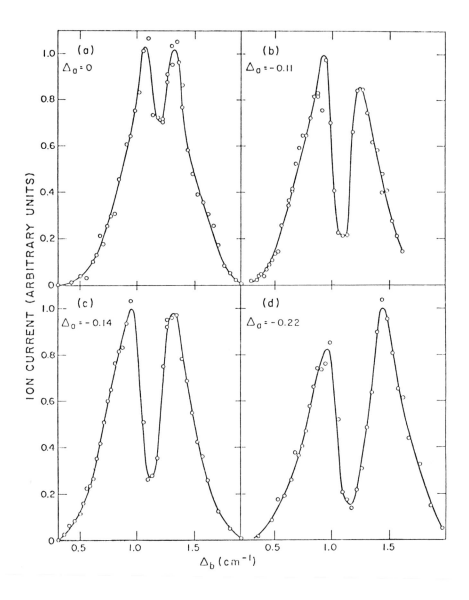

Fig. 5. Double-resonance measurements in three-photon ionization, for zero detuning and three increasingly negative detunings of the pump laser.

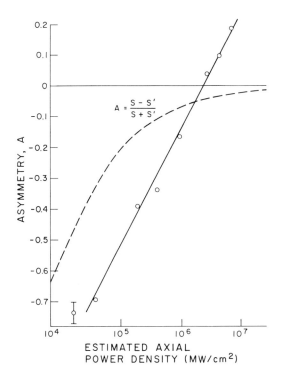

$$A = \frac{S - S'}{S + S'}$$

Fig. 6. Observed asymmetry (A) (defined as shown where S is the peak height of the component of the Rabi doublet reached with the lower probe laser frequency, S' that for the higher probe laser frequency) is plotted against log of the axial intensity of the pump laser.

The authors are indebted to Drs. David Nitz and Arlee Smith for many very helpful comments and suggestions on the content of this paper, as well as to Dr. Marc Levenson for invaluable discussions.

References

1. P. Lambropoulos, paper in this Conference.
2. P. B. Hogan, S. J. Smith, A. T. Georges and P. Lambropoulos, Phys. Rev. Lett. 41, 229 (1978).
3. A. T. Georges, P. Lambropoulos and P. Zoller, Phys. Rev. Lett. 42, 1609 (1979).
4. B. W. Shore, Phys.Rev. A 17, 1739 (1978).
5. A. Smith and S. Geltman, private communication.
6. P. Agostini, A. T. Georges, S. E. Wheatley, P. Lambropoulos and M. D. Levenson, J. Phys. B 11, 1733 (1978).
7. E. Courtens and A. Szöke, Phys. Rev. A 15, 1588 (1977).

Saturation and Stark-Splitting of Resonant Transitions in Stochastically Fluctuating Laser Fields of Arbitrary Bandwidth

A.T. Georges, P. Lambropoulos, and P. Zoller[1]

Department of Physics, University of Southern California
Los Angeles, CA 90007, USA

The interaction of strong electromagnetic radiation with matter usually leads to nonlinear processes [1-23]. In fact, the presence of such nonlinearities may usefully be taken as the definition of "strong" radiation, at least in the sense the term is used in this paper. Examples are: Multi-photon excitation or ionization, harmonic generation or wave mixing, as well as the saturation of a bound-bound transition, even if a single-photon transition, as it is the saturation that introduces the non-linearity. A process that depends on the field non-linearly cannot be described in terms of the average intensity and spectrum of the radiation (first order correlation function) because in general higher order correlation functions are also involved [24]. The electron responds to more than just the total radiation; it "sees" the fluctuations of the field.

Perhaps the simplest case of a non-linear process is the non-resonant N-photon ionization of an atom, or more generally the N-photon transition from a (sharp) bound state to a continuum [25]. Throughout this paper, "non-resonant" implies the absence of resonances with intermediate states. Under these conditions, the process is proportional to the Nth order field correlation function. As a result, N-photon ionization induced by a chaotic field (CF) - or equivalently by a laser with many independent modes - is larger by a factor of N! than that induced by a purely coherent field. This has been known for quite some time now and has been verified experimentally, the most dramatic demonstration having been given by the Saclay group in 11-photon ionization of Xe [26]. Theoretically this result is obtained easily because the absence of intermediate resonances allows one to treat the field as monochromatic. This leads to a separation of field from atomic variables and the process is found to depend on a single correlation function. Thus only the intensity fluctuations affect the transition probability.

If the frequency of the field, or an integral multiple of it, is resonant with a transition to an intermediate state, the field bandwidth also plays a role if it is comparable to or larger than the natural width of that state. In the simplest case, only the spectral line-shape of the source will matter, as in the weak-field limit, for example. If in addition the field is sufficiently strong to saturate the resonant transition, not only the bandwidth (a feature of the first order correlation function) but also higher order correlation functions affect the process. The precise effect will then depend on the detailed stochastic properties of the field. As discussed subsequently, whether the field undergoes amplitude or simply phase fluctuations makes a tremendous difference. Thus two particular

---

[1] on leave of absence from the University of Innsbruck, 6020 Innsbruck, Austria

models representing two extreme cases are commonly employed in related studies: The phase diffusion (PD) field [24] whose amplitude is assumed to be constant while its phase undergoes fluctuations of a diffusion type and the chaotic field (CF) [24] whose complex amplitude undergoes Gaussian fluctuations.

The role of field fluctuations and of the associated bandwidth in the resonant interaction of strong radiation with matter has been an extremely active subject in the last three years or so [1-16]. The interest has been stimulated by recent experiments on resonance fluorescence [17-18] and double optical resonance [20]. A number of interesting theoretical results have been obtained under the assumption of a bandwidth due entirely to phase fluctuations [6,8,10-13,16]. The PD model is expected to be adequate for the interpretation of experiments with well stabilized CW lasers such as those used in recent experiments on resonance fluorescence under strong fields [17,18]. There is however another type of experiment [20] - such as multiphoton excitation and ionization - performed with considerably stronger, multimode, pulsed lasers whose amplitude undergoes substantial fluctuations, often comparable to - if not stronger than - those of a CF. This poses a more general and far more significant problem: How does a strong, stochastically fluctuating field - with both amplitude and phase fluctuations - affect the saturation, the associated Stark splitting and the overall observed signal in a non-linear process involving a resonant transition?

The essential mathematical problem - which can be formulated in more than one way [21-23] - can be reduced to the solution of the equations of motion of a two-level system (TLS) strongly coupled to a stochastically fluctuating field. If the TLS is described in terms of its density matrix, one ultimately has to solve a set of stochastic differential equations. For phase diffusion, the task is facilitated substantially because the atomic variables can at the appropriate stage of the calculation be decorrelated rigorously from the field variables [16]. The decorrelation here refers to the stochastic averaging over the fluctuating variables. As a result, the stochastic differential equations are reduced to differential equations obeyed by the averaged atomic variables. For amplitude fluctuations, however, the decorrelation is not valid and if used as an approximation one does not know the magnitude of the error, unless, of course, the solution of the complete problem is known.It is only in the weak field limit that the decorrelation is valid for all fields.

In this paper, we present results from the solution of the problem for a chaotic field which, as is well known, undergoes Gaussian amplitude fluctuations. These results have been obtained with two parallel approaches based on formulations whose details have been published elsewhere [21-23] and will only be discussed here very briefly. In both approaches, the equations of motion of the density matrix of the TLS are coupled to a stochastic field.

In one of the approaches [21-22] the CF is assumed to be Markovian and is represented by its Fokker-Planck equation. As has been shown elsewhere, if one is interested in certain one-time atom-field averages, the stochastic density matrix equations can be reduced to an infinite set of differential equations for these averages. If in addition, one considers the stationary limit of the averaged density matrix of the TLS, the equations reduce to an infinite system of linear algebraic equations which can be used to obtain solutions in terms of continued fractions.

Let $\rho_{11}(t)$ and $\rho_{22}(t)$ be the diagonal matrix elements (populations) of the density matrix of the TLS where $|2\rangle$ and $|1\rangle$ are the upper and lower states, respectively, with energies $\hbar\omega_2$ and $\hbar\omega_1$. These matrix elements are of course coupled to $\rho_{12}(t)$. The stationary limit of the average (indicated by angular brackets) population difference $\langle n(t)\rangle = \langle\rho_{22}(t)\rangle - \langle\rho_{11}(t)\rangle$ corresponds to $t \to \infty$ and can be written as a continued fraction involving the averaged Rabi frequency Ω, the detuning Δ from resonance, the field bandwidth b and the spontaneous decay width κ_2 of the upper state. This continued fraction has been shown [21] to converge for all values of the above parameters. In general, it must be calculated numerically but the convergence is very rapid which enables one to obtain results for arbitary field intensities and bandwidths. The approach has also been generalized [22] to calculate the spectrum of resonance fluorescence of a TLS driven by a strong chaotic field. Then we find that the stochastic average of the spectrum obeys an inhomogeneous three-term recursion relation which we have solved numerically.

In the second approach [23], the chaotic field $\varepsilon(t)$ is written as a complex Gaussian stochastic process described by the infinite sequence of its field-correlation functions. Although such as process is not necessarily Markovian, we do here assume a Markovian chaotic field with first order correlation function $\langle\varepsilon^*(t_1)\varepsilon(t_2)\rangle = \varepsilon^2\, e^{-b|t_1-t_2|}$ where $\varepsilon^2 = \langle|\varepsilon(t)|^2\rangle$, and the resulting spectrum is obviously Lorentzian. The correlation functions of the chaotic field obey well known relations that can be found, for example, in Ref.24. Again using the density matrix equations, an integral equation for n(t) can be obtained. But in attempting to calculate $\langle n(t)\rangle$ one encounters correlations of the form $\langle\varepsilon^*(t_1)\varepsilon(t_2)n(t_2)\rangle$ in which the field and atom variables can be decorrelated only for a phase-diffusion field. To solve the problem for the chaotic field, we have used the correlation functions of the field to obtain a series expansion for $\langle\varepsilon^*(t_1)\varepsilon(t_2)n(t_2)\rangle$. When substituted into the integral equation for $\langle n(t)\rangle$ and after an iteration procedure whose details will be presented elsewhere [23], we were able to obtain a series integral equation which can be written in a diagramatic form. Taking the Laplace transform, one can then express $\langle n(t=\infty)\rangle$ in terms of a continued fraction equivalent to the one of the first approach.

Before presenting some of the results obtained with the above approaches, it is worth discussing an analytical result that illustrates much of the physics involved. For a chaotic field of zero bandwidth, the continued fraction simplifies considerably and $\langle n(\infty)\rangle$ can be written as

$$\langle n(\infty)\rangle^{CH} = \langle n(\infty)\rangle^{PD}(1+\tfrac{1}{S})e^{1/S}\, E_1\,(1/S) \tag{1}$$

where $S = 1/2\ \Omega^2/(\Delta^2+1/4\ \kappa_2^2)$ is the usual saturation parameter of the TLS in a monochromatic field, $\Omega=2\hbar^{-1}\mu_{12}\varepsilon$ is the average Rabi frequency, and E_1 is the exponential integral. Here ^{CH} indicates chaotic and PD phase-diffusion. Note that $\langle n(\infty)\rangle^{PD} = -(1+S)^{-1}$. From the properties of the exponential integral one finds that $\langle n(\infty)\rangle^{CH}$ and $\langle n(\infty)\rangle^{PD}$ are equal to first order in S (weak field); for large S, $\langle n(\infty)\rangle^{CH} \approx -\ell n S/S$ while $\langle n(\infty)\rangle^{PD} \approx -1/S$. As expected, in both cases $\langle n(\infty)\rangle$ tends to zero as $S \to \infty$, but it approaches zero more slowly, for a chaotic field is less effective than the coherent field in saturating a one-photon transition. As we will see below, this turns out to be a basic feature of the chaotic field that persists for arbitrary bandwidth.

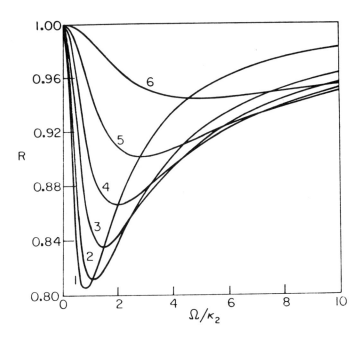

<u>Fig. 1</u> Saturation of a two-level system under a strong stochastic field. The ratio $R = \langle\rho(\infty)\rangle^{CH}/\langle\rho_{22}(\infty)\rangle^{PD}$ is shown as a function of Ω/κ_2 for $\Delta=0$. The curves numbered 1 to 6 correspond to field-bandwidths $b=0$, $0.2\kappa_2$, $0.5\kappa_2$, κ_2, $2\kappa_2$ and $5\kappa_2$

Let us now consider some of the results illustrating the effects of field fluctuations for chaotic fields of arbitrary bandwidth. In Fig. 1 we plot the ratio $R = \langle\rho_{22}(\infty)\rangle^{CH}/\langle\rho_{22}(\infty)\rangle^{PD}$ as a function of field intensity for various field bandwidths b. As we progress from curve 1 (corresponding to b=0) to curve 6 (corresponding to $b=5\kappa_2$) the bandwidth increases. For zero field, the ratio R is unity, but as the field increases, the ratio drops to a minimum value which depends on the bandwidth. Also the field strength at which the minimum occurs depends on the bandwidth. Curve 1 represents the analytical results of Eq.(1). The other curves have resulted from the numerical calculation of the continued fraction. Curve 2, for example, shows that for a bandwidth $b=0.2\kappa_2$ the minimum occurs at a field strength for which $\Omega\approx\kappa_2$. Clearly the results illustrated by these curves prove an assertion made earlier: The chaotic field is always less effective than a phase-diffusion field in saturating a resonant transition. It also implies that the decorrelation approximation will be inaccurate - by a maximum of 20% - if used for a chaotic field. The difference between chaotic and phase-diffusion fields decreases as the bandwidth increases becoming about 6% for b = $5\kappa_2$.

We have found similar behavior also when $|1>$ and $|2>$ are coupled through an N-photon transition. For weak fields, the chaotic field is more efficient than the coherent field in raising the system to state $|2>$. This is due to the well known enhancement [25] of the transition probability for a chaotic field by a factor of N!. As the field increases and saturation begins to set in, the chaotic field eventually becomes less efficient than the coherent. The weak field limit of R is N! in this case, while the strong field limit is again unity.

In Fig. 2 we present results for resonance fluorescence in a strong chaotic field. The dashed curves correspond to a phase-diffusion field and the solid curves to a chaotic field. In both cases, the calculation is for fields exactly on resonance and with bandwidth b =κ_2. Recall that for a monochromatic coherent field, the spectrum of the spontaneously emitted photons ω_k has the well known triplet structure with the central peak at the frequency of the strong field, and the sidebands shifted by an amount equal to the Rabi frequency. The overall structure is preserved in a phase-diffusion field of finite bandwidth. Under a chaotic field of the same band-width, the spectrum undergoes a dramatic change. The triplet structure is still there, but much less pronounced with only the central peak being clearly visible. The main reason for the near obliteration of the triplet structure is that the sidebands are smeared because the amplitude fluctuation causes the fluctuation of the Rabi frequency. The central peak is affected somewhat by the amplitude fluctuations, but it broadens mainly because of the finite bandwidth. Thus we have found that for a chaotic field of zero bandwidth, the central peak remains essentially unchanged while the triplet structure is smeared out. In that case, the splitting turns out to be equal to $\Omega/\sqrt{2}$. As is evident in Fig. 2, the on-resonance splitting for a chaotic field is smaller than that for a phase-diffusion field of the same bandwidth and intensity. The triplet structure of reso-nance fluorescence does therefore persist in a resonant chaotic field but is much less pronounced and perhaps not easily detectable. It does become slightly more visible with increasing average Rabi frequency as is evident in Fig. 2.

A few comments on the case of off-resonant excitation are of interest at this point. It is known that, for excitation by a detuned PD field, the spectrum becomes asymmetric (sidebands of unequal heights), the sideband closer to the atomic transition frequency being higher. For a CF of finite bandwidth, we again find asymmetric sidebands, but owing to the intensity fluctuations of the field, these sidebands are broader than they are for a phase-diffusion field. In addition, the central line is weaker than it would be for a PD field and its width is essentially determined by the bandwidth of the CF. Many more interesting differences in detail appear as one progresses from the PD field to the CF of zero and then of finite band-width. Space however does not permit their full discussion here. A complete analysis will be published elsewhere [22]. As is evident from Fig. 2 and the above discussion, the most dramatic effect of amplitude fluctuations on resonance fluorescence is the severe suppression of the sidebands. This suppression had been anticipated qualitatively by Avan and Cohen-Tannoudji [13] but the triplet structure is not washed out as their argument had suggested.

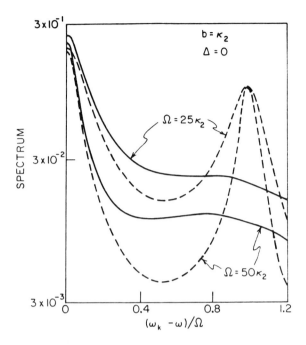

Fig. 2   The spectrum of resonance fluorescence under a strong stochastic field exactly on-resonance ($\Delta=0$) and of bandwidth $b=\kappa_2$. The dashed curves correspond to a PD field and the solid curves to a CF. The spectrum of the fluorescent photons ω_k is plotted as a function of $(\omega_k-\omega)/\Omega$ where ω is the atomic frequency and Ω the averaged Rabi frequency as defined in the text. Only half of the symmetric spectrum is shown.

We finally trun to the effects of field fluctuations on Stark splitting in double optical resonance (DOR). The TLS is again coupled to a near-resonant strong stochastic field $(\varepsilon(t)e^{i\omega t} + c.c.)$ while a second weak stochastic field $(\varepsilon'(t)e^{i\omega't} + c.c.)$ is near-resonant with a second trans-ition $|2> \rightarrow |3>$. In Fig. 3 we show results for strong chaotic fields of finite bandwidth and of two different intensities. In the same figure, re-sults corresponding to phase-diffusion fields of the same bandwidth and intensities are also shown. The most striking feature in Fig. 3 is the sub-stantial broadening of the peaks caused by the amplitude fluctuations of the CF. That this is not a result of the finite bandwidth only, becomes evident by noticing the difference from the corresponding curve for the PD field. The situation is reminiscent of and related to the broadening of the sidebands in Fig. 2. Our calculation corresponds to an experiment on Stark splitting in 3-photon ionization that we have analyzed in a recent paper [20] using a PD model. That analysis has left unanswered the question of the origin of the substantial broadening (beyond the laser bandwidth and ionization widths) that was evident in the experimental data. Our pre-sent results, as illustrated in Fig. 3, provide the answer. The field of the experiment did have amplitude fluctuations, although it probably was not chaotic. It must be stressed here that this additional broadening is

present even for a chaotic field of zero bandwidth since it is an effect
basically related to intensity fluctuations.

If the strong field is detuned from resonance, the Stark-splitting peaks
become asymmetric. As we have shown in previous work [16], the finite bandwidth of a PD field causes the reversal of this asymmetry. Qualitatively,
this can be understood as an enhancement of the transition $|1> \leftrightarrow |2> \leftrightarrow |3>$
due to the overlap of the atomic frequency ω_{21} with photons in the tail of
the Lorentzian spectrum of the strong field. The reversed asymmetry, being
a bandwidth effect, is also found in excitation with a CF of finite bandwidth.

A new feature, due to the CF, is that on resonance the effective Stark
splitting is less than it would be for a coherent field of the same intensity and bandwidth. This is apparent in Fig. 3 and can be shown to be
independent of the bandwidth. Again it basically is an effect related to
intensity fluctuations.

Another interesting aspect is the effect of the laser line shape. Hogan
et.al. [20] have reported experimental evidence for a reversal of the peak-
asymmetry in the doublet spectrum of DOR due to the finite bandwidth of
the exciting laser light. In that experiment, however, the asymmetry was
reversed for detunings of a few laser line widths, reverting back to
normal for larger detunings. The theory, on the other hand, which is based
on a Lorentzian laser line shape, predicts a reversed asymmetry for arbitrary detuning [16]. Physically, the reversed asymmetry originates from the
enhancement of the two-step process - caused by the overlap of the wing of
the spectrum with the resonant transition - in comparison with the off-
resonant two-photon absorption line of the doublet [16]. Thus the discrepancy between theory and experiment can be explained by the fact that the
actual laser line shape was not Lorentzian. Its wings certainly decreased

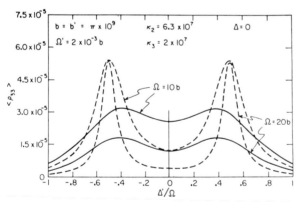

Fig. 3 AC Stark splitting under a resonant ($\Delta=0$) strong stochastic field
in double optical resonance. The averaged population $<\rho_{33}>$ of the third
level is plotted as a function of the detuning Δ' of the probe field
measured in units of Rabi frequency Ω. The dashed curves correspond to a
PD field and the solid curves to a CF. The primed quantities Ω' and b'
correspond to the probe field, and κ_3 is the spontaneous decay width of
level $|3>$. All widths are given in sec^{-1}.

much faster than a Lorentzian would. As a result such a spectrum is expected to lead to a reversed asymmetry only within a certain range of detunings.

Even laser theory predicts a cut-off for the Lorentzian laser line shape [27]: the phase $\phi(t)$ of a single mode laser can be shown to obey

$$\ddot{\phi}(t) + \beta\dot{\phi}(t) = F(t) . \tag{2}$$

$F(t)$ is a random Gaussian force with $\langle F(t)F(t')\rangle = 2\beta^2 b\delta(t-t')$. The spectrum of the laser described by (2) is Lorentzian with width b having a cut-off at frequencies $\beta \gg b$. Note that the commonly employed approximation, where $\ddot{\phi}$ is neglected in (2), corresponds to the limit $\beta \to \infty$ (phase-diffusion model). In this limit the lineshape is purely Lorentzian. We have calculated the spectrum of DOR for the laser described by (2) [28,29]. If the strong laser is tuned off-resonance from the transition $|1\rangle - |2\rangle$, which is probed by a second weak laser inducing population in level $|3\rangle$, the relative heights h_+ of the two-photon absorption line $|1\rangle\nleftrightarrow|3\rangle$ and h_- of the two-step line $|1\rangle \to |2\rangle \to |3\rangle$ are given by

$$h_+ \sim 1/(\kappa_2 + 2b)$$
$$h_- \sim [\tfrac{1}{4} \frac{\Omega^2}{\Delta^2} (\kappa_2 + 2b \frac{\beta^2}{\Delta^2 + \beta^2}) + 2b \frac{\beta^2}{\Delta^2 + \beta^2}] /\kappa_2(\kappa_2 + \kappa_3). \tag{3}$$

κ_2 and κ_3 are the natural widths of level $|2\rangle$ and $|3\rangle$. Ω is the Rabi frequency $|1\rangle - |2\rangle$ and Δ denotes the detuning of the intense laser from resonance. Equation (3) we derived under the assumption $\Omega \ll \Delta$ and $\beta \gg b$. For $2b < \kappa_2$ (3) predicts $h_+ \gtrless h_-$ in accordance with monochromatic theory. If $2b > \kappa_1$ and $\Delta \ll \beta$, i.e., the overlap of the laser with the transition $|1\rangle - |2\rangle$ is well within the Lorentzian part of the spectrum, the asymmetry is reversed ($h_+ < h_-$). For detunings larger than the cut-off of the laser line ($\Delta \gg \beta$), the asymmetry reverts back to normal ($h_+ > h_-$) in qualitative agreement with the experiment of Hogan et al [20]. Since the reversed peak asymmetry is basically a line shape effect, the present conclusions are expected to be valid even in the presence of amplitude fluctuations.

Acknowledgment

This work was supported by the National Science Foundation through Grant No. PHY78-23812. One of us (Peter Zoller) gratefully acknowledges support through the Max Kade Foundation and the Österreichische Fonds Zur Förderung der Wissenschaftl. Forschung under contract No. 2097.

References

1. P.A. Apanasevich, G.I. Zhovna, and A.P. Khapalyuk, J. Appl. Spectrosc. 8, 14 (1968).
2. L.D. Zusman and A. I. Burshtein, Sov. Phys. JETP 34, 520 (1972).
3. S.G. Przhibelskii and V.A. Khodovoi, Opt. Spectrosc. 32, 125 (1972).
4. S.G. Przibelskii, Opt. Cpectrosc. 35, 415 (1973).
5. Yu. S. Oseledchik, J. Appl. Spectrosc. 25, 1036 (1976).
6. G.S. Agarwal, Phys. Rev. Lett. 37, 1383 (1976), Phys. Rev. A 18, 1490 (1978).

7. H. J. Carmichael and D.F. Walls, J. Phys. B: Atom. Molec. Phys. $\underline{9}$, 1199 (1976).
8. J.H. Eberly, Phys. Rev. Lett. 37, 1387 (1976): J.L.F. de Meijere and J.H. Eberly, Phys. Rev. A $\underline{17}$, $\overline{14}16$ (1978).
9. W.A. McClean and S. Swain, $\overline{J}$. Phys. B: Atom. Molec. Phys. $\underline{10}$, L143 (1977).
10. H.J. Kimbel and L. Mandel, Phys. Rev. A $\underline{15}$, 689 (1977).
11. P. Zoller, J. Phys. B $\underline{10}$, L321 (1977); ib$\overline{id}$. 11, 805 (1978).
12. P. Zoller and F. Ehlotzky, J. Phys. B $\underline{10}$, 302$\overline{3}$ (1977).
13. P. Avan and C. Cohen-Tannoudji, J. Phys. B $\underline{10}$, 155 (1977).
14. S.G. Przhibelskii, Opt. Spectrosc. $\underline{42}$, 8 (19$\overline{77}$).
15. P.V. Elyutin, Opt. Cpectrosc. 43, 3$\overline{18}$ (1977).
16. A.T. Georges, P. Lambropoulos, $\overline{P}$hys. Rev. A $\underline{18}$, 587 (1978).
17. H. Walther, in Proceedings of the Second Laser Spectroscopy Conference, Megeve, France, 1975 (Springer-Verlag, Berlin 1975); see als review in Multiphoton Processes edited by J.H. Eberly and P. Lambropoulos (John Wiley and Sons, New York, 1978).
18. F.Y. Wu, R.E. Grove and S. Ezekiel, Phys. Rev. Lett. $\underline{35}$, 1426 (1975); see also review by S. Ezekiel and F.Y. Wu in Multiphoton Processes edited by J.H. Eberly and P. Lambropoulos (John Wiley and Sons, New York, 1978).
19. P. Agostini, A.T. Georges, S.E. Wheatley, P. Lambropoulos and M.D. Levenson, J. Phys. B $\underline{11}$, 1733 (1978).
20. P.B. Hogan, $\overline{S}$.J. Smith, A.T. Georges and P. Lambropoulos, Phys. Rev. Lett. $\underline{41}$, 229 (1978).
21. $\overline{P}$. Zoller, Phys. Rev. A $\underline{19}$, 1151 (1979) and in press.
22. P. Zoller, Phys. Rev. A $\overline{in}$ press.
23. A.T. Georges and P. Lambropoulos, Phys. Rev. A in press.
24. R.J. Glauber in "Quantum Optics and Electronics", edited by C. DeWitt et al. (Gordon and Breach Publishers, New York, 1965).
25. P. Lambropoulos in "Advances in Atomic and Molecular Physics", Vol 12, 87 (Academic Press, New York, 1976).
26. C. Lecompte, G. Mainfray, C. Manus and F. Sanchez, Phys. Rev. A $\underline{11}$, 1009 (1975).
27. H. Haken in Handbuch der Physik XXV/2c, edited by S. Flügge (Springer Verlag, Berlin, 1970).
28. P. Zoller and P. Lambropoulos, submitted to J. Phys. B.
29. S. Dixit, P. Lambropoulos and P. Zoller, to be published.

Part VI

Coherent Transients, Time Domain Spectroscopy

Two-Pulse Spectroscopy with Picosecond Laser Pulses

W. Kaiser, J.P. Maier, and A. Seilmeier

Physik Department der Technischen Universität München
D-8000 München, Fed. Rep. of Germany

Introduction

With the advent of powerful bandwidth limited light pulses of a
few picosecond duration it became possible to measure very short
population lifetimes of vibrational modes of polyatomic molecules
in the liquid phase./1/ Tunable infrared pulses or stimulated
Raman scattering were used successfully to generate an excess
population in certain molecular vibrations. Interrogation with
delayed probe pulses allowed to study the instantaneous degree
of the excitation. Two probing techniques have been applied in
a series of investigations: /1/ (i) Spontaneous anti-Stokes
scattering of the probe light gives direct information which
vibrational mode is excited and how the excess population develops as a function of time. Unfortunately, the anti-Stokes
signals are difficult to measure on account of the small Raman
scattering cross sections and the small excess population
generated with ultrashort pulses. Experiments were restricted
to neat liquids or highly concentrated systems. (ii) Quite
different is the situation when the photons of the probe pulse
are adjusted in frequency to make transitions from the excited
vibrational states to a fluorescent electronic state. Short
time-delays between the exciting IR-pulse and the (visible) probe
pulse allow to study the time behavior of the vibrationally excited molecule. This probing technique has the advantage of much
higher sensitivity making the investigation of highly diluted
systems possible. On the other side, the observed fluorescence
signal gives less specific information.

The fluorescence signal at fixed frequencies ν_1 and ν_2 does not
tell us accurately the probed vibrational energy state. It will
be shown in this note that the combination of frequency dependent
studies and time resolved measurements gives valuable new information on the vibrational modes observed and on the vibrational
lifetime involved.

Our experimental system is depicted in Fig. 1. Two pulses of
frequency ν_1 and ν_2, tunable over a wide frequency range, are
generated by two different parametric systems. A single pulse of
a mode-locked glass laser at 9455 cm^{-1} is divided into two pulses
via a beam splitter. One pulse produces the infrared pulse at

378

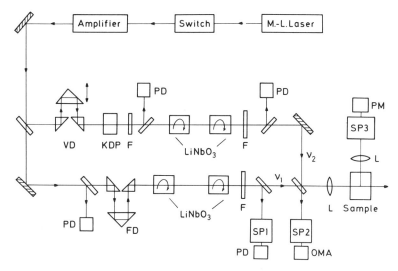

Fig. 1 Schematic of experimental set-up. A single
pulse of a Nd-glass laser at ν_L=9455 cm^{-1} generates
a tunable infrared (ν_1) and a tunable visible (ν_2)
pulse in two parametric generators consisting of
two LiNbO$_3$ crystals each. The two pulses excite
the sample selectivity. The fluorescence is monitored with the help of spectrometer SP3.

frequency ν_1 via parametric amplification in LiNbO$_3$ crystals./2/
The second pulse is first upconverted to the second harmonic
frequency at ν_{2L} = 18,910 cm^{-1} before it traverses two other
LiNbO$_3$ crystals for the generation of the pulse with frequency
ν_2, usually in the red part of the spectrum. Fig. 2 summarizes
several important parameters of the two parametric processes.
Frequency tuning is possible over several thousand cm^{-1}. The
bandwidth of the pulses depends on the frequency; in favorable
cases (e.g. around ν_1 = 3000 cm^{-1}) and under carefully chosen
experimental conditions the frequency width amounts to approximately 8 cm^{-1} for pulses of 4 ps duration.

A two-pulse system allows a variety of investigations (see Fig.3).
Tuning the IR-frequency ν_1 and holding the value of ν_2 constant
changes the excited vibrational states in the electronic ground
state and gives transitions to different Franck-Condon states in
S_1. The situation is simpler when the total energy $h\nu_1 + h\nu_2$ is
held constant and when both frequencies are changed to investigate different vibrational states in S_0. In a third set of experiments the IR-frequency is held constant and the frequency ν_2
is tuned through the bottom of S_1.

The transition rate from the initial to the final state is proportional to the product of the infrared dipole moment $<\chi_1^0 \, Q\chi_m^0>$
which connects the initial with the intermediate level and the
Franck-Condon factor $<\chi_m^0 \, \chi_f^1>$ which determines the transition
from the intermediate to the final state in the electronic state

Pulse Properties

	IR	VIS
Tuning Range	2500 to 7000 cm^{-1}	12500 to 17000 cm^{-1}
Frequency Width	~8 cm^{-1}	~7 cm^{-1}
Pulse Duration	~4 ps	~4 ps
Divergence	3 mrad	3 mrad
Intensity	10^9 W/cm^2	10^9 W/cm^2

Fig. 2 Important parameters of our para-
metrically generated pulses.

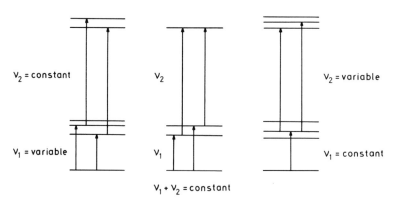

Fig. 3 Schematic of three types of investigations
using the two-pulse technique.

S_1. These two factors vary widely for different transitions. Two
examples relevant for the following investigations should be
mentioned here. In the IR range around 3000 cm^{-1} the C-H stretching modes have their prominent dipole moments. The corresponding
Franck-Condon (F-C) factors are known to be considerably smaller
than the F-C factors of the C-C ring modes which couple quite
well to the π-electrons of the chromophore. Frequently, the v=2
overtones of the C-C modes have the largest F-C factors. These
modes have their frequencies around 3000 cm^{-1} with smaller IR
dipole moments. The total transition via excited modes around
3000 cm^{-1} is a sum over different components. The many modes
with small IR dipole moment and small F-C factors give a broad

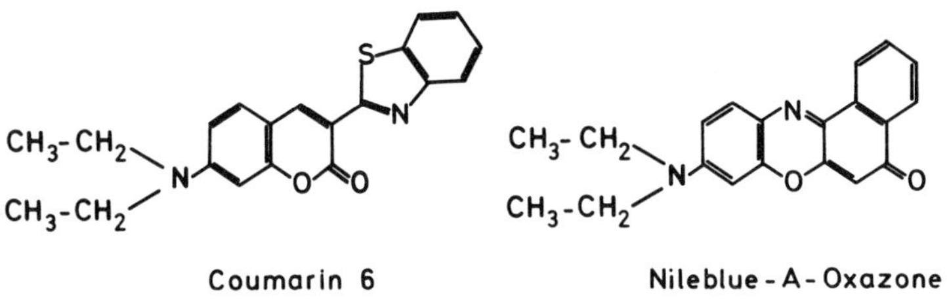

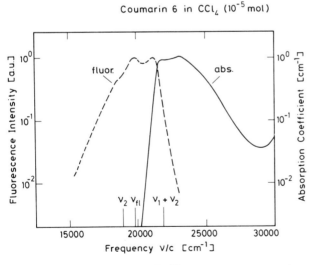

Coumarin 6 **Nileblue-A-Oxazone**

Fig. 4 The two molecules investigated here.

frequency independent contribution. The C-C ring modes should show up on account of their very large F-C factors and the C-H modes are candidates of spectroscopic prominence due to their large dipole moment. The experimental results presented below bear out these considerations.

Two molecules discussed in this paper are depicted in Fig.4. We note the CH_3 and CH_2 groups which will be seen in the following IR spectra.

In Fig. 5 the broad absorption and fluorescence spectra of Coumarin 6 in CCl_4 are presented. These one-photon spectra are typical for molecules of this size (42 atoms); they provide little information on the molecule and its interaction with the surrounding. It is our goal to learn more about medium size molecules in liquid solutions at room temperature. It will be shown in this paper that the ultrafast two-pulse technique allows us to obtain new information on these large molecules.

Fig. 5 Absorption and fluorescence spectra of Coumarin 6 (10^{-5} M in CCl_4).

It should be mentioned at this point that low temperature studies in solid matrices and in jet streams have offered new possibilities of higher resolution spectroscopy in medium size molecules. /3,4/ However, multiple-sites effects in solids and clustering in the jet stream require careful interpretation of the observed data. In Fig. 6 we present a two-pulse spectrum of nileblue-A-oxazone./5/ The frequency of the first pulse was held at $\nu_1 = 2935$ cm^{-1} while the frequency of the second pulse was tuned from 15,600 cm^{-1} to 16,150 cm^{-1}. The detailed two-pulse spectrum of Fig. 6b should be compared with the smooth one-photon absorption spectrum of Fig. 6a taken over the same total energy

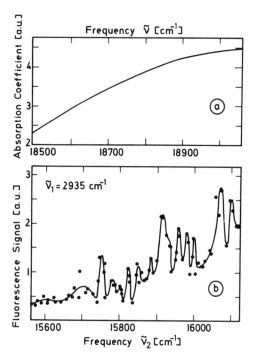

Fig. 6 (a) Expanded part of conventional absorption spectrum of nileblue-A-oxazone in the visible. (b) Ultrafast two-pulse spectrum of nileblue-A-oxazone (10^{-4}M in CCl$_4$). The observed fluorescence is plotted versus ν_2, the frequency of the red pulse; the frequency ν_1 of the infrared pulse is held constant at $\nu_1 = 2935$ cm^{-1}. Note the shift of the abscissa in Figs. a and b for ready comparison of equal values of the total excitation energy.

range. The structure of Fig. 6b is tentatively assigned to transitions involving mixed intermediate states consisting of skeletal modes with large F-C factors and low frequency bending modes where large dipole moments are borrowed. Experiments have shown that the observed structure in the spectrum is quite sensitive to the molecular surrounding. Next, a three-dimensional picture, signal versus ν_1 and ν_2, has to·be measured for a more detailed analysis of the spectral information.

A different spectrum is obtained when the total energy, $\nu_1+\nu_2$, is held constant and the infrared frequency ν_1 is tuned through the 3000 cm^{-1} range./6/ Fig. 7a shows distinct maxima and minima superimposed on a·background. Comparison with Fig. 7b depicting the standard one-photon infrared spectrum shows little common features. The prominent CH$_3$, CH$_2$, and CH modes in the IR spectrum

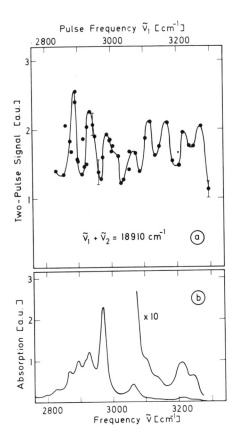

Pulse Frequency $\tilde{\nu}_1$ [cm^{-1}]

$\tilde{\nu}_1 + \tilde{\nu}_2 = 18910$ cm^{-1} (a)

Frequency $\tilde{\nu}$ [cm^{-1}]

Frequency $\tilde{\nu}_1$ [cm^{-1}]

$t_D = 0$ (a)

$t_D = 7$ ps (b)

(c)

$\times 10$

Frequency $\tilde{\nu}$ [cm^{-1}]

Fig. 7 (a) Ultrashort two-pulse spectrum versus frequency ν_1 of infrared pulse. The frequencies of ν_1 and ν_2 were adjusted to make the sum constant, $\nu_1 + \nu_2 = 18,910$ cm^{-1}. Nileblue-A-oxazone in CCl$_4$, 10^{-4} M. (b) Conventional infrared spectrum.

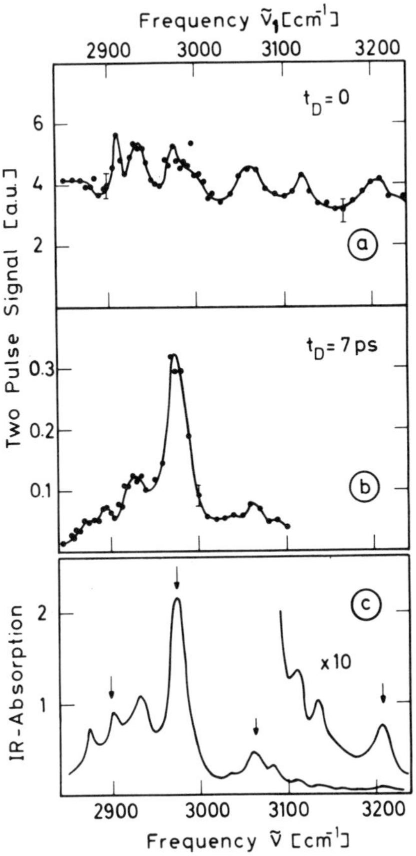

Fig. 8 (a) Two-pulse spectrum versus infrared frequency ν_1; $\nu_2 = 18,910$ cm^{-1}. No time delay t_D between both pulses. (b) Two-pulse spectrum measured with $t_D = 7$ ps. Note the similarity to 8c. (c) Conventional IR spectrum in the same frequency range.

do not show up in Fig. 7a. There is indication that overtones of the skeletal modes (factor 10 smaller in Fig. 7b) are related to the peaks in Fig. 7a. Supporting evidence for this notion comes from a series of investigations briefly summarized in Fig. 8 and 9. Coumarin 6 in CCl$_4$ was investigated with $\nu_2 = 18,910$ cm^{-1} and ν_1 tuned between 2850 cm^{-1} and 3250 cm^{-1}. The two-pulse IR spectrum is depicted in Fig. 8a. Again, the peaks of this spectrum do not correlate with the magnitudes of the one-photon IR spectrum of Fig. 8c. We have made time resolved measurements at various distinct IR frequencies marked by arrows in Fig. 8c. In Fig. 9 the corresponding curves are presented for ν_1 = 2894 cm^{-1}, 2976 cm^{-1}, 3065 cm^{-1}, and 3210 cm^{-1}. The pulse of frequency ν_1 interacts with the molecules. The excitation is interrogated by a pulse of frequency ν_2 after a time interval t_D. Fig. 9·shows four different curves. For three frequency settings we find two time constants, a fast decay of the signal and a subsequent slower one with time values of 6 ps and 8 ps. At a frequency ν_1 = 3210 cm^{-1}, on the other hand, the fluorescence signal decays quite rapidly, the curve represents the time resolution of our measuring system. It is important to note that at the frequencies ν_1 = 2894 cm^{-1}, 2976 cm^{-1}, and 3065 cm^{-1} we have CH$_2$, CH$_3$, and CH stretching modes, respectively, while at ν_1 = 3210 cm^{-1} an overtone of the skeletal modes is located (see Fig. 8c). It appears from Fig. 9 that the longer time constants of several picoseconds are related to the CH-stretching modes. Next, we have remeasured the IR two-pulse spectrum with a fixed delay time of t_D = 7 ps. After this time, the fast component of Fig. 9 has disappeared and we look at the signals with longer time constants. The delayed IR spectrum is depicted in Fig. 8b.

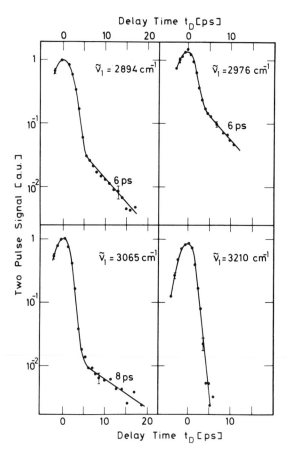

Fig. 9 Fluorescence signal versus delay time t_D between IR pulse at ν_1 and visible pulse at ν_2. Measurements were taken at four different IR frequencies of 2894, 2974, 3065, and 3210 cm^{-1}. The longer relaxation times correspond to CH stretching modes (see arrows in Fig. 8c).

Vibration	$\tilde{\nu}$ [cm^{-1}]	Direct T_1 [ps]	Linewidth $\Delta\tilde{\nu}$ [cm^{-1}] $\to$ T [ps]	
C-H Stretch	2894	6 ± 1	15	~ 0.7
v = 1	2976	6 ± 1	20	~ 0.5
	3065	8 ± 2	20	~ 0.5
C-C Ring				
v = 2	3210	≤ 0.5	35	~ 0.3
Combination Modes	2895 to 3065	≤ 0.5		

Fig. 10 Summary of observed time constants and data estimated from conventional line widths.

There is a striking similarity with the one-photon spectrum of Fig. 8c; the CH_2, CH_3, and CH are clearly resolved in the two-pulse spectrum. Obviously, the signals with longer time constants are proportional to the infrared dipole moment. We interpret our findings as follows. The ultrashort IR pulses around 3000 cm^{-1} lead to two distinctly different excitation processes. Interaction with the overtones of the C-C modes gives large total transition rates whith short time constants while the excitation of the CH-stretching modes gives smaller transition rates which show up on account of their longer time constant. We take Fig. 8b as a convincing prove that we have indeed excited the different CH-stretching modes.

In Fig. 10 we present a summary of our time constants and compare those with the linewidths of the one-photon IR spectrum. The bands of the CH modes are made up of inhomogeneous broadening and dephasing processes. The calculated time constants of less than one picosecond do not give information on the population lifetime measured by the two-pulse technique. The time constants of the overtones and combination modes are much shorter; an upper limit is given by the time resolution of 0.5 ps of our present system. This value is consistent with the broad IR bands. As an example the mode at 3210 cm^{-1} is listed.

We have made preliminary measurements on other molecules of similar size but different structure and obtained related spectra and time constants. The two-pulse technique appears to be a valuable tool to improve our knowledge of medium size molecules at room temperature. New experimental data become available which may not be obtained with standard spectroscopic methods.

/1/ A. Laubereau, W. Kaiser, Review of Modern Physics $\underline{50}$, 607 (1978)

/2/ A. Seilmeier, K. Spanner, A. Laubereau, W. Kaiser, Optics Commun. $\underline{24}$, 237 (1978)

/3/ E.V. Shpol'skii, T.N.Bolotnikova, Pure Appl. Chem. $\underline{37}$,183 (1974)

/4/ D.H. Levy, L. Wharton, R.E. Smalley, in Chemical and Biochemical Applications of Lasers, ed. C.B.Moore, Acad.Press (1977)

/5/ A. Seilmeier, W. Kaiser, A. Laubereau, S.F.Fischer, Chem.Phys. Lett. $\underline{58}$, 225 (1978)

/6/ A. Seilmeier, W. Kaiser, A. Laubereau, Optics Commun. $\underline{26}$, 441 (1978)

Spectroscopy of Photoreaction Centres by Tunable Picosecond Parametric Oscillators

S.A. Akhmanov[1], A.Yu. Borisov[1], R.V. Danielius[2], R.A. Gadonas[2], V.S. Kozlovskij[1], A.S. Piskarskas[2], and A.P. Razjivin[1]

[1]Moscow State University, Moscow, USSR
[2]Vilnius State University, SU-232054, Vilnius, USSR

1. Introduction

At present photosynthesis is the greatest solar energy converting system. Although its efficiency is rather poor (0.2 – 0.5 %), the amount of energy conserved in green biomass is immense, i.e. $\sim 3 \cdot 10^{21}$ J per year, which is ten times as much as the energy consumed by mankind in 1970. It is important to note that low efficiency is not the principal feature of photosynthesis. It is rather a consequence of the traditional organization of rural economy. In optimal physiological and illumination conditions this efficiency may be easily increased by one order of magnitude (!) and may reach 3–5 % (see review [1]).

It has been confirmed in many works that the input part of the photosynthesis apparatus acts as a photoelectric device, i.e. the light photon absorbed by Chl usually generates a pair: strong oxidant + strong reductant separated by a biomembrane. This pair is the precursor of subsequent endergonic biochemical processes, viz. electron and proton transport via photosynthetic charge transfer chain, formation of electrochemical transmembrane potential, formation of adenosine triphosphate etc. As shown in ref. [2], energetic evaluation concerning the efficiency of solar energy transformation in the primary stages : solar radiation ⟶ electrochemical energy which is performed by the chlorophyll-protein input part of the photosynthesis machinery, was carried out. The 10 % efficiency was derived. According to BOLTON [3] , the maximum possible efficiency of model cells of such type may reach even 15 – 17 % and 22 – 25 % for a single and double layer systems absorbing light in different optical regions, respectively.

The above said undoubtedly emphasizes the importance of detailed investigation of the molecular structure and function of photosynthesis, especially of its energy-collecting and transforming chlorophyllous part. The important fact that photoelectrochemistry of natural photosynthesis does without superpurity and supercrystallinity of light-absorbing Chl gives us hope that their technological artificial models may be elaborated on the basis of stable red-absorbing dyes and light-resistant polymers.

The light-conversion machinery in photosynthesis includes :
a) Chl-protein complexes that are responsible for light absorption and efficient funneling of singlet excited states to RCs ;

Abbreviations used below : Chl – chlorophyll;
BChl – bacteriochlorophyll; RC – reaction centre.

b) RCs which convert this energy into electrochemical energy of the redox pair separated in space by the biomembrane and stabilized in time to the level making it accessible to slow biochemical reactions limited by diffusion.

The process usually starts with photon absorption by antenna Chl molecule thus converting it into a singlet excited state. A great variety of dye molecules in various model systems (Chl among them) were shown, in numerous works, to exist in photoexcited singlet state for $10^{-8} - 10^{-9}$ seconds. First investigations of this important function for Chls in vivo provided data in the same region (1 - 2 ns) for purple, green bacteria and Chl a of plants (see re-view [4]) which was the consequence of restricted abilities of the phase fluorometric technique used. Then a time-lever method was elaborated for phase-fluorometry. This made it possible to discover, for the first time, the short-living components of Chl - fluorescence [5,6] which were usually masked by the nanosecond ones obser-ved earlier. The fluorescence yield and the lifetime of these photo-synthetic components were proved to correlate with the state of RCs and in active photosynthesis, reached 10 - 100 picoseconds for purple and green bacteria [5] and $\lesssim 30 - 70$ picoseconds for photosystem - I of plants [6] . It means that several dozens of picoseconds are sufficient for excitation energy to be funneled from 50 - 1000 antenna molecules to active RC and to be trapped with the efficiency of about 90 % (!). There has been no precedent of this kind in physics and photochemistry for random systems. Shortly afterwards, laser methods were involved for the solution of this prob-lem. The first data were suprisingly similar - approximately 30 - 50 picosecond fluorescence lifetimes in cases with different samples. Later, CAMPILLO and SHAPIRO discovered a serious source of artifacts in them due to the interaction between multiple excited states in Chls [7] . After a drastical reduction of the power of excitation monopulses, they succeeded in obtaining correct data for Chl fluor-escence lifetimes in vivo of the order of 100 picoseconds for two wild strains of purple bacteria Rh. sphaeroides [8] . As shown in ref. [9] , this lifetime was determined to be $\simeq 200$ ps for Rh. rubrum bacteria. Besides, trains of ps pulses were used in a num-ber of earlier works. This mode of work is associated with addition-al fluorescence quenching, as has been demonstrated in MAUZERALL and GEACINTOV laboratories [10, 11] .

On the other hand, succesful isolation by biochemists of pure photoactive RC particles from some purple bacteria made it possible to study the kinetics of excitation energy conversion into what now is widely adopted as P^F state [12 - 14] . According to the concep-tion currently accepted, it is a photoinduced state of RC with one electron being transferred to bacteriopheophytine (BPhe) molecule from the BChl dimer P 870 [13] . The risetime is shorter than 7 pico-seconds [14] . The lifetime approximates 120 - 250 ps [15,16] and reflects subsequent electron withdrawel from BPhe to the quinone molecule of RC [14] providing its essential separation from the parent BChl dimer and the corresponding stabilization in time.

It was accepted that multiple excitons created by the excitation laser pulse cannot change the photosynthesis process the way it had been known in luminescence studies, because the first photon successfully absorbed by RC should convert it into P^F state that is safe from interaction with the next ones.

Our preliminary experiments with picosecond kinetics at different wavelengths, howe ver, revealed their marked dependence on the quantity of light photons per RC illuminated by the picosecond laser pulse [17]. That is why we have undertaken a detailed study of this problem.

2. Experimental set-up

Experimental data of picosecond investigations of the primary stage of photosynthesis in RC presented below are obtained by picosecond absorption spectrometer produced and applied by the Department of Astronomy and Quantum Electronics of Vilnius State University. In contrast with picosecond absorption spectrometers known from earlier works [15, 16], the above mentioned spectrometer enabled us to obtain a wide wavelength tuning range of exciting and probing pulses. This achievement is due to the application of picosecond parametric oscillators (PPO) presented and described in our earlier publications [18 - 21].

Fig. 1 presents the main experimental sketches of PPO.

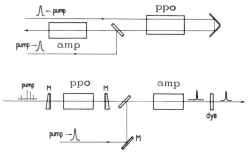

Fig. 1   Traveling wave and synchronously pumped single pulse PPO s

The upper scheme represents the travelling-wave parametric oscillator directly pumped by the single picosecond pulse. Spectral narrowing and energy increase take place in the second crystal, pumped independently.

The lower scheme represents synchronously pumped cavity PPO in which the pumping train of regular picosecond pulses has been used. Single pulse of tunable wavelength is selected in the parametric amplifier, independently pumped by the strobing pulse.

The main characteristics of the investigated PPO are selected and presented in Table 1. The choice of a definite PPO depends on the peculiarities of the point in question. To cover the visible region, KDP PPO pumped by the third harmonic (λ = 355 nm) or ADP PPO pumped by the fourth harmonic (λ = 266 nm) give optimal issues. Using the fourth harmonic with the single crystal scheme, the broad spectral band radiation covering almost all visible range is obtained. The results may be of importance in the registration of ultrafast absorption changes that take place simultaneously in broad spectral range. The main advantage of the broad band PPO radiation in comparison with picosecond continuum excited in solutions is a smooth spectral envelope and high reproducibility. $LiNbO_3$ crystal PPO directly pumped by YAG : Nd^{3+} laser radiation enables to obtain continuous wave length tuning in near infrared region up to 4.8μ. The PPO synchronously pumped by the second harmonic of picosecond

Table 1   Powerful two-crystal picosecond parametric oscillators

Nr.	Crystal	Matching type	Pump wave length [nm]	Energy conversion [%]	Tuning range [μ]	Spectral band-width [cm⁻¹]	Pulse-width [ps]	Energy [m]] ($\varkappa$ = 0,4)
1	KDP	eoe	532	51	0,78-1,4	6-8	20-25	5-7
2	KDP	eoe	355	15	0,45-1,4	6-8	25	0,8-1,5
3	LiNbO₃	eoo	532	3,5	0,66-2,7	10	25	0,5-0,8
4	LiNbO₃	eoo	1064	17	1,4-4,8	50	33	8-10
5	ADP	eoo	266	10	0,45-0,65	20-25	25	0,2-0,5
6*	ADP	eoo	266	5	visible	5300	20-25	0,1-0,2
7**	α - HJO₃	eoe	530	7	0,7-2,2	5-6	3-4	0,2
8	KDP	eoe	530	30	0,78-1,4	6-8	3-4	1-2

*   Single crystal PPO
**   PPO sinchronously pumped by the second harmonic train of picosecond pulses obtained from phosphate glass : Nd[3+]   laser with p.r.r. 2 Hz.

phosphate glass : Nd[3+]   laser is of great interest because of low oscillation threshold, spectral band-limited pulses and beam divergency near diffraction limits.

The absorption spectrometer scheme used in our picosecond measurements in RC is presented in Fig. 2   and discussed below. Passive mode-locked YAG : Nd[3+]   laser with intracavity single pulse selector was used as the driving oscillator. Pulse repetition rate of up to 25 Hz was attainable. The energy of the single pulse has been amplified to 40-60 mJ by two cascade amplifiers. Frequency doubling took place in KDP crystal (ℓ =   1 cm) with conversion efficiency of 40-50 %. The beam of the second harmonic was split then into two

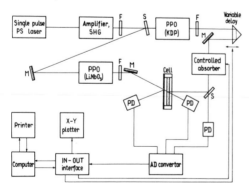

Fig. 2   Experimental set-up of picosecond spectrometer

beams with intensities ratio 5 : 1. The more intensive beam pumped the PPO of excitation channel built in by two-crystal (tandem) scheme. The second beam pumped two-step travelling-wave PPO of the probing channel. In this case $LiNbO_3$ crystals have been used and placed in the oven for the temperature tuning of wave length. The exciting and the probing beams were incidental on the cell to its opposite sides. The probing beam was limited by the diafragm of 0.3 - 0.5 mm in diameter. The exciting beam was much broader what insured homogeneous excitation of the volume under probe. The automatic delay line in the exciting channel provided time scanning with the accuracy of 1 ps in the 500 ps range.

The measurement of the cell absorption and the intensity of the excitation was carried out by the registration set-up which involved photodiodes and amplitude analysers AI-256-6 and provided for data storing and its averaging according to the fixed quantity of the laser flashes.

In order to provide for a higher absorption resolution (0.005 opt. units) and the accuracy of the measurement of ultrafast absorption changes, a computerized system with data storing and processing was employed. Mini computer D-3-28 controled the three step-motors which drove both the optical delay line, the attenuator of excitation energy as well as the monochromator. The final results of the experiment were displayed by the printer or the x-y plotter.

The presented absorption spectrometer enabled the accomplishment of picosecond measurements at the excitation and probing regions of 790-1600 nm and 660-2700 nm, respectively, with the time resolution of 20 ps. The possible operation regimes may be used :
a) automatic recording of ps kinetics of absorption changes,
b) automatic recording of photobleaching curves,
c) recording of photoinduced absorption spectrum with ps time resolution,
d) experiments with tandem pulse excitation.

3. Results and Discussion

In most earlier works on laser application in photosynthesis, investigated objects were illuminated by 530 nm pulses. We were the first to realize the selective excitation and the probing of RC preparations in their major absorption maxima [22] . The dependence of transmittance in 870 nm band on the pulse energy is shown in Fig. 3 (solid points).

The absolute quantum yield φ_p of primary P870 photooxidation was calculated from these data as the ratio of P870 oxidized to the number of photons absorbed (Fig. 3 , open circles). The dependence becomes more sensitive being represented versus PRC parameter (photons per reaction centre during one excitation pulse). One can see that φ_p tends to reach the unity if PRC ≤ 1, but drops drastically in region PRC > 1. We believe that fast biexcitation processes emerge in these conditions and become prevailing. The principal difference between signals induced by weak and strong excitation monopulses (PRC ≤ 1 and PRC >> 1 correspondingly) is also demonstrated in the kinetics of fast absorption changes at 802 nm (compare solid points and open circles in Fig. 4). The absorption of quantum by RC results in charge separation state which vanishes with $\tau \sim$ 200 ps (solid points). In the case of strong laser pulse, excited RC can absorb some more quanta just after the first one due to the remaining absorption (≈ 10-20 %) at 870 nm. This process is

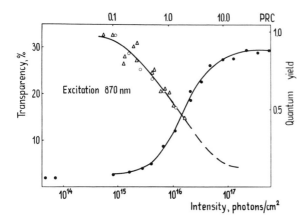

Fig. 3   Optical transmittance of RC at 870 nm (●) and absolute
quantum yield of primary RC photooxidation (o) vs   870 nm exciting
pulse energy and PRC parameter. Initial absorption of the sample
≈ 1.7 at optical pathlength 1 mm

accompanied by absorption increase at 802 nm. These two differing
kinetics favor the idea that absorption of additional quantum converts
RC into a state different from the initial one natural for photosynthesis. The comparison of photobleaching curves around 870, 805 and
750 nm (Fig. 5) strengthen this point of view.

The kinetics at 748 and 810 nm were also shown to depend upon
PRC parameter. The above data put forward the question concerning
the nature of RC state responsible for the absorption of   subsequent
photons.

We believe that interaction of excitations belonging to different
RCs is hardly probable because at least four conditions should be
maintained at the same time : a) RC close association; b) P870
and I⁻ being dopped not deeper than ≈ 7-8 Å in the protein matrix;
c) specific adjoining of reaction centres in pairwise fashion with
[P870 I] containing open parts; d) simultaneous existing of excited states in adjoining centres.

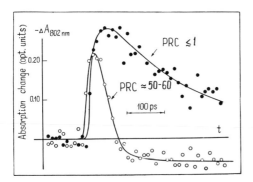

Fig. 4   Kinetics of absorption changes of RC at 802 nm induced by
weak (●) and strong (o) 870 nm exciting pulses

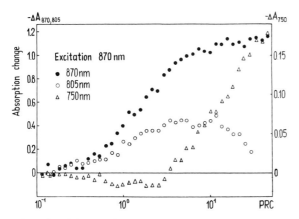

$-\Delta A_{870,805}$ ... $-\Delta A_{750}$

Excitation 870 nm
● 870nm
○ 805nm
△ 750nm

Absorption change

1.2 1.0 0.8 0.6 0.4 0.2 0

0.15 0.10 0.05 0

10^{-1} 10^0 10^1 PRC

Fig. 5   Absorption changes of RC at 870 nm (●), 805 nm (o) and 750 nm (△) vs PRC

Four general schemes may be proposed for events within the RC particle following the second photon absorption :

a) $\quad [\text{P870 I}] \xrightarrow{h\nu_1} [\text{P870* I}] \xrightarrow{h\nu_2} [\text{P870** I}]$

b) $\quad [\text{P870 I}] \xrightarrow{h\nu_1} [\text{P870* I}] \xrightarrow{k_e} [\text{P870}^+ \text{ I}^-] \xrightarrow{h\nu_2} [\text{P870}^+ \text{ I}^-]$ *

c) $\quad [\text{P870 I}] \xrightarrow{h\nu_1} [\text{P870* I}] \xrightarrow{k_e} [\text{P870}^+ \text{ I}^-] \xrightarrow{h\nu_2} [\text{P870}^+ (\text{I}^-)*]$

d) $\quad [\text{P870 I}] \xrightarrow{h\nu_1} [\text{P870* I}] \xrightarrow{k_e} [\text{P870}^+ \text{ I}^-] \xrightarrow{h\nu_2} [(\text{P870}^+)* \text{ I}^-]$

Here $h\nu_1$ and $h\nu_2$ are the first and second quanta absorbed by the same RC ; * signifies excited state ; ** is higher excited state due to biphoton excitations; I, I⁻ — primary porphyrinous electron acceptor in normal and photoreduced state; P870, P870* — BChl dimer of RC in normal and photoexcited state; P870⁺, (P870⁺)* — the cation of P870 in normal and photoexcited state; (I⁻)* — the anion of I in photoexcited state; [P870⁺ I⁻] , usually designated in literature as P^F state.

Besides, one should bear in mind that the lifetime of P870* is restricted within 7 ps according to [14] and that the duration of monopulses used in this work ⋞ 30 ps is shorter than the lifetime of P^F state.

Photoreaction a) requires either the lifetime of P870* being near the upper limit (5~7 ps) or, as was pointed out by Dr. GODIK, energy gap between [P870* I] and [P870⁺ I⁻] states within 50—60 meV which will poise quasi equilibrium between them with the portion of [P870* I] ⋟ 10 %. The data of [9] for fluorescence life time of RC equal to 15 ± 8 ps favors this version, but recent data for above energy gap obtained by SHUVALOV (≃ 120 meV [23]) contradict it. Therefore (a) cannot yet be rejected, especially for powerful pulses with PRC ⋟ 100. It is quite probable that in such conditions more than one quantum may be successfully absorbed by photoexcited reaction centre during one pulse. For moderate PRC ⋞ 10, however, one of photoreactions (b), (c) and (d) should evidently work because the extinction in 870 nm band drops after $h\nu_1$ absorption and the lifetime of P^F state exceeds greatly that for P870*.

The following experiments lend direct support to the conception of $h\nu_2$ interaction with RC in P^F state. After a weak 870 nm pulse the second powerful one comes with delay characteristic of P^F state lifetime (50—500 ps). It results in essential changes of 802 nm transmittance thus proving that the interaction with P^F (no P870* can survive after such delay) converts reaction centre into some different state (see Fig. 6).

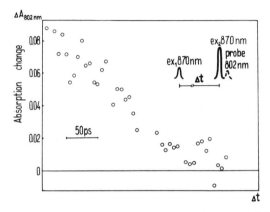

Fig. 6   Absorption changes of RC at 805 nm vs delay time between 870 nm exciting pulses in double pulse tandem excitation scheme

Only a weak absorption increase was observed in bacteriopheophytine band at 748 nm (see open circles in Fig. 8) for PRC values near the unity. Stronger monopulses (PRC $\stackrel{>}{\sim}$ 10), however, induced noticeable transparency (see solid points in Fig. 8). The same conclusion may be derived after analyzing the corresponding light curve in Fig. 5. Thus it appears that the signal which was interpreted by many authors as due to bacteriopheophytine photoreduction in the primary photoact reflects multiexcitation conversion of RC pigment complex. No similar phenomena were observed at 870 and 800 nm bands. A considerable part of 870 nm band still undergoes bleaching at PRC $\simeq$ 1 (see Fig. 3).

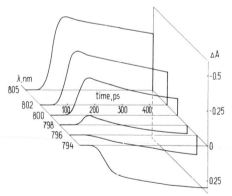

Fig. 7   3-dimensional pattern of absorption changes in P800 band of RC vs   time and wavelength when excited by weak 870 nm pulses

In Fig. 7 , a series of kinetics within 800 nm band is presented in a form of three-dimensional pattern. It is seen that this band exhibits a blue shift and partial photobleaching just after the photon absorption. The latter decays in several hundreds of picoseconds.

4. Main Conclusions

1. Illumination of photosynthetic reaction centres with picosecond laser pulses may result in multistep absorption by separate RC what leads to nonlinear physical processes competing with the primary charge separation of photosynthesis. These processes differ from the previous one in spectral responses, so that one should be careful not to ascribe them to the natural sequence of photosynthesis events.

In our opinion, these nonlinear processes may also provide useful information concerning the mutual localization and interaction of 6 porphyrine molecules involved into RC.

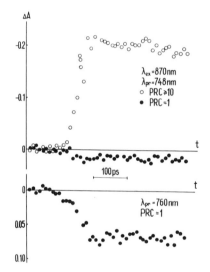

Fig. 8 Kinetics of absorption changes of RC in 760 nm band induced by weak (o) and strong (o) 870 nm exciting pulses

2. In the light of point 1 above, the PRC (photons per reaction centre) parameter should be introduced into laser absorption studies of picosecond photosynthesis processes. It means the quantity of photons absorbed per separate reaction centre illuminated during one picosecond monopulse.

3. The picosecond data from variuos laboratories may be comparable only when obtained either at approximately equal PRC or at PRC $\lesssim$ 1 which is the best because it excludes possibility of nonlinear processes.

4. In laser picosecond studies of reaction centres excitation of 870 nm absorption band should be undoubtfully preferred because this band bleaches after the absorption of the first photon, thus protecting

RC from the others. In this case more powerful pulses can be used without involving nonlinear processes.

5. It was demonstrated that under conditions when RC in P^F state absorbs one more 87o nm quantum (PRC $\geqslant$ 10) it converts into some state involving bacteriopheophytine molecule. Corresponding spectral response is similar to those observed in refs. [11, 14].

6. Our experiments with reaction centres from Rh. rubrum have shown that fast changes of optical density around 750 nm (which were currently ascribed to bacteriopheophytine reduced in the course of P^F formation) nearly disappeared after we succeeded in lowering PRC parameter down to the unity. At the same time the absorption changes at 870 nm, which manifest primary electron photoexpelling from P870, survived. The above said supports an important conclusion viz. that it is not the bacteriopheophytine molecule which acquires this photoelectron.

The former conception should be reconsidered. Experiments aiming at the identification of the primary electron acceptor in bacterial photosynthesis are in progress.

Acknowledgement

We are grateful to Drs. GODIK and SHUVALOV for valuable discussion of the material presented.

References

1. D.O. Hall, FEBS Letters, 64, p.6 (1976).
2. A.Yu. Borisov, in "Photosynthesis in relation to model systems", Elsavier, Y.Barber ed., v.3, pp.1-26 (1979).
3. J.R.Bolton, Science, 202, p.705 (1978).
4. A.Yu.Borisov, V.I.Godik, Biochim.Biophys.Acta, 301, p.227 (1973).
5. a) A.Yu.Borisov, V.I.Godik, Biochim.Biophys.Acta, 223, p.441 (1970).
   b) A.Yu.Borisov, V.I.Godik, J.Bioenerget., 3, p.211 (1972).
6. A.Yu.Borisov, M.D.Il'ina, Biochim.Biophys.Acta, 305, p.264 (1973).
7. A.J.Campillo, V.H.Kollman, S.L.Shapiro, Science, 193, p.227 (1976).
8. A.J.Campillo, R.C.Hyer, T.G.Monger, W.W.Parson, S,L.Shapiro, Proc. Natl.Acad.Sci.U.S., 74, p.1997 (1977).
9. V.Z.Paschenko, A.A.Kononenko, S.P.Protasov, A.B.Rubin, L.B.Rubin, N.Y.Uspenskaya, Biochim.Biophys.Acta, 461, p.403 (1977).
10. D.Mauzerall, Biophys.J., 16, p.87 (1976).
11. N.E.Geacintov, J.Breton, Biophys.J., 17, p.1 (1977).
12. W.W.Parson, R.K.Clayton, R.J.Cogdell, Biochim.Biophys.Acta, 387, p.265 (1975).
13. J.Fajer, D.C.Brune, M.S.Davis, A.Forman, L.D.Spaulding, Proc.Natl.Acad. Sci.U.S., 72, p.4956 (1975).
14. K.J.Kaufman, K.M.Petty, P.L.Dutton, P.M.Rentzepis, Biochem.Biophys. Res.Comm., 70, p.839 (1976).
15. K.J.Kaufman, P.L.Dutton, T.L.Netzel, J.S.Leigh, P.M.Rentzepis, Science, 188, p.1301 (1975).
16. M.G.Rockley, M.W.Windsor, R.J.Cogdell, W.W.Parson, Proc.Natl.Acad. Sci.U.S., 72, p.2251 (1975).

17. S.A.Akhmanov,A.Yu.Borisov,R.V.Danielius,V.S.Kozlovskij,A.S.Pis-karskas,A.P.Razjivin, in "Picosecond phenomena", Berlin,C.V.Shank ed. pp.134-139 (1978).

18. K.P.Burneika,M.V.Ignatavichius,V.J.Kabelka,A.S.Piskarskas,A.Yu. Stabinis, Pisma v JETP,16,p.365 (1972).

19. R.Danielius,G.Dikchius,M.Ignatavichius,V.Kabelka,N.J.Koroteev and A.Piskarskas, Opt.Commun.18,p.22 (1976).

20. R.Danielius,G.Dikchius,V.Kabelka,A.Piskarskas,A.Stabinis,J.Jasevich-iute, Kvantovaja elektronika,4,p.2379 (1977).

21. R.Danielius,G.Dikchius,V.Kabelka,A.Piskarskas,A.Stabinis,J.Jasev-ichiute, Litovskij fizitcheskij sbornik,18,p.93 (1978).

22. S.A.Akhmanov,A.Yu.Borisov,R.V.Danielius,A.S.Piskarskas,A.P.Raz-jivin,V.D.Samuilov (1977) Pisma v JETP,26,655.

23. V.A.Shuvalov,V.V.Klimov, Biochim.Biophys.Acta,440,p.587 (1976).

Ultraslow Optical Dephasing of Pr^{3+}: LaF_3[1]

R.G. DeVoe, A. Szabo[2] S.C. Rand, and R.G. Brewer

IBM Reserach Laboratory, 5600 Cottle Road
San Jose, CA 95193, USA

Abstract: Optical free induction dephasing times as long as 16 μsec, corresponding to an optical homogeneous linewidth of 10 kHz, have been observed for the $^3H_4 \leftrightarrow {}^1D_2$ transition of Pr^{3+} ions in LaF_3 at 2°K. Measurements are facilitated by a frequency-locked cw dye laser and a new form of laser frequency switching. Zeeman studies reveal a Pr-F dipole-dipole dephasing mechanism where the Pr nuclear moment is enhanced in both $^1D_2$ and $^3H_4$.

In this Letter, we report a new advance in the observation of extremely long optical dephasing times in a low temperature solid. Coherently prepared Pr^{3+} impurity ions in a LaF_3 host crystal exhibit optical free induction decay (FID) where the dephasing times correspond to an optical linewidth of only 10 kHz half-width half-maximum and a spectral resolution of 5×10^{10}. At this level of resolution, which represents a fifty fold increase over our previous measurements [1], it is now possible to perform detailed optical studies of magnetic Pr-F dipole-dipole interactions in the ground and optically excited states. Heretofore, such weak relaxation effects could be detected only in the ground state by spin resonance techniques [2-4] or radio-frequency optical double resonance [5,6].

The Pr^{3+} transition $^3H_4 \leftrightarrow {}^1D_2$ monitored at 5925Å involves the lowest crystal field components of each state. These are singlet states where the 2J+1 degeneracy is lifted by the crystalline field due to the low Pr^{3+} site symmetry, perhaps C_2 or C_{2v}. The nuclear quadrupole interaction [7] of Pr^{3+} (I=5/2) splits each Stark level into three hyperfine components which are each doubly degenerate ($\pm I_z$), and to a first approximation, three equally probable optical transitions connecting these states occur, namely, $I_z'' \leftrightarrow I_z' = \pm5/2 \leftrightarrow \pm5/2$, $\pm3/2 \leftrightarrow \pm3/2$, and $\pm1/2 \leftrightarrow \pm1/2$. All three transitions overlap and can be excited simultaneously by a monochromatic laser field since the Pr^{3+} hyperfine splittings of order 10 MHz are considerably less than the inhomogeneous crystalline strain broadening of ~5 GHz. Weaker transitions of the type $5/2 \leftrightarrow 1/2$, $5/2 \leftrightarrow 3/2$, ... also occur among these hyperfine states because of a nonaxial field gradient at the Pr^{3+} nucleus which mixes the $|I_z\rangle$ wavefunctions slightly. As noted previously [1,8], the weaker transitions redistribute the ground state

---

[1]Work supported in part by the U.S. Office of Naval Research.
To appear in Physical Review Letters June 4, 1979.

[2]On leave from the National Research Council of Canada, Ottawa.

hyperfine population distribution drastically in an optical pumping cycle, and play an important role in the optical dephasing measurements reported here.

BLEANEY [9] has shown that when an electronic singlet of a rare earth ion admixes with close lying Stark split levels of a given J manifold, it produces in second order a pseudo-quadrupole moment and an enhanced nuclear magnetic moment

$$m_i = (g_N\beta_N - 2g_J\beta\Lambda_{ii})I_i \ ,$$

(1)

where the notation is that of TEPLOV [4]. Here, the principal axes are labeled i=x,y,z, the nuclear and electronic g values are g_N and g_J, the electronic matrix element $\Lambda_{ii}=\sum_{n\neq 0} A_J |<0|J_i|n>|^2/(E_n-E_0)$ connects the lower state $|0>$ with an excited state $|n>$ removed in energy by E_n-E_0, and A_J is the Pr^{3+} hyperfine constant. Now imagine that a fluctuating local magnetic field $\tilde{H}_z$ exists at the Pr^{3+} site due to distant pairs of F nuclei participating in mutual spin flips, and ignore other dephasing mechanisms for the moment. This field modulates the optical transition frequency randomly through a Pr-F dipole-dipole interaction and produces a HWHM homogeneous optical linewidth

$$\Delta\nu = |\gamma_z'' I_z'' - \gamma_z' I_z'| \tilde{H}_z/2\pi$$

(2)

where γ_z'' and γ_z' are the Pr^{3+} *enhanced* gyromagnetic ratios ($\gamma_z=m_z/hI_z$) of $^3H_4$ and $^1D_2$. Because the Pr nuclear wavefunctions are mixed to some extent [7], rigorously I_z is not a good quantum number. Nevertheless, to a good approximation [5,8] $I_z''\sim I_z'$ and as already mentioned, we expect three strong optical transitions $|\pm 5/2>\rightarrow|\pm 5/2>$, $|\pm 3/2>\rightarrow|\pm 3/2>$, and $|\pm 1/2>\rightarrow|\pm 1/2>$. Therefore, from (2) three different decay times should appear in an optical FID. We shall see that this idea is supported and that γ_z' for $^1D_2$ can be obtained since γ_z'' is known [5] and $\tilde{H}_z=2\pi\Delta\nu_{rf}/\gamma_z''$ can be deduced from an rf-optical double resonance linewidth [6] of the $^3H_4$ state. Furthermore, these experiments offer a new way of testing ab initio calculations [10] of Λ_{ii} as well as the Pr^{3+} site symmetry, which remains controversial [10,11].

The technique adopted for observing optical FID relies on laser frequency switching [12], but in a new form. A cw dye laser radiates a beam at 5925Å which is linearly polarized at a power of ~4 mW. The beam passes through a lead molybdate acousto-optic modulator which is external to the laser cavity and oriented at the Bragg angle. The Bragg diffracted beam is focused to a 200 micron diameter in a 7×7×10 mm^3 crystal of Pr^{3+}:LaF_3 (0.1 or 0.03 atomic % Pr^{3+}) which is immersed in liquid helium at 2°K, and the emerging laser and FID light, which propagates parallel to the crystal c axis, then strikes a PIN diode photodetector. The Pr^{3+} ions are coherently prepared while the modulator is driven continuously and efficiently at 110 MHz. FID follows when the rf frequency is suddenly shifted (100 nsec rise time) from 110 to 105 MHz, the duration of the switching pulse being 40 µsec. Note that the laser is switched 500 homogeneous linewidths. Figure 1 shows FID signals produced in this way where the dephasing time $T_2/(1+\sqrt{1+\chi^2 T_1 T_2})\sim T_2/2$ is independent of power broadening since $\chi^2 T_1 T_2 <<1$, χ being the Rabi frequency. The anticipated heterodyne beat of 5 MHz frequency is readily observed because the shifted laser and FID beams overlap due to the change in the Bragg angle (0.4 mrad) being less than the beam divergence (7 mrad). This type of extra-cavity laser frequency switching is compatible with laser frequency locking which we now consider.

To detect ultraslow dephasing times by FID, the laser frequency must remain fixed within the sample's narrow homogeneous linewidth $\Delta\nu=1/(2\pi T_2)$ for an interval $\sim T_2$ - a stability condition which is less stringent than in a linewidth measurement. In the present work, a frequency stability of $\sim$10 kHz in a time of $\sim$16 μsec is required. To this end, our laser is locked to an external reference cavity which provides an error signal in a servo loop of high gain for correcting slow frequency drift and high frequency jitter. The noise spectrum as seen from the error signal or a spectrum analyzer is not flat but is dominated by isolated jumps of 30 to 100 kHz in a 10 μsec period. At such times, the sample is prepared at two (or more) discrete frequencies which result in a deeply modulated FID pattern. This behavior agrees with a computer simulation of FID which assumes a bimodal spectrum. However, at other times frequency jumps do not occur, and the free induction decays monotonically as in Fig. 1. Under these conditions, a laser jitter of <10 kHz permits a reliable decay time measurement of these *single events* which are considerably longer-lived than the time-averaged value of many decays. These signals are captured with a Biomation 8100 Transient Recorder and then reproduced on an X-Y chart recorder.

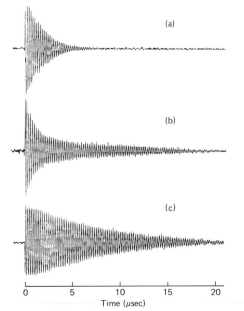

Fig. 1 Free induction decay of 0.1 atomic % Pr^{3+} in LaF_3 at 2°K in the presence of an external magnetic field $H_0 \propto c$ axis. H_0 equals (a) 0.5G (earth's field), (b) 19G and (c) 76G. The optical heterodyne beat frequency is 5.005 MHz. Cases (a) and (c) are plotted in Fig. 2

A key feature of the measurement is an optical pumping absorption-emission cycle which transfers population from any given hyperfine level of the $^3H_4$ ground state to its two neighbors, for example from $|3/2\rangle$ to

|5/2> and |1/2> within the same inhomogeneous packet. As a result, each of the three $^3H_4$ hyperfine states excited (three packets) will be depleted and FID cannot be detected. However, by sweeping the laser frequency at a slow rate of ≤10 kHz/16 µsec so as not to influence the decay rate, the pumping cycle can be reversed [1] and the hyperfine population partially restored. The $^3H_4$ hyperfine population distribution which results depends on the sweep rate and the relative transition probability among the hyperfine states as they decay from $^1D_2$ to $^3H_4$ via intermediate states. Therefore, the pumping cycle dictates which of the three strong transitions can be prepared to yield FID.

In Fig. 1, a dramatic variation in the FID occurs when a weak external field H_0 is applied perpendicular to the crystal c axis. The T_2 dephasing times for the three cases are (a) 3.6 µsec at H_0=0.5G (earth's field), (b) 3.5 and 15.6 µsec at H_0=19G, and (c) 15.8 µsec at H_0=76G. Note that case (c) corresponds to a 10 kHz HWHM linewidth which appears to be the *narrowest homogeneously broadened optical transition detected in a solid.* Its magnitude is comparable to NMR linewidths [2,4,6] which result from a magnetic dipole-dipole dephasing process. Cases (a) and (c) are single exponentials (Fig. 2), the ratio of the two decay times being 4.6. The intermediate case (b) is dominantly a biexponential and displays precisely the same two decay times found in (a) and (c). It is significant that the decay time ratio approximates 5 and that the magnitude of these decay times is essentially independent of magnetic field. These results are consistent with (2) where we expect three decay times in the ratio 5:3:1, and we conclude that case (a) represents dephasing due to the |5/2> state, case (c) to the |1/2> state, and case (b) to both of these states with possibly a small contribution from |3/2> as well. We conclude that application of a weak magnetic field modifies the optical pumping cycle and the $^3H_4$ population distribution in a sensitive way by mixing the nuclear wavefunctions |I_z> further since the $^1D_2$ Zeeman and quadrupole energies [8] can be comparable. This model is also consistent with the

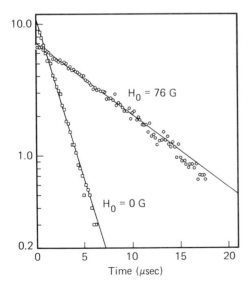

Fig. 2 Semilog FID plots of the data of Figs. 1(a) and (c) showing a simple exponential decay

401

zero field rf-optical double resonance observation [6,8] that the $^3H_4$
quadrupole transition $|5/2\rangle\leftrightarrow|3/2\rangle$ is more intense than the $|3/2\rangle\leftrightarrow|1/2\rangle$
More detailed calculations of the nuclear wavefunctions are needed to test
these ideas further and will require determining the orientation of the
principal axes x, y, z for both $^3H_4$ and $^1D_2$.

We now turn to (2) to determine the $^1D_2$ enhanced gyromagnetic ratio
γ_z'. A fluctuating local dipolar field of $\bar{H}_z=0.41G$ at the Pr^{3+} site due
to the fluorine nuclei can be deduced from the ground state value [5]
$\gamma_z''/2\pi=23$ kHz/G and a ground state linewidth [6] of 9.5 kHz for the $^3H_4$
quadrupole transition $|5/2\rangle\leftrightarrow|3/2\rangle$ at $H_0=0G$. The same local field
modulates the optical transition frequency producing a considerably broader
linewidth of 44 kHz ($I_z=5/2$) at $H_0=0G$. Therefore, we find from (2) that
$\gamma_z'/2\pi=20\pm4$ kHz/G where we have taken the enhanced moments of $^3H_4$ and $^1D_2$
to be of opposite sign. This quantity is bounded by $1.29<\gamma_z'/2\pi<19$ kHz/G,
the lower limit being derived from the first term of (1), i.e., with no
enhancement. The upper limit follows from the second term of (1) where
we assume in Λ_{zz} the maximum matrix element $\langle1|J_z|0\rangle=2$, the lowest Stark
level of $^1D_2$ mixes with the first excited state where $E_1-E_0=23$ cm^{-1}, $g_J=1$,
and $A\sim1.093\times10^9$ Hz. If γ_z'' and γ_z' are assumed to be of the same sign,
$\gamma_z'/2\pi=66$ kHz/G which exceeds the upper limit. In addition, ab initio
calculations [10] of $\langle J_z\rangle$ are in serious disagreement with our experimental
results.

Other broadening mechanisms we have considered appear to be negligible.
They include a $^1D_2$ radiative decay time of 0.5 msec [13] (0.16 kHz) and
phonon processes [8] (0.8 kHz). Our linewidths are also independent of
Pr^{3+} concentration in the range 0.03 to 0.1 atomic % so that Pr^{3+}-Pr^{3+}
interactions are excluded. Since the width is independent of laser power
and a nutation signal is not detected, we estimate that the optical
transition matrix element $\mu_{ij}\leq4.5\times10^{-5}$ Debye. This implies that only 10^{-5}
of the $^1D_2$ ions return directly by radiative decay to the ground $^3H_4$ state;
the remainder radiate to excited Stark split states of $^3H_4$ and other states
[13] followed by rapid spontaneous phonon emission processes to the ground
state. Clearly, the optical pumping cycle is not simple. The contribution
of laser frequency jitter to the linewidth appears to be small since the
decay time varies with external magnetic field in a predictable manner.
We expect that a significantly higher spectral resolution can be achieved
in the near future and will further improve precision measurements of this
kind where ultraslow optical dephasing processes occur.

We are indebted to D. Horne for the design and construction of the
laser frequency locking circuit and to K. L. Foster for technical
assistance. We are pleased to acknowledge conversations with E. Wong,
L. E. Erickson, C. S. Yannoni, I. D. Abella, E. L. Hahn, W. B. Mims, and
A. Wokaun.

References

[1]A. Z. Genack, R. M. Macfarlane and R. G. Brewer, Phys. Rev. Lett. 37,
1078 (1976); R. M. Macfarlane, A. Z. Genack, S. Kano and R. G. Brewer,
Journal of Luminescence 18/19, 933 (1979).

[2]K. Lee and A. Shir, Phys. Rev. Lett. 14, 1027 (1965).

[3] W. B. Mims, *Electron Paramagnetic Resonance*, ed. S. Geschwind (Plenum, NY, 1972), p. 263.

[4] M. A. Teplov, Soviet Phys. JETP **26**, 872 (1968).

[5] L. E. Erickson, Opt. Comm. **21**, 147 (1977).

[6] R. M. Shelby, C. S. Yannoni and R. M. Macfarlane, Phys. Rev. Lett. **41**, 1739 (1978).

[7] T. P. Das and E. L. Hahn, *Nuclear Quadrupole Resonance Spectroscopy* (Academic, 1958).

[8] L. E. Erickson, Phys. Rev. **16B**, 4731 (1977).

[9] B. Bleaney, Physica **69**, 317 (1973).

[10] S. Matthies and D. Welsch, Phys. Status Solidi B **68**, 125 (1975).

[11] E. Y. Wong, O. M. Stafsudd and D. R. Johnston, J. Chem. Phys. **39**, 786 (1963); V. K. Sharma, J. Chem. Phys. **54**, 496 (1971).

[12] R. G. Brewer and A. Z. Genack, Phys. Rev. Lett. **36**, 959 (1976); A. Z. Genack and R. G. Brewer, Phys. Rev. **17A**, 1463 (1978).

[13] M. J. Weber, J. Chem. Phys. **48**, 4774 (1978).

Optical Coherence Storage in Spin States

J.B.W. Morsink and D.A. Wiersma

Picosecond Laser and Spectroscopy Laboratory
Department of Physical Chemistry, University of Groningen
Nijenborgh 16, 9747 AG Groningen, The Netherlands

ABSTRACT

The formation of a new type photon echo is described which is stimulated from optically induced nuclear spin polarization in the electronic ground state. This anomalous photon echo was observed in the molecular mixed crystal triphenylmethyl in triphenylamine and in the system Pr^{3+} doped into LaF_3. In the latter system an echo lifetime was obtained that exceeds by a factor of 10^8 the fluorescence lifetime.

1. INTRODUCTION

Photon echo phenomena sofar have been associated with an induced macroscopic polarization at optical frequencies. In this paper we will demonstrate that in a multilevel system a new type photon echo may be observed whose lifetime solely depends on the lifetime of the levels in the electronic ground state. A prerequisite to observation of this echo is that optical branching occurs. The anomalous photon echo formation is examined by study of the time evolution of the density matrix of a three level system which serves as a model system. The interesting conclusion that follows from the calculations is that optical coherence may be stored for a long time in the electronic ground state in the form of a frequency dependent nuclear spin polarization. It is further shown that the echo appearance and lifetime may be manipulated by the application of a magnetic field. Finally the echo is demonstrated in some systems which indicates that the echo formation is a general feature of multilevel systems.

2.1. CLOSED THREE LEVEL SYSTEM

In this section we derive expressions for the two pulse echo (2PE) and the three pulse stimulated echo (3PSE) for a closed three level system. It is our aim to show that, in such a system, it is possible to produce a 3PSE at a probe delay-time that far exceeds the fluorescence lifetime.

   In a multilevel system this anomalous 3PSE effect can occur if the following conditions hold:
1) Decay processes *within* the multilevel structure of the electronic ground state must be much slower than the decay processes that occur at the optical transition.
2) Several (at least two) levels of the electronic groundstate must be optically connected to the same level in the excited state (optical branching). A three level model system such as shown in fig. 1a is the simplest one that still has all physical properties of interest. A complex multilevel system is here approximated by taking out of the groundstate multiplet only two levels and of the excited state just one.

   The ensemble averaged properties of the echo system can be described by

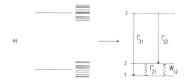

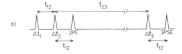

Fig. 1. (a) Three level model-system as an approximation of a multi-level system. Γ_{31} and Γ_{32} are the decay rates at the optical transitions, Γ_{21} and W_{12} are the decay rates between the hyperfine levels.
(b) Optical pulse sequence used to generate the 2PE and the 3PSE.

the time evolution of a density matrix. For an inhomogeneously broadened line the echo intensity can be calculated by averaging this density matrix over the lineshape. The density matrix $\rho(t)$ of the model system must obey the following equation:

$$\dot{\rho}(t) = \frac{1}{i\hbar}[H,\rho(t)] - \tfrac{1}{2}\{\Gamma,\rho(t)\} + F. \tag{1}$$

Whereby $H = H_0+V$ denotes the hamiltonian of the system including the interaction ($V = -\vec{\mu}.\vec{E}\cos\omega t$) with the radiation field. The brackets and curly brackets stand for commutator and anticommutator respectively. Γ and F are the decay and feeding matrix [1]:

$$\Gamma = \begin{pmatrix} W_{12} & 0 & 0 \\ 0 & \Gamma_{21} & 0 \\ 0 & 0 & \Gamma_{31}+\Gamma_{32} \end{pmatrix}, \quad F = \begin{pmatrix} \Gamma_{31}\rho_{33}+\Gamma_{21}\rho_{22} & 0 & 0 \\ 0 & \Gamma_{32}\rho_{33}+\Gamma_{12}\rho_{11} & 0 \\ 0 & 0 & 0 \end{pmatrix}. \tag{2}$$

The eigenvalues of the unperturbed hamiltonian H_0 are: $\hbar(\omega_1+\delta_1)$, $\hbar(\omega_2+\delta_2)$, $\hbar(\omega_3+\delta_3)$. The deltas account for the inhomogeneous contribution to the energy of each level. The decay paths shown in fig. 1a assure a closed system: $\rho_{11}(t)+\rho_{22}(t)+\rho_{33}(t)=1$. Rapidly oscillating terms of eq. (1) can be eliminated by transforming the off-diagonal elements of the optical transitions:

$$\tilde{\rho}_{23} = \rho_{23}\,e^{-i\omega t}, \qquad \tilde{\rho}_{13} = \rho_{13}\,e^{-i\omega t},$$

$$\tilde{\rho}_{32} = \tilde{\rho}_{23}^{\,*}, \qquad \tilde{\rho}_{31} = \tilde{\rho}_{13}^{\,*}. \tag{3}$$

This leads, with the rotating wave approximation, to the following equations:

$$\dot{\rho}_{11} = i\alpha(\tilde{\rho}_{31}-\tilde{\rho}_{13})+\Gamma_{31}\rho_{33}-W_{12}\rho_{11}+\Gamma_{21}\rho_{22}, \tag{4a}$$

$$\dot{\rho}_{22} = i\beta(\tilde{\rho}_{32}-\tilde{\rho}_{23})+\Gamma_{32}\rho_{33}+W_{12}\rho_{11}-\Gamma_{21}\rho_{22}, \tag{4b}$$

$$\dot{\rho}_{33} = i\beta(\tilde{\rho}_{23}-\tilde{\rho}_{32})+i\alpha(\tilde{\rho}_{13}-\tilde{\rho}_{31})-\Gamma_f\rho_{33}, \tag{4c}$$

$$\dot{\rho}_{12} = i\Delta_{21}\rho_{12}+i\alpha\tilde{\rho}_{32}-i\beta\tilde{\rho}_{13}-\tfrac{1}{2}\Gamma_s\rho_{12}, \tag{4d}$$

$$\dot{\tilde{\rho}}_{13} = i\Delta_{31}\tilde{\rho}_{13}+i\alpha(\rho_{33}-\rho_{11})-i\beta\rho_{12}-\tfrac{1}{2}(\Gamma_f+W_{12})\tilde{\rho}_{13}, \tag{4e}$$

$$\dot{\tilde{\rho}}_{23} = i\Delta_{32}\tilde{\rho}_{23}+i\beta(\rho_{33}-\rho_{22})-i\alpha\rho_{21}-\tfrac{1}{2}(\Gamma_f+\Gamma_{21})\tilde{\rho}_{23}. \tag{4f}$$

with the definitions:

$$\Delta_{21} = \omega_2-\omega_1+\delta_2-\delta_1, \quad \Delta_{31} = \omega_3-\omega_1-\omega+\delta_3-\delta_1, \quad \Delta_{32} = \omega_3-\omega_2-\omega+\delta_3-\delta_2,$$

$$\Gamma_f = \Gamma_{31}+\Gamma_{32}, \quad \Gamma_s = \Gamma_{21}+W_{12}, \quad \alpha = \mu_{13}E_o/2\hbar, \quad \beta = \mu_{23}E_o/2\hbar,$$

whereby was assumed that:

$$\mu_{13} = \mu_{31} \text{ and } \mu_{23} = \mu_{32}. \tag{5}$$

We used for solution of the differential eqs. 4a-4f different techniques depending on whether the exciting laser is on or off. For solution of the time-evolution of the density matrix during the exciting pulses we used the general method as developed by Brewer in his description of the Raman beat [2]. Free decay time-evolution can be partly solved by simple Laplace transformation techniques.

2.2. FREE TIME EVOLUTION of the DENSITY MATRIX

Laplace transforming the coupled diagonal equations leads to the following solutions:

$$\rho_{33}(t) = \rho_{33}(0)e^{-\Gamma_f t}, \tag{6a}$$

$$\rho_{11}(t) = \rho_{11}(0)\{Ae^{-\Gamma_s t}+(1-A)\}+\rho_{22}(0)\{(A-1)e^{-\Gamma_s t}+(1-A)\}$$

$$\qquad +\rho_{33}(0)\{Be^{-\Gamma_s t}+Ce^{-\Gamma_f t}+D\}, \tag{6b}$$

$$\rho_{22}(t) = 1 - \rho_{11}(t) - \rho_{33}(t), \tag{6c}$$

with the definitions:

$$A = \frac{W_{12}}{\Gamma_s}, \qquad B = \frac{W_{12}\Gamma_{31}}{\Gamma_s(-\Gamma_s+\Gamma_f)} - \frac{\Gamma_{32}\Gamma_{21}}{\Gamma_s(-\Gamma_s+\Gamma_f)},$$

$$C = \frac{\Gamma_{31}}{(-\Gamma_f+W_{12})} - \frac{\Gamma_{21}\Gamma_{32}}{\Gamma_f(-\Gamma_f+\Gamma_s)} - \frac{\Gamma_{21}\Gamma_{31}W_{12}}{\Gamma_s(-\Gamma_f+\Gamma_s)(-\Gamma_f+W_{12})},$$

$$D = 1 - A.$$

The off-diagonal elements have the trivial solutions:

$$\rho_{12}(t) = \rho_{12}(0) \, e^{-\frac{1}{2}\Gamma_s t + i\Delta_{21} t}, \tag{7a}$$

$$\tilde{\rho}_{13}(t) = \rho_{13}(0) \, e^{-\frac{1}{2}(\Gamma_f + W_{12})t + i\Delta_{31} t}, \tag{7b}$$

$$\tilde{\rho}_{23}(t) = \rho_{23}(0) \, e^{-\frac{1}{2}(\Gamma_f + \Gamma_{21})t + i\Delta_{32} t}. \tag{7c}$$

2.3. DRIVEN TIME EVOLUTION of the DENSITY MATRIX

The solution of this problem was obtained using the following assumptions:
a) neglegible decay
b) on resonance excitation
c) the Rabi frequences α and β far exceed the ground state splitting:
$\omega_{21} = \omega_2 - \omega_1$.
From conditions a, b and c it follows that *during* excitation:

$$\Delta_{21}, \ \Delta_{31} \text{ and } \Delta_{32} \approx 0, \quad \Gamma_{21}, \ W_{12}, \ \Gamma_s \text{ and } \Gamma_f \approx 0.$$

Pairwise combinations of the modified eqs. 4a-4f and their c.c. in a manner as shown by Brewer [2] leads to "Bloch like" equations. The initial values of the density matrix elements however will be different after each free decay evolution. After some algebraic manipulations we find for the diagonal elements:

$$\rho_{33}(t) = \rho_{33}(0) + \frac{(1-\cos(\varepsilon t))}{2\delta^2}\left\{\alpha^2(\rho_{11}(0)-\rho_{33}(0))+\beta^2(\rho_{22}(0)-\rho_{33}(0))\right.$$
$$\left. +\alpha\beta(\rho_{12}(0)+\rho_{21}(0))\right\}$$
$$-\frac{\sin(\varepsilon t)}{2i\delta}\left\{\alpha(\tilde{\rho}_{13}(0)-\tilde{\rho}_{31}(0))+\beta(\tilde{\rho}_{23}(0)-\tilde{\rho}_{32}(0))\right\}, \tag{8a}$$

$$\rho_{11}(t) = \frac{(\alpha^2-\beta^2)}{\delta^2}\rho_{33}(0) + \frac{\beta^2}{\delta^2}\left\{\rho_{22}(0)+\rho_{11}(0)\right\}+\left\{\frac{\alpha^2(1+\cos(\varepsilon t))-2\beta^2}{2\delta^4}\right\}$$
$$\cdot\left\{\alpha^2(\rho_{11}(0)-\rho_{33}(0))+\beta^2(\rho_{22}(0)-\rho_{33}(0))+\alpha\beta(\rho_{12}(0)+\rho_{21}(0))\right\}$$
$$+ \sin(\varepsilon t)\left(\frac{\alpha^2}{2i\delta^3}\right)\left\{\alpha(\tilde{\rho}_{13}(0)-\tilde{\rho}_{31}(0))+\beta(\tilde{\rho}_{23}(0)-\tilde{\rho}_{32}(0))\right\}$$
$$- \frac{\cos(\delta t)}{\delta^2}\left\{(\alpha^2-\beta^2)\rho_{33}(0)+\beta^2\rho_{22}(0)-\alpha^2\rho_{11}(0)+\left(\frac{\alpha^2-\beta^2}{\delta^2}\right)\right.$$
$$\left.\cdot\left[\alpha^2(\rho_{11}(0)-\rho_{33}(0))+\beta^2(\rho_{22}(0)-\rho_{33}(0))+\alpha\beta(\rho_{12}(0)+\rho_{21}(0))\right]\right\}$$
$$+ \sin(\delta t)\left(\frac{i\alpha\beta}{\delta^3}\right)\left\{\beta(\tilde{\rho}_{13}(0)-\tilde{\rho}_{31}(0))-\alpha(\tilde{\rho}_{23}(0)-\tilde{\rho}_{32}(0))\right\}, \tag{8b}$$

$$\rho_{22}(t) = 1 - \rho_{11}(t) - \rho_{33}(t). \tag{8c}$$

For the off-diagonal elements we obtain:

$$\rho_{12}(t) = \frac{3\alpha\beta}{\delta^2} \left\{ \rho_{33}(0) - \frac{1}{3} + \frac{2\alpha^2}{\epsilon^2} (\rho_{11}(0) - \rho_{33}(0)) + \frac{2\beta^2}{\epsilon^2} (\rho_{22}(0) - \rho_{33}(0)) \right.$$

$$\left. + \frac{2\alpha\beta}{\epsilon^2} (\rho_{12}(0) + \rho_{21}(0)) \right\}$$

$$+ \cos(\delta t) \left\{ \rho_{12}(0) - \frac{3\alpha\beta}{\delta^2} \rho_{33}(0) + \frac{\alpha\beta}{\delta^2} - \frac{2\alpha^3\beta}{\delta^4} (\rho_{11}(0) - \rho_{33}(0)) \right.$$

$$\left. - \frac{2\alpha\beta^3}{\delta^4} (\rho_{22}(0) - \rho_{33}(0)) - \frac{2\alpha^2\beta^2}{\delta^4} (\rho_{12}(0) + \rho_{21}(0)) \right\}$$

$$+ \sin(\delta t) \left(\frac{i}{\delta^3} \right) \left\{ \alpha(\alpha^2 \tilde{\rho}_{32}(0) + \beta^2 \tilde{\rho}_{23}(0)) - \beta(\alpha^2 \tilde{\rho}_{31}(0) + \beta^2 \tilde{\rho}_{13}(0)) \right\}$$

$$+ \cos(\epsilon t) \left(\frac{\alpha\beta}{2\delta^4} \right) \left\{ \alpha^2 (\rho_{11}(0) - \rho_{33}(0)) + \beta^2 (\rho_{22}(0) - \rho_{33}(0)) \right.$$

$$\left. + \alpha\beta(\rho_{12}(0) + \rho_{21}(0)) \right\}$$

$$+ \sin(\epsilon t) \left(\frac{\alpha\beta}{2i\delta^3} \right) \left\{ \alpha(\tilde{\rho}_{13}(0) - \tilde{\rho}_{31}(0)) + \beta(\tilde{\rho}_{23}(0) - \tilde{\rho}_{32}(0)) \right\} \quad , \quad (8d)$$

$$\tilde{\rho}_{13}(t) = \cos(\delta t) \left(\frac{\beta}{\delta^2} \right) \left\{ -\alpha\tilde{\rho}_{23}(0) + \beta\tilde{\rho}_{13}(0) \right\}$$

$$+ \sin(\delta t) \left(\frac{i\beta}{\delta^3} \right) \left\{ -\alpha\beta(\rho_{11}(0) - \rho_{22}(0)) - \beta^2\rho_{12}(0) + \alpha^2\rho_{21}(0) \right\}$$

$$+ \frac{\alpha^2}{2\delta^2} (\tilde{\rho}_{13}(0) + \tilde{\rho}_{31}(0)) + \frac{\alpha\beta}{2\delta^2} (\tilde{\rho}_{32}(0) + \tilde{\rho}_{23}(0))$$

$$- \sin(\epsilon t) \left(\frac{i\alpha}{2\delta^3} \right) \left\{ \alpha^2 (\rho_{11}(0) - \rho_{33}(0)) + \beta^2 (\rho_{22}(0) - \rho_{33}(0)) \right.$$

$$\left. + \alpha\beta(\rho_{12}(0) + \rho_{21}(0)) \right\}$$

$$+ \cos(\epsilon t) \left(\frac{\alpha}{2\delta^2} \right) \left\{ \alpha(\tilde{\rho}_{13}(0) - \tilde{\rho}_{31}(0)) + \beta(\tilde{\rho}_{23}(0) - \tilde{\rho}_{32}(0)) \right\} \quad , \quad (8e)$$

$$\tilde{\rho}_{23}(t) = - \cos(\delta t) \left(\frac{\alpha}{\delta^2} \right) \left\{ -\alpha\tilde{\rho}_{23}(0) + \beta\tilde{\rho}_{13}(0) \right\}$$

$$- \sin(\delta t) \left(\frac{i\alpha}{\delta^3} \right) \left\{ -\alpha\beta(\rho_{11}(0) - \rho_{22}(0)) - \beta^2\rho_{12}(0) + \alpha^2\rho_{21}(0) \right\}$$

$$+ \frac{\alpha\beta}{2\delta^2} (\tilde{\rho}_{13}(0) + \tilde{\rho}_{31}(0)) + \frac{\beta^2}{2\delta^2} (\tilde{\rho}_{23}(0) + \tilde{\rho}_{32}(0))$$

$$- \sin(\varepsilon t)\left(\frac{i\beta}{2\delta^3}\right)\{\alpha^2(\rho_{11}(0) - \rho_{33}(0)) + \beta^2(\rho_{22}(0) - \rho_{33}(0))$$

$$+ \alpha\beta(\rho_{12}(0) + \rho_{21}(0))\}$$

$$+ \cos(\varepsilon t)\left(\frac{\beta}{2\delta^2}\right)\{\alpha(\tilde{\rho}_{13}(0) - \tilde{\rho}_{31}(0)) + \beta(\tilde{\rho}_{23}(0) - \tilde{\rho}_{32}(0))\} . \tag{8f}$$

whereby : $\delta = \sqrt{\alpha^2 + \beta^2}$ and $\varepsilon = 2\delta$.

Repetitive use of eqs. 6a-c, 7a-c and 8a-f then leads to calculation of the density matrix at the "echo-time" τ. Inspection of the expression for the polarization $P(\tau)$, from which the echo can be derived, shows that we only have to calculate $\tilde{\rho}_{13}(\tau)$ and $\tilde{\rho}23(\tau)$:

$$P(\tau) = \mu_{13} \rho_{31}(\tau) + \mu_{31}\rho_{13}(\tau) + \mu_{32} \rho_{23}(\tau) + \mu_{23} \rho_{32}(\tau)$$

$$= \mu_{13} 2\text{Re}[\tilde{\rho}_{13}(\tau)e^{i\omega\tau}] + \mu_{23} 2\text{Re}[\tilde{\rho}_{23}(\tau)e^{i\omega\tau}].$$

In fig. 1b we show the pulse sequence used for the generation of the 2PE and 3PSE. Under the following, physically very reasonable, conditions:

$$\Delta t_1, \Delta t_2, \Delta t_3 << t_{12}, t_{23} \qquad \Gamma_{21}, W_{12} << \Gamma_{31}, \Gamma_{32}$$

$$\text{and } \Gamma_{31}^{-1}, \Gamma_{32}^{-2} << t_{23}$$

we have calculated for the 2PE and the anomalous 3PSE the optical off-diagonal elements. It seems worthwhile to note here that it is *not* possible for a three level system to define $\pi/2$ and π pulses as in the case of a two-level system. We have therefore chosen to derive the general expressions. It is further assumed that the groundstate splitting ($\hbar\omega_{21}$) is much smaller than kT. Levels 1 and 2 are then initially equally populated at t = 0, therefore:

$$\rho_{11}(0) = \rho_{22}(0) = \tfrac{1}{2}. \tag{9}$$

All other elements of the density matrix are zero. In deriving $\tilde{\rho}_{13}(\tau)$, we note that analogous manipulations lead to a solution for $\tilde{\rho}_{23}(\tau)$.

2.4. The 2PE CALCULATIONS

This analysis mainly serves to demonstrate how to perform the calculations of the matrix elements. From eqs. 7b-c and their c.c. we conclude that the rephasing of $\tilde{\rho}_{13}(t)$ (with $\exp(i\Delta_{31}t_{12})$) between the second pulse and the echo can compensate the dephasing between the first and second pulse of $\tilde{\rho}_{31}(t)$ and also of $\tilde{\rho}_{32}(t)$ if we have the condition

$$|(\delta_2 - \delta_1)| << t_{12}^{-1} . \tag{10}$$

Expression (10) tells that the difference in inhomogeneous contribution to level 1 and 2 in the time span t_{12} is neglegible. With eq. (9) the c.c. of eqs. 8e-f, 7b-c expressions can be deduced for $\tilde{\rho}_{31}(\Delta t_1 + t_{12})$ and $\tilde{\rho}_{32}(\Delta t_1 + t_{12})$ just prior to the second pulse:

$$\tilde{\rho}_{31}(\Delta t_1 + t_{12}) = e^{-\frac{1}{2}\Gamma_f t_{12}} \cdot e^{-i\Delta_{31} t_{12}} \cdot \frac{i\alpha}{4\delta} \sin(\varepsilon\Delta t_1), \tag{11a}$$

$$\tilde{\rho}_{32}(\Delta t_1 + t_{12}) = e^{-\frac{1}{2}\Gamma_f t_{12}} \cdot e^{-i\Delta_{32} t_{12}} \cdot \frac{i\beta}{4\delta} \sin(\varepsilon\Delta t_1). \tag{11b}$$

The second pulse makes $\tilde{\rho}_{13}(t)$ $(t > \Delta t_1 + t_{12} + \Delta t_2)$ dependent of $\tilde{\rho}_{31}(\Delta t_1 + t_{12})$ and $\tilde{\rho}_{32}(\Delta t_1 + t_{12})$ (see eq. 8e). After a second free decay period t_{12} the expressions for $\tilde{\rho}_{13}(t)$ $(t = \tau = \Delta t_1 + t_{12} + \Delta t_2 + t_{12})$ becomes:

$$\tilde{\rho}_{13}(\tau) = e^{-\Gamma_f t_{12}} \left[\left\{ \frac{i\alpha^3}{4\delta^3} \sin(\delta\Delta t_2) \cdot \sin(\varepsilon\Delta t_1) \right\} \right.$$

$$\left. + e^{i\omega_{21} t_{12} + i(\delta_2 - \delta_1) t_{12}} \left\{ \frac{i\alpha\beta^2}{4\delta^3} \sin^2(\delta\Delta t_2) \cdot \sin(\varepsilon\Delta t_1) \right\} \right], \tag{12a}$$

and analogous:

$$\tilde{\rho}_{23}(\tau) = e^{-\Gamma_f t_{12}} \left[\left\{ \frac{i\beta^3}{4\delta^3} \sin(\delta\Delta t_2) \cdot \sin(\varepsilon\Delta t_1) \right\} \right.$$

$$\left. + e^{-i\omega_{21} t_{12} - i(\delta_2 - \delta_1) t_{12}} \left\{ \frac{i\alpha^2\beta}{4\delta^3} \sin^2(\delta\Delta t_2) \cdot \sin(\varepsilon\Delta t_1) \right\} \right]. \tag{12b}$$

It is interesting to note that eqs. 12a-b lead to the same result as obtained by Schenzle et al. [3] (Eq. 24 of their paper) when the appropriate adaptions are made.

2.5. The ANOMALOUS 3PSE CALCULATIONS

Obviously the *normal* 3PSE can be calculated, but because of lack of space, we will present only the results of the anomalous 3PSE calculation. Again we conclude from eq. 7b that the optical off-diagonal element $\tilde{\rho}_{13}(t)$ will get a positive phase-factor $\Delta_{31} t_{12}$ in the time interval t_{12} after the third pulse. There are already elements in the expression of $\tilde{\rho}_{13}(t)$, that have the phase factor $-\Delta_{31} t_{12}$ and $-\Delta_{32} t_{12}$ from $\tilde{\rho}_{31}(\Delta t_1 + t_{12})$ and $\tilde{\rho}_{32}(\Delta t_1 + t_2)$ respectively. From eqs. 8a-f we derive, that this coupling is caused by the second and third excitation pulses. The remaining important question to answer now is: which elements have still an appreciable value after the long free decay period t_{23}? It may be seen from eqs. 6a-c and 7a-c that only $\rho_{11}(\Delta t_1 + t_{12} + \Delta t_2 + t_{23})$, $\rho_{22}(\Delta t_1 + t_{12} + \Delta t_2 + t_{23})$ and $\rho_{12}(\Delta t_1 + t_{12} + \Delta t_2 + t_{23})$ (plus c.c.) will be left. $\rho_{12}(\Delta t_1 + t_{12} + \Delta t_2 + t_{23})$ plus c.c. however can also be excluded, as this element does not have the correct phase factor to cancel the factor $\Delta_{31} t_{12}$. At this stage we come to the following important conclusion: the optical diagonal elements $\tilde{\rho}_{13}(\tau)$ and $\tilde{\rho}_{23}(\tau)$ at the time τ of the anomalous 3PSE formation have been induced by the third pulse from the ground state polarization values of $\rho_{11}(t)$ and $\rho_{22}(t)$ at the time t just prior to this pulse. With these data, we deduce from eq. 8e:

$$\tilde{\rho}_{13}(\tau) = e^{-\frac{1}{2}\Gamma_f t_{12}} \cdot e^{i\Delta_{31}t_{12}}$$

$$\cdot \left[\rho_{11}(\Delta t_1 + t_{12} + \Delta t_2 + t_{23}) \cdot \left(-\frac{i\alpha\beta^2}{\delta^3} \sin(\delta\Delta t_3) - \frac{i\alpha^3}{2\delta^3} \sin(\epsilon\Delta t_3) \right) \right.$$

$$\left. + \rho_{22}(\Delta t_1 + t_{12} + \Delta t_2 + t_{23}) \cdot \left(\frac{i\alpha\beta^2}{\delta^3} \sin(\delta\Delta t_3) - \frac{i\alpha\beta^2}{2\delta^3} \sin(\epsilon\Delta t_3) \right) \right]. \quad (13)$$

The groundstate diagonal elements contain terms with phase factors $-\Delta_{31}t_{12}$ and $-\Delta_{32}t_{12}$. An analogous expression results for $\tilde{\rho}_{23}(\tau)$:

$$\tilde{\rho}_{23}(\tau) = e^{-\frac{1}{2}\Gamma_f t_{12}} e^{i\Delta_{32}t_{12}}$$

$$\cdot \left[\rho_{11}(\Delta t_1 + t_{12} + \Delta t_2 + t_{23}) \cdot \left(\frac{i\alpha^2\beta}{\delta^3} \sin(\delta\Delta t_3) - \frac{i\alpha^2\beta}{2\delta^3} \sin(\epsilon\Delta t_3) \right) \right.$$

$$\left. + \rho_{22}(\Delta t_1 + t_{12} + \Delta t_2 + t_{23}) \cdot \left(\frac{-i\alpha^2\beta}{\delta^3} \sin(\delta\Delta t_3) - \frac{i\beta^3}{2\delta^3} \sin(\epsilon\Delta t_3) \right) \right]. \quad (14)$$

Whereby only the following terms out of the diagonal elements lead to rephasing at $t = \tau$:

$$\rho_{11}(\Delta t_1 + t_{12} + \Delta t_2 + t_{23}) : e^{-\Gamma_s(t_{12}+t_{23})} \cdot \sin(\epsilon\Delta t_1)$$

$$\cdot \left[e^{-i\Delta_{31}t_{12}} \frac{i\alpha}{4\delta} \left\{ (A)\left(\frac{i\alpha^3}{2\delta^3} \sin(\epsilon\Delta t_2) + \frac{i\alpha\beta^2}{\delta^3} \sin(\delta\Delta t_2) \right) \right. \right.$$

$$+ (A-1)\left(\frac{i\alpha\beta^2}{2\delta^3} \sin(\epsilon\Delta t_2) - \frac{i\alpha\beta^2}{\delta^3} \sin(\delta\Delta t_2) \right) \quad (15)$$

$$\left. + (B)\left(-\frac{i\alpha}{2\delta} \sin(\epsilon\Delta t_2) \right) \right\}$$

$$+ e^{-i\Delta_{32}t_{12}} \frac{i\beta}{4\delta} \left\{ (A)\left(\frac{i\alpha^2\beta}{2\delta^3} \sin(\epsilon\Delta t_2) - \frac{i\alpha^2\beta}{\delta^3} \sin(\delta\Delta t_2) \right) \right.$$

$$+ (A-1)\left(\frac{i\beta^3}{2\delta^3} \sin(\epsilon\Delta t_2) + \frac{i\alpha^2\beta}{\delta^3} \sin(\delta\Delta t_2) \right)$$

$$\left. \left. + (B)\left(-\frac{i\beta}{2\delta} \sin(\epsilon\Delta t_2) \right) \right\} \right] .$$

It can easily be shown from the relation:

$$\rho_{22}(\Delta t_1 + t_{12} + \Delta t_2 + t_{23}) = 1 - \rho_{11}(\Delta t_1 + t_{12} + \Delta t_2 + t_{23}), \quad (16)$$

that expression (15) with reversed sign gives the important elements of $\rho_{22}(\Delta t_1+t_{12}+\Delta t_2+t_{23})$. Similar to the 2PE calculations, there are terms with the phase factor $(\Delta_{31}-\Delta_{32})t_{12} = (\omega_{21}+(\delta_2-\delta_1))t_{12}$. Depending on whether or not relation (10) holds, these terms will lead to modulation in the anomalous 3PSE intensity as a function of t_{12}.

The following physical picture emerges from the calculations for the description of the anomalous echo formation. The second pulse transforms the dephasing pattern, that has been created during the first interval t_{12} into a *frequency dependent* optical polarization (FDOP) within the inhomogeneous lineshape. Simultaneously however, a frequency dependent spin polarization (FDSP) in the levelstructure of the groundstate is created. Both polarizations have a one to one correspondence in every particle of the ensemble. The FDOP will decay completely (with Γ_f) and the FDSP will decay slowly if the time interval t_{23} is comparable with Γ_s^{-1}. The third excitation pulse then transforms the remaining FDSP into a *new* FDOP. During the last time interval t_{12} rephasing then occurs which results in the anomalous 3PSE formation.

3.1. EXPERIMENTAL RESULTS

We studied the anomalous 3PSE in two multilevel systems. The first one is the molecular radical system triphenylmethyl (TPM) in triphenylamine (TPA)[1]. The second one is the inorganic system Pr^{3+}/LaF_3. We have chosen also this latter system, because the level structure is much simpler as in the former case and the ground state parameters are well known [4]. References [5, 6, 7] give a detailed description of our results. The experimental equipment used is described in ref. [8]. We now proceed by discussing the experimental results in the short-hand style.

3.2. The $^2A_1 \leftarrow {}^2A_1$ TRANSITION of TPM in TPA

The results of 3PSE measurements on the $^2A_1 \leftarrow {}^2A_1$ doublet-doublet transition of the energetically lowest site of TPM in TPA are shown in fig. 2. (The experiments were performed on a crystal with a guest concentration of 10^{-3} mol % at 1.5 K).

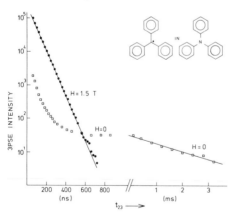

Fig. 2.

Intensity of the 3PSE of TPM in TPA at 1.5 K versus second-third excitation pulse delay in zero field and in a magnetic field of 1.5 T. The time separation between the first and second pulse was 40 ns.

---

[1] We note that the experiments on this system were performed in cooperation with Wim H. Hesselink (ref. 6).

In zero magnetic field the normal 3PSE exhibits a non-exponential decay and is followed by the anomalous 3PSE decay on msec. time scale. The multi-level system is formed by the interaction of the radical electronspin with the nuclear spin of the phenyl hydrogen atoms. The non-exponential decay may arise from cross relaxation processes in the manifold of nuclear spin states. Upon deuteration of the TPM the initial 3PSE decay, up to 400 nsec, becomes exponential. This confirms our earlier supposition [5], that the decay is greatly influenced by *intramolecular* spin-spin interaction processes.

In a high magnetic field (H = 1.5 T, see also fig. 2) only a fast purely exponential decay remains. Optical branching no longer occurs because the electron- and nuclear spins in this situation are quantized along the magnetic field direction. The 3PSE decay time (65 nsec) is in excellent agreement with the expected decay time of a two level system: $\frac{1}{2}T_1$, where T_1 is the fluorescence lifetime (131±3 nsec). The multilevel structure in a high magnetic field therefore merely causes an additional source of inhomogeneous broadening.

3.3. The $^3P_0 \leftarrow {}^3H_4$ TRANSITION in Pr^{3+}/LaF_3

We also studied photon echo phenomena in the $^3P_0 \leftarrow {}^3H_4$ transition of Pr^{3+} doped (ca 1 at %) into LaF_3 at 2 K. Next to the anomalous 3PSE measurements

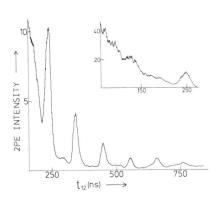

Fig. 3

Intensity of the 2PE versus excitation pulse delay in the $^3P_0 \leftarrow {}^3H_4$ transition of Pr^{3+} in LaF_3 at 2 K.

we also wish to mention here 2PE and normal 3PSE measurements. Fig. 3 shows the 2PE decay beat pattern and in fig. 4 we show the 3PSE decay modulation

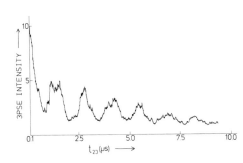

Fig. 4.

Intensity of the 3PSE versus second-third excitation pulse delay at 2 K in the $^3P_0 \leftarrow {}^3H_4$ transition of Pr^{3+} in LaF3. The separation between the first two excitation pulses was 50 ns.

on a time scale where the 2PE signal was no longer detectable. From these results we were able to calculate the ground- and excited state level splittings and the parameters of the excited state spin hamiltonian.

We now proceed by discussing the anomalous echo effect. From the work of Chen et al. [9, 10], we conclude that optical branching occurs in the $^3P_0 \leftarrow ^3H_4$ transition. This can also be derived from the "sharp" recurrences (beats) in the 2PE decay (see fig. 3), resulting from strong interference effects [7]. Further it is known from Erickson's work [4] that the lifetime of the groundstate hyperfine levels at 4.2 K is 0.5 sec., which far exceeds the fluorescence lifetime ($\sim 47\mu sec$ [11]). These data satisfy the earlier formulated condition for anomalous 3PSE formation. The background in fig. 4 at $\sim 9\mu sec$ is due to the anomalous 3PSE effect. In zero magnetic field we measured for this component a lifetime of ~ 0.5 sec. In fig. 5 the magnetic

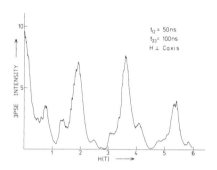

Fig. 5

Intensity of the 3PSE versus magnetic field in the $^3P_0 \leftarrow ^3H_4$ transition of Pr^{3+} in LaF_3 at 2 K. The separation between the excitation pulses is shown in the figure. The magnetic field was perpendicular to the crystal c-axis.

field dependence of the normal 3PSE is shown. The modulation pattern demonstrates, that in a high magnetic field (perpendicular to the crystal c-axis) optical branching still occurs. In fig. 6 we show a picture of the anoma-

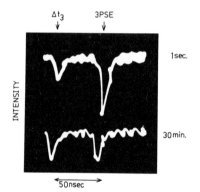

Fig. 6

Anomalous 3PSE trace obtained after the third probe pulse in a magnetic field of 4.8 T at 2 K (the first pulse on the left is the probe pulse). The upper trace was taken at 1 second after the second pulse. The lower trace was taken at 30 minutes after the second pulse. The first two excitation pulses were separated by 50 ns.

lous echo in a magnetic field of 4.8 T. We were able to observe an echo even 30 minutes! after the second excitation pulse. The increase of the anomalous 3PSE-lifetime in a magnetic field may be due to suppression of fluor and/or lanthanium spin cross relaxation processes. This is a point of further investigation.

4. SUMMARY

We have shown that in a multilevel system, where optical branching occurs, optical coherence may be stored for a long time in the electronic ground state.

The mechanism for this optical storage system was elucidated and it was shown that the formation of a frequency dependent nuclear spin polarization in the electronic ground state is the key element in the echo formation-mechanism.

The observed effect of spin diffusion on the anomalous echo-lifetime stimulates further research in this area.

ACKNOWLEDGEMENTS

We are grateful to Johan Kemmink for help in the experiments on Pr^{3+}/LaF_3, and to Marianne A. Morsink for her courtesy to type this paper.

REFERENCES

1) W.G. Breiland, M.D. Fayer and C.B. Harris, Phys. Rev. A13 383 (1976). H. de Vries and D.A. Wiersma, J. Chem. Phys. 70 issue: June 15th (1979).
2) R.G. Brewer in "Frontiers in Laser Spectroscopy", vol. I, Les Houches 1975 (ed. R. Balian, S. Haroche and S. Liberman), (North Holland Publ. Co., Amsterdam, 1977).
3) A. Schenzle, S. Grossman and R.G. Brewer, Phys. Rev. A13 1891 (1976).
4) L.E. Erickson, Opt. Commun. 21, 147 (1977).
5) W.H. Hesselink and D.A. Wiersma, Chem. Phys. Lett. 50 51 (1977).
6) J.B.W. Morsink, W.H. Hesselink and D.A. Wiersma, Chem. Phys. Lett. 63 issue: June 15th (1979).
7) J.B.W. Morsink and D.A. Wiersma, submitted for publication in Chem. Phys. Lett.
8) J.B.W. Morsink, T.J. Aartsma and D.A. Wiersma, Chem. Phys. Lett. 50 51 (1977).
9) Y.C. Chen and S.R. Hartmann, Phys. Lett. 58A 201 (1976).
10) Y.C. Chen, K.P. Chiang and S.R. Hartmann, Opt. Commun. 26 269 (1978).
11) M.I. Weber, J. Chem. Phys. 48 4774 (1968).

Time-Resolved Infrared Spectral Photography

D. S. Bethune, J. R. Lankard, M. M. T. Loy and P. P. Sorokin

IBM Thomas J. Watson Research Center

Yorktown Heights, New York 10598 USA

Introduction

A new technique was recently developed[1] which allows broadband infrared absorption spectra to be photographed with $\approx$5nsec time resolution, even under some conditions, in a single-shot. Experimental details and theoretical aspects of this method are discussed at length in [2]. Briefly, the technique utilizes third order nonlinearities of alkali metal vapors to achieve two objectives: (1) generation of a pulsed, broadband, infrared continuum beam (ν_{ir}) which can probe the absorption of a sample and (2) subsequent conversion of this i.r. beam, with its encoded spectral information about the sample, to the visible, where it can be photographically or photoelectrically recorded. The two principal nonlinear processes involved are diagrammed in Fig. 1.

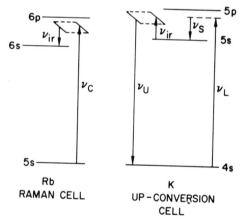

Figure 1 Level diagrams for the Raman generation and four-wave upconversion processes.

Generation of ν_{ir} is accomplished by *broadband* stimulated electronic Raman scattering (SERS) in a first alkali metal vapor heat-pipe oven (HP1 in Fig. 2). A broadband ($\Delta\nu\sim1000cm^{-1}$) visible dye laser beam ν_c, originates from a mirrorless, superfluorescent dye cell (C_1 in Fig. 2) and is subsequently amplified to the level of a few millijoules and focused into HP1. This visible beam, the primary beam for the SERS process, is converted to an infrared Stokes beam having approximately the same spectral width and pulse duration ($\approx$5nsec) as ν_c. The i.r. spectral range spanned by ν_{ir} depends on both ν_c and the energy of the final electronic state involved in the Raman transition. For Rb, with ν_c in the vicinity of the 6p resonance lines, the infrared generated falls in the 2.7μm range. With the use of dye laser continua spectrally located near other alkali resonance lines, generation of broadband i.r. in several different regions, extending out to at least 30μm, should be possible.[1]

The second nonlinear process diagrammed in Fig. 1 is responsible for the upconversion that occurs in the second heat-pipe oven (HP2 in Fig. 2). A narrow band laser beam ν_L, injected into HP2 via a silicon wafer (Si$_2$), generates by SERS a narrow-band Stokes beam ν_S. Beams ν_L, ν_S and ν_{ir} beat together in HP2 to produce an upconverted beam ν_U by means of the resonantly enhanced four-wave mixing process shown in Fig. 1. The intensity of a given spectral component of ν_U is linearly proportional to its corresponding ν_{ir} component. Thus infrared spectra are simply translated to the visible by the fixed Raman energies of the alkali metal atoms in HP2. This procedure allows one to take advantage of the fact that detectors with high inherent spectral resolution exist for the visible region.

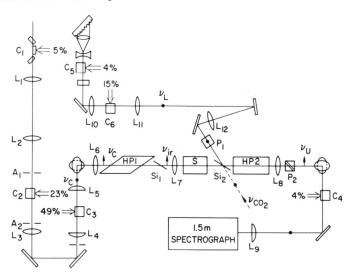

Figure 2 Diagram of experimental apparatus.

Experimental

The dye cells depicted in Fig. 2 were all transversely pumped by the third harmonic of a Quanta-Ray Nd^{3+}:YAG laser. Up to 100mj of pulse energy is available at this frequency. This was apportioned to the various dye cells as indicated in the diagram. The basic pulse width was ~5nsec.

The polarizations of the various beams were chosen as shown in Fig. 2. This allowed the narrow-band laser beam ν_L to be conveniently nulled by polarizer P_2, while the orthogonally polarized upconverted beam is transmitted.

For most of our work thus far we have utilized the Rb $5s \rightarrow 6p \rightarrow 6s$ Raman scheme (shown in Fig. 1) to generate ν_{ir}. Using dyes such as POPOP, Bis-MSB, and dimethyl POPOP in p-dioxane, and Stilbene 420 in mixtures of water and methanol, i.r. continua covering the range from $\sim 4,000 cm^{-1}$ to $\sim 2600 cm^{-1}$ have been produced. At the longer i.r. wavelengths, the cross-section for the Raman process shown in Fig. 1 becomes less resonantly enhanced, requiring ν_c pump energies of a few millijoules to exceed SERS thresholds.

For the Raman-driven upconverter we have exclusively employed K vapor and the scheme shown in Fig. 1. DPS in p-dioxane is used in both the narrow-band oscillator and amplifier cells (C_5 and C_6). In some instances an upconverted beam amplifier cell C_4 was also utilized.

Discussion of Spectra

The upconverted spectra shown in Fig. 3 are typical of our first results.

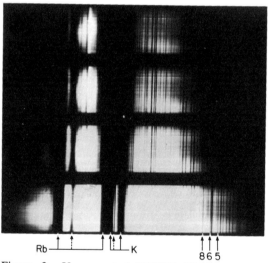

Figure 3   Upconverted POPOP Rb-K spectrum (second order, sample cell removed); increasing exposures (top to bottom). The numbers 8,6,5 refer to numbered peaks in the IUPAC calibration spectrum given in Fig. 4.

Here the combination of a 5 torr Rb cell and a continuum beam based on solutions of POPOP in p-dioxane was used to generate an i.r. continuum, and the sample cell S was completely removed. Dark bands correspond to $4s \rightarrow 5p_{3/2,1/2}$ resonance absorptions of K atoms in the second cell and to absence of i.r. light at the exact $6p_{3/2,1/2} \rightarrow 6s$ frequencies of Rb. Between each of these absorption line pairs is a dark line, due to interference nulls in $\chi^{(3)}_{xxyy}$ for K in the upconverter cell and in $\chi^{(3)}_{xxxx}$ for Rb in the Raman cell.[1]

The myriad of sharp, dark lines on the high frequency side of the spectrum results from residual water vapor in the optical path between the two vapor cells. A densitometer trace of a portion of the spectrum in Fig. 3 is shown in the upper portion of Fig. 4. The correspondence with a published IUPAC water vapor spectrum[3] (lower trace) is apparent.

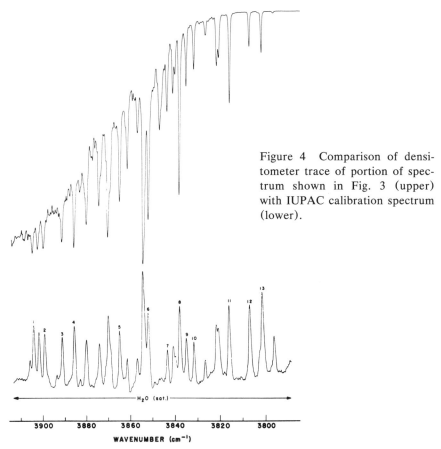

Figure 4 Comparison of densitometer trace of portion of spectrum shown in Fig. 3 (upper) with IUPAC calibration spectrum (lower).

The spectra of Fig. 3 represent superpositions of dozens of shots taken with the lower power apparatus of [1]. One result of this averaging is that a random spectral noise that we have since repeatedly encountered (side infra) is effectively averaged out, leaving well-resolved i.r. bands with apparently correct intensity distributions.

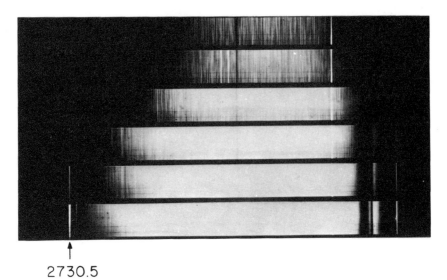

↑
2730.5

Figure 5   Upconverted dimethyl POPOP Rb-K spectra with accumulated shots per exposure of 1,3,5,7,9 and 20 (top to bottom).   No upconverted beam amplifier was used.

In Fig. 5 the wide range of i.r. covered with the use of dimethyl POPOP in cells C_1, C_2, and C_3 is demonstrated.   The bright line at the left of the lower spectra results from upconversion of a 2730.5cm^{-1} lasing transition $(6s_{1/2} \rightarrow 5p_{1/2})$ in the K cell HP2.   The bright line seen at the right of the upper spectra is the narrow-band laser beam ν_L, which remains fixed in frequency in all of the spectra shown in Fig. 5.   From spectra such as those shown in Fig. 5, we observe that phase-matching appears to play no important role in the Raman-driven upconverter, since there is no obvious intensity peaking of upconverted light at the wavelength (ν_L) that corresponds to the exact phase-matching i.r. wavelength.   The very striking broadband nature of the upconversion process, together with its observed high efficiency[1], is further evidence of this.   A possible explanation for this is that the SERS process has very high gain (several hundred cm^{-1}).   This may give a short "active region", where the Stokes wave ν_S grows rapidly while the intensity of the pump wave ν_L falls.   If this region were only a few millimeters long, fairly large wavevector mismatching could be tolerated.   Tuning of ν_L over a wide range is observed to have no strong effect on the upconverted spectra, in line with the above comments.

At the right of the bottom spectrum of Fig. 5 are seen again diffuse dark lines corresponding to the Rb $6p \rightarrow 6s$ transitions and also a sharp dark line between them representing the Rb Raman null.   The latter occurs approximately at 3607cm^{-1}.   Thus, a spectral region $\sim$1000cm^{-1} in width can be spanned with a single continuum dye solution, provided several shots per plate exposure are allowed. Because of insufficient exposure, the single-shot spectrum covers only approximately $600cm^{-1}$.

The top few spectra display the granular spectral character of the upconverted light, mentioned earlier. To obtain a reasonably accurate spectrum comprising narrow absorption lines generally requires averaging ten shots or so. Even so, in Fig. 5 a dark line at $\sim3208cm^{-1}$ is clearly evident in the top most (single-shot) exposure. This dark line appears in virtually all spectra covering this region. Its frequency coincides with the K 4d-5f transitions. Potassium atoms in the 4d states are probably produced by photodissociation of K_2 dimers as the beam ν_L passes through HP2.

Figure 6 shows increasing exposures of upconverted spectra with ~300 torr of CH_4 in the 18cm long sample cell. The CH_4 Q-branch is marked. The P and R branches are also clearly seen. Upconverted spectra have also been recorded with NH_3 and various other organics in the sample cell.[2]

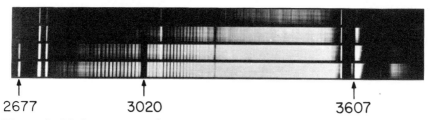

$\uparrow$ 2677 $\uparrow$ 3020 $\uparrow$ 3607

Figure 6 Methane spectra (second order, 300 torr). Shots per exposure are 10,20,30 and 50 (top to bottom). Cells C_1, C_2 and C_3 contain 5×10^{-4} molar solutions of Stilbene 420 in 1:1 H_2O-CH_3OH.

Example of a Time-Resolved Spectrum: $CH_3NC \rightarrow CH_3CN$

To demonstrate the *transient* capability of the new technique, the isomerization of methyl isocyanide to methyl cyanide was studied. This well-known isomerization, exothermic by 23.7kcal/mole[4], has been suggested as an ideal unimolecular reaction for testing thermal explosion theories.[5] Recently we observed that this thermal isomerization can be initiated by single pulses of CO_2 TEA laser radiation, tuned to coincide with the ν_4 band of the CH_3NC molecule.[6]

The ν_1 band of CH_3NC, associated with the symmetric stretch of the CH_3 group, lies near $2966cm^{-1}$. The top strip in Fig. 7 shows the upconverted spectrum of CH_3NC in this region. In agreement with an earlier study by Thompson and Williams[7], the P, Q, R branch contours of the ν_1 band are clearly shown, but the P and R lines are not resolved. The sharp lines on the high-frequency side of the R branch of the ν_1 band belong to the ν_5 band. The isomerization product CH_3CN also has a ν_1 band centered at $2955cm^{-1}$, but the absorption band is too weak to be observed here.

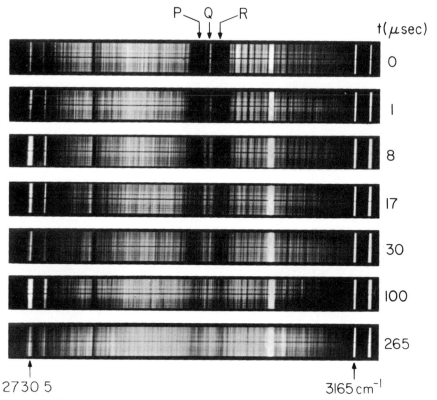

P Q R

t(μsec)

0

1

8

17

30

100

265

2730 5   3165 cm^{-1}

Figure 7 Changes in the upconverted spectra of 100 torr CH_3NC as it thermally explodes (isomerizes), various times after the application of an unfocused 0.5J pulse of CO_2 TEA laser·radiation ($\nu_{CO_2}=960cm^{-1}$). Each spectrum represents 16 superimposed shots.

Figure 7 shows time-resolved upconverted spectra of CH_3NC (100 torr initial pressure) taken at intervals of 1,8,17,30,100 and 265μsec, respectively, after the unfocused 0.5J CO_2 laser pulse, applied to the sample cell via Si_2 (Fig. 2), has initiated the thermal isomerization. Two effects can be clearly seen from these spectra. First, the intensity of the ν_1 absorption band weakens as time increases. Second, the absorption band contours of the P and R branches broaden substantially, with maximum broadening occurring 30μsec after the laser pulse. This broadening is due to the increase in temperature from the energy released in the isomerization process. As shown by Gerhard and Dennison[8], for a parallel band transition in a symmetric rotator (e.g., the ν_1 band of CH_3NC), the separation between the maxima of the P and R branch contours is proportional to the square root of the absolute temperature. Using this relation and the measured separations between P and R bands, we obtain the instantaneous temperature as a function of time, plotted in Fig. 8.

422

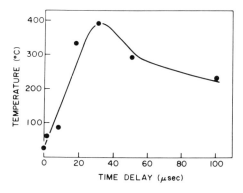

Figure 8 Temperature of the reacting CH_3NC gas as a function of time, determined from the separation of the P and R branches.

About 30μsec after the laser pulse, the temperature rises to about 673K: it then gradually cools. The maximum measured temperature of 673K indicates this to be the approximate "threshold temperature" for thermal isomerization to CH_3CH; this is in reasonable agreement with thermal data.[9] Molecules of CH_3NC hotter than this threshold temperature rapidly isomerize to CH_3CN and are thus not detected by the i.r. probe. A true determination of the kinetic temperature could be made by seeding the CH_3NC gas with a small amount of stable foreign gas (such as HCN) that has strong i.r. bands in a convenient range.

Acknowledgements

This work is partly supported by the U. S. Army Research Office and the Office of Naval Research. We would like to thank our colleague, Dr. J. Ors, for preparation of the CH_3NC sample.

References

1. D. S. Bethune, J. R. Lankard and P. P. Sorokin, Optics Lett. **4**, 103 (1979).
2. D. S. Bethune, J. R. Lankard, M. M. T. Loy and P. P. Sorokin, IBM J. Res. Develop. (to be published).
3. IUPAC Commission on Molecular Structure and Spectroscopy, Tables of Wave Numbers for the Calibration of Infrared Spectrometers (Pergamon, London, 1977), p. 98.
4. M. H. Baghal-Vayjooee, J. L. Collister and H. O. Pritchard, Can. J. Chem. **55**, 2634 (1977).
5. H. O. Pritchard and B. J. Tyler, Can. J. Chem. **51**, 4001 (1973).
6. D. S. Bethune, J. R. Lankard, M. M. T. Loy, J. Ors and P. P. Sorokin, Chem. Phys. Lett. **57**, 479 (1978).
7. H. W. Thompson and R. L. Williams, Trans. Faraday Soc. **48**, 502 (1952).
8. S. L. Gerhard and D. M. Dennison, Phys. Rev. **43**, 197 (1933).
9. F. W. Schneider and B. S. Rabinovitch, J. Amer. Chem. Soc. **84**, 4215 (1962).

Part VII

Optical Bistability, Superradiance

Theory of Optical Bistability

R. Bonifacio, L.A. Lugiato, and M. Gronchi

Istituto di Fisica dell'Università, Via Celoria, 16
I-20133 Milano, Italy

1. Introduction

Optical bistability is a phenomenon which arises in the trans
mission of light by an optical cavity filled with a resonant
medium. Precisely, we consider a c-w laser beam which is injected
into a resonant cavity tuned or nearly tuned to the incident
light. The incident beam is partially transmitted, partially
reflected and partially scattered by the cavity. When the cavity
is empty the transmitted intensity is proportional to the incid
ent intensity, where the proportionality constant depends on the
cavity detuning. On the other hand, when the cavity is filled
with material resonant or nearly resonant with the incident field
the transmitted intensity is a nonlinear function of the incident
intensity. In particular, under suitable conditions, the transmitted intensity varies discontinuously and exhibits a hysteresis
cycle. This is the so-called Optical Bistability (OB). This phenomenon was predicted by SZÖKE and coworkers in 1969 [1]. Some
years later McCALL [2] performed a numerical treatment of OB in
a Fabry-Perot cavity, which suggested the experiments of GIBBS,
McCALL and VENKATESAN in Na and Ruby, in which the bistable
response was nicely observed for the first time [3, 4]. These
experiments have shown that this system is the basis for quite
a number of device applications as optical memory, optical trans
istor, clipper, limiter etc. These crucial results stimulated a
very active theoretical research which bifurcated into two direc
tions. The first channel was mainly interested in the device
aspects of the phenomenon and in particular investigated the
feasibility of electro-optical systems to produce bistability and
related phenomena. A review of this approach can be found in [5].
The second channel [6 - 34] studies optical bistability as a
fundamental chapter of the interaction between multiatomic systems
and radiation. Precisely, it considers the optical bistable system
as the passive counterpart of the laser. The semiclassical mean
field of OB has been developed in [6 - 12], the quantum statistical mean field theory in [13 - 23], the exact semiclassical

treatment in a ring cavity in $[24 - 25, 27 - 29]$, the numerical
analysis in a Fabry-Perot cavity in $[2, 30 - 34]$. In particular
the theory of OB developed in $[6 - 8, 10, 12 - 13, 18, 21, 24 -$
$25, 27 - 29]$ has the following characteristics: it is completely
microscopic, it is analytical, it takes propagation effects fully
into account, it has been developed both at a semiclassical level,
i.e. neglecting fluctuations and correlations, and at a fully
quantum statistical level. In this paper we shall illustrate only
the exact semiclassical theory in a ring cavity formulated in
$[24 - 25, 27 - 29]$.

2. Semiclassical Treatment of the Steady State

In order to describe theoretically this phenomenon, it is
easier to consider a ring cavity (Fig.1) than a Fabry-Perot, because in the ring cavity one has to deal with propagation only
in one direction, thus avoiding standing wave difficulties. For
simplicity we assume that mirrors 3 and 4 have 100% reflectivity.
We call R and T (with R + T = 1) the reflectivity and transmitt-
ivity of mirrors 1 and 2.

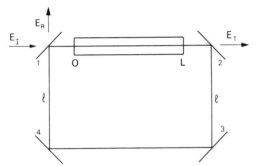

Fig.1.   Ring cavity. E_I, E_T and E_R are the incident, transmitted
and reflected field respectively.

We describe the dynamics of the coupled system atoms + radiation
field by the well known one-sided Maxwell-Bloch equations, which
are here written in general for an inhomogeneously broadened
system of N atoms

$$\frac{\partial S_\omega}{\partial t} = \frac{\mu}{\hbar} E D_\omega - \left[\gamma_\perp + i(\omega - \omega_o)\right] S_\omega , \qquad (1.1)$$

$$\frac{\partial D_\omega}{\partial t} = -\frac{\mu}{2\hbar}\left(E S_\omega^* + E^* S_\omega\right) - \gamma_\parallel \left[D_\omega - \frac{\mathcal{N}(\omega)}{2}\right] , \qquad (1.2)$$

427

$$\frac{\partial E}{\partial t} + c \frac{\partial E}{\partial z} = - g \int d\omega \, S_\omega .$$

$$(1.3)$$

In Eq. (1) $\mathcal{N}(\omega)d\omega$ is the number of atoms with frequency between ω and $\omega+d\omega$. One has of course $\int d\omega \, \mathcal{N}(\omega) = N$. We call S_ω and D_ω the polarization and the population difference (precisely, one half the difference between the populations of the lower and of the upper level) associated to the isochromat of frequency ω. E is the slowly varying envelope of the electric field. μ is the modulus of the dipole moment of the atoms and g is a coupling constant given by

$$g = 4\pi \omega_0 \mu / V,$$

$$(2)$$

where ω_0 is the frequency of the incident field and V the volume of the atomic sample. $\gamma_{\shortparallel}$ and $\gamma_\perp$ are the inverse of the atomic relaxation times T_1 and T_2 respectively. Of course the field equation is supplemented by some proper boundary conditions (see Fig.1):

$$E_T(t) = \sqrt{T} \, E(L,t) ,$$

$$(3.1)$$

$$E(0,t) = \sqrt{T} \, E_I + RE(L,t-\Delta t) \, e^{-i\theta T}.$$

$$(3.2)$$

In (3) E_I and E_T are the incident and transmitted field amplitudes respectively, $\Delta t = (2l+L)/c$ is the time the light takes to travel from mirror 2 to mirror 1 and θ is the cavity detuning parameter

$$\theta = (\omega_c - \omega_0)/K , \qquad K = cT/\ell ,$$

$$(4)$$

where ω_c is the frequency of the cavity that is nearest to resonance and $\ell = 2(L+l)$ is the total length of the cavity. K is the empty cavity width. The second contribution in the r.h.s. of (3.2) describes a feedback mechanism, arising from the mirrors, which is essential for the rise of bistability.

Let us first consider the steady state $(\partial E/\partial t = \partial S_\omega/\partial t = \partial D_\omega/\partial t = 0)$. It is suitable to introduce the adimensional amplitudes (y is taken real for definiteness)

$$F(z) = \frac{\mu E(z)}{\hbar \sqrt{\gamma_\perp \gamma_\parallel}} \quad , \qquad y = \frac{\mu E_I}{\hbar \sqrt{\gamma_\perp \gamma_\parallel T}} \quad . \tag{5}$$

One easily derives from (1) the following equation for the stationary field:

$$\frac{\partial F}{\partial z} = -\alpha F \chi \left(|F|^2 \right) , \tag{6}$$

where α is the linear absorption coefficient

$$\alpha = \mu g N / 2 \hbar c \gamma_\perp$$

and χ is the complex nonlinear susceptibility

$$\chi \left(|F|^2 \right) = \int d\omega \, \frac{\mathcal{P}(\omega)}{N} \left(1 - i \, \frac{\omega - \omega_o}{\gamma_\perp} \right) \frac{1}{1 + |F|^2 + \frac{(\omega - \omega_o)^2}{\gamma_\perp^2}} . \tag{7}$$

Eq. (6) is popular in the community of spectroscopists, but it contains only 50% of the physics of optical bistability. The other 50% is contained in the boundary conditions (3). In fact the bistability arises from the very fact that the field at z = 0 is in general quite different from the incident field because there is the feedback contribution. To describe the response of the system we must derive the relation which expresses the transmitted intensity $X = |F(L)|^2$ as a function of the incident intensity $Y = y^2$. By combining the solution of Eq. (5) with the boundary condition (3.2) we find a parametric representation for this function:

$$X = \frac{2}{\rho^2 - 1} \left[\alpha L - \left(1 + \Delta^2 \right) \ln \rho \right] ,$$

$$Y = \frac{2}{T^2} \frac{\alpha L - \left(1 + \Delta^2 \right) \ln \rho}{\rho^2 - 1} \left[\rho^2 + R^2 \right. \tag{8}$$

$$\left. - 2 R \rho \, \cos \left(\Delta \ln \rho - \theta T \right) \right] ,$$

where

$$\rho = |F(0)|/|F(L)| \quad, \quad \Delta = (\omega_A - \omega_o)/\gamma_\perp \ . \tag{9}$$

Δ is the atomic detuning parameter and ω_A is the central frequency of the atomic line. At resonance $(\Delta = \theta = 0)$ Eq. (8) reduces to [24]:

$$\ln\left[1 + T\left(\frac{y}{x} - 1\right)\right] + \frac{x^2}{2}\left\{\left[1 + T\left(\frac{y}{x} - 1\right)\right]^2 - 1\right\} = \alpha L , \tag{10}$$

where $x = \sqrt{X}$. Eq. (3) holds for a homogeneously broadened system and includes all propagation effects. Also in the case of a in-homogeneously broadened system we have obtained an analytical solution assuming a lorentzian atomic line:

$$\mathcal{N}(\omega) = \left(N/\pi T_2^*\right)\left[(\omega - \omega_A)^2 + (T_2^*)^{-2}\right]^{-1}. \tag{11}$$

The latter solution is more complicated and will be reported elsewhere.

Before analyzing the exact solution, let us state a theorem which shows that the solution drastically simplifies in the limit of small mirror transmissivity. Precisely, let us consider the following triple limit (mean field limit):

$$T \to 0 \quad, \quad \alpha L \to 0, \quad (\omega_c - \omega_o)\mathcal{L}/c \to 0, \tag{12}$$

with $C \equiv \alpha L/2T$ constant, $\theta = (\omega_c - \omega_o)\mathcal{L}/cT$ constant. In this limit the exact solution reduces to the state equation of the Mean Field Theory (MFT) of OB [6, 8, 12]:

$$Y = X\left\{\left[1 + \chi_1(X)\right]^2 + \left[\theta - \chi_2(X)\right]^2\right\}, \tag{13}$$

$$\chi_1(X) = 2C \operatorname{Re}\chi(X) = \frac{\sigma + \sqrt{1+X}}{\sqrt{1+X}} \frac{2C}{\Delta^2 + (\sigma + \sqrt{1+X})^2} , \tag{14.1}$$

$$\chi_2(X) = 2C \operatorname{Im}\chi(X) = \frac{2C\Delta}{\Delta^2 + (\sigma + \sqrt{1+X})^2}, \qquad (14.2)$$

where $\sigma = (\chi_1 T_2^*)^{-1}$. In particular, (13, 14) for $\sigma = 0$ follow from (8) performing the mean field limit (12). The advantages of the MFT are that it is simple and allows a fully quantum statistical treatment. On the basis of the MFT we have shown that OB is a relevant example of first order phase transition in a system far from thermal equilibrium. In particular we predicted a critical slowing down which has been recently observed in a hybrid device [35]. Furthermore we have given a description of the spectra of transmitted and fluorescent light [6,7,13,21] thereby establishing the connection between OB and resonance fluorescence. The mean field theorem (12), (13) shows that the MFT works very well when T is small enough. Note that for αL small the field inside the sample becomes more and more uniform. Furthermore the conditions

$$(\omega_c - \omega_o)\ell/c \ll 1 \quad , \quad (\omega_c - \omega_o)\ell/cT \quad \text{finite}$$

mean that the cavity detuning must be much smaller than the free spectral range, but of the same order of magnitude of the cavity width. A relation of type (13) between incident and transmitted field was first given in [3] on the basis of phenomenological arguments. Our approach derives this formula from first principles as an analytical solution of the Maxwell-Bloch equations with boundary conditions , pointing out its limit of validity and giving explicit expressions for $\chi_{1,2}$ in the case of the in-homogeneously broadened lorentzian lineshape (11). The homogeneous case is trivially obtained by putting $\sigma = 0$ in (13). On the other hand (8) contains propagation effects which must be measurable far from the mean field limit (12), when (13) does not hold. These propagation effects, as we shall see, induce an anomalous transient behavior (self-pulsing) which is unpredictable in the framework of the mean field theory. Fig. 2 shows the shape of the curves of transmitted vs incident light when there is bistability. The part of the curves with negative slope are unstable, as we shall see later, so that one finds a hysteresis cycle. Curve e) in Fig. 2.1 is the mean field result for C = 50, $\Delta = \theta = \sigma = 0$ (purely absorptive case); curve e) in Fig. 2.2 is the mean field result for C = 50, $\Delta = 10$, $\theta = 2.25$, $\sigma = 0$ (dispersive case). In both Figs. 2.1, 2.2 ... the curves a,b,c,d show the exact solution (8) for different values of αL and of the transmissivity, chosen in such a way that C = αL/2T is constant equal to 50. For high values of αL and T, as in curve a) there is no bistability

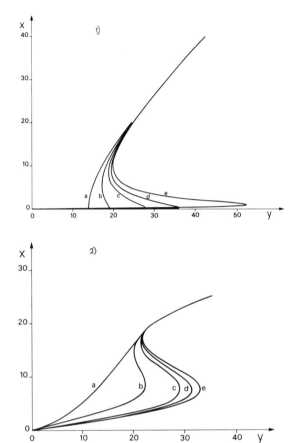

Fig. 2 Plot of the transmitted amplitude $x = \sqrt{X}$ vs the incident
amplitude y in the homogeneously broadened case. In both Figs.2.1
and 2.2 curves a), b), c), d) show the exact stationary solution;
curve e) is the mean field result. In Fig. 2.1 $C = 50$, $\Delta = \Theta = 0$;
in Fig. 2.2 $C = 50$, $\Delta = 10$, $\Theta = 2.25$. For curves a) $\alpha L = 100$
$T = 1$; for curves b) $\alpha L = 50$; $T = .5$; for curves c) $\alpha L = 20$
$T = .2$; for curves d) $\alpha L = 10$; $T = .1$.

whereas the bistable behavior increases by decreasing αL and T.
In this way one approaches the mean field result which is already
a good approximation for $\alpha L \simeq 1$.

Let us now consider in detail the mean field state equation (13).
When $\Delta, \Theta \ll 1$ one can neglect the term $[\Theta - \chi_2]^2$ so that one reduces
to the equation for purely absorptive optical bistability, in

432

which only the absorptive part of the susceptibility is important. On the contrary, for $\Delta \gg 1$ the absorptive part χ_1 is negligible and one has the equation for purely dispersive optical bistability, in which only the dispersive part of the polarization plays a role. The state equation (13), in the case of homogeneous broadening ($\sigma = 0$) has been analyzed in [8, 9, 12]. In this case one can analytically derive the conditions which ensure a bistable response. One has two conditions:

$$2C > \Delta\Theta - 1 \, , \tag{15 a}$$

$$(2C - \Delta\Theta + 1)^2 (C + 4\Delta\Theta - 4) > 27 C (\Delta + \Theta)^2. \tag{15 b}$$

Condition (15a) guarantees that the curve Y(X) given by (13) has an inflection point $X_{inf} > 0$; condition (15b) ensures that the slope of the curve at X_{inf} is negative. The analysis of the bistability conditions (15) leads to the following results. First, in the case of purely absorptive OB one has bistability for $C > 4$ as it is well known [6]. Second, for $C < 4$ not only absorptive but also dispersive bistability is impossible. Finally, when $C > 4$ the hysteresis cycle is largest for $\Delta = \Theta = 0$, i.e. in the absorptive case. The conclusion is that in the case of a homogeneously broadened system purely absorptive bistability is more suitable than dispersive OB, at least when the mean field approximation holds.

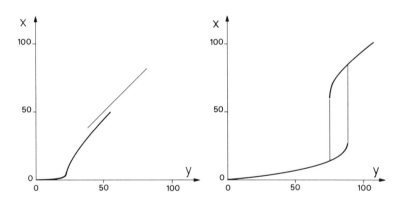

Fig. 3   Mean field curves for the transmitted amplitude $X = \sqrt{X}$ vs the incident amplitude y in the case of inhomogeneous broadening (lorentzian atomic line). For $\Delta = \Theta = 0$ (absorption case) one has no bistability, whereas for $\Delta = 30$, $\Theta = 1$ one has a hysteresis cycle. In both curves C = 180 and $\sigma = 15$.

On the other hand let us consider the case of inhomogeneous broadened systems. In general, for Δ, θ and σ fixed one has bistability when C is larger than a suitable threshold value $C_{min}(\Delta, \theta, \sigma)$ which increases rapidly with σ. When $\sigma \gg 1$ one finds values of C such that absorptive bistability is impossible, whereas dispersive bistability is possible. The situation is illustrated in Fig. 3. The conclusion is that in the case of inhomogeneously broadened systems dispersive bistability is more suitable than absorptive bistability.

3. Linear Stability Analysis

Let us now discuss the stability of the stationary solutions and for definiteness let us consider the purely absorptive case and assume homogeneous broadening. To check the stability of the steady state, we must consider an initial condition in which the system is slightly displaced from the stationary state and investigate whether the system returns to the stationary state or not. This regression to the steady state is described by the linearized Maxwell-Bloch equations for the deviation $\delta E(z,t)$ of the field from the stationary state. The time evolution of this deviation is a linear combination of exponentials

$$\delta E(z,t) = \sum_n c_n(z) \exp\left[-\lambda_n t\right] . \tag{16}$$

The index n labels the modes of the cavity. The complex eigenvalue λ_n characterizes the regression of the nth mode to its stationary value. In particular, Re λ_n gives the damping rate of the nth mode. By using the boundary conditions (3) we derived from the linearized Maxwell-Bloch equation the exact eigenvalue equation [25]

$$\lambda_n = i\Delta\omega_n + \frac{c}{\ell} \ln \frac{1}{1-T}$$

$$+ 2CK\gamma_\perp \frac{1}{L} \int_0^L dz \frac{1}{1+F^2(z)} \frac{\lambda_n + \gamma_{||}\left[1 - F^2(z)\right]}{(\lambda_n + \gamma_\perp)(\lambda_n + \gamma_{||}) + \gamma_\perp\gamma_{||} F^2(z)} , \tag{17}$$

where $\Delta\omega_n = \omega_n - \omega_0$ is the difference between the frequency of the nth mode and the frequency of the incident field, which is also a frequency of the cavity. Eq. (17) can be solved by iteration and one finds that λ_n has the structure

$$\lambda_n = i \Delta \omega_n + \frac{c}{\ell} \ln \frac{1}{1-T} - f(F, C, \Delta \omega_n) , \qquad (18)$$

so that defining $G = \mathrm{Re}\, f$

$$\mathrm{Re}\, \lambda_n = \frac{c}{\ell} \ln \frac{1}{1-T} - G(F, C, \Delta \omega_n) . \qquad (19)$$

Note that when G is positive (19) is a loss-minus-gain form. For an empty cavity $f = 0$, so that all the modes are equally damped. For a filled cavity under suitable conditions the gain can exceed the loss, so that the nth mode is no longer damped but amplified and the steady state is unstable. A simple analysis can be performed in the mean field limit $T \ll 1$, $\alpha L \ll 1$ with $\alpha L/2T = C$ fixed. In this limit one finds [27, 29]

$$G = -2C k \gamma_\perp \frac{1}{1+X} \frac{\gamma_\parallel (1-X) + i \Delta \omega_n}{(\gamma_\parallel + i \Delta \omega_n)(\gamma_\perp + i \Delta \omega_n) + \gamma_\perp \gamma_\parallel X} . \qquad (20)$$

Assume for simplicity that $\gamma_\perp = \gamma_\parallel = \gamma$. For states in the low transmission branch all the modes are stable. For states on the part with negative slope the resonant mode $n = 0$ (for which $\Delta \omega_n = 0$) is such that $\mathrm{Re}\, \lambda_o < 0$ and is therefore unstable. Hence these states are always unstable as we anticipated. The states in the high transmission branch such that $X = \sqrt{X} < C/2$ are unstable provided some off-resonance modes lie in the symmetrical ranges of the variable $\Delta \omega_n$ in correspondence of which $\mathrm{Re}\, \lambda_n < 0$ (Fig. 4). Finally for $X = \sqrt{X} > C/2$ the stationary state is always stable. Note that the point $x = C/2$ lies in the high transmission branch only for $C > 2(1 + \sqrt{2}) > 4$. Hence the instability in the high transmission branch can arise only when there is a hysteresis cycle. Note also that the detuning $\Delta \omega_n$ of the unstable modes is always smaller than the RABI frequency of the transmitted field.

4.   Self-Pulsing in Optical Bistability

One asks what happens in correspondence to the instability points in the high transmission branch. The answer is given in [27]. One has two possibilities. The first is that the system simply precipitates to the stable low transmission state which corresponds to the same value of the incident field. (Fig. 5.2).

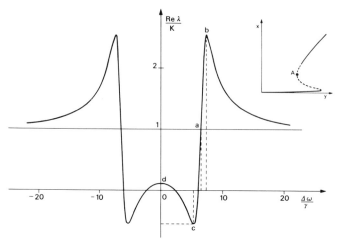

Fig. 4. Plot of the real part of the eigenvalue λ_n vs $\Delta\omega_n = \omega_n - \omega_0$ in the limit $T \ll 1$, $\alpha L \ll 1$ with $\alpha L/2T = C = 20$, $\gamma_\perp = \gamma_\parallel = \gamma$, $x = 6.1$. One has $(\Delta\omega/\gamma)_a = \sqrt{x^2 - 1}$ $(x > 1)$; $(\Delta\omega/\gamma)_b = \sqrt{x^2 + 2x - 1}$ $(x > \sqrt{2} - 1)$; $(\Delta\omega/\gamma)_c = \sqrt{x^2 - 2x - 1}$ $(x > \sqrt{2} + 1)$; $(Re\,\lambda/K)_c = 1 - C/2x$; $(Re\,\lambda/K)_d = dy/dx$. On the hysteresis cycle $x = 6.1$ is point A.

The other possibility is that the system approaches a limit cycle, in which the transmitted light is given by an infinite sequence of pulses (Fig. 5.1). In this case one has spontaneous generation of pulses. This behaviour is similar to that one finds in the so-called "second threshold" of the laser [36, 37]. In our case, the self-pulsing behaviour is very interesting from a practical view-point because it suggests a device to transform c-w light into pulsed light. This possibility has been independently suggested by McCALL on the basis of a completely different mechanism [30]. When there is only one symmetrical couple of unstable modes, the frequency of the oscillations in the limit cycle is roughly equal to the detuning $|\Delta\omega_n|$ of the unstable modes. In this case, one can even analytically calculate the limit cycle by using HAKEN's theory of phase transitions in systems far from thermal equilibrium [38]. This method has been generalized in [29] in order to be applied to OB. The agreement between analytical and numerical data for $T = 0.01$ is remarkable [39].

When T and αL are not small enough, one must consider the exact eigenvalue equation (17) instead of the approximate expression (20). The numerical solution of (17) [40] shows that, increasing T and αL with $C = \alpha L/2T$ fixed, the ranges of the variables $\Delta\omega_n$ such that $Re\,\lambda_n < 0$ become smaller and smaller. For T large enough all instabilities - as well as the bistability - disappear.

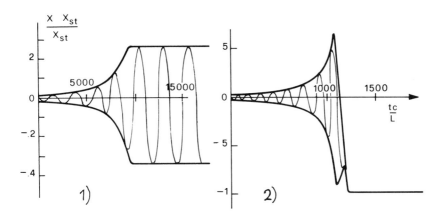

Fig. 5 Time evolution of the transmitted field when the stationary state on the high transmission branch is unstable. The time
width of the pulses has been increased by a factor ~ 300.

5. Connection Between Optical Bistability, Saturation Spectroscopy
   and the Laser with Injected Signal

5.1 Saturation Spectroscopy

Let us consider an atom driven by a strong pump field. Under
suitable conditions, a weak probe field which is incident on the
atom can experience gain. This effect has been both theoretically
predicted [41, 42] and experimentally observed [43, 44]. A similar
effect arises in an extended medium, under suitable conditions the
absorption coefficient of the probe field can become negative,
thereby producing gain. For definiteness, we call "probe gain" the
probe absorption coefficient changed of sign:

$$ g_p \left(F_s^2, \Delta\omega \right) = - \alpha_p \left(F_s^2, \Delta\omega \right) , \qquad (21) $$

where $\Delta\omega$ is the difference between the frequencies of the probe
field and of the saturator field. Clearly, the probe gain is a
gain in space. On the other hand the problem of optical bistability
is basically different, because first we have not a probe field
and second we have a cavity. The mirrors give rise to boundary
conditions which imply that the field inside the medium is quite
different from the incident field. Using these boundary conditions
we derived the exact eigenvalue equation (17) which rules the
stability of the stationary states. As we have seen the eigen-

values have the loss-minus-gain structure (19), where the correct expression of the loss term comes from the boundary conditions. We call G for definiteness "effective gain"; G is a gain in time. It is a different object from the probe gain g_p and has in general a quite different analytical structure. However one proves the following relevant theorem: in the mean field limit one has a simple proportionality relation between G and g_p, provided in the expression of g_p one takes as saturator intensity the effective driving intensity X and not the incident intensity Y

$$G\left(X, \Delta\omega_n\right) = c\left(L/\ell\right) g_p\left(F_s^2 = X, \Delta\omega_n\right). \qquad (22)$$

We recall that X is a discontinuous function of Y which can be obtained only by taking into account the boundary conditions. Finally, we have shown that under suitable conditions the effec tive gain exceeds the losses, thereby producing the instability. An important point is that this instability arises only in bi-stable situations. When the system shows self-pulsing it works like a parametric oscillator which transfers energy from the external signal to some unstable mode whose detuning is always smaller than the RABI frequency of the transmitted field.

5.2   Laser with Injected Signal

The steady state properties of our system in the mean field limit (when propagation effects can be neglected) appear similar to those of the laser with injected signal [47 - 50], in the sense that the outgoing intensity is formally a multivalued function of the incident intensity. However the physical behaviour of the two systems is radically different. In fact a stability analysis shows that the laser with injected signal is not bistable [48], [49, 50] , i.e. the transmitted intensity varies continuously with the incident intensity.

Acknowledgements

We thank V. Benza, P. Meystre and M. Sargent III for their precious collaboration and for stimulating discussions.

References

1. A.Szöke, V.Daneu, S.Goldhar and N.A.Kurnit, Appl.Phys.Lett. 15, 376 (1969). See also H.Seidel, U.S. Patent N. 3, 610, 731 (0ct. 5, 1971)
2. S.L.McCall, Phys.Rev. A9, 1515 (197 4).
3. H.M.Gibbs, S.L.McCall and T.N.C. Venkatesan, Phys.Rev.Lett. 36, 1135 (1976).

4. T.N.C.Venkatesan and S.L.McCall, Appl.Phys.Lett.3 C, 282 (1977)

5. J.Marburger and J.Garmire, in <u>Proceedings of the Conference on the Physics of Fiber Optics</u>, Rhode Island, June 1978, Plenum Press.

6. R. Bonifacio and L.A. Lugiato, Opt.Commun. 19, 172 (1976)

7. R. Bonifacio and L.A. Lugiato, Phys.Rev. A 18, 1129 (1978)

8. R. Bonifacio and L.A. Lugiato, Lett.Nuovo Cimento 21,517 (1978)

9. S.S.Hassan, P.D.Drummond and D.F.Walls, Opt.Commun. 27, 480 (1978).

10. R.Bonifacio and P.Meystre, Opt.Commun. 27, 147 (1978), 29, 131 (1978).

11. P.Schwendimann, Journal of Physics 12 A L 39 (1979).

12. R.Bonifacio, M.Gronchi and L.A.Lugiato, to appear on Nuovo Cimento B.

13. R.Bonifacio and L.A.Lugiato, Phys.Rev.Lett. 40, 1023 (1978).

14. G.S.Agarwal, L.M.Narducci, D.H.Feng and R.Gilmore, in L.Mandel and E.Wolf (eds.), <u>Coherence and Quantum Optics IV</u>, Proc. 4th Conference Rochester, U.S.S.; Plenum Publishing Corporation, New York, 1978.

15. G.S.Agarwal, L.M.Narducci, R.Gilmore and D.H.Feng, Opt.Lett. 2, 88 (1978).

16. G.S.Agarwal, L.M.Narducci, R.Gilmore and D.H.Feng, Phys.Rev. A 18, 620 (1978).

17. C.R.Willis, Opt.Commun. 23, 151 (1977), 26, 62 (1978).

18. R.Bonifacio, M.Gronchi and L.A.Lugiato, Phys.Rev. A 18, 2266 (1978).

19. A. Schenzle and H.Brandt, Opt.Commun. 27, 85 (1978).

20. W.C.Schieve, to be published.

21. L.A. Lugiato, Nuovo Cimento B 50, 89 (1979).

22. F.T.Arecchi and A.Politi, to appear on Opt. Commun.

23. P.D. Drummond and F.D.Walls, to be published.

24. R.Bonifacio and L.A. Lugiato, Lett.Nuovo Cimento 21,505 (1978).

25. R.Bonifacio and L.A.Lugiato, Lett.Nuovo Cimento 21,510 (1978).

26. J.Marburger and F.S.Felber, Phys.Rev. A 17, 335 (1978).

27. R.Bonifacio, M.Gronchi and L.A.Lugiato, to appear Opt.Commun.

28. R.Bonifacio, M.Gronchi, L.A.Lugiato and P.Meystre, to be published.

29. V.Benza and L.A.Lugiato, to appear Zeits f.Physik B.

30. S.L.McCall, Appl.Phys.Lett., 32, 284 (1978).

31. P.Meystre, Opt.Commun. 26, 277 (1978).

32. H.J.Carmichael, to appear Optica Acta.

33. E.Abraham, R.K.Bullough and S.S.Hassan, to appear Opt.Commun.

34. J.A.Hermann, to appear Optica Acta.

35. E.Garmire, J.H.Marburger, S.D.Allen and H.G.Winful, Appl. Phys.Lett. 34, 374 (1979).

36. H.Risken and K.Nummedal, Journ.Appl.Phys. 49, 4662 (1968).

37. R.Graham and H.Haken, Z.Phys. 213, 420 (1968).

38. H.Haken, Zeits f.Physik B 21, 105 (1975) B 22, 69 (1975).

39. V.Benza, L.A.Lugiato and P.Meystre, to be published.

40. V.Benza, R.Bonifacio, M.Gronchi, L.A.Lugiato and M.Sargent III, to be published.

41. S.G.Rautian and I.I.Sobel'man, Sov.Phys. JETP 14, 328 (1962)

42. B.R.Mollow, Phys.Rev. A 5, 2217 (1972).

43. D.W.Aleksandrov,A.M.Bonch-Bruevich, V.A.Khodovoi and N.A. Chigir, JETP Letters 18, 58 (1973).

44. F.Y.Wu, S.Ezekiel, M.Ducloy and B.R.Mollow, Phys.Rev.Lett. 38, 1077 (1977).

45. S.Haroche and F.Hartmann, Phys.Rev. A6, 1280 (1972)

46. M.Sargent III and P.E.Toschek, Appl.Phys. 11, 107 (1976).

47. V.Degiorgio and M.O.Scully, Phys.Rev. A2, 1170 (1970).

48. M.B.Spencer and W.E.Lamb Jr., Phys.Rev. A5, 884 (1972).

49. W.W. Chow, M.O.Scully and E.W. Van Stryland, Opt. Commun. 15, 6 (1975).

50. L.A.Lugiato, Lett.Nuovo Cimento 23, 609 (1978).

Controlling Light with Light: Optical Bistability and Optical Modulation

H. M. Gibbs, S. L. McCall, A. C. Gossard, A. Passner, and W. Wiegmann

Bell Laboratories, Murray Hill, New Jersey 07974

T. N. C. Venkatesan

Crawford Hill Laboratories Holmdel, New Jersey 07733

Abstract

A simple GaAs etalon has been used to observe optical bistability, pulse shaping, and modulation of one light beam by another. Such devices may lead to all-optical, high-speed computation and communication; all-optical signal processing may be appropriate with the advent of low-loss, high bit-rate optical transmission systems. The etalon consists of a sandwich of 4 μm of GaAs between 0.2 μm layers of $Al_{0.42}Ga_{0.58}As$ grown by molecular beam epitaxy. Nonlinear laser spectroscopy was used to identify appropriate non-linearities essential to the operation of this device.

1. Introduction

Recently, much progress has been made toward long-lived light sources, high quantum-efficiency detectors, and lossless, high-bandwidth optical fibers. Optical signal processing using all-optical circuitry seems relevant at this point. The success of an all-optical logic system requires an active element in which one light beam controls another. Optical bistable devices have this capability. Since the first observation in 1974 [1] of optical bistability in the transmission of a Fabry-Perot interferometer containing Na vapor, much progress has been made toward practical devices. The latest advance, a 5 μm GaAs-GaAlAs trilayer, will be described after the fundamentals of optical bistability are illustrated and essentials from nonlinear spectroscopy of GaAs are presented. Optical hysteresis in lasers is not discussed here (see *Note in Proof* at end of text).

2. Controlling Cavity Transmission Via Intracavity Intensity-Dependent Refractive Index

A. Optical Modulation Using a Fabry-Perot Interferometer

Summation of the multiply-reflected beams transmitted by a Fabry-Perot results in the well-known transmission equation

$$I_T = I_I / \left[1 + \frac{4R}{(1-R)^2} \sin^2\phi/2 \right] \tag{1}$$

where I_I is the incident light intensity, I_T is the transmitted intensity, R is the reflectivity of each of the lossless mirrors having transmission T = 1-R, and the phase factor ϕ is given by

$$\phi = (4\pi nL/\lambda)\cos\theta \tag{2}$$

where n is the refractive index of the material of length L between the reflecting surfaces, λ is the wavelength, and θ is the angle (inside the medium) between the beam and the etalon normal. Taking θ to be zero, the maxima occur for

$$2nL = m\lambda \tag{3}$$

where the integer m is called the order number. The importance of these formulae for this discussion is simply that *the transmission of the Fabry-Perot depends upon the light intensity if the optical pathlength nL depends upon the light intensity*. Intensity dependence of nL can occur in many ways including the following mechanisms already used to demonstrate optical bistability: index changes resulting from optical pumping of atoms from one state to another (Na, [1]); index changes from partial saturation of a resonant transition changing index contributions from much stronger but far-away bands (ruby, [2]); length and index changes from temperature changes caused by optical absorption (color filters, [3]; GaAs, [4]); index changes induced by saturating the free exciton resonance (GaAs, [4]); index changes produced by feeding back an electrical signal from an output detector to an intracavity phase shifter (hybrid devices, [5]). The transmission of the Fabry-Perot can also be intensity-dependent as a result of saturation of intracavity absorption. Historically [6], this was discussed and sought first, but all of the observed bistable systems [1-5] function primarily dispersively. Hence the discussion here is restricted to intensity-dependent dispersion rather than absorption.

The scanning of the wavelengths of peak transmission of a Fabry-Perot by an intensity-dependent intracavity optical pathlength is easily illustrated by a Corning 4-74 color filter polished down to form a 57-μm etalon and coated to yield 80% reflectivity [7]. A 632.8 nm He-Ne laser signal beam and a 647.1 nm control beam are focussed to the same 15-μm diameter spot on the etalon. The 11%/pass absorption of the control beam heats the etalon, increasing its length and, to a lesser extent, increasing its refractive index. The increase in optical pathlength can sweep the Fabry-Perot peak through the signal wavelength. Or, if the signal is initially transmitted through the etalon as in Fig.1a, the signal transmission is reduced by a control pulse and then returns as the etalon cools after the pulse is removed. Of course, a strong signal pulse is able to sweep the Fabry-Perot without a control pulse; this is shown in Figs.1b and 1c.

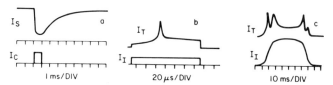

Fig.1 Sweeping of the frequency of a Fabry-Perot's peak transmission caused by thermal effects induced by optical absorption. (a) Transmission of a 1 mW 632.8 nm signal beam I_s through a 57 μm color filter (4-74 Corning) is greatly reduced by thermal scanning of the etalon by a 1 W 647.1 nm control beam I_c. (b) The 1 W 647.1 nm control beam can also alter its own color filter transmission changing a square top input pulse into a much shorter pulse on a pedestal. (c) Self-thermal scanning of a 100 mW beam through two peaks of a GaAs etalon [3 and 4].

B. *Optical Bistability and Differential Gain*

In the small phase shift approximation, sin $\phi/2 \approx \phi/2$, and (1) is approximated by

$$I_1 \approx \left[1 + R\phi^2/T^2\right]I_T. \tag{4}$$

A simple model [1,8] for bistability and AC gain emerges if one makes the not unreasonable approximation that the dependence of the optical pathlength on intensity is linear:

$$\phi = \beta I_T - \phi_0 \tag{5}$$

where ϕ_0 is the zero-intensity detuning between laser and Fabry-Perot cavity and β quantifies the strength of the nonlinearity. Substitution of (5) into (4) yields an inflection point $(d^2I_1/dI_T^2 = 0)$ for $I_T = 2\phi_0/3\beta$. If at that point $0 < dI_1/dI_T < 1$, which requires

$$0 < |\phi_0| < \sqrt{3}T/\sqrt{R}, \qquad \text{AC GAIN} \qquad (6)$$

then an AC gain G results:

$$G = \frac{dI_T}{dI_I} = \left(1 - R\phi_0^2/3T^2\right)^{-1}. \qquad (7)$$

For larger detunings, i.e.,

$$|\phi_0| > \sqrt{3}T/\sqrt{R}, \qquad \text{OPTICAL BISTABILITY} \qquad (8)$$

I_T of (4) becomes a double-valued function of I_I, see Fig.2. The reflectivity sets the scale for the required initial detuning. In fact, the required detuning for bistability $\sqrt{3}T/\sqrt{R}$ is $\sqrt{3}/2$ times the instrument width of the Fabry-Perot; *for bistability the detuning must be close to an instrument width*, ensuring that the low intensity transmission is low. The strength of the nonlinearity β sets the scale for the required intensity. Thus *the larger the nonlinearity, the lower the threshold intensity*, suggesting the use of resonance enhancement. The examples which follow may not strictly obey (5), but the essential points of the model apply: the required intensity is reduced by anything which increases the dependence of optical pathlength on light intensity and by increased reflectivity (provided the finesse is dominated by the reflectivity finesse, not by nonflatness, losses, etc.).

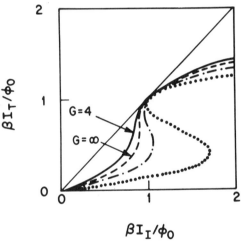

Fig.2 Characteristic transfer function calculated from the simple dispersive model of (4) and (5) for R = 0.9. The values of ϕ_0 are 0.158, 0.183, 0.224, and 0.3216, respectively, from left to right. The G=∞ curve separates the AC gain and bistable regions [1 and 8].

Returning to the simple color filter etalon, note the optical bistability and limiter action in Fig.3. The etalon transmission passes through a maximum both in switching on (the high spike) and just before switching off (the rounded knee, upper center). The transmission function of Fabry-Perot, ordinarily a symmetrical, periodic function of the phase when the optical pathlength is intensity-independent, is highly asymmetrical under conditions for bistability; see Fig.4.

C. Device Considerations

The remainder of this article concentrates on a 5-μm GaAs all-optical device [4] which we regard as the closest to a practical device so far. The first optical bistable system, a 11-cm plane Fabry-Perot contain-

443

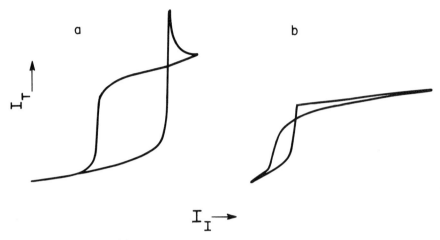

Fig.3 Optical bistability (a) and limiter action (b) in a 57-μm color filter. Input is 0.1 Hz triangular wave. The minimum power for bistability was 18 mW for optimum focussing (2.5 cm focal length) [3].

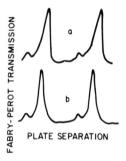

Fig.4 (a) Asymmetry of a Fabry-Perot scanned piezoelectrically under bistability conditions via an intracavity anti-reflection coated, 2 mm-thick Corning 3131 filter. (b) Symmetric scan in low-intensity linear regime. Small peak arises from improper mode matching [3].

ing Na vapor, was slow (≈ μs), sensitive to small MHz shifts in laser frequency, and not easily miniaturized [1]. The first room-temperature solid to exhibit bistability, ruby, was even slower (≈ ms) [2]. The color filter devices utilizing nonresonant thermal changes in optical pathlength require large energies and are slow (μs to ms), although low-temperature devices might switch in a ns or less [3]. The hybrid devices, for example Fig.5, are typically a cm in length to provide sufficient phase shift for feedback signals of several volts [5]; since they are nonresonant and some even cavity-less, either long phase shifters or external amplifiers are essential. As an all-optical device, the GaAs etalon is superior to all of these except that it has not been operated above 120°K. Its short length makes it the least sensitive to wavelength instabilities (except for no-cavity hybrid devices). Its small size should lead to high packing density when adequately heatsunk. Present switching energies are 0.6 nJ measured for a switch-beam diameter five times the cw diameter, implying 24 pJ actually used; sub-pJ energies are expected for smaller future devices. The wavelength of operation is convenient for AlGaAs diode lasers. And since GaAs is the nonlinear material, optical integrated circuitry including lasers, detectors, modulators, bistable memory elements, clippers, limiters, and amplifiers is a real possibility. All of these important device advantages were anticipated and motivated the choice of GaAs as a good candidate for a practical device.

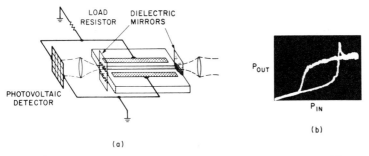

(a)   (b)

Fig.5 (a) Schematic view of self-contained, integrated bistable device. (b) Experimental optical hysteresis observed with the device shown in (a) [5c].

3. Nonlinear Optical Spectroscopy of Semiconductors

Construction of low-power optical devices requires resonantly enhanced nonlinear susceptibilities. Lasers have been used to study optical nonlinearities of many materials. In fact, the directionality of lasers minimizes the power required for a given intensity and permits spatial probing of samples with resolution approaching one wavelength. The monochromaticity possible with lasers is far narrower than most solid-state resonances; higher-power, multi-mode operation is often sufficient. However, most of the nonlinear optical studies have been non-resonant; for example; saturation of band-to-band transitions was found to require about 100 kW/cm^2 [9]. Studies of the changes in bandedge absorption in the vicinity of the exciton resonance resulting from intense above-bandgap pumping revealed a much lower saturation intensity of less than 1 kW/cm^2 to eliminate the narrow exciton resonance [10].

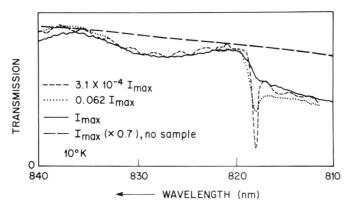

Fig.6 Nonlinear transmission of 500 ns pulses through an antireflection-coated trilayer consisting of 0.42 μm of GaAs between 2.38 and 3.33 μm layers of $Al_{0.24}Ga_{0.76}As$. $I_{max} \approx 43$ kW/cm^2 in the center of the focal spot on the sample [11].

For optical bistability and self-modulation, the nonlinear transmission of a single laser beam at wavelengths in the vicinity of the exciton line is of primary concern. The nonlinear absorption [11] of the free exciton resonance of GaAs is shown in Fig.6, taken with the apparatus of Fig.7. The sample was similar to that of Fig.8, but the GaAs layer was thinner and the AlGaAs windows thicker. In reasonable agreement with the above-bandgap results [10], a saturation intensity $I_s = 150$ W/cm^2 was found [11] where the exciton absorption α was found to saturate as

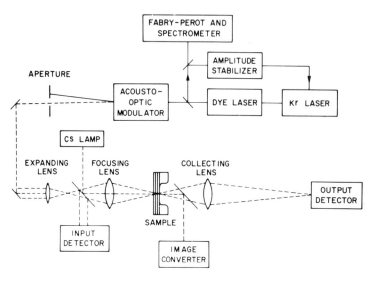

Fig.7 Experimental apparatus for nonlinear spectroscopy and optical bistability observations in GaAs [11].

$$\alpha = \alpha_B + \alpha_0/(1+I/I_s).$$

This saturation behavior is abrupt, as desired, but it saturates only down to a background absorption α_B. SZOKE et al. [6] showed that $\alpha L/T$ must exceed 8 for absorptive bistability. Since background absorption is similar to a loss via mirror transmission, $\alpha_0 L/(T+\alpha_B L)$ must exceed 8. The largest ratio α_0/α_B that we obtained was 6.6, consistent with our failure to observe absorptive optical bistability. The difficulty experimentally is obtaining narrow, strong exciton features simultaneously with flat, parallel surfaces [11].

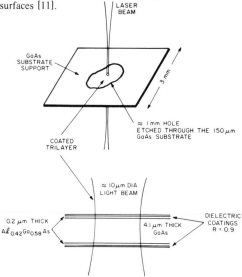

Fig.8 Trilayer sandwich grown by molecular beam epitaxy and used for optical bistability and optical modulation [4 and 7].

446

Even though saturation becomes more difficult off-resonance, for different reasons all of the intrinsic bistable systems operate predominantly by off-resonance refractive index changes rather than absorptive changes. From Fig.6, one can note that the background absorption α_B becomes smaller as the wavelength increases away from the exciton peak. By operating in this region with thicker samples, large enough index changes were obtained to permit bistability. These nonlinear index changes were detected by the resultant modulation of the etalon transmission as described in the next section.

4. Optical Modulation and Bistability in GaAs

The sample, grown by molecular beam epitaxy, is diagrammed in Fig.8. The best exciton features were observed in samples with low Al concentration $(x < 0.25)$. The highest finesse (flattest surfaces) were found with high Al concentration windows $(x > 0.3)$ which stop the selective etch of the substrate more abruptly. The sample of Fig.8 had a ratio of exciton peak absorption to band-to-band absorption of only 2.5 compared with values as high as 7 observed on low x samples. But the finesse for 90% coatings was 16, well below the bandgap, compared with only 8 in low x samples; nonflatness presumably accounts for not achieving the theoretical reflectivity finesse of 30. The reduced peak absorption was considered of lesser importance for off-resonance operation, and indeed, the 4-μm thickness of the sample and high finesse combined to produce Fabry-Perot shifts approaching one instrument width as needed for significant amplitude modulation and optical bistability.

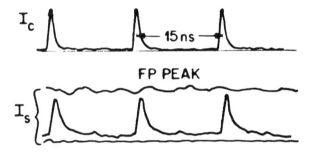

Fig.9 Control of a 10-mW cw dye laser beam I_s by 2.5-W 514.5 nm 200 ps mode-locked Ar pulses I_c. Turn-on of the signal follows the integrated control pulse which sweeps the Fabry-Perot peak onto the signal wavelength $(\lambda_s = 821.3$ nm). The 2-ns recovery time is related to the carrier lifetime. (10°K; $\lambda_{exciton} \approx 818$ nm) [7].

Figure 9 illustrates the control of an 821 nm cw dye laser beam by a train of 514.5 nm, 200 ps pulses; ps pulses could be synchronized in this manner. The integrated absorption of the control pulse creates enough carriers to destroy the excitonic contribution to the refractive index. This is shown in Fig.10. The absorption (Fig.10a) and intracavity phase shift (Fig.10b) are modified appreciably by the control pulse. If the laser frequency ν_L is at c in Fig.10c, the etalon transmission is low until the control pulse shifts the etalon peak onto the laser frequency. Such modulation studies could lead to nonlinear refractive index measurements with a careful deconvolution of the Fabry-Perot instrument function. A crude analysis revealed a maximum etalon shift of about one instrument width about 4 or 5 nm below the exciton resonance. The shift dropped off roughly as $1/\Delta\lambda$ for long λ. Absorption greatly reduced the transmission close to the exciton resonance. These measurements were made with a weak infrared cw dye laser beam so that it served only as a probe; all of the nonlinearities originated from the 514.5 nm control beam.

With no probe beam but with a strong infrared beam, optical bistability was observed as shown in Fig.11g. This excitonic bistability was observed up to 120°K (see Fig.11i for 107°K operation); at this temperature, kT already exceeds the 4 meV exciton binding energy. Also, thermal effects become

447

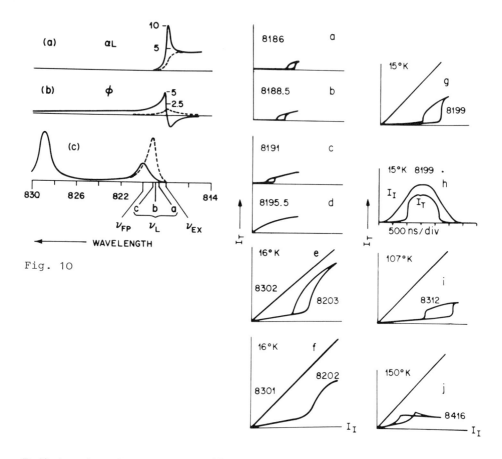

Fig. 10

Fig.10 Approximate GaAs absorption αL (a), roundtrip phase shift ϕ (b), and Fabry-Perot transmission under conditions of optical bistability (c). The solid curves are for zero input intensity where the exciton absorption is fully effective. The dashed curves are for intensities which are high enough to saturate the exciton feature, but low enough to leave the band-to-band contributions to α and ϕ unaffected. Relative frequency positions of laser ν_L, Fabry-Perot ν_{FP}, and free exciton ν_{EX} for observed optical bistability are shown in (c). The values a,b,c of ν_L refer to Fig.11a,b,c, respectively [4].

Fig.11 Optical bistability in 4 μm long GaAs between 0.2 μm $Al_{0.42}Ga_{0.58}As$ windows with R = 0.9 coatings. All figures show excitonic effects except (j) which shows thermal. Unitless numbers are wavelengths in Å. The laser-Fabry-Perot detuning depends upon position on the sample and was not determined. (a) to (d): dependence of bistability on laser tuning for a fixed position on the sample at 10°K. (e) and (f): bistability and differential gain compared with the next Fabry-Perot transmission at longer wavelength. (g) and (h) bistability as seen in x-y display (g) and time display (h) under the same conditions; in (h), I_I is divided by 2 relative to (g). (i) Excitonic bistability at "high" temperature. (j) Thermal bistability with 20 ms pulses.

more important at higher temperature, permitting thermal optical bistability to be observed with pulses of several ms (Fig.11j). The low-temperature bistability is not thermal in origin. In fact, excitonic and thermal optical bistability have initial detunings of opposite sign because their dn/dI have opposite signs. From Fig.11a-11c corresponding to ν_L at points a to c in Fig.10c, one can deduce that the excitonic dn/dI is negative whereas it is positive for heating.

The switching times for a triangular input are both about 40 ns; see Fig.11h. Much faster switch-on was achieved using a 589 nm, 200 ps pulse with the infrared intensity held midway between the turn-off and turn-on intensities; see Fig.12. A detector-limited switch-on time of about 1 ns was observed. Recent measurements [12] show that the exciton absorption is reduced to the band-to-band value in less than one ps by a sub-ps, above-bandgap saturation pulse. This suggests that an optical bistability switch-on time of a few ps should be possible.

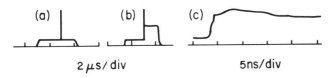

Fig.12 Switch-on of GaAs optical bistability by 590 nm, sub-ns pulse. The experimental parameters were adjusted so that switching did not occur in (a), but did in (b). The switch-on in (b) is expanded in (c), revealing a 1-ns, detector-limited switch-on time. The spikes in (a) and (b) are absent if the red probe beam is absent, i.e., they are gated 820 nm light and not leakage from the 590 nm control pulse.

Much remains to be done based on this first observation of optical bistability in semiconductors. Switch-on times of 10 ps or less should be demonstrated. Methods of sub-ns switch-off should be found. Dimensions should be reduced as much as possible toward the 230 nm wavelength of the light in the medium. Room-temperature operation should be sought, perhaps using other excitons with larger binding energies. Nonlinear laser spectroscopy of semiconductors should contribute to this search for the ultimate optically bistable element.

Note in Proof. A bistable laser was proposed by G. J. Lasher, Solid-State Electron. *7*, 707 (1964) in which either electrical or light pulses could switch states. Laser output hysterisis as a function of cavity detuning or excitation current was seen many years ago; e.g., R. L. Fork, W. J. Tomlinson, and L. J. Heilos, Appl. Phys. Lett. *8*, 162 (1966) reported hysteresis from polarization competition and V. N. Lisitsyn and V. P. Chebotaev, JETP Lett. *7*, 1 (1968) reported hysteresis from saturation in an intracavity absorption cell (theory in A. P. Kazantsev, S. G. Rautian, and G. I. Surdutovich, Sov. Phys. JETP *27*, 756 (1968). Recently hybrid optical bistability has been observed by inserting an electro-optic tuner inside a dye laser cavity: K. H. Levin and C. L. Tang, Appl. Phys. Lett. *34*, 376 (1979). We believe that the simplicity, smallness, and speed of intrinsic devices containing no gain medium give them important advantages over bistable lasers.

References

1. H. M. Gibbs, S. L. McCall, and T. N. C. Venkatesan, Phys. Rev. Lett. *36*, 1135 (1976).

2. T. N. C. Venkatesan and S. L. McCall, Appl. Phys. Lett. *30*, 282 (1977).

3. S. L. McCall and H. M. Gibbs, J. Opt. Soc. Am. *68*, 1378 (1978). S. L. McCall, H. M. Gibbs, W. P. Greene, and A. Passner, unpublished.

4. H. M. Gibbs, S. L. McCall, T. N. C. Venkatesan, A. C. Gossard, A. Passner, and W. Wiegmann, CLEA 1979 and unpublished.

5. (a) A. A. Kastal'skii,Sov. Phys. - Semicond. *7*, 635 (1973). (b) P. W. Smith and E. H. Turner, Appl. Phys. Lett. *30*, 280 (1977). P. W. Smith, E. H. Turner, and P. J. Maloney, IEEE J. Quantum Electron. QE-14, 207 (1978). (c) P. W. Smith, I. P. Kaminow, P. J. Maloney, and L. W. Stulz, Appl. Phys. Lett. *33*, 24 (1978) and *34*, 62 (1979). (d) E. Garmire, S. D. Allen, J. Marburger, and C. M.Verber, Opt. Lett. *3*, 69 (1978). (e) A. Feldman, Opt. Lett. *4*, 115 (1979).

6. A. Szöke, V. Daneu, J. Goldhar, and N. A. Kurnit, Appl. Phys. Lett. *15*, 376 (1969). S. L. McCall, Phys. Rev. A *9*, 1515 (1974). E. Spiller, J. Opt. Soc. Am. *61*, 669 (1971) and J. Appl. Phys. *43*, 1673 (1972). J. W. Austin and L. G. DeShazer, J. Opt. Soc. Am. *61*, 650 (1971).

7. H. M. Gibbs, T. N. C. Venkatesan, S. L. McCall, A. Passner, A. C. Gossard, and W. Wiegmann, Appl. Phys. Lett. *34*, 511 (1979).

8. H. M. Gibbs, S. L. McCall, and T. N. C. Venkatesan in F. T. Arecchi, R. Bonifacio, and M. O. Scully, eds., *Coherence in Spectroscopy and Modern Physics* (Plenum, NY, 1978), p. 111.

9. J. Shah, R. F. Leheny, and C. Lin, Solid State Comm. *18*, 1035 (1976). P. D. Dapkus, N. Holonyak, Jr., R. D. Burnham, and D. L. Keune, Appl. Phys. Lett. *16*, 93 (1970).

10. J. Shah, R. F. Leheny, and W. Wiegmann, Phys. Rev. B *16*, 1577 (1977) and R. F. Leheny and J. Shah, Solid State Electron. *21*, 167 (1978).

11. H. M. Gibbs, A. C. Gossard, S. L. McCall, A. Passner, W. Wiegmann and T. N. C. Venkatesan, Solid State Comm. *30*, 271 (1979).

12. C. V. Shank, R. L. Fork, R. F. Leheny, and J. Shah, Phys. Rev. Lett. *42*, 112 (1979).

Editors' Note

The following contribution to the topic of optical bistability was elicited: S.D. Smith and D.A.B. Miller (Heriot-Watt University, Edinburgh) reported recent original advances on bistability in the semiconductor InSb. The first observations of optical bistability in a one element device above 1st order were demonstrated and used to produce "optical transistor" action with differential signal gain. These results are in print for Appl. Phys. Lett. (Nov. 1979), Opt. Commun. (Nov. 1979) and Opt. Lett. (Oct. 1979).

Theory of Superfluorescence[1]

F. Haake

Fachbereich Physik, Gesamthochschule Essen
D-4300 Essen, Fed. Rep. of Germany

On the basis of a one dimensional model we show that superfluorescent pulses fluctuate strongly in shape even though each
pulse has a classical dynamical behavior. Our results are compatible with existing experimental data.

Superfluorescent light pulses are radiated by collections of
many atoms which are excited such that, at some initial time,
there is neither a macroscopic electric polarization nor an
external electric field present. Such an initial state would be
infinitely long-lived would not spontaneous emission events,
i.e. quantum fluctuations initiate its decay.

When no collisions between the atoms nor any other dephasing
mechanisms are effective the radiation process is cooperative.
The light pulse generated can then have a peak intensity I_{max}
proportional to N^2, the square of the number of atoms, and a
line width proportional to N. In practice, both I_{max} and the
line width will be much larger than is characteristic for
incoherent radiation pulses.

The cooperativity of the process suggests that the atoms behave in effect nearly classically - N classical dipoles radiating in phase with one another would produce a peak intensity
proportional to N^2. However, the fact that the radiation is
initiated by quantum fluctuations leads one to expect that the
process is essentially quantum mechanical in character.

In fact, the classical and quantum aspects of the process are
complementary. I shall argue in the following that

(i)   each radiated superfluorescent pulse corresponds to a
       solution of a classical problem, if the number of atoms
       is large.

---

[1]In this talk I report results of an investigation done by
R. Clauber, myself, J. Haus, F. Hopf, H. King, and G. Schröder, see ref. [1,2,3,4].

(ii)   In spite of its classical dynamics no such pulse can be
        predicted in its precise time dependence. The reason for
        that is that the initial values for the electric polari-
        zation of the atoms is random due to the quantum mecha-
        nical uncertainty the polarization has in the initial
        state.

(iii)   Different pulses differ greatly in their shapes. The
        peak intensity I_{max}, the time t_{max} of its appearance,
        the width of the first lobe, and other parameters
        characterizing the pulse shape fluctuate from one
        pulse to the next. These fluctuations are macroscopic
        manifestations of the initial randomness of the atomic
        polarization.

The features of superfluorescent pulses just described are
in accord with what one generally expects for macroscopic
systems in which an initially prepared state of unstable equilibrium decays due to the influence of microscopic fluctuations. Other wellknown examples for such decay processes are
the switch-on of a laser, the abrupt switch-on of a temperature
gradient in a Bénard cell, and spinodal decomposition in a
magnet.

I should also point out that the features of superfluorescent
pulses mentioned have been predicted on the basis of the so-
called single-mode model of Bonifacio, Schwendimann, and Haake
[5] which accounts for the atoms by means of a macroscopic
dipole and for the electric field by means of a single harmonic
oscillator and simulates propagation effects by a loss mechanism for the oscillator. While this model - nor its refinements
[6] - cannot be expected to be quantitatively correct it has,
because of its simplicity, provided physical insight and, most
importantly, permitted to find the conditions under which
superfluorescence can be and has been observed experimentally.

In describing the radiating atoms and their radiation field
we are facing a quantum mechanical problem in three spatial
dimensions. However, in order to reduce the complexity of
calculations to a manageable degree we have chosen to investigate a one dimensional model in which the operators for the
fields we use to describe the system depend on a single spatial
variable x. Such a one dimensional model is an oversimplifica-
tion with respect to any real radiating system. It should, however, share the most important qualitative features of its
dynamical and statistical behavior with systems in which the
atoms are confined to a cylindrical volume of Fresnel number F
near unity. Systems with $F \approx 1$ represent the best possible
compromise between the two conflicting requirements that the
radiation go preferentially along the cylinder axis and that
the electric field have a weak dependence on the radial coordinates and suffer little diffraction losses.

In order to simplify the problem even further we assume that
the number density of atoms n is large enough that there are
many atoms in every disk-shaped section of the cylinder extend-

ing over one wavelength λ along the axis. It is then possible to treat the dynamics of the system in terms of smoothly varying field operators for the atomic polarization and inversion.

Our model allows for propagation of electromagnetic waves in both directions along the cylinder axis. We represent the electric field by envelope operators $E_R^{(\pm)}(x,t)$ and $E_L^{(\pm)}(x,t)$ which describe the space-time modulation of the right- and leftgoing waves, respectively. The superscripts distinguish the positive and negative frequency parts. Similarly, we let the atomic polarization density be represented by the envelope operators $R_\pm(x,t)$ and $L_\pm(x,t)$ for the right- and leftgoing waves. Finally, we denote the atomic inversion density by $Z(x,t)$. I choose these operators as well as the space- and time coordinates to be dimensionless. Lengths are referred to the length ℓ of the cylinder and times to the socalled superfluorescence time τ, defined by

$$1/\tau \;=\; 3\lambda^2\, n\ell/8\pi\tau_0 \;=\; \ell c/\ell_c^2 \quad . \tag{1}$$

Here τ_0 is the natural lifetime of the excited atomic state and ℓ_c is the cooperation length introduced by ARECCHI and COURTENS. We normalize the field operators such that the inversion operator Z has the eigenvalues $+1$ and -1 in the initially prepared excited state and the ground state, respectively and such that we have the commutation rules

$$[R_+(x),R_-(x')] \;=\; (4/N)\,\delta(x-x')\,Z(x)$$

$$[Z(x),R_\pm(x')] \;=\; \pm(2/N)\,\delta(x-x')\,R_\pm(x) \tag{2}$$

$$[E_R^{(+)}(x),E_R^{(-)}(x')] \;=\; (4/\alpha N)\,\delta(x-x') \quad .$$

Similar rules hold for the envelopes of the leftgoing fields. Note that all commutators are of order $1/N$ and that the fields will therefore tend to behave classically. The electric field commutator involves the parameter

$$\alpha \;=\; (\ell/\ell_c)^2 \;=\; \ell/c\tau \tag{3}$$

which I will require to be smaller than unity below.

The dynamical behavior of our field operators is governed by their Heisenberg equations of motion which read, in the rotating wave approximation

$$\partial/\partial t\, R_\pm \;=\; Z\, E_R^{(\mp)} \quad , \qquad \partial/\partial t\, L_\pm \;=\; Z\, E_L^{(\mp)}$$

$$\partial/\partial t\, Z \;=\; -\tfrac{1}{2}\{E_R^{(+)}\, R_+ + E_L^{(+)}\, L_+ + \text{h.c.}\} \tag{4}$$

$$(\partial/\partial x + \alpha\,\partial/\partial t)\, E_R^{(\pm)} \;=\; R_\mp$$

$$(-\partial/\partial x + \alpha\,\partial/\partial t)\, E_L^{(\pm)} \;=\; L_\mp \quad .$$

These equations, although of quantum mechanical nature, are identical in appearence with the Maxwell-Bloch equations well-known from classical radiation theory.

The initial state with respect to which we have to evaluate expectation values of our operators is the vacuum of the electric field and the excited state for each atom,

$$|0\rangle \quad = \quad |vac, \{\uparrow\}\rangle \qquad . \qquad (5)$$

At t = 0, therefore, all normally ordered expectation values of electric field operators vanish. Likewise, all expectation valuex involving an odd number of polarization operators vanish. The nonvanishing even-order moments of the polarization have, if $N \gg 1$, Gaussian behavior and can all be built up as sums of products of the second-order moments

$$\langle 0| \ R_+(x,0) \ R_-(x',0) \ |0\rangle \qquad =$$

$$= \langle 0| \ L_+(x,0) \ L_-(x',0) \ |0\rangle \qquad = \qquad (4/N) \ \delta(x-x') \qquad . \ (6)$$

These second-order moments express the quantum mechanical uncertainty the atomic polarization must have in an eigenstate of the inversion since the corresponding operators do not commute.

Obviously, we can, at t = 0, calculate all normally ordered expectation values, i.e. moments like

$$\langle 0| \ E_R^{(-)} \ E_R^{(+)} \ E_L^{(-)} \ E_L^{(+)} \ R_+ \ L_+ \ Z \ L_- \ R_- \ |0\rangle \qquad , \qquad (7)$$

as if the fields were classical random fields with a probability distribution which assigns the sharp values zero and unity to the electric field and the atomic inversion, respectively, at each point x. The polarizations, on the other hand, would then have, at each x, a Gaussian distribution with the small width (4/N).

The quasiprobabilistic picture of the statistical behavior of our fields just given for t = 0 can also be used at later times. During the early stages of the radiation process, when the depletion of the atomic excited state is still negligable, we can linearize the equations of motion (4) by putting $Z(x,t) = 1$. The time-dependent fields $E_{R,L}^{(\pm)}(x,t)$, $R_\pm(x,t)$, and $L_\pm(x,t)$ are then easily found [1] as linear functionals of their initial values. Because of this linearity the normally ordered expectation values retain the same factorization properties they had initially.

The solutions of the linearized equations of motion are identical in structure with the solutions we would find were we concerned with the calssical Maxwell Bloch equations. Therefore, we can calculate all normally ordered expectation values as if the radiation pulse were triggered by an initially random polarization and had classical deterministic dynamics afterwards.

For later times, when the pulse intensity has risen to a macroscopic level and when, eventually, the nonlinearity in the equations of motion becomes effective, the above classical

454

picture becomes applicable in yet another, more direct sense. While up to such times we can make statements about ensembles of pulses only we can now say, in view of the correspondence principle, that each observable pulse has classical dynamics if N ≫ 1. However, since each such pulse was initially triggered by a random polarization, different pulses must be expected to differ in shape and the time dependence of no individual pulse can be predicted.

According to the above argument an ensemble of solutions of the classical Maxwell Bloch equations arising from a Gaussian ensemble of initial polarization configurations can be identified with an ensemble of sufficiently many experimental pulses. Unfortunately, because of the nonlinearity of the Maxwell Bloch equations we have no analytic solutions available and must calculate the classical trajectories numerically.

The question immediately arises whether it is possible to calculate by numerical means so many classical solutions that the function space available to the initial polarization configurations is reasonably well represented. Serious doubts may arise from the fact that the Gaussian probability distribution for the polarization does not, according to eq. (6), suppress spatial inhomogeneities and thus allows for spatial fluctuations of arbitrarily short wavelengths. However, for systems shorter than the cooperation length, i.e. with

$$\alpha = (\ell/\ell_c)^2 < 1 \qquad\qquad\qquad (8)$$

the Maxwell Bloch equations themselves suppress short-wavelength fluctuations in the initial polarizations relative to long-wavelength ones. This dynamical preponderance of long wavelengths can most easily be checked by Fourier-analyzing the linearized Maxwell Bloch equations. Such an analysis reveals that only a few low order Fourier components of the fields get significantly amplified if $\alpha < 1$. Consequently, the radiated intensity will not depend sensitively on the initial values of short wavelength Fourier components of the polarizations.

We have calculated, for various fixed values of the parameters α and N, large numbers of classical intensity trajectories. The initial polarization configurations were fixed either in terms of Fourier series truncated at some low order or, in most calculations, by Gaussian random numbers at each point along a spatial grid along the cylinder.

The ensembles of intensity trajectories so generated display large fluctuations. For instance, for $\alpha = 0.32$, $N \approx 1,5 \times 10^8$ the hight I_{max} and the delay time t_{max} of the first intensity peak have, in an ensemble of ~ 3500 trajectories based on random number realizations of the initial polarization, root mean squares of ~ 14 % and ~ 11 % of the respective means. Histograms of the distributions of t_{max} and I_{max} are shown in Figs. 1 and 2, respectively. It is expected that these results would not change if the number of trajectories were

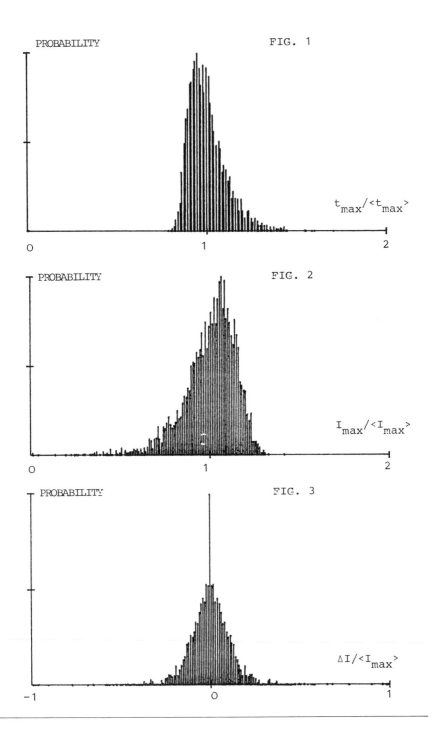

PROBABILITY   FIG. 1

$t_{max}/<t_{max}>$

O   1   2

PROBABILITY   FIG. 2

$I_{max}/<I_{max}>$

O   1   2

PROBABILITY   FIG. 3

$\Delta I/<I_{max}>$

-1   0   1

increased since the statistics was stabilized after a few hundred trajectories already. These statistical results are compatible with estimates based on the linearized treatment of the early-stage regime [7]. In a paper to be presented at this conference Dr. VREHEN will report experimental data on the statistics of superfluorescent pulses which are in amazingly good agreement with the above predictions.

We have also calculated the mean delay time as a function of the number density of atoms. The result is compatible with the experimental results of GIBBS, VREHEN, and HIKSPOORS [8] and VREHEN and SCHUURMANS [9].

Another result of ours should be easily comparable to experimental data. While there is a considerable statistical spread in the energy radiated to either side of the cylinder we find that the total atomic energy is preferentially radiated symmetrically to both sides. Fig. 3 shows the probability of finding given values for the difference of the first intensity peaks of the right- and leftgoing pulses.

The sizable magnitude of the fluctuations we predict, the apparent success in explaining some of the few available experimental data, as well as the failure of our one dimensional model to account for the absence of ringing in the experiments of GIBBS and VREHEN make it seem quite desirable to extend our investigation to three dimensions and, on the other hand, to accumulate more experimental data.

We would like to thank R. BONIFACIO, D. POLDER, M.F.H. SCHUURMANS, and Q.H.F. VREHEN for fruitful discussions and correspondence.

We gratefully acknowledge financial support by a NATO Research Grant, No. 1445 and by the U.S. Department of Energy under Contract No. EY 76-O2-3064.

References

1. R. Glauber and F. Haake, Phys. Lett. 68 A, 29 (1978).

2. F. Haake, H. King, G. Schröder, J. Haus, R. Glauber, and F. Hopf, Phys. Rev. Lett., to appear in the June 25, 1979 issue.

3. F. Haake, H. King, G. Schröder, J. Huas, and R. Glauber, Phys. Rev. A, to be published.

4. F. Hopf, Phys. Rev. A, to be published.

457

5. R. Bonifacio, P. Schwendimann, and F. Haake, Phys. Rev. A $\underline{4}$, 302 (1971); Phys. Rev. A $\underline{4}$, 854 (1971).

6. R. Bonifacio and L.A. Lugiato, Phys. Rev. $\underline{A11}$,1507 (1975); $\underline{A12}$, 587 (1975).

7. D. Polder, M.F.H. Schuurmans, and Q.H.F. Vrehen, J. Opt. Soc. Am. $\underline{68}$, 699 (1978), and Phys. Rev. A, to be published.

8. H.M. Gibbs, Q.H.F. Vrehen, and H.M.J. Hikspoors, Phys. Rev. Lett. $\underline{39}$, 547 (1977).

9. Q.H.F. Vrehen and M.F.H. Schuurmans, Phys. Rev. Lett. $\underline{42}$, 224 (1979).

Quantum Theory of Superfluorescence

M.F.H. Schuurmans and D. Polder

Philips Research Laboratories, Eindhoven, The Netherlands

Abstract

A quantum-mechanical description of the initiation of super-fluorescence in an extended pencil-shaped medium of N two-level atoms is presented. Propagation effects along the pencil axis are fully included. Transverse effects are neglected. The theory can be extended beyond the initiation regime, where decreasing inversion becomes important. The quantum noise initiating superfluorescence consists of the zero-point fluctuations of both matter and field. The effective initial tipping angle $\theta_0 \simeq 2/\sqrt{N}$. SF can also be described by classical Maxwell-Bloch equations with field and matter noise sources. In this stochastic variables description, the SF output due to one single shot excitation of the inversion corresponds to one single element of the ensemble of noise source functions. Averaged over many repeated experiments, the mean delay time $\langle \tau_D \rangle = (\tau_R/4) [\ln \sqrt{2\pi N}]^2$ and the relative standard deviation $\Delta \tau_D = 2.3/(\ln N)$, where τ_R is the collective decay time.

1. Introduction

Superfluorescence (SF) is the *cooperative* emission from a large number of initially inverted atoms, *without initial macroscopic dipole moment*.

SF was predicted by DICKE [1] in 1954. It was first observed by the MIT group of FELD and co-workers [2] in 1973. Since then several observations of SF on different atomic and molecular transitions have been reported [3,4]. All observations pertain to extended systems, i.e. samples with linear dimensions large compared to the wavelength λ_0 of the emitted radiation.

This paper summarizes our contribution to the theory of SF. Our work has been greatly stimulated and influenced by the experimental work of GIBBS and VREHEN in this laboratory [4]. These experiments were conducted on a transition in atomic Cs which is well suited for a study of SF under a wide range of conditions. Our theoretical work started from the state of the art at the Rochester conference on Coherence and Quantum Optics in 1977. We will briefly summarize the situation at that time. Two main theories of SF from pencil-shaped extended samples existed : (i) the quantum-mechanical mean field or one mode theory of BONIFACIO and co-workers [5] and (ii) the semi-classical Maxwell-Bloch theory of MACGILLIVRAY and FELD [6]. The mean field theory neglects the spatial variation of the field envelope throughout the sample, whereas the semi-classical theory includes the propagation effects along the main axis of the sample. The importance of propagation effects was already indicated by the 1973 MIT experiments. Both theories do not include transverse effects. The semi-classical theory describes the initiation of SF by a fluctuating polarization source. However, proper treatment of the initiation requires a fully quantum-mechanical theory. At the Rochester conference, papers were presented [7] which attempted to combine the semi-classical and quantum-mechanical treatments, but none of them led to a completely quantum-mechanical theory with full inclusion of the propagation effects. The strength of the quantum-noise initiating SF was under debate [5,6,8] and the nature of the noise was not fully understood. The interpretation of the SF output originating from one single-shot excitation and the shot-to-shot fluctuations in the SF outputs, was obscure. Finally, semi-classical theory predicts ringing of SF outputs, but GIBBS and VREHEN have observed single-pulse SF [4].

Our theoretical work pertains to a pencil-shaped sample with cross-sectional area $S >> \lambda_0^2$, length L and Fresnel number $F \equiv S/(\lambda_0 L) = 1$, containing N two-level atoms with transition wavelength λ_0 and transition dipole moment μ transverse to the pencil axis. The atomic density $\rho = N/(SL)$ is so large that the characteristic time for collective decay along the pencil axis $\tau_R = (8\pi\tau_n/3)/(\rho\lambda_0^2 L)$ is much shorter than the natural lifetime τ_n of a single atom. The atomic density should be so small that dipole-dipole interaction can be neglected, i.e. $\rho(\lambda_0/2\pi)^3 << 1$. Atomic motion as well as collisions are disregarded, but the atoms occupy random positions. The inversion is produced by swept-pulse excitation and only emission along the direction of excitation is considered.

Our results can be summarized as follows :

(i) In the linear regime, when decreasing atomic inversion can be ignored, SF is described quantum-mechanically with full inclusion of the propagation effects along the pencil axis. After the initiation SF becomes classical in the linear regime. The theory can be further extended into the non-linear regime.

(ii) The quantum noise initiating SF consists of the zero-point fluctuations of both matter and field. Consideration of normally ordered field correlation functions leads to the GH theory of GLAUBER and HAAKE [9], which ascribes the initiation to the matter zero-point fluctuations. On the other hand, consideration of antinormally ordered atomic polarization correlation functions leads to the PSV theory presented by VREHEN and the present authors [10], which ascribes the initiation to zero-point fluctuations of the vacuum field.

(iii) For each given operator ordering there is a stochastic variables description of the initiation of SF in terms of classical Maxwell-Bloch equations with a corresponding noise source: a field source for atomic polarization measurements pertaining to antinormal ordering (PSV theory) and a polarization source for field measurements pertaining to normal ordering (GH theory).

(iv) The mean squared effect of the field or polarization noise can be simulated by a coherent pulse of area Θ_0 and an effective initial tipping angle Θ_0 of the collective Bloch vector, respectively. We find $\Theta_0 \simeq 2/\sqrt{N}$, in agreement with the prediction of the quantum-mechanical theory of BONIFACIO and LUGIATO [5] but distinctively different from the predictions of MACGILLIVRAY and FELD [6] and EBERLY and REHLER [8]. The value $\Theta_0 \simeq 2/\sqrt{N}$ has been confirmed by recent experiments done by VREHEN [11].

(v) In a stochastic variables description the response to a single-shot excitation of the inversion exhibits the action of one single element of the corresponding ensemble of noise source functions. This interpretation, which was first explicitly introduced by VREHEN and the present authors [10], and the extension of the theory into the non-linear regime, opens the way for calculations of single-shot SF emission outputs. Numerical calculations of this type for the emission intensity have been done by HAAKE et al [12] and by HOPF [13]. The outputs of such non-linear, one-dimensional, calculations exhibit ringing. We suspect that transverse effects and inhomogeneous broadening are responsible for the observed absence of ringing [4]. Note that the absence of ringing cannot be explained by referring to quantum-mechanical averages since these averages relate to many repeated experiments.

(vi) The stochastic initial motion of the collective Bloch vector, in a linear Maxwell-Bloch description with a field noise source, is completely described by a bivariate Gaussian probability distribution function. The knowledge of this distribution function has enabled us to calculate analytically the spread in delay times as it would be measured over many repeated experiments. Within a 10 % accuracy limit, the relative standard deviation $\Delta\tau_D = 2.3/(\ln N)$. Recent experimental results reported by VREHEN are consistent with this prediction [14].

The principal assumption in our theory is that the initiation, propagation and collective growth of the SF emission are governed by the interaction of atoms with field modes inside two small solid angles λ_0^2/S around the pencil axis. This treatment in terms of two end-fire modes is made possible by our explicit assumption of Fresnel number $F=1$, i.e. the diffraction angle $\lambda_0/\sqrt{S}$ equals the geometric angle $\sqrt{S}/L$. The reason for singling out the end-fire modes is, of course, the largest logarithm of gain along the pencil axis. The gain of field modes outside the specified solid angles is ignored.

The remaining part of this paper elucidates the results (i) - (vi).

2. Quantum Description of the Linear Regime

In this section the operator Maxwell-Bloch equations describing the initiation of SF will be introduced. A detailed derivation can be found in the PSV paper [10].

The two-level atom j at the position $\underline{r}_j$ is described by the raising operator R_j^+, the lowering operator R_j^- and the inversion operator R_j^3. Each R_j^3 has eigenvalues $\pm\frac{1}{2}$ and the set $\{R_j^+,^3\}$ obeys spin $\frac{1}{2}$ angular momentum commutation relations. The atom-field system is described by the Heisenberg atomic operators $\{R_j^+,^3(t)\}$ and the Heisenberg electric field operator component $E(\underline{r},t)$ parallel to the transition dipole moment μ. The Heisenberg equations of motion are derived with the aid of the electric-dipole approximation.

We would like to find an approximate description of SF using as a guiding principle the fact that the gain of waves travelling through the active medium is largest for end-fire modes of propagation. In fact we will assume that, since $F=1$, a plane wave treatment of the two end-fire modes is sufficient for the description of the gross features of SF. We need not bother about the radiation reaction field part of E. This part accounts for lateral incoherent emission on the time scale $\tau_n \gg \tau_R$ and will be disregarded. The end-fire modes couple to certain collective atomic operators. We divide the pencil into slices of thickness d, oriented perpendicular to the pencil (x) axis. The center position of the slice is denoted by the discrete variable x. Since $\rho(S\lambda_0/2\pi)\sim\tau_n/\tau_R$ and $\tau_n \gg \tau_R$ we may choose d such that each slice contains many atoms and is still thin compared with $\lambda_0/2\pi$. The collective atomic operators are defined by

$$R_x^{\pm,3}(t) = \frac{1}{N_s} \sum_{j\in\{j\}_x} R_j^{\pm,3}(t), \tag{1}$$

where N_s is the mean number of atoms in a slice and $\{j\}_x$ denotes the collection of atoms with (fixed) positions $\underline{r}_j$, where $x-d/2 < x_j < x + d/2$. In the spirit of the plane wave end-fire mode assumption we now require the collective operators to be essentially independent of the particular random positions in the slice. Accordingly, the collective atomic operators interact with a slice-averaged electric field E(x,t), defined by integrating $E(\underline{r}_j,t)$ over $\underline{r}_j$ in a slice and dividing the result by the slice volume Sd. The slice-averaged electric field corresponds to an equiphase plane average of the "microscopic" electric field. Such equiphase plane averaging is in order because the mean interatomic distance is assumed to be larger than $\lambda_0/2\pi$, cf. KRAMERS [15]. The slice-averaged field E(x,t) consists of modes in the two specified solid angles around the pencil axis. When diffraction losses are disregarded, E(x,t) corresponds to the Maxwell field in the plane wave end-fire modes. By the very introduction of the collective atomic operator variables $\{R_x^{\pm,3}(t)\}$ and the operator Maxwell field

E(x,t) in the Heisenberg equations of motion for $\{R_j^{\pm,3}(t)\}$ and
E(r_j,t), and by formally adopting x as a continuous variable,
we find that $\{R^{\pm,3}(x,t)\}$ and E(x,t) satisfy operator valued
Maxwell-Bloch equations. These equations can be further simplified by the use of the rotating wave approximation (RWA)
and the introduction of slowly varying field E_L, E_R and polarization P_L, P_R envelope amplitudes of left (L; towards -x) and
right running (R) end-fire modes of the positive frequency
parts $E^{(+)}$ and $\mu^* R_2^-$ respectively. Similarly, the hermitean con-
jugated operators $E_L^\dagger$, $E_R^\dagger$ and $P_L^\dagger$, $P_R^\dagger$ correspond to the negative
frequency parts $E^{(-)}$ and μR^+, respectively. The use of the RWA
inevitably means that the dipole-dipole interaction between
the atoms is not accounted for adequately. However, since
$\rho(\lambda_0/2\pi)^3 \ll 1$ the dipole-dipole dephasing time $T_\Delta \simeq \tau_n/[\rho(\lambda_0/2\pi)^3]$
is much longer than τ_n so that dipole-dipole interaction can
be left out of account.

Having thus described schematically the derivation of operator Maxwell-Bloch equations for SF, we now specialize to the
initiation of SF. We approximate $2R^3$ by the identity operator,
thereby neglecting decreasing atomic inversion. Left and right
travelling waves then decouple and the dimensionless field operator $E \equiv -i\tau_R \mu E_R/\hbar$ and the dimensionless polarization operator $P \equiv P_R/\mu^*$ obey the *linear* Maxwell-Bloch equations

$$\frac{\partial P}{\partial T} = E, \tag{2}$$

$$\frac{\partial E}{\partial X} = P, \tag{3}$$

where $X \equiv x/L$ and $T \equiv \tau/\tau_R = (t-x/c)/\tau_R$. Similar equations hold
for the L-envelope amplitudes. We must supplement the Maxwell-
Bloch equations with initial and boundary conditions. The
atomic system is inverted by swept excitation at $\tau = t-x/c = 0$.
The initial state vector $|\Psi\rangle$ consists of the product of the
field vacuum state (no photons) and the matter state with all
atoms inverted, i.e. $2R^3|\Psi\rangle = |\Psi\rangle$. We only consider emission in
the direction of excitation. The appropriate equations of mo-
tion are thus (2) and (3).

The initial operator polarization

$$P(X,T=0) = P_0(X), \tag{4}$$

where, by definition, $P_0^\dagger|\Psi\rangle = 0$ and

$$[P_0^\dagger(X), P_0(X')] = \delta(X-X')/N, \tag{5}$$

exhibiting the Bose character of P_0. The commutator describes
the uncertainty in the initial atomic polarization. More specifically, it describes the incoherent addition of matter zero-
point fluctuation waves in the (R) - end-fire solid angle λ_0^2/S.

Since we consider SF emitted to the right, the electric
field at the left end-face is equal to the vacuum field E_0
incident on it, i.e.

$$E(X=0,T) = E_0(T). \tag{6}$$

By definition, the Bose operator E_0 satisfies $E_0|\Psi> = 0$ and

$$[E_0(T), E_0^{\dagger}(T')] = \delta(T-T')/N. \tag{7}$$

The commutator describes the addition of running waves of the vacuum-field zero-point fluctuations in the (R)-solid-angle λ_0^2/S. The result (7) is obvious when it is rewritten in the form

$$[E_0(T), E_0^{\dagger}(T')] = \frac{3}{2} \frac{\lambda_0^2}{4\pi S} \frac{\delta(t-t')}{\tau_n}, \tag{8}$$

where $\delta(t-t')/\tau_n$ is twice the well-known strength of the zero-point fluctuations initiating single-atom spontaneous emission, $\lambda_0^2/(4\pi S)$ is the fraction of solid angle involved in the initiation of (R) - SF and the factor 3/2 compensates for the linear dipole description of the atomic transition.

Eqs. (2) - (4) and (6) describe hyperbolic differential equations with initial and boundary conditions prescribed on the characteristic lines $T = 0$ and $X = 0$, respectively. Greens function propagator analysis then shows that the unique solution (E,P) preserves the Bose character of P along the characteristic lines X = constant and X' = constant and for $T' = T$, i.e.

$$[P^{\dagger}(X,T), P(X',T)] = \delta(X-X')/N. \tag{9}$$

Similarly the Bose character of E is preserved along the characteristic lines T = constant and T' = constant and for $X' = X$, i.e.

$$[E(X,T), E^{\dagger}(X,T')] = \delta(T-T')/N. \tag{10}$$

The ordering of the operators in the commutation relations (9) and (10) is different. E is a Bose operator on the field *ground* state and P is a Bose operator on the *excited* state of the atoms. This will be of prime importance in the following.

Eqs. (2) - (10) give a complete quantum-mechanical description of the initiation of SF, with inclusion of the propagation effects along the pencil axis. It is interesting to note that we could have started from the classical linear Maxwell-Bloch equations as well, and then introduce quantum mechanics by invoking the Bose character of E and P as described above. The present derivation has the advantage of elucidating in greater detail the physics involved and the assumptions needed.

The difference in character of the Bose operators E and P becomes explicit in the GH theory and the PSV theory of the initiation of SF. The GH theory calculates normally ordered field correlation functions, and since expectation values of normally ordered products of E_0 and $E_0^{\dagger}$ vanish, it ascribes the initiation of SF to the initial uncertainty in the atomic polarization as described by (5). The PSV theory calculates antinormally ordered polarization correlation functions, and since antinormally ordered products of P_0 and $P_0{}^\dagger$ have vanishing expectation values, it ascribes the initiation of SF to the zero-point fluctuations of the vacuum field as described by (7). Our eqs. (2) - (10) allow the calculation of arbitrary field-polarization correlation functions. A complete quantum-mechanical description of SF thus involves the zero-point fluctuations of both field and matter. If the appropriate commutation relations are used, the GH and PSV theory yield identical results.

3. Stochastic Variables Descriptions and Classical Behavior

We will show that the statistical behavior of our system can be conveniently represented in terms of classical stochastic variables. Depending on the choice of the desired correlation functions, the stochastic variables description is different.

In order to obtain vanishing initial and boundary variables we introduce the matter field $\Omega = E - E_0(T)$ and the collective Bloch vector component $M = P - P_0(X)$. The Maxwell-Bloch equations (2) and (3) then read

$$\frac{\partial M}{\partial T} = \Omega + E_0(T), \tag{11}$$

$$\frac{\partial \Omega}{\partial X} = M + P_0(X). \tag{12}$$

All variables are still operators with the appropriate commutation rules. Since $P_0{}^\dagger|\Psi\rangle = 0$ and $E_0|\Psi\rangle = 0$, we have $\langle MM^\dagger\rangle = \langle PP^\dagger\rangle$ and $\langle\Omega^\dagger\Omega\rangle = \langle E^\dagger E\rangle$. Thus $4\langle MM^\dagger\rangle$ equals the mean squared tipping angle of the collective Bloch vector and $N\hbar\omega_0\langle\Omega^\dagger\Omega\rangle/\tau_R$ is the mean field intensity emitted.

The stochastic variables description suited for the calculation of *antinormally* ordered correlation functions is obtained from (11) and (12) by treating Ω, $\Omega^\dagger$, M and $M^\dagger$ as complex valued c-numbers $\tilde\Omega$, $\tilde\Omega^*$, $\tilde M$ and $\tilde M^*$, by considering $(E_0, E_0{}^\dagger)$ as a classical fluctuating field source $(\tilde E_0, \tilde E_0{}^*)$ and by putting $P_0 = 0$ since the operator P_0 does not contribute to antinormally ordered correlation functions. Moreover, quantum-mechanical average and the average over stochastic variables are to be identified. This stochastic variables description pictures the atom-field system as being driven by a bivariate field noise source $(\tilde E_0, \tilde E_0{}^*)$, which is Gaussian since E_0 is a Bose operator. All correlation functions involving an odd number of functions $\tilde E_0$ and $\tilde E_0{}^*$ vanish. In view of (7) and $E_0|\Psi\rangle = 0$, the second order correlation function

$$\langle \tilde E_0(T)\, \tilde E_0{}^*(T')\rangle = \delta(T-T')/N. \tag{13}$$

All even higher order correlation functions factorize according to

$$\langle \tilde{E}_0(T_1) \ldots \tilde{E}_0(T_p) \tilde{E}_0^*(T_1') \ldots \tilde{E}_0^*(T_p') \rangle = \sum_p \prod_{j=1}^p \langle \tilde{E}_0(T_j) \tilde{E}_0^*(T_{Pj}) \rangle, \qquad (14)$$

where the sum is over all possible permutations of times.

In this stochastic variables picture, the behavior of the collective Bloch vector in each individual experiment is determined by one representative out of all possible noise source functions $\tilde{E}_0(T)$:

$$\tilde{M}(X,T) = \int_0^T dT' \; I_0(2\sqrt{X(T-T')}) \tilde{E}_0(T'), \qquad (15)$$

where I_0 is the modified Bessel function of zeroth order. The average over the ensemble of stochastic variables must be understood as an average over many repeated experiments. For instance, the mean squared tipping angle evolves according to

$$\langle \tilde{M} \tilde{M}^* \rangle (X,T) = \frac{1}{N} \int_0^T I_0^2(2\sqrt{XT'}) dT'. \qquad (16)$$

The above stochastic variables description is particularly well suited for the calculation of the results of polarization measurements. However, measurements on the electric field detect *normally* ordered field correlation functions. The stochastic variables description for the calculation of these correlation functions is obtained from (11) and (12) by treating Ω, $\Omega^\dagger$, M and $M^\dagger$ as complex valued c-numbers $\tilde{\Omega}$, $\tilde{\Omega}^*$, $\tilde{M}$ and $\tilde{M}^*$, and by considering $(P_0, P_0^\dagger)$ as a classical fluctuating polarization source $(\tilde{P}_0, \tilde{P}_0^*)$. In this case E_0 is put equal to zero since that operator does not contribute to normally ordered correlation functions. The stochastic initiation of SF is now due to a bivariate Gaussian (P_0 is a Bose operator) polarization noise source $(\tilde{P}_0, \tilde{P}_0^*)$ satisfying

$$\langle \tilde{P}_0^*(X) \tilde{P}_0(X') \rangle = \delta(X-X')/N. \qquad (17)$$

The behavior of the field in each individual experiment is determined by one representative out of all possible noise source functions :

$$\tilde{\Omega}(X,T) = \int_0^X I_0(2\sqrt{(X-X')T}) \; P_0(X') dX'. \qquad (18)$$

The emission intensity as measured over many repeated experiments evolves according to

$$\langle \tilde{\Omega}^* \tilde{\Omega} \rangle (X,T) = \frac{1}{N} \int_0^X I_0^2(2\sqrt{X'T}) dX'. \qquad (19)$$

Note that

$$\frac{\partial}{\partial T} \langle \tilde{M} \tilde{M}^* \rangle = \frac{\partial}{\partial X} \langle \tilde{\Omega}^* \tilde{\Omega} \rangle, \qquad (20)$$

which expresses the conservation of energy in the mean.

For increasing field and polarization amplitudes, it is reasonable to assume that the operator character of both atomic

466

and field variables becomes less and less important. In fact, polarization-polarization, polarization-field and field-field commutators governing the quantum-mechanics are fixed in value, while polarization and field themselves grow exponentially due to amplification as indicated by the large argument behavior of I_0 in (15) and (18). In this classical regime which begins after a few τ_R the noise sources in the stochastic variables descriptions can be ignored for the calculation of the further development of the system. In other words, once the amplitudes of Ω and $\tilde{M}$ become large compared to the noise amplitudes, SF essentially follows its deterministic course. The system enters a non-linear regime roughly at the delay time τ_D of the SF pulse. Since τ_D is much larger than τ_R, cf. (30), the solutions in the linear and classical regime can be continued in the non-linear regime by the non-linear Maxwell-Bloch equations

$$\tau_R \frac{\partial n}{\partial t} = -4 \ \mathrm{Re} \ [\tilde{M}_L \tilde{\Omega}_L^* + \tilde{M}_R \tilde{\Omega}_R^*], \tag{21}$$

$$\tau_R \frac{\partial \tilde{M}_{L,R}}{\partial t} = n\tilde{\Omega}_{L,R}, \tag{22}$$

$$\frac{L}{c}(\frac{\partial}{\partial t} \pm c \frac{\partial}{\partial x}) \ \tilde{\Omega}_{L,R} = \tilde{M}_{L,R}, \tag{23}$$

where n is the dimensionless atomic inversion (+1 initially) and the + and - signs correspond to (R) and (L) waves, respectively. The coupling between (R) and (L) waves as described by (21) is a slowly varying envelope approximation.

The stochastic variables descriptions given above are suited for the calculation of either the field evolution in terms of a polarization noise source or the Bloch vector evolution in terms of a field noise source. However, if the operator ordering can be ignored, i.e. in the classical regime, either of the two stochastic variables descriptions can be used to calculate both field and polarization. The ensemble of classically behaving single shot outputs $\tilde{\Omega}$ and $\tilde{M}$ is generated equally well by the various representations of the one or the other noise source.

4. Initial Bloch Vector Motion

This section gives a brief survey of the PSV results for the initiation of SF. Due to the field noise, the collective Bloch vector starts jittering in a sort of two-dimensional Brownian motion about its upright position. The mean polarization is zero ($<\tilde{M}>=<\tilde{M}^*>=0$) since there is no preferential direction of motion for the Bloch vector. The T<<1 expansion of (16) yields

$$<\tilde{M} \ \tilde{M}^*>^{\frac{1}{2}} = (T/N)^{\frac{1}{2}} \tag{24}$$

and shows that the mean squared tipping angle initially grows by diffusion. The atoms still radiate as single entities. Using the definition of τ_R and (24), the decrease in upper level population is given by

$$\Delta p = \frac{3}{2} \frac{\tau}{\tau_n} \frac{\lambda_0^2}{4\pi S}. \tag{25}$$

This is precisely the result expected for single atom dipole emission in the solid angle λ_0^2/S. After τ_R, amplification becomes predominant. The drift motion of the Bloch vector due to the matter field $\tilde{\Omega}$ results in an exponentially growing tipping angle. For the mean squared behavior this is shown by the $XT\gg1$ expansion of (16) :

$$<\tilde{M}\ \tilde{M}^*>^{\frac{1}{2}} = \exp[\ 2(XT)^{\frac{1}{2}}]\ /(8\pi NX)^{\frac{1}{2}}\ . \tag{26}$$

A complete description of the initial motion of the collective Bloch vector is furnished by the probability density Π defined as

$$\Pi(\alpha,\alpha^*,X,T) = <\delta(\alpha-\tilde{M})\delta(\alpha^*\tilde{M}^*)>. \tag{27}$$

All ensemble averages can be expressed as moments of this function

$$<\tilde{M}^p\ \tilde{M}^{*q}>\ = \int\alpha^p\alpha^{*q}\ \Pi(\alpha,\alpha^*,X,T)d^2\alpha, \tag{28}$$

where the integral is over the entire complex α-plane. Using the Gaussian nature of the bivariate field noise, one obtains with the method of linked moments [16]

$$\Pi(\alpha,\alpha^*,X,T) = \frac{e^{-|\alpha|^2/<\tilde{M}\ \tilde{M}^*>}}{\pi<\tilde{M}\ \tilde{M}^*>}\ . \tag{29}$$

We finally consider the delay time of the SF pulse. Following the PSV theory, it is defined as the time τ_D at which $4\ \tilde{M}\ \tilde{M}^*$, the squared tipping angle in the linear theory, becomes equal to 1 at the end-face x=L. From (26) we find the mean delay time

$$<\tau_D>= \frac{\tau_R}{4}(\ln\sqrt{2\pi N})^2. \tag{30}$$

If we introduce an effective initial tipping angle Θ_0 in a classical Maxwell-Bloch theory to simulate the mean behavior of the Bloch vector, we find from equating delay times that Θ_0 must be given by

$$\Theta_0 = \frac{2}{\sqrt{N}}\ [\ \ln(2\pi N)^{1/8}]^{\frac{1}{2}}\ . \tag{31}$$

Analytic calculation of the relative standard deviation $\Delta\tau_D = [(<\tau_D^2>/<\tau_D>^2)-1]^{\frac{1}{2}}$ using (29) yields within a 10 % accuracy and for $10^5 < N < 10^{40}$ the expression

$$\Delta\tau_D = 2.3/(\ln N). \tag{32}$$

Recently, VREHEN has shown that our theoretical predictions for Θ_0 and $\Delta\tau_D$ are consistent with experiment, at least for $N \sim 10^8$ Cs atoms [14].

4. Conclusion

We have given a quantum-mechanical description of the initiation of superfluorescence in an extended pencil-shaped medium with Fresnel number $F = 1$. Propagation effects along the pencil axis are fully included. Transverse effects are disregarded. Stochastic variables descriptions in terms of classical linear Maxwell-Bloch equations with a field or matter noise source are given for the calculation of antinormally and normally ordered correlation functions, respectively. A single shot experimental result is associated with one representative of the noise source functions. The theory can be extended beyond the linear regime if it is assumed that the evolution becomes classical in that regime, i.e. noise is of minor importance for the further evolution of the system. Field and matter solutions are continued into the non-linear regime by using the non-linear Maxwell-Bloch equations.

HAAKE et al [12] and HOPF [13] have numerically calculated single-shot field intensities using the non-linear Maxwell-Bloch equations with the polarization noise source of the preceding section. They used conditions realized in the experiments of VREHEN [4]. Classical behavior before τ_D is then likely since $2\sqrt{\tau_D/\tau_R} \sim 10$ in these experiments. Their numerical results for the delay time mean and variance agree with our analytical predictions. It is of interest to add here that onset of classical behavior long before τ_D implies that the relative standard deviation in the peak amplitudes of pulses selected to have equal delay times is much smaller than $\Delta\tau_D$.

The way in which classical behavior is reached clearly deserves further discussion. Other problems that deserve attention are (i) the inclusion of transverse effects, (ii) the extension of the theory to Fresnel numbers different from 1 and (iii) the inclusion of inhomogeneous broadening. Recently, we have given a unified theory of superfluorescence and amplified spontaneous emission by including homogeneous broadening in the linear theory [17].

References

1. R.H. Dicke, Phys. Rev. 93, 99 (1954).
2. N. Skribanowitz, I.P. Herman, J.C. MacGillivray and M.S. Feld, Phys. Rev. Lett. 30, 309 (1973).
3. M. Gross, C. Fabre, P. Pillet and S. Haroche, Phys. Rev. Lett. 36, 1035 (1976); A. Flusberg, T. Mossberg and S.R. Hartmann, Phys. Lett. 58A, 373 (1976); A.T. Rosenberger, S.J. Petuchowski, and T.A. DeTemple; *Cooperative Effects in Matter and Radiation*, C.M. Bowden, D.W. Howgate and H.R. Robl, eds. (Plenum, New York, 1977); J. Okada, K. Ikeda and M. Matsuoka, Opt. Commun. 27, 321 (1978).
4. Q.H.F. Vrehen, H.M.J. Hikspoors and H.M. Gibbs, Phys. Rev. Lett. 38, 764 (1977); H.M. Gibbs, Q.H.F. Vrehen and H.M.J. Hikspoors, Phys. Rev. Lett. 39, 547 (1977); H.M. Gibbs; *Cooperative Effects in Matter and Radiation*, C.M. Bowden, D.W. Howgate and H.R. Robl, eds. (Plenum, New York, 1977)

p. 61; Q.H.F. Vrehen ibid p. 79.

5. R. Bonifacio and L.A. Lugiato, Phys. Rev. A11, 1507 (1975); A12, 587 (1975); G. Banfi and R. Bonifacio, Phys. Rev. Lett. 33, 1259 (1974); Phys. Rev. A12, 2068 (1975).

6. J.C. MacGillivray and M.S. Feld, Phys. Rev. A14, 1169 (1976).

7. S. Haroche; *Coherence and Quantum Optics IV*, L. Mandel and E. Wolf eds. (Plenum, New York, 1978), p. 539; E. Ressayre and A. Tallet, ibid p. 799.

8. J.H. Eberly and N.E. Rehler, Phys. Lett. 29A, 142 (1969); N.E. Rehler and J.H. Eberly, Phys. Rev. A3, 1735 (1971).

9. R. Glauber and F. Haake, Phys. Lett. 68A, 29 (1978).

10. M.F.H. Schuurmans, D. Polder and Q.H.F. Vrehen, J. Optical Soc. Am. 68, 699 (1978); D. Polder, M.F.H. Schuurmans and Q.H.F. Vrehen, Phys. Rev. A19, 1192 (1979).

11. Q.H.F. Vrehen and M.F.H. Schuurmans, Phys. Rev. Lett. 42 224 (1979).

12. F. Haake, H. King, G. Schröder, J. Haus and R. Glauber, submitted to Phys. Rev. A; Same authors and F. Hopf, submitted to Phys. Rev. Lett.

13. F. Hopf, private communication.

14. Q.H.F. Vrehen, to be published in the FICOLS 1979 proceedings.

15. H.A. Kramers, *Die Grundlagen der Quantentheorie* (Longmans, London, 1964).

16. M. Lax; *Brandeis University Summer Institute in Theoretical Physics, 1966, Statistical Physics*, Vol. 2, eds. M. Chretien, E.P. Gross and S. Deser (Gordon and Breach, New York, 1968).

17. M.F.H. Schuurmans and D. Polder, Phys. Lett., to be published.

Experiments on the Initiation and Coherence Properties of Superfluorescence

Q.H.F. Vrehen

Philips Research Laboratories, Eindhoven, The Netherlands

1. Introduction

In the first experiment on superfluorescence (SF) SKRIBANOWITZ et al [1] observed ringing. This ringing was ascribed to axial variations of the atom and field variables, the evolution of which was calculated with the semiclassical Maxwell-Bloch equations [2]. In the first experiment that strictly obeyed the BONIFACIO and LUGIATO [3] conditions for "pure" SF, except that longitudinal pumping was used, GIBBS et al [4] discovered regimes both of single-pulse and of multiple-pulse emission. VREHEN et al [5] have demonstrated experimentally that the multiple pulses in the cesium experiment are related to transverse variations in the field intensities. GIBBS et al [6] have suggested that the absence of ringing in the single-pulse emission might result from dynamic transverse effects. In SF those effects cannot be avoided by taking a sample of large cross-section; a large Fresnel number leads to emission in many transverse modes [5].

Recently GLAUBER and HAAKE [7] (GH) and POLDER et al [8] (PSV) have developed completely quantum mechanical theories that fully include propagation effects. These theories, just as the semiclassical treatment of MACGILLIVRAY et al [2], are one-dimensional in the sense that transverse variations in the atom and field variables are neglected.

According to PSV two stages may be distinguished in the SF evolution. In the first stage the inversion is nearly constant, the equations of motion can be linearized and the motion is strongly affected by quantum noise. In the second stage fields and polarization are nearly classical, the quantum noise can be neglected and eventually the motion becomes non-linear. The quantum noise present in the first stage makes itself felt in the second stage in macroscopic fluctuations in the pulse parameters. PSV have estimated analytically the variance in the delay time. HAAKE et al [9] have studied numerically the fluctuations in delay time, pulse shape and intensity. It seems to us that of these parameters the delay time may be least affected by the transverse effects mentioned above.

In this paper three experiments will be discussed, all designed to investigate some basic property of SF emission. In the first experiment a direct measurement has been made of the average strength of the noise source that initiates SF (section 2). In the second experiment information is obtained about the temporal and spatial coherence of SF from the interference between the beams emitted by two different, simultaneously pumped SF cells (section 3). The final experiment has yielded the first preliminary data on the intrinsic quantum fluctuations in the SF delay times (section 4).

2. The Strength of the Quantum Noise

The SF emission is initiated by quantum noise. The strength of the noise source is a quantity of fundamental interest in itself. Moreover, both the delay time of the SF pulse and the amount of ringing depend on that noise strength according to the semiclassical and quantum mechanical theories that include propagation effects. In semiclassical calculations the noise source is often replaced [2, 6, 10] by a short coherent pulse of area θ which propagates along the axis of the sample and rotates the Bloch vector away from its upright position over the small angle θ, the so called initial tipping angle. An effective initial tipping angle θ_0 may now be defined as the tipping angle that yields a delay time equal to the average delay time resulting from the noise. So defined θ_0 provides a convenient measure of the quantum noise power.

The magnitude of θ_0 has been under debate. PSV have derived an explicit expression $\theta_0^{PSV} = (2/\sqrt{N}) \sqrt{(\ln 2\pi N)/8}$. For the cesium experiment one finds $\theta_0^{PSV} = 2.3 \times 10^{-4}$. Comparable values are obtained from GH and from BONIFACIO and LUGIATO [3]. A much smaller value, $\theta_0^{MF} = 6.7 \times 10^{-6}$, is predicted from the work of MACGILLIVRAY and FELD [2] and a much larger one, $\theta_0^{RE} = 2.7 \times 10^{-2}$ from that of REHLER and EBERLY [11].

Recently SCHUURMANS and the present author [12] have reported a direct measurement of θ_0. The measurement does not depend on any assumption about the SF evolution after the initial stage and it is not affected by possible errors in the measurement of atomic density. From the experiment it is concluded that $1 \times 10^{-4} < \theta_0 < 2.5 \times 10^{-3}$. Clearly, therefore, the experiment gives support to the theories of GH and PSV.

In earlier work the observation of single-pulse SF from cesium atoms has been reported. The possibility has then been entertained that the absence of ringing could at least partly be attributed to very large value of θ_0 ($\theta_0 \approx 10^{-2}$). That possibility can now be excluded. Indeed, for the observed value of θ_0 the one-dimensional theories [2, 7, 8] predict a fair amount of ringing. The small measured value of θ_0 strengthens the conclusion from the single-pulse experiment that transverse effects need to be incorporated in the theory.

In the semiclassical theory based on the Maxwell-Bloch equations the delay time τ_D depends on θ_0 and on the atomic density through the SF characteristic time τ_R. From an analytical approximation to the solution of the Maxwell-Bloch equations

by HERMANN [13] one finds $\tau_D \simeq (\tau_R/3)[\ln(\theta_0/2\pi)]^2$ for θ_0 of
the order of 1×10^{-4}. With $1 \times 10^{-4} < \theta_0 < 2.5 \times 10^{-3}$ the result is
$\tau_D/\tau_R = 30 \pm 10$. Experimentally the ratio τ_D/τ_R varies from 20
for long delay times (15-20 ns) to 45 for very short delay
times (5-7 ns), with an uncertainty of (-40 %, + 60 %) due to
possible errors in the measured density. Thus the observed
delay times are consistent with the Maxwell-Bloch theory.

3. Temporal Coherence

In SF experiments little information has been obtained so far
about variations of the phase of the electromagnetic field.
Two extreme possibilities can be envisioned. First, it may be
that during the non-linear stage of pulse evolution the phase
is constant; the pulse then is coherent and its spectral width
corresponds to the Fourier transform of its shape. Note that
the phase may vary randomly from shot to shot in the interval
$[0,2\pi]$. Alternatively it may be that the phase varies rapidly
during the whole pulse; the pulse is then temporally incoherent
and its spectral width is much larger than the Fourier transform of its shape. It would be of interest to measure the
spectral content of the SF emission. No spectra have been reported so far.

In their work on quantum beats in SF VREHEN et al [14] described beats between two transitions with different, inco-
herently excited, upper levels and different lower levels. Such
beats cannot be understood from interferences within the atom
but they can be readily explained as interferences between two
classical coherent fields. A similar argument holds for the
beats observed by GROSS et al [15] generated by different velocity groups of atoms and those reported by MAREK [16] from
different isotopes in rubidium. Those experiments lend support
to the view that a classical coherent field is emitted.

Here we wish to report an experiment in which beats are
observed between two different SF samples emitting independently [17]. The experimental apparatus is shown in fig. 1. Two
identical 5 cm cesium cells are simultaneously pumped and their
infrared output brought to interference at the detector. The
transverse magnetic fields are adjusted so that in one cell the
atoms are excited to the level $7P^{3/2}$ ($m_J = -3/2$, $m_I = -5/2$) and
in the other cell to $7P^{3/2}$ ($m_J = -3/2$, $m_I = -3/2$). Depending
on the precise setting of the magnetic fields the two SF transitions have frequencies differing by 300 to 600 MHz. Optimum
overlap between the two beams requires that the virtual image
of the pumped volume in cell 1 coincides precisely with the
pumped volume in cell 2. The experimental technique used to
realize that condition will not be discussed here. Let us
assume that each beam has temporal coherence during the pulse.
The visibility of the beats, i.e. the modulation depth, may
then be smaller than one for a number of reasons. First, the
alignment of the beams may be imperfect. Second, the frequency
response of the detector may be too small. Third, the uncorrelated fluctuations in the delay times, peak intensities and
pulse shapes of the two pulses may reduce the modulation depth.

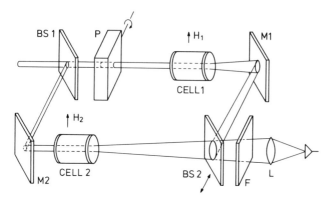

Fig.1 Apparatus for the observation of beats between the outputs from two independent SF cells. $M_{1,2}$ mirrors; $BS_{1,2}$ beam-
splitters; F filter; L lens.

And finally the spatial distribution of amplitude and phase
may be different for the two beams, and will change from shot
to shot. Consider a plane just in front of the sample and normal to its axis. If the beams have complete temporal coherence
then in each point of the plane the phase of each of the beams
is constant during the pulse. However, the point-to-point variations in amplitude and phase may be different for the two
beams. The diameter of the source is 430 µm, which, according
to the van Cittert-Zernike theorem guarantees coherence within
a full angle of 4.4 mrad. In the experiment a full emission
angle of 15 mrad is accepted by the detection system. Most of
the SF radiation is emitted within this angle.

In fig. 2 an example is presented of the interference of
the beams together with a Fourier spectrum of the pulse. The
spectrum has a central peak and a side lobe around the modulation frequency. The modulation depth m equals the ratio of

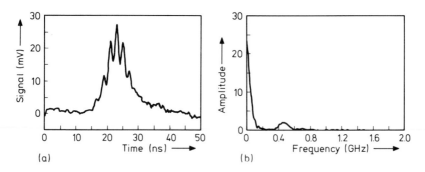

Fig.2 a) Beats resulting from the interference between two
independently emitted SF pulses b) the Fourier spectrum of the
pulse.

474

the areas under side lobe and central peak. It varies strongly
from shot to shot. After correction for the detector response
m fluctuates between 0 and 0.5. The beat frequency agrees well
with that calculated from the atomic parameters and the values
of the magnetic fields. In the example shown the FWHM pulse
width is 7.5 ns and the FWHM spectral width of the side lobe
is 130 MHz. The pulse is thus close to transform limited. The
result of this experiment can most easily be interpreted as
follows. The shot to shot variations in the modulation depth
are due to differences between the spatial distributions of
amplitude and phase in the two beams. The SF emission has a
high degree of temporal coherence. The occasional occurrence
of large phase fluctuations as considered by HOPF [18] cannot
be excluded.

4. Delay Time Fluctuations

In the following an experiment is described to measure quantum
fluctuations in the SF delay times. In principle such a measurement could be made by firing many shots successively and
measuring the delay time of each. Unfortunately the laser pump
pulse varies from shot to shot and thus the sample preparation
is not sufficiently reproducible. In our experiment this problem is circumvented by pumping two cells simultaneously and
studying the delay time differences. The setup is sketched in
fig. 3. The cells are placed in corresponding positions in the
beam after it has been divided in two equal parts by a beam
splitter. Both cells have the same length (5 cm) and the
atomic densities and magnetic fields are made equal as closely
as possible. The electrical signal from one of the cells is
given an additional delay of 55 ns so that for each shot both
pulses can be displayed on the same trace of the Transient
Digitizer, and their delay times can be measured. In an experiment pulses are recorded at a rate of 5 Hz and about 500
pulses can be stored on one disk for analysis afterwards. The
analysis is made with the help of a computer which determines
the delay time as the time at which the pulse reaches its
maximum intensity. Superimposed on the signal is noise, which
introduces some randomness in the measurement of the delay

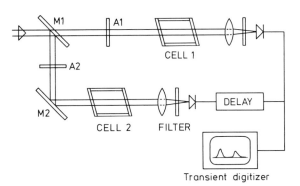

Fig.3 Setup for the study of delay time fluctuations.

time. This effect has been found to be small and will be neglected in the following. Unfortunately, on part of the shots the excitation is so weak that no SF pulse can be observed from either one or both of the cells.

As expected the delay times in the two cells are partly correlated. The correlation can be ascribed to variations in the pumping process, while the uncorrelated part is supposedly due to the quantum fluctuations. For the analysis the following assumptions are made : 1. The two cells are identical in all respects. Delay time differences for the same shot are thus entirely due to quantum fluctuations. 2. Shot to shot variations in the excitation can be described fully as variations in the excited state density. No variations in the transverse or longitudinal excitation profile are considered. 3. The quantum fluctuations in the delay time for a given density are proportional to the average delay time at that density. 4. Otherwise the quantum fluctuations in the delay time and the density fluctuations are uncorrelated.
The first assumption is justified if the delay time distributions of the two cells are identical within statistical limits. Experimentally the average delay times are usually slightly different for the cells. A correction is made by multiplying all delay times of one cell by the appropriate factor. After that correction the criterion is applied that the second and third moments of the distributions should be the same. The second assumption is hard to evaluate. It is known that an increase of the cross-sectional area of the pumped volume with a factor 15 leads to a reduction in the delay time by a factor 1.5 to 2.[5]. It is estimated that the cross-section might vary by a factor of 2 from shot to shot and some delay time variations will result. However, as long as the relative quantum fluctuations are not affected such variations are equivalent to variations in the density. The third assumption is in agreement with present theoretical understanding. The last assumption can be justified at least qualitatively from the "dot diagram" of the number pairs $\{(\tau_{1i} + \tau_{2i})/2, (\tau_{1i} - \tau_{2i})\}$ where τ_{1i} and τ_{2i} are the delay times of cells 1 and 2 in the

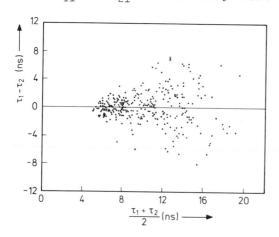

Fig.4 A plot of the delay time difference $\tau_1-\tau_2$ versus the arithmetic mean $(\tau_1+\tau_2)/2$ for 323 pulses.

i-th pulse. Such a diagram is shown in fig. 4 from which it can appreciated that little correlation exists except the one described in the third assumption.

Each shot can be characterized by three quantities, the (unknown) density ρ and the measured delay times τ_1 and τ_2 for the two cells. Let τ be the average delay time at the fixed density ρ. Each shot can then also be characterized by the numbers (τ_1, τ_2, τ). Let $\delta = \tau_1 - \tau_2$ and let brackets $<>$ indicate an average over many shots. From the assumptions it follows that $<\tau> = <\tau_1> = <\tau_2>$. One would like to derive from the experiments the variance of τ_1 at some fixed value of τ, e.g., $<\tau>$. That variance will be denoted $\sigma(\tau_1, <\tau>)$. Since for a given τ, i.e. for a given density, the quantities τ_1 and τ_2 are uncorrelated $\sigma(\delta, <\tau>) = 2\sigma(\tau_1, <\tau>)$. Let us define the following probability distributions:

$P(\delta, \tau)d\delta.d\tau$: probability that δ and τ take on their respective values within the interval $d\delta.d\tau$,
$Q(u)du$: probability that for a given τ $\delta \equiv u\tau$ takes on a value within τdu
$R(\tau)d\tau$: probability for τ to take on a value within the interval $d\tau$.

Note that the conditional probability $Q(u)$ can be defined because of our third assumption. With the definitions

$$\sigma(\delta, <\tau>) = <\tau>^2 \int u^2 Q(u)du \qquad (1)$$

Let us define furthermore the quantity $\sigma(\delta)$ as

$$\sigma(\delta) = \int\int P(\delta, \tau)\delta^2 d\delta.d\tau \qquad (2)$$

It follows
$$\sigma(\delta) = \int u^2 Q(u)du \int \tau^2 R(\tau)d\tau = \sigma(\delta, <\tau>)(1 + \sigma(\tau)/<\tau>^2) \qquad (3)$$

From equation (3) one sees that $\sigma(\delta, <\tau>)$, the variance of δ for a given $\tau = <\tau>$ equals $\sigma(\delta)$ the variance of δ for all values of τ apart from the correction factor $(1 + \sigma(\tau)/<\tau>^2)$. $\sigma(\delta)$ can be obtained from the experiment immediately. The parameter τ is not measured. However, since τ reflects only density fluctuations whereas τ_1 reflects both density fluctuations and quantum fluctuations it is obvious that $\sigma(\tau)/<\tau>^2$ $<\sigma(\tau_1)/<\tau_1>^2$. Experimentally it is found that $\sigma(\tau_1)/<\tau_1>^2$ is of the order of 0.1 only, so that the correction factor differs only slightly from 1. In fact, to calculate that factor we have used the distribution of $(\tau_1 + \tau_2)/2$ instead of the unknown distribution of τ itself.

In fig. 5 the distribution is shown of delay time differences $\tau_1 - \tau_2$ for the same set of data as used for fig. 4. Some further statistical data for that experiment are collected in Table I. From the experiments performed so far a preliminary value can, be quoted for the relative standard deviation at a given density as (13 ± 3) %.

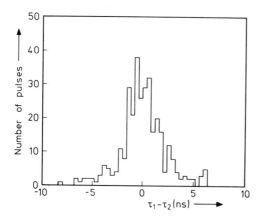

Fig.5 Distribution of the delay time differences $\tau_1 + \tau_2$ for a set of 323 pulses. Average delay time is 1Q2 ns.

The experimental value can be compared with theoretical predictions. DEGIORGIO [19] was the first to report theoretical results. He found a relative standard deviation of 1.2/lnN in the framework of the mean-field theory. It corresponds to 6.5% for the present experiment. POLDER et al [8] have made an analytical estimate of the delay time fluctuations. These authors arrive at a relative standard deviation of 2.3/lnN, amounting to 12.5 % for the cesium case. Finally HAAKE et al [9] numerically obtained a value of 12 % using the parameters of the 'cesium experiment. It may be concluded that the preliminary experimental result on the fluctuations in delay time agrees reasonably well with the GH and PSV theories. Further experiments are in progress.

5. Summary

The experiments described in this paper give support to several aspects of the recent theories of PSV and GH which treat SF quantum mechanically with full inclusion of propagation effects. The strength of the noise source as characterized by the measured effective initial tipping angle is in good a-

Table 1 Statistics of delay times in an experiment of 500 pulses. Number of pulses: total 500, with signal on both detectors 323, on first detector only 21, on second detector only 40, without signal on either detector 116. Delay times as observed for cell 1 have been multiplied by .934, those of cell 2 by 1.071 to make the averages equal. In parentheses values as observed. All times in ns.

	Average value	Standard deviation
τ_1	10.15(10.87)	3.43(3.67)
τ_2	10.15(9.44)	3.43(3.19)
$(\tau_1 + \tau_2)/2$	10.15	3.24
$\tau_1 - \tau_2$	0	2.24

greement with those theories, and so are the first measurements of the quantum fluctuations in the delay times. The measured delay times agree within (the fairly large) experimental error with the calculated ones. The observation of beats when two beams interfere is consistent with the emission of essentially classical fields during much of the pulse evolution. The most important limitation of the GH and PSV theories seems to be their neglect of transverse variations in the atom and field variables. The development of a three-dimensional theory of SF is now called for, but it may prove to be a formidable task.

Acknowledgement

The author would like to thank Mr. A.N. Andersen for his co-operation in the experiment on temporal coherence and Mr. J.J. der Weduwe for assistance with the measurements. Stimulating discussions with Prof. D. Polder and Dr. M.F.H. Schuurmans are gratefully acknowledged.

References

1. N. Skribanowitz, I.P. Hermann, J.C. MacGillivray and M.S. Feld, Phys. Rev. Lett. 30, 309 (1973).
2. J.C. MacGillivray and M.S. Feld, Phys. Rev. A14, 1169 (1976).
3. R. Bonifacio and L.A. Lugiato, Phys. Rev. A11, 1507 (1975); A12, 587 (1975).
4. H.M. Gibbs, Q.H.F. Vrehen and H.M.J. Hikspoors, Phys. Rev. Lett. 39, 547 (1977).
5. Q.H.F. Vrehen, H.M.J. Hikspoors and H.M. Gibbs in Proceedings of the Fourth Rochester Conference on Coherence and Quantum Optics, L. Mandel and E. Wolf editors (Plenum, New York, 1978).
6. H.M. Gibbs, Q.H.F. Vrehen and H.M.J. Hikspoors, in Proceedings of the Third International Conference on Laser Spectroscopy, J.L. Hall and J.L. Carlsten editors, (Springer, Berlin, 1977).
7. R. Glauber and F. Haake, Phys. Lett. 68A, 29 (1978).
8. D. Polder, M.F.H. Schuurmans and Q.H.F. Vrehen, Phys. Rev. A19, 1192 (1979); J. Opt. Soc. Am. 68, 699 (1978).
9. F. Haake, H. King, G. Schröder, J. Haus and R. Glauber, submitted to Phys. Rev.; same authors and F. Hopf, submitted to Phys. Rev. Lett.
10. R. Saunders, S.S. Hassan and R.K. Bullough, J. Phys. A9, 1725 (1976).
11. N.E. Rehler and J.H. Eberly, Phys. Rev. A3, 1735 (1971).
12. Q.H.F. Vrehen and M.F.H. Schuurmans, Phys. Rev. Lett. 42, 224 (1979).
13. J.A. Hermann, Phys. Lett. 69A, 316 (1979).
14. Q.H.F. Vrehen, H.M.J. Hikspoors and H.M. Gibbs, Phys. Rev. Lett. 38, 764 (1977).
15. M. Gross, J.M. Raimond and S. Haroche, Phys. Rev. Lett. 40, 1711 (1978).
16. J. Marek, private communication.
17. Q.H.F. Vrehen, A.N. Andersen and J.J. der Weduwe, to be published.
18. F.A. Hopf, private communication.
19. V. Degiorgio, Opt. Comm. 2, 362 (1971).

Coupled Transitions in Superradiance

A. Crubellier, C. Brechignac, P. Cahuzac, and P. Pillet

Laboratoire Aimé Cotton, Centre National de la Recherche Scientifique
F-91405 Orsay, France

In the theoretical study of superradiance, the atoms or molecules are
generally considered as two-level systems. However in many experimental
situations, this simple assumption is not valid. It is the case, for example, in all experiments exhibiting superradiant beats [1-3]. Moreover,
these beats are sometimes due to a specific property of superradiance : the
existence of interferences between different atoms. Other examples of multilevel superradiance are encountered each time one observes several superradiant transitions sharing a common level [3,4]. In addition, superradiance
is generally observed between degenerate levels, and it has been already
shown experimentally that the level-degeneracy influences the properties of
the superradiant light, in particular its polarization characteristics [5].

In all these experimental situations, the conditions for cooperative
emission are fulfilled for several "coupled" transitions that cannot be
treated independently. Typical cases of two coupled transitions are shown
in Fig. 1 . In such cases, the mutual influence of the different coupled
transitions is different in the cooperative emission compared to the non-
cooperative spontaneous emission. In superradiance , new features can be
theoretically predicted. The aim of this paper is to present various experimental evidences for the specific properties of the coupling between superradiant transitions. A theoretical study of coupled transitions can be
found in [6,7]. In the experiments, because of the duration of the excitation, one can observe superradiant pulses that coexist with the exciting
pulse. Then, a "pumping" effect, which is somewhat analogous to an optical
pumping, is added to the coupling between the transitions and it can strongly modify the characteristics of the emitted light.

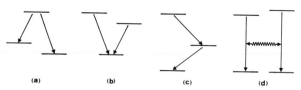

(a)   (b)   (c)   (d)

Fig. 1 : Typical level-diagrams of coupled transitions ; in case
(d) the two transitions are coupled only if they are
able to interfere or to beat

In this paper we present and qualitatively interpret the results of two experimental studies of superradiance, which manifest the existence of a cooperative coupling between different transitions and of a pumping effect. The first study has been realized by A. CRUBELLIER, S. LIBERMAN and P. PILLET and preliminary results can be found in [5]. In these experiments, superradiance is observed between (degenerate) levels of atomic Rb and the polarization characteristics of the emitted light are determined. The second experimental study is due to C. BRECHIGNAC and P. CAHUZAC and its concerns the observation of superradiant emission, in the visible range, on a three-level cascade of atomic Sr , whose lowest level is the ground state itself.

I Superradiance Between Degenerate Levels. Polarization Characteristics

The influence of level-degeneracy in superradiance manifests, in particular, in the polarization properties of the emitted light. In this section is presented an extensive study of the polarization characteristics of superradiant light, which has been realized with Rb atoms and for several $J \rightarrow J'$ transitions.

The experimental set-up is shown in Fig. 2 ; it is essentially the same as in [5] , where a detailed description can be found.

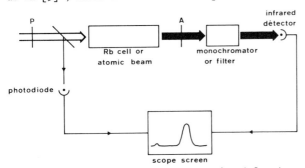

Fig. 2 : General scheme of the experimental set-up

Concerning the Rb atoms, two different experimental devices have been utilized : either a temperature-regulated cell, or a special oven , described in [5] , which provides a multiple atomic beam ; in this latter case, the influence of the Doppler effect can be considerably reduced. The Rb atoms are excited by a short N_2- pumped dye laser pulse (about 5 ns duration). The atoms are illuminated through a polarizer P , which provides either linearly or circularly polarized light. Infrared superradiant pulses are detected through a suitable filter and through an analyzing polarizer A . The signals are recorded using a transient analyser.

The infrared superradiant transitions observed when either the $6p \, ^2P_{1/2}$ or the $6p \, ^2P_{3/2}$ level is initially excited are shown in Fig. 3 .

The experimental observations concerning the polarization characteristics of the superradiant pulses are summarized in Table I .

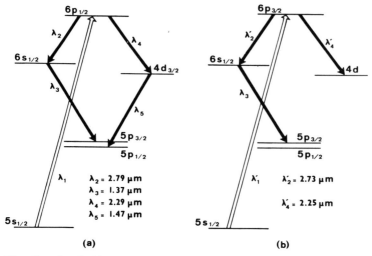

Fig. 3 : Level diagrams showing the analyzed superradiant transitions ; case (a) refers to $5s_{1/2} \to 6p_{1/2}$ (λ_1 = 421.5 nm)
excitation , whereas case (b) refers to $5s_{1/2} \to 6p_{3/2}$
(λ_1' = 420.2 nm) excitation. In the latter, the $4d \to 5p$
transition has been observed without monochromatisation,
so that no assignment can be proposed.

For most of the transitions, the polarization appears to be either linear
with the same polarization direction as the exciting light or purely circular. In these cases, the analysis is easy, despite the shortness of the
superradiant pulses. For the other transitions, a first analysis using one

Table I : Summary of experimental observations. In the case of
linearly polarized superradiant emission, one observed
either a polarization direction $\vec{\varepsilon}$ identical to that
of the exciting light or a polarization direction $\vec{\varepsilon'}$
which changes randomly from pulse to pulse

excitation super- radiant transitions	$6p_{1/2}$		$6p_{3/2}$	
	linear $\vec{\varepsilon}$	σ^+	linear $\vec{\varepsilon}$	σ^+
$6p \to 6s$	linear $\vec{\varepsilon'}$	σ^-	linear $\vec{\varepsilon}$	σ^-
$6s \to 5p$	linear $\vec{\varepsilon'}$	σ^-	linear $\vec{\varepsilon'}$	σ^+
$6p \to 4d$	linear $\vec{\varepsilon'}$	σ^+	linear $\vec{\varepsilon}$	σ^+
$4d \to 5p$	linear $\vec{\varepsilon'}$	σ^-	linear $\vec{\varepsilon}$	σ^-

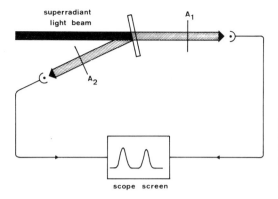

superradiant
light beam

A₁

A₂

scope screen

Fig.4: Detail of the experimental set-up used for
analyzing the cases of Table I
where no privileged polarization is found with the set-up of
Fig.2

polarizer A allows one to conclude only that, in average, no privileged
polarization , neither linear nor circular, can be observed. In order to
better investigate the polarization of these transitions, the experimental
set-up has been slightly changed, as shown in Fig. 4 .

The superradiant light beam is separated in two parts by a beam-splitter and
the two beams are detected separately through analyzing polarizers A_1 and
A_2 . One of the two detected signals is delayed by about 20 ns and the two
signals are calibrated, added and then recorded. A typical recording, shown
in Fig. 5 , displays the two identical pulses that are observed on the
$6s_{1/2} \rightarrow 5p_{3/2}$ transition, when getting off both A_1 and A_2 .

If A_1 and A_2 correspond to two orthogonal circular polarizations, the
two recorded pulses have always equal heights. On the contrary, if A_1 and
A_2 correspond to two orthogonal linear polarizations, the signal appears to
be quite different from pulse to pulse (see Fig. 6).

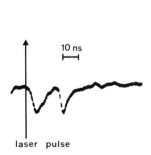

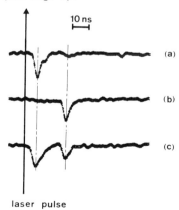

Fig. 5 : Typical recording obtained
with the experimental set-up
of Fig. 4, when getting off
both polarizers A_1 and A_2

Fig. 6 : Typical recordings obtained
with the set-up of Fig. 4 ,
when the two beams are
analyzed by two orthogonal
polarizers

All these results strongly suggest that the light is in fact totally linearly polarized, but with a polarization direction which changes randomly from
pulse to pulse. The different transitions exhibiting this property can be
found in Table I.

Complete interpretation of these results has been done, using a semi-
classical theoretical model in which the beginning of the emission is described quantum-mechanically, using the Markovian approximation [8,9] ; the
level-degeneracy has been introduced in bóth quantum and semi-classical
periods of the evolution of the system. However it is possible to have some
physical insight which gives a simple understanding of the phenomena.

It is to be noted first that, for instantaneous excitation, the coupling
between 6p → 6s and 6p → 4d should almost completely inhibit the emission
on the cascade 6p → 4d → 5p [7]. Superradiant emission on this cascade is
observed only when the light pulse emitted on 6p → 6s appears before the
end of the exciting pulse ; it is therefore due to a pumping effect [3,4].

Some situations of Table I are described in Fig. 7. Since we deal with
Zeeman sublevels, it is necessary to choose a quantization axis. For sake
of simplicity, it has been taken along the direction of propagation of the
exciting light. In that condition, superradiant light is to be seen either
in $\sigma^{\pm}$ polarization or in a superposition of σ^+ and σ^- polarizations.
In addition, it has been shown that, after a short quantum-mechanical period,
the semi-classical approximation is valid [8,9] and one can expect only a

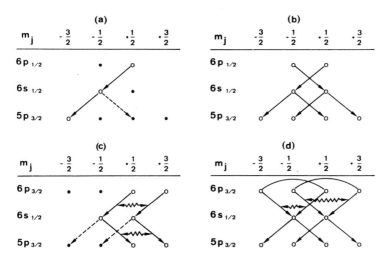

Fig. 7 : Schematic representation of the observed polarization characteristics in four chosen cases ; cases (a) and (c) (resp. (b) and (d)) correspond to circularly σ^+ (resp. linearly) polarized excitation. Points and
circles represent, respectively, non-populated and populated Zeeman sublevels.
Density matrix coherences between Zeeman sublevels are pointed out by curved
lines (case (d)). Phase relations between the emitted fields are represented
by wavy lines. Solid arrows indicate the effective superradiant transitions;
broken arrows indicate transitions on which superradiance is inhibited

coherent superposition of σ^+ and σ^- polarizations, which agrees with the observation of totally polarized light.

Case (a) of Fig. 7 , for instance, shows a typical example of an inhibition process [7] . The circular σ^+ polarized exciting light populates only the $m_J = 1/2$ sublevel of $6p\ ^2P_{1/2}$. The emission on $6p \rightarrow 6s$ is therefore purely σ^- polarized and populates only the $m_T = -1/2$ sublevel of $6s\ ^2S_{1/2}$. The two possible Zeeman transitions from this level to sublevels of $5p\ ^2P_{3/2}$ are coupled as in case (a) of Fig. 1 . The observed σ^- polarization of the light emitted on $6s \rightarrow 5p$ shows clearly the inhibition of the less probable transition.

Case (c) of Fig. 7 gives an example of coupling between two interfering Zeeman transitions (case (d) of Fig. 1). The σ^+ polarized exciting light populates the $m_J = 3/2$ and $m_J = 1/2$ sublevels of $6p\ ^2P_{3/2}$. The two possible Zeeman transitions from these sublevels to sublevels of $6s\ ^2S_{1/2}$ have obviously the same σ^- polarization and the same frequency . If they were not coupled, one would observe two successive pulses, the delay time of the second pulse being three times larger than the delay time of the first one. This has been never observed. In fact, one can show that the two transitions are coupled via interatomic interferences, which appear not only on each transition but also between the two transitions. The consequence of this coupling is the initiation of the emission on the less probable transition, this emission appearing always simultaneously with the emission on the other transition.

Cases (d) and (b) of Fig. 7 are very interesting from the point of view of the symmetry properties of the light. In both cases the exciting light is linearly polarized. In case (d) the superradiant light emitted on $6p \rightarrow 6s$ has the same linear polarization and therefore the same symmetry properties as the exciting light. In case (b) , on the contrary , the superradiant light emitted on $6p \rightarrow 6s$ does not present, in average, any privileged polarization, the pulses having a randomly distributed linear polarization direction. In this latter case, the symmetry properties of the superradiant light are therefore quite different from those of the exciting light. The explanation of the phenomenon in both cases is simple. The exciting light can be considered as a coherent superposition of σ^+ and σ^- polarized light and the polarization direction is determined by the relative phase of the σ^+ and σ^- fields. In case (d), the excitation will create non-zero off-diagonal density matrix elements between the sublevels $m_J = 3/2$ and $m_J = -1/2$ (respectively $m_J = 1/2$ and $m_J = -3/2$) of $6p\ ^2P_{3/2}$; the complex argument of these coherences is fixed by the polarization direction of the excitation. The symmetry of the exciting light is therefore transmitted to the atomic system. The complex argument of the coherences fixes in turn the relative phase of the σ^+ and σ^- emitted superradiant fields, which explains the observed result. In case (b) the situation is quite different. The excitation provides a statistical mixing of the sublevels $m_J = 1/2$ and $m_J = -1/2$ of $6p\ ^2P_{1/2}$. The atomic system does not keep any memory of the polarization direction of the exciting light. In fact, the atomic state just after the excitation has a cylindrical symmetry and the observation of a random polarization direction appears as an example of spontaneous symmetry breaking. This phenomenon represents a new manifestation of the fluctuations [10,11] that are a characteristics of superradiance. At the beginning of the emission, when the atom-field system behaves purely quantum-mechanically, the fluctuations affects in particular the relative phase of the σ^+ and σ^- fields. The polarization for all further emission is fixed by the value of this relative phase at the end of the quantum-mechanical period and this value, because of the fluctuations, changes randomly from pulse to pulse.

II Superradiance in a Three-level System Cascading to the Ground Level

In this section is presented and interpreted the observation of super-radiant emission, in the visible range, on a three-level cascade of atomic Sr, whose lowest level is the ground state itself. Let us mention that the observation of superradiance in the visible range has been achieved only very recently, using Eu atoms [12,13].

The experimental set-up is shown in Fig. 8. It includes two pulsed dye-lasers pumped by the same nitrogen laser. They provide pulses with 5-6 ns time duration and a few kilowatt of peak power. The Sr vapour is obtained in a heat-pipe type absorbing cell. Both laser pulses are simul-taneous and copropagating. They are focused in the cell and the laser flux density is of the order of 5 MW/cm^2. The active region is approximately a cylinder with 1 cm length and 0.4 mm diameter. The backward fluorescence is observed through a monochromator. Time-resolved studies are performed using a boxcar integrator following a fast photomultiplier. The time-resolution limit of the detection is of the order of 1 ns.

The diagram of the relevant atomic levels is shown in Fig. 9.

Either the $^1S_0$ or the $^1D_2$ level is populated by resonant two-photon absorption from the ground level. The detuning between the blue laser frequency and the frequency of the lower transition is of a few angströms. One

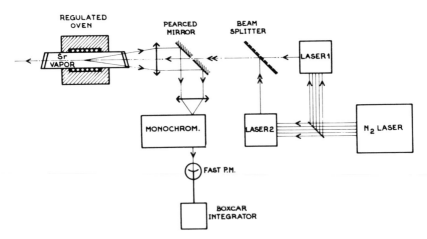

Fig. 8 : Experimental set-up

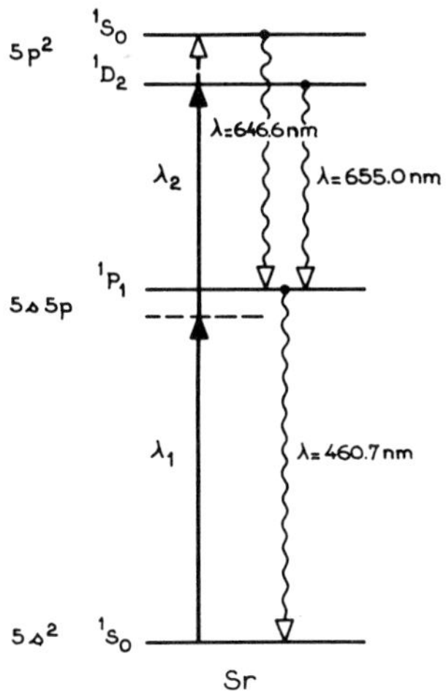

Fig. 9 : Relevant energy levels

$5p^2$ — $^1S_0$, $^1D_2$

$\lambda = 646.6$ nm

λ_2 $\lambda = 655.0$ nm

$5s\,5p$ — $^1P_1$

λ_1 $\lambda = 460.7$ nm

$5s^2$ $^1S_0$

Sr

is therefore ensured that the $^1P_1$ level is not populated by absorption of "blue" photons.

The observations have been made with constant laser intensities and variable temperature, i.e. variable atomic density. Dealing with the upper transition ($\lambda = 646.6$ nm or $\lambda = 655.0$ nm) and at rather low atomic densities, one observes a signal which decreases exponentially with a decay constant equal to the lifetime of the upper level. Above a threshold value of the atomic density ($n_0 \simeq 3.5 \; 10^{13}$ atoms/cm^3), it appears suddenly a very sharp peak, directively emitted, whose height increases rapidly and whose width becomes quickly equal to the resolution limit of the detection. Dealing with the lower transition ($\lambda = 460.7$ nm), an analogous phenomenon is observed in both cases of excitation, but with an atomic density threshold slightly higher than for the red lines. Two typical recordings of the blue emission, below and above the threshold, are shown in Fig. 10 ; Fig. 11 is a photograph of the blue and red emitted light such as it appears in a screen placed before the detector (for a temperature higher than both thresholds).

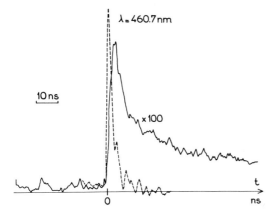

$\lambda = 460.7$ nm

10 ns

×100

Fig. 10: Temporal condensation for the blue emission ($\lambda = 460.7$ nm); the dashed line (resp. full line) represents the recorded signal at 580°C, i.e. $n_0 \sim 4.4 \; 10^{14}$ atoms/cm^3 (resp. at 510°C, i.e. $n_0 \sim .8 \; 10^{14}$ atoms/cm^3)

487

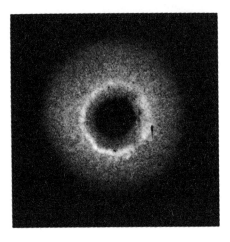

The experimental values of the thresholds are in agreement with the threshold condition for superradiance given in [14]. The delay times of the
different light pulses with respect to the center of the two exciting pulses are always small (less than 4 ns) but they are inversely proportional
to the atomic density. Let us finally mention that the blue pulse always
appears after the red one, the delay time between them being small (1- 2 ns)
but significant.

All these observations allow us to conclude that, above a given atomic
density threshold, the two successive light pulses that are emitted on the
two transitions of the cascade correspond to cooperative emission. As far
as the upper transition is concerned, this result is not surprising since
a complete population inversion is produced on this transition by the two-
photon excitation. The obtainment of a large population inversion between
the intermediate level and the ground level is somewhat more surprising.
However, the red pulses always appear before the end of the exciting laser
pulses and the existence of a pumping effect must be taken into account.
Actually one can show that the observation of a light pulse corresponding
to the lower transition can be well interpreted as a consequence of this
pumping effect. Other phenomena, for example a four-wave mixing process,
could be considered, but this simple interpretation accounts, at least qualitatively, for all of the experimental observations. Calculations have
been made using a semi-classical model including the duration of the excitation. However the result can be easily understood without calculations.
In a very simplified description, the evolution of the system consists in
alternated periods of pumping and of deexcitation. At the beginning of the
two exciting pulses, all of the N atoms are in their ground level. The
pumping transfers quickly $N/2$ atoms in the upper level ($^{1}S_{0}$ or $^{1}D_{2}$).
It provides therefore a complete population inversion on the upper transition. Cooperative emission then appears on this transition, after a time
delay. It transfers the $N/2$ atoms of the upper level on the intermediate
one. The upper level is therefore empty again and, if the excitation still
lasts, the pumping transfers in this level half of the atoms that are in
the ground level, i.e. $N/4$ atoms. One obtains therefore an important
population inversion between the intermediate level ($N/2$ atoms) and the

ground level (N/4 atoms) and cooperative emission can occur on the lower transition of the cascade. Notice that in this very simplified description the levels are considered as nondegenerate. The semi-classical calculations, including a correct treatment of level-degeneracy, lead to similar results.

The results of these two experimental studies illustrate two interesting aspects of the study of coupled transitions in superradiance. First the presence of coupled transitions leads to various specific properties of the superradiant emission. As an example of the new experimental possibilities afforded by the observation of superradiance in the visible range, the experiments with Sr atoms show that cooperative emission down to a ground level is possible. The experimental study using Rb atoms demonstrates the specific polarization properties of the superradiant light, which are due to the coupling between the Zeeman transitions. Secondly, the coupling between superradiant transitions reveals some fundamental properties of the cooperative emission. For example, the superradiant light emitted by the Rb atoms has been shown to be always totally polarized ; this is a confirmation of the validity of the semi-classical approximation. In addition, the observation of a spontaneous symmetry breaking provides a new manifestation of the fluctuations that are inherent to superradiance . The study of coupled superradiant transitions can therefore, in this way, help the understanding of the phenomenon of superradiance itself.

[1] Q. H. F. Vrehen, H. M. J. Hikspoors and H. M. Gibbs, Phys. Rev. Lett. 38, 764 (1977).

[2] M. Gross, J. M. Raimond and S. Haroche, Phys. Rev. Lett. 40, 1711 (1978).

[3] J. Marek, J. Phys. B 12, L229 (1979).

[4] M. Gross, C. Fabre, P. Pillet and S. Haroche, Phys. Rev. Lett. 36, 1035 (1976) ; S. Haroche, C. Fabre, M. Gross and P. Pillet, in Atomic Physics 5, edited by R. Marrus, P. Prior and H. Shugart (Plenum, New York, 1977).

[5] A. Crubellier, S. Liberman and P. Pillet, Phys. Rev. Lett. 41, 1237 (1978).

[6] A. Crubellier, Phys. Rev. A 15, 2430 (1977).

[7] A. Crubellier and M. G. Schweighofer, Phys. Rev. A 18, 1797 (1978).

[8] P. Pillet, Thèse de 3ème cycle, Paris (1977).

[9] S. Haroche, M. Gross and P. Pillet, in Proceedings of the Fourth Rochester Conference on Coherence and Quantum Optics , edited by L. Mandel and E. Wolf (Plenum, New York, 1978), p. 539.

[10] D. Polder, M. F. H. Schuurmans and Q. H. F. Vrehen, Phys. Rev. A 19, 1192 (1979).

[11] F. Haake and R. Glauber, to be published.

[12] C. Bréchignac and P. Cahuzac, J. de Physique Paris 40, L-123 (1979).

[13] P. Cahuzac, H. Sontag and P. E. Toschek, to be published in Opt. Comm..

[14] J. C. McGillivray and M. S. Feld, Phys. Rev. A 14, 1169 (1976).

Part VIII

Laser Spectroscopic Applications

Phase Conjugate Optics[1]

J. AuYeung and A. Yariv

California Institute of Technology, Pasadena, CA 91125, USA

1. Introduction

A new research area in coherent optics has emerged and has been receiving increasing attention from many scientists as its important applications are recognized. Phase conjugate optics is the name which seems to have attached itself to this new field. The main feature of phase conjugate optics is the generation of an electromagnetic wave with a phase distribution which is, at each point in space, the reversal of that of an arbitrary incoming monochromatic wave. The wavefront, after being generated, proceeds to propagate in the opposite direction, retracing in reverse the path of the incoming wave. Thus, the phase reversal or conjugation process results in what is frequently called a time-reversed replica of the incident wave. If we consider, as an example, an incoming spherical wavefront which, diverging from a point, has a radius of curvature R, its conjugate-replica will be an outgoing spherical wavefront converging toward the same point and with a radius of curvature -R. Phase conjugation techniques have been used in the past for imaging through phase distorting media; well known examples can be found in holography [1] and adaptive optical systems [2]. The new and attractive feature, which differentiates phase conjugate optics from the previous techniques, is the use of nonlinear optical mixing to generate in real time without the need for intermediate electronics, and with amplification if desired, a time-reversed replica of an incident wave.

Before discussing in detail the techniques that can be used to produce such conjugate wavefronts, it is appropriate to define exactly what a conjugate wavefront is, and to describe its properties. Consider an incident electromagnetic wave

$$E_i(\vec{r},t) = \text{Re}\{A(\vec{r}) \exp[i(\omega t - \vec{k}\cdot\vec{r})]\} \tag{1}$$

which is propagating with a wave vector $\vec{k}$ and oscillating at an optical frequency ω. The complex amplitude $A(\vec{r})$ contains both amplitude and phase information of the field. The conjugate of this field is defined to be

$$E_r(\vec{r},t) = \text{Re}\{A^*(r) \exp[i(\omega t + \vec{k}\cdot\vec{r})]\} \tag{2}$$

It is clear that E_r is obtained by taking the complex conjugate of the spatial part of E_i, and it describes a wave propagating in the opposite direction to that of E_i. The field E_i, as it propagates inside a lossless

---

[1]Research supported by the Army Research Office, Durham, N.C.

dielectric medium with a real permittivity $\varepsilon(\vec{r})$, obeys the scalar wave equation

$$\nabla^2 E_i + \omega^2 \mu \varepsilon(\vec{r}) E_i = 0 \tag{3}$$

If the field is also assumed to be traveling in the positive z direction, (3) implies that

$$\nabla^2 A + [\omega^2 \mu \varepsilon(\vec{r}) - k^2]A - 2ik \frac{\partial A}{\partial z} = 0 \tag{4}$$

The complex conjugate of (4) is

$$\nabla^2 A^* + [\omega^2 \mu \varepsilon(\vec{r}) - k^2]A^* + 2ik \frac{\partial A^*}{\partial z} = 0 \tag{5}$$

and can be recognized as the wave equation describing the propagation of the conjugate wave E_r which is therefore a legitimate solution of Maxwell's equations. In general, in a lossless dielectric medium, there are two solutions to Maxwell's equations; they are two counterpropagating waves which are the conjugate of each other. When the conjugate wave is generated at a plane $z = z_0$, it propagates in the negative z direction and retraces the original path of the incident wave, evolving as the conjugate of the incident wave at all points for $z < z_0$. The phase fronts of the two waves are identical in shape except that they move in opposite directions.

A frequently quoted application of phase conjugation is its ability to compensate for phase distortions. For example, if an incident plane wave passes through a phase distorting medium such as a poor quality optical component, it will acquire an undesirable phase $\exp[i\phi(x,y)]$. The conjugate of this distorted wave, after being generated, will have a phase of $\exp[-i\phi(x,y)]$, which, upon retracing the original path through the phase distorting medium, will emerge as a plane wave with the phase aberration exactly cancelled.

2. Phase-Conjugate Wavefront Generation by Nonlinear Optical Mixing

The techniques proposed to date for phase conjugation include stimulated scattering and nonlinear optical mixing. The most widely used method [3-7] is now known as degenerate four-wave mixing. All optical fields involved in the mixing process are of the same frequency, ω, hence the name *degenerate*. The geometry of the interaction is shown in Fig.1 [7]. A nonlinear medium,

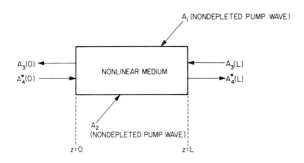

Fig.1 Geometry of four-wave mixing

for example, carbon disulphide (CS_2), is illuminated by two intense pump beams, A_1 and A_2, both at frequency ω. These waves are chosen to be plane waves

$$E_1(\vec{r},t) = \frac{1}{2} A_1 \exp[i(\omega t - k_1 \cdot \vec{r})] + c.c. \tag{6}$$

$$E_2(\vec{r},t) = \frac{1}{2} A_2 \exp[i(\omega t - \vec{k}_2 \cdot \vec{r})] + c.c. \tag{7}$$

which are counterpropagating such that the sum of their wave vectors $\vec{k}_1(\omega) + \vec{k}_2(\omega) = 0$. Upon incidence, along an arbitrary direction, of an arbitrary input wave

$$E_4(\vec{r},t) = \frac{1}{2} A_4 \exp[i(\omega t - \vec{k}_4 \cdot \vec{r})] + c.c. \tag{8}$$

at the same frequency ω but with a wave vector $\vec{k}_4(\omega)$, a third order non-linear polarization

$$P_{NL}^{\omega+\omega-\omega} = \frac{1}{2} \chi^{(3)} A_1 A_2 A_4^* \exp\{i[(\omega+\omega-\omega)t - (\vec{k}_1+\vec{k}_2-\vec{k}_4)\cdot\vec{r}]\} + c.c. \tag{9}$$

is formed. It radiates to give a field

$$E_3(\vec{r},t) = \frac{1}{2} A_3 \exp[i(\omega t - \vec{k}_3 \cdot \vec{r})] + c.c. \tag{10}$$

also at the frequency ω, but with a wave vector $\vec{k}_3(\omega) = \vec{k}_1 + \vec{k}_2 - \vec{k}_4$ that is equal to $-\vec{k}_4(\omega)$. $\chi^{(3)}$ is a third order nonlinear susceptibility tensor element of the medium. The backward going wave E_3, with an amplitude proportional to A_4^*, in turn combines with the two pump waves to induce a third order nonlinear polarization,

$$P_{NL}^{\omega+\omega-\omega} = \frac{1}{2} \chi^{(3)} A_1 A_2 A_3^* \exp\{i[(\omega+\omega-\omega)t - (\vec{k}_1+\vec{k}_2-\vec{k}_3)\cdot\vec{r}]\} + c.c. \tag{11}$$

which radiates into E_4. The phase matching condition is automatically satisfied in this process, since $\vec{k}_3 = -\vec{k}_4$. The two waves, without loss of generality, can be assumed to be counterpropagating along the z direction. The evolution of the amplitudes, $A_3(z)$ and $A_4(z)$, is obtained by solving the wave equation

$$\nabla^2 E - \frac{\varepsilon}{c^2} \frac{\partial^2}{\partial t^2} E = \frac{4\pi}{c^2} \frac{\partial^2}{\partial t^2} P_{NL} \tag{12}$$

for each wave with its appropriate driving term given in (9) and (11), respectively. Neglecting the depletion of the pump beams, E_1 and E_2, and using the approximation

$$\left| \frac{d^2 A_j}{dz^2} \right| \ll \left| k_j \frac{dA_j}{dz} \right| \quad \text{for} \quad j = 3,4$$

two coupled equations can be obtained from the wave equation:

$$\frac{dA_3}{dz} = i\kappa^* A_4^* \tag{13}$$

494

$$\frac{dA_4^\star}{dz} = i \kappa A_3 \tag{14}$$

where $\kappa^\star = \frac{2\pi\omega}{cn} \chi^{(3)} A_1 A_2$, and n is the linear index of refraction of the non-linear medium. The solutions, when expressed in terms of the boundary values $A_3(L)$ and $A_4(0)$ are

$$A_3(z) = \frac{\cos|\kappa|z}{\cos|\kappa|L} A_3(L) + \frac{i\kappa^\star \sin|\kappa|(z-L)}{|\kappa|\cos|\kappa|L} A_4^\star(0) \tag{15}$$

$$A_4^\star(z) = \frac{i|\kappa|\sin|\kappa|z}{\kappa^\star \cos|\kappa|L} A_3(L) + \frac{\cos|\kappa|(z-L)}{\cos|\kappa|L} A_4^\star(0) \tag{16}$$

z = 0 and z = L specify the boundaries of the interaction region as shown in Fig.1. In particular, when there is only an incident wave $A_4(0)$ at z = 0 with $A_3(L) = 0$, (15) yields

$$A_3(0) = -i(\frac{\kappa^\star}{|\kappa|} \tan|\kappa|L) A_4^\star(0) \tag{17}$$

while at the other end (z = L)

$$A_4(1) = \frac{A_4(L)}{\cos|\kappa|L} \tag{18}$$

The nonlinear medium, used under the condition prescribed above, can be viewed as a mirror, reflecting an input wave E_4 to give a backward wave $E_3 \propto E_4^\star$. The difference between such a conjugate mirror and a conventional one lies in the fact that the reflected wave is a time-reversed replica and retraces the exact path of the incident wave; whereas the reflected wave from a conventional mirror obeys the conventional law of reflection. The intensity reflectivity of the conjugate mirror, when derived from (17), is $\tan^2(|\kappa|L)$ which depends on the nonlinear properties of the material and the pump power. When $|\kappa|L > \pi/4$, $\tan|\kappa|L > 1$ and amplified reflection will occur; and at $|\kappa|L = \pi/2$, an infinite reflectivity, corresponding to self-oscillation, is predicted.

In the above analysis, E_3 and E_4 are plane waves. However, the result obtained is also applicable to waves of arbitrary phase front or transverse spatial dependence. This is because the solutions given in (15) to (18) are independent of the propagating direction and thus describe plane waves with any wave vector $\vec{k}$. An incident wave with a particular phase front and transverse spatial dependence can be expressed as a linear superposition of plane waves with different wave vectors. Each of these plane waves will give rise to its own conjugate plane wave with identical nonlinear reflectivity. A conjugate replica of the incident wavefront is obtained when the individual conjugate plane waves are superposed together.

An alternate viewpoint of four-wave mixing has a direct analogy in holography [4,5,8]. The incident (object) wave E_4 interferes with a (reference) pump beam, say E_1, to form a grating which then diffracts the other counter-propagating (reconstruction) pump beam, E_2, to give the conjugate image wave, E_3. At the same time, the interference of E_4 and E_2 forms a hologram which diffracts E_1 to yield E_3. The backward conjugate wave in turn forms a grating with either pump to diffract the other, resulting in the amplifi-

cation of the incident object wave. Four-wave mixing, however, differs from
the conventional holographic experiments in the simultaneity of the grating
formation and the reconstruction steps. This makes possible the processing
of information in real time. The name, real-time holography, has been frequently used to refer to this kind of nonlinear optical means of wavefront
generation.

3. Experimental Demonstration of Conjugate Wavefront Generation

In this section, an experiment which demonstrates the generation of a phase-
conjugate wavefront by degenerate four-wave mixing is described. The experimental set-up is shown in Fig.2. The laser source was a Q-switched ruby

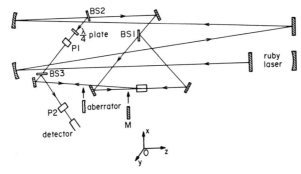

Fig.2. Experimental set-up

laser operated in a single longitudinal and transverse mode. The linearly
polarized (s-polarization) laser pulse had a duration of about 25 nsec and
a typical energy of 10 millijoules. The intensity of the beam varied transversely as $\exp(-r^2/w_0^2)$ with a beam radius w_0 of 0.33 mm. The laser pulse
was divided by a beam splitter (BS1) into two beams which passed through
the sample cell in opposite directions, to provide the two pump beams,
$E_1(\vec{r},t)$ and $E_2(\vec{r},t)$, required for the four-wave mixing interaction. The
field at the plane of the laser output mirror was imaged by two 6m radius-
of-curvature mirrors onto a plane at the center of the sample cell. The
path length between the laser and the cell was therefore 12 m, and this provided a round trip length of 24 m, or equivalently, 80 nsec delay to prevent
feedback of the pulse back to the laser. A small portion of the laser energy was diverted by a beam splitter (BS2) to serve as the signal wave,
$E_4(\vec{r},t)$. Its polarization was converted by a quarter-wave plate and a
polarizer (P1) to be orthogonal (p-polarized) to those of the two pump
waves. The signal beam entered the cell almost parallel to the pump waves.
The angle was limited to about ten milliradians to allow for a large overlap of the beams, and hence a large interaction region.

The nonlinear medium was p-methoxy-benzylidene p-n-butylaniline (MBBA)
contained in a 5.7 cm long quartz cell. This is an opaque nematic liquid
crystal at room temperature. It turns transparent at temperatures above
$T_c = 42.5°C$, at which point the material passes into the isotropic phase.
Its properties at temperatures near and above the transition point, T_c, have
been investigated elsewhere [9], and it has been found to possess a very
high nonlinear susceptibility at these temperatures, thus making it a suitable material for phase-conjugate wavefront generation.

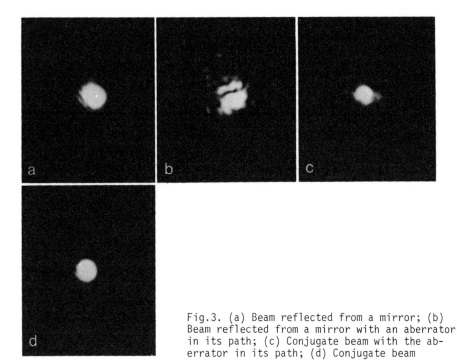

Fig.3. (a) Beam reflected from a mirror; (b) Beam reflected from a mirror with an aberrator in its path; (c) Conjugate beam with the aberrator in its path; (d) Conjugate beam

The two s-polarized counterpropagating pump waves and the p-polarized signal wave induced a nonlinear polarization inside the MBBA which radiated a p-polarized conjugate wave, $E_3(\vec{r},t)$. The conjugate beam propagated backward along the path of the signal wave, and was directed by a beam splitter (BS3) to a detector. The polarizer P2 in Fig.2 passed only the p-polarized conjugate wave and rejected any s-polarized light that was caused by scattering of the intense pump waves.

To verify the conjugate nature of the radiated field and its ability to compensate for phase distortions, the following tests were carried out. First, a mirror M, as shown in Fig.2, was placed next to the sample cell to retroreflect the signal beam, and Fig.3a shows the spatial quality of the beam. Then a phase aberrator was inserted in the path of the signal beam and the picture of the reflected beam from the mirror M was taken. The beam had therefore passed through the aberrator twice. Fig.3b shows such a picture in which the aberrator was a piece of transparent double-stick Scotch tape. The mirror was then removed to allow the four-wave mixing to take place. Fig.3c shows the generated conjugate beam which passes through the same aberrator before incidence onto the film. Fig.3d finally shows the picture of the conjugate beam with the aberrator removed. Since the conjugate beam had a different intensity than the mirror-reflected signal beam, Figs. 3a and 3b were taken with the appropriate neutral density filter in front of the film to simulate the intensity of the conjugate beam. The correction for phase distortion by the conjugate beam is quite evident by comparing Figs. 3b and 3c.

497

4. Phase Conjugation by Photon Echoes

All optical mixing schemes for phase conjugation require the simultaneous illumination of a medium with several intense optical beams. There is, however, another phase reversal technique which uses the finite "memory" time of the atoms in the "mixing" medium. This technique does not require the optical fields to be present simultaneously.

Heer and McManamon [10] first pointed out that wafefront correction and phase conjugation can take place in a photon-echo geometry involving forward traveling noncoincident four-wave mixing. Shiren [11] later showed that a backward conjugate echo can result when two of the noncoincident input pulses are opposite to each other.

These ideas will be explored here and a perturbation treatment will be used [12] to analyze the problem in a similar way as the perturbational description of cw nonlinear optical mixing [13]. It also includes the effects of collisional relaxation (T_1, T_2) on the process of conjugate echo formation. In the previous sections, the nonlinear susceptibility was taken to be a phenomenological constant. The analysis presented in this section will illustrate how the nonlinear susceptibility of an atomic system can be derived quantum mechanically.

Consider an ensemble of two-level atoms with energy levels E_s and E_m $(E_m > E_s)$. The atoms are subjected to three pulses of duration δ of optical radiation of the same frequency ω but traveling along arbitrary directions

$$E_i(\vec{r},t) = \frac{1}{2} \vec{E}_i(t) \exp[i(\omega t - \vec{k}_i \cdot \vec{r})] + \text{c.c.}, \qquad i = 1,2,3 \tag{19}$$

The pulses may or may not coincide, and the pulse areas are taken to be much smaller than π. The induced dipole moment can be obtained as

$$\langle \mu_j \rangle = \text{trace}(\hat{\rho}\,\hat{\mu}_j) = \rho_{12}(\hat{\mu}_j)_{21} + \rho_{21}(\hat{\mu}_j)_{12} \tag{20}$$

where $\hat{\rho}$, the density matrix operator, obeys

$$\frac{d\hat{\rho}}{dt} = \frac{1}{i\hbar} [\hat{H}, \hat{\rho}] \tag{21}$$

The total Hamiltonian $\hat{H}$ is the sum of the unperturbed value $\hat{H}_0$ and the perturbation $\hat{V}(t)$, where

$$\hat{V}(t) = -\hat{\mu} \cdot \vec{E}(\vec{r},t) \tag{22}$$

We shall assume $(\hat{\mu}_j)_{ms} = \mu_{jms}$ for $m \neq s$, and zero otherwise. Eq. (21) can be solved by expressing $\hat{\rho}$ as a perturbation series, i.e., $\rho_{ij} = \rho_{ij}^{(0)} + \rho_{ij}^{(1)} + \rho_{ij}^{(2)} + \cdots$. Furthermore, the effect of the population relaxation and dephasing collisions can be included by adding two phenomenological time constants to (21). Specifically, $\rho_{ij}^{(1)}$ obeys the following equation of motion [14]

$$\frac{\rho_{ij}^{(1)}}{dt} - i\omega_{ji}\,\rho_{ij}^{(1)} + \frac{\rho_{ij}^{(1)}}{T_{ij}} = \frac{-i}{\hbar} [\hat{V}(t), \hat{\rho}^{(0)}]_{ij}, \qquad i,j = 1,2 \tag{23}$$

For $i = j$, T_{ij} equals T_1, and $\rho_{ii}^{(0)}$ is determined by the ith energy level thermal equilibrium population. For $i \neq j$, T_{ij} equals T_2, and $\rho_{ij}^{(0)} = 0$.

We shall look for a term in ρ_{12} that is proportional to the product of the three applied fields. Upon the incidence of the first pulse, only the off-diagonal density matrix elements, ρ_{12} and ρ_{21}, are perturbed to first order in E_1. The driving term on the right hand side of (23) is proportional to $(\rho_{11}^{(0)} - \rho_{22}^{(0)})E_1(t)$. After the perturbation due to the first pulse, ρ_{12} evolves as $\exp[t(i\omega_{ms} - 1/T_2)]$ and in the limit of small δ ,

$$\rho_{12}^{(1)} (t_3 > t > t_1) = -i \frac{\mu_{1sm}}{2\hbar} \delta(\rho_{11}^{(0)} - \rho_{22}^{(0)}) E_1 \times$$

$$\exp[i(\omega t_1 - \vec{k}_1 \cdot \vec{r})] \exp[(i\omega_{ms} - \frac{1}{T_2})(t-t_1)] \qquad (24)$$

The second pulse affects the diagonal elements, ρ_{11} and ρ_{22}, giving rise to perturbation terms proportional to $E_1^* E_2$. In this case, the driving term in (23) is the product of $E_2(t)$ and the $\rho_{12}^{(1)}(t)$ given in (24). ρ_{11} and ρ_{22} then evolve as $\exp(-t/T_1)$ after the second pulse,

$$\rho_{11}^{(1)}(t > t_2) = \frac{-\mu_{1ms}\mu_{2sm}}{4\hbar^2} \delta^2 (\rho_{11}^{(0)} - \rho_{22}^{(0)})E_1^* E_2 \exp[i\omega(t_2 - t_1)$$

$$- i(\vec{k}_2 - \vec{k}_1) \cdot \vec{r}] \exp[-(i\omega_{ms} + \frac{1}{T_2})(t_2 - t_1)] \exp[- \frac{1}{T_1}(t-t_2)] \qquad (25)$$

When the third pulse arrives, ρ_{12} and ρ_{21} are affected by driving terms proportional to $E_3(t)(\rho_{11}^{(1)}(t) - \rho_{22}^{(1)}(t))$ when the on-diagonal density matrix elements are given in (25). Thus ρ_{12} and ρ_{21} are proportional to $E_1^* E_2 E_3$ and evolve as $\exp[t(i\omega_{ms} - 1/T_2)]$ for $t > t_3$. Setting $t_1 = 0$, the final expression for the induced dipole moment is

$$<\mu_j> = \frac{i\delta^3}{4\hbar^3} \mu_{1ms}\mu_{2sm}\mu_{3sm}\mu_{jms} E_1^* E_2 E_3 \exp\{i[\omega t-(-\vec{k}_1 + \vec{k}_2 + \vec{k}_3) \cdot \vec{r}$$

$$+ (\omega_{ms}-\omega)(t-t_2-t_3)] \} \exp[- \frac{1}{T_2}(t+t_2-t_3) - \frac{1}{T_1}(t_3-t_2)] \qquad (26)$$

where $\omega_{ms} \equiv \omega_m - \omega_s$. Let the resonance offset parameter be $\Delta \equiv (\omega_{ms} - \omega)$. An inhomogeneous distribution of resonant frequencies is described by a normalized lineshape function $g(\Delta)$. The induced polarization P_j is obtained by summing $<\mu_j>$ over all Δ. The result is

$$P_j = N \int_{-\infty}^{\infty} <\mu_j> g(\Delta) d\Delta$$

$$= \frac{i\delta^3}{4\hbar^3} N\mu_{1ms}\mu_{2sm}\mu_{3sm}\mu_{jms}E_1^* E_2 E_3 \exp\{i[\omega t - (-\vec{k}_1 + \vec{k}_2 + \vec{k}_3) \cdot \vec{r}] S(t-t_2-t_3)$$

$$\times \exp[- \frac{1}{T_2}(t+t_2-t_3) - \frac{1}{T_1}(t_3-t_2)] \qquad (27)$$

where $S(t) = \int_{-\infty}^{\infty} g(\Delta) e^{i\Delta t} d\Delta$ is the Fourier transform of the normalized lineshape function $g(\Delta)$, and N is the dipole moment density.

For an even $g(\Delta)$ the function $S(t-t_2-t_3)$, hence P_j, is essentially zero except for a duration $\sim 2\pi/\Delta_{1/2}$ centered on

$$t = t_2 + t_3 \qquad\qquad\qquad (28)$$

where $\Delta_{1/2}$ is the width of $g(\Delta)$. During this period the sample will radi-ate a photon echo, provided the phase matching conditions are satisfied. If the pulses E_2 and E_3 are exactly opposite, $\vec{k}_2 + \vec{k}_3 = 0$, then the radiated echo at $t = t_2 + t_3$ will propagate according to (27) in the direction $-k_1$, i.e., the reverse direction to E_1. These results have been noted by Shiren [11], who applied the formalism of photon echoes. Our perturbation approach brings out specifically the field dependence $E_1^* E_2 E_3$ which is in a form identical to that used to describe four-wave conjugation in conventional nonlinear optics terminology [13]. It should be recalled, however, that here $E_1(t)$, $E_2(t)$, and $E_3(t)$ do not necessarily coincide in time.

By comparing (27) with (9) or (11), it is obvious that the induced polar-ization given in (27) can be expressed as a product of a third order non-linear susceptibility $\chi^{(3)}$ and the applied fields. The analysis presented in this section shows how $\chi^{(3)}$ can be derived. The formalism is general and is applicable to describe the interaction of other systems with pulsed or continuous optical waves.

The output wave radiated by the polarization (27) is the conjugate of the first wave $E_1(t)$. In holographic terms we can describe the process as one whereby a "hologram" is written into the atomic medium by pulses $E_1(t)$ and $E_2(t)$ which do not coincide in time. An interrogation pulse at t_3 opposite in direction to $E_2(t)$ gives rise to the backward readout echo at $t = t_3 + t_2$ which is the conjugate replica of $E_1(t)$. If pulses E_2 and E_3 are parallel to each other as well as to pulse E_1, i.e., $\vec{k}_1 = \vec{k}_2 = \vec{k}_3$, then the induced echo has, according to (27), a spatial dependence of $\exp[i(\omega t - \vec{k}_1 \cdot \vec{r})]$. It is thus radiated in the forward direction.

The special case of $\vec{E}_2(\vec{r},t) = \vec{E}_3(\vec{r},t)$ is of interest. The pulse formed is proportional to $E_1^* E_2^2 \exp\{i[\omega t - (2\vec{k}_2 - \vec{k}_1) \cdot \vec{r}]\}$ and occurs, according to (28), at $t = 2t_2$. If $\vec{k}_1 = \vec{k}_2$, then the pulse is radiated in the forward direction and can be recognized as the usual forward photon echo [15]. It is interesting to note that although only two pulses are applied, the phenom-enon involves a third order mixing, except that two of the fields (E_2 and E_3) are degenerate. The demonstrations of phase conjugation by photon echo have been reported recently [16,17].

5. A Phase-Conjugate Resonator

It has been mentioned that a nonlinear material, when used in the four-wave mixing geometry, can be treated as a conjugate mirror. If such a mirror is used to replace one of the two conventional mirrors that form a laser cavity, it will be shown that there is no stability condition necessary for oscilla-tion to occur.

Consider an incident Gaussian field

$$E_i = E_i(\vec{r}) \exp[i(\omega t - kz - \frac{kr^2}{2\rho}) - \frac{r^2}{w^2}] + c.c. \qquad\qquad (29)$$

500

with a complex amplitude $E_i(\vec{r})$, a radius of curvature ρ, and a spot size w, which is also frequently written as

$$E_i = E_i(\vec{r}) \, \exp[i(\omega t - kz - \frac{kr^2}{2q_i})] + c.c. \tag{30}$$

The complex radius of curvature q_i in (30) is defined as

$$\frac{1}{q_i} = \frac{1}{\rho} - \frac{i\lambda}{\pi w^2} \tag{31}$$

The reflected field from the phase-conjugate mirror is

$$E_r \propto E_i^*(\vec{r}) \, \exp[i(\omega t + kz + \frac{kr^2}{2\rho}) - \frac{r^2}{w^2}] + c.c. \tag{32}$$

which can also be expressed as

$$E_r \propto E_i^*(\vec{r}) \, \exp[i(\omega t + kz - \frac{kr^2}{-q_i^*})] + c.c. \tag{33}$$

Using the conventional ray matrix formalism which describes the effects of optical components on Gaussian beam propagation, the phase-conjugate mirror can be represented by a "ray" matrix

$$\underset{\approx}{M} = \begin{bmatrix} A & B \\ C & D \end{bmatrix} = \begin{bmatrix} 1 & 0 \\ 0 & -1 \end{bmatrix} \quad . \tag{34}$$

with the output and input complex radii of curvature related by

$$q_r = \frac{Aq_i^* + B}{Cq_i^* + D} \tag{35}$$

Consider an optical resonator which is bounded on one end by a mirror having a radius of curvature, R. The resonator contains arbitrary intracavity optical components described collectively by an $A'B'C'D'$ matrix, M', for optical propagation from left to right, and again by an $A''B''C''D''$ matrix, $\underset{\approx}{M''}$, for propagation from right to left. The resonator is also bounded on the other end by a phase-conjugate mirror. Choosing a plane to the immediate right of the real mirror, we trace a beam that propagates to the right and get, after one round trip, the following matrix product:

$$\underset{\approx 1}{M_1} = \begin{bmatrix} A_1 & B_1 \\ C_1 & D_1 \end{bmatrix} = \begin{bmatrix} 1 & 0 \\ \frac{-2}{R} & 1 \end{bmatrix} \begin{bmatrix} A'' & B'' \\ C'' & D'' \end{bmatrix} \begin{bmatrix} 1 & 0 \\ 0 & -1 \end{bmatrix} \begin{bmatrix} A' & B' \\ C' & D' \end{bmatrix}$$

$$= \begin{bmatrix} 1 & 0 \\ \frac{-2}{R} & 1 \end{bmatrix} \begin{bmatrix} 1 & 0 \\ 0 & -1 \end{bmatrix} \tag{36}$$

where we have used the relation

$$\underset{\approx}{M''}\underset{\approx}{M} = \underset{\approx}{M}(\underset{\approx}{M'})^{-1} \tag{37}$$

which can be shown straightforwardly, using the reciprocity property of the group of optical elements represented by M' (or M''). Self-consistency demands the field to reproduce itself at the aforementioned plane after one round trip, and hence,

$$q = \frac{A_1 q^* + B_1}{C_1 q^* + D_1} \qquad (38)$$

Upon solving for the beam parameters it is found that $\rho = -R$, that is, the radius of curvature of the Gaussian beam matches that of the real mirror at the plane of the mirror, which can either be a concave or a convex mirror. There is no constraint on the spot size of the beam and also no stability condition for such a cavity. The result obtained is independent of the cavity length and the intracavity optical components.

In conventional resonators, all allowed eigenmodes are obtained by demanding a single round trip self-consistent solution. However, in the case of a phase-conjugate resonator, it is possible to have allowed modes which will reproduce themselves after two round trips. Using the same procedure as before, the ray matrix describing the beam propagation for two round trips is $(M_1)^2 = I$ where I is the identity matrix. Hence, in contrast to the single round trip case, any complex radius of curvature, (i.e., both ρ and w) at the initial plane will yield a self-consistent solution. The resonator is stable for any real convex or concave mirror.

In the above analysis, the frequency ω of the Gaussian beam is the same as that of the pump waves illuminating the nonlinear medium where four-wave mixing takes place. If the incident wave is at a frequency $\omega+\delta$ while the pump waves are at frequency ω, the induced nonlinear polarization and hence the reflected conjugate wave is at a frequency $\omega+\omega-(\omega+\delta) = \omega-\delta$. Using the same self-consistent formalism, it can be shown in a straightforward way that no solution exists for the one-round-trip case, and a unique solution exists for the two-round-trip case, leading to interesting longitudinal and transverse mode spectra [18]. However, the reflectivity of the conjugate mirror decreases rapidly with increasing $|\delta|$ due to phase mismatching and hence these spectra are not easily observable and will probably be of limited practical importance.

The phase conjugate resonator is useful mainly when used in the degenerate case when all oscillating modes are locked to the same frequency as that of the pump beam. It will correct for any intracavity phase aberrations and has been experimentally found to allow oscillation even in a conventional unstable cavity configuration [18].

The most frequently quoted example for the application of phase conjugation is the correction of aberrations that arise when an optical wave passes through a phase distorting medium. Other applications of phase conjugate optics include optical filtering, compensation for group velocity dispersion in communication channels, optical gating, and others. Active research is going on in this relatively new area. Many promising applications are yet to be explored.

Acknowledgment

We wish to acknowledge the significant contribution of D. Fekete and D.M. Pepper to this work.

References

1. R. J. Collier, C. B. Burckhardt, and L. H. Lin, Optical Holography (Academic Press, New York, 1971).
2. J. Opt. Soc. Amer. 67 (1977) (special issue on adaptive optics).

502

3. B. I. Stepanov, E. V. Ivakin, and A. S. Rubanov, Soviet Phys.-Dokl. 16, 46 (1971).
4. J. P. Woerdman, Opt. Comm. 2, 212 (1970).
5. A. Yariv, Appl. Phys. Lett. 28, 88 (1976); A. Yariv, J. Opt. Soc. Amer. 66, 301 (1976).
6. R. W. Hellwarth, J. Opt. Soc. Amer. 67, 1 (1977).
7. A. Yariv and D. M. Pepper, Opt. Lett. 1, 16 (1977).
8. A. Yariv, Opt. Comm. 25, 23 (1978).
9. G.K.L. Wong and Y. R. Shen, Phys. Rev. A 10, 1277 (1974).
10. C. V. Heer and P. F. McManamon, Opt. Comm. 23, 49 (1977).
11. N. S. Shiren, Appl. Phys. Lett. 33, 299 (1978).
12. A. Yariv and J. AuYeung, IEEE J.Q.E. QE-15, 224 (1979).
13. J. A. Armstrong, N. Bloembergen, J. Ducuing, and P. S. Pershan, Phys. Rev. 127, 1918 (1962).
14. A. Yariv, Quantum Electronics, 2nd ed. (Wiley, New York, 1975).
15. I. D. Abella, N. A. Kurnit, and S. R. Hartmann, Phys. Rev. 141, 319 (1966).
16. N. C. Griffen and C. V. Heer, Appl. Phys. Lett. 33, 865 (1978).
17. M. Fujita, H. Nakatsuka, H. Nakanishi, and M. Matsuoka, Phys. Rev. Lett. 42, 974 (1979).
18. J. AuYeung, D. Fekete, D. M. Pepper, and A. Yariv, to appear in IEEE J.Q.E.

Laser-Induced Selective Multistep Processes: Impact on Nuclear Physics, Chemistry and Biology

V.S. Letokhov

Institute of Spectroscopy, Academy of Sciences USSR
SU-142092 Moscow, Troitzk, Podol'skii rayon, USSR

In this report I should like to discuss some features of multistep processes of selective action on atoms and molecules by laser radiation which make them rather efficient when they are applied to various fields – beginning with nuclear physics up to biology. This discussion will be concrete enough. I will begin with the simplest object – hydrogen atom. On this example I will discuss the ways of controlling the proton beam characteristics by laser radiation. Then I will proceed to the problem of selective detection of single atoms by the laser selective multistep ionization and to the process of laser selective multistep photodissociation of polyatomic molecules and its applications. I will finish the discussion with DNA molecule that is a complicated object for us.

Of course, my report due to such a large scope will be inevitably superficial and fragmentary. This drawback is eliminated to some extent by two other reports in which two concrete problems of this broad program are being discussed in more detail. They are: 1) laser detection of single atoms by selective multistep photoionization method – the report by G.I.Bekov [1] ; 2) selective action on DNA and RNA components by picosecond laser pulses by Angelov et al. [2] .

2. General Remarks

Multiple-step excitation of a multilevel quantum system (atom, molecule) by multifrequency laser radiation may be effective when the rate of induced transitions at each intermediate level W_{Kn} is much larger than the relaxation rate of each intermediate level $1/T_1^{(k)}$ and $1/T_1^{(n)}$. When a multilevel system is simultaneously irradiated by a set of laser pulses with the same duration $\tau_p \ll T_1^{(k)}$ (k=i,.... , f) at resonance frequencies ω_{Kn} of successive quantum transitions this case is considered optimal (Fig.1). If the energy fluence $\mathcal{E}(\omega_{Kn})$ of each of such pulses complies with the condition of saturation

$$\mathcal{E}(\omega_{Kn}) \gg \mathcal{E}_{sat}^{Kn} = \frac{\hbar\omega_{Kn}}{6(\omega_{Kn})} ,\qquad (1)$$

where $6(\omega_{Kn})$ is the cross-section of the resonance transition k $\rightarrow$ n, under the action of multifrequency laser radia –

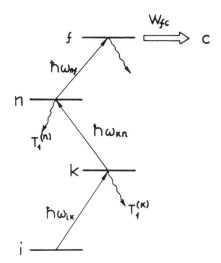

Fig.1 General scheme
of multistep excita-
tion of multilevel
quantum system by a
several laser pulses
with different fre-
quencies.

tion these quantum systems become distributed evenly, on the
average, on the initial, all intermediate and final levels [1]:

$$\frac{n_i}{g_i} \ldots = \frac{n_k}{g_k} = \ldots \frac{n_f}{n_f} , \qquad (2)$$

where g_k is the statistical weight of a k-th state. If the
statistical weight rises fairly with increasing number of quantum state an overwhelming portion of quantum systems may be
concentrated in highly excited states. Vibrational excitation
of polyatomic molecules is an excellent example of this (see
below Fig.8).

If the conditions of saturation of intermediate quantum
transitions (1) are met with a sufficient reserve the even
distribution of quantum systems on intermediate levels (2) can
take place in a time τ_{exc} which is shorter than the irradiation time τ_p:

$$\tau_{exc} \simeq \sum_K W_{Kn}^{-1} \lesssim \tau_p ; \quad W_{Kn} = \delta_{Kn} I(\omega_{Kn}). \qquad (3)$$

A highly excited quantum system may transfer into continuum either spontaneously or under irradiation or any additional perturbation, for example, due to ionization or dissociation. If the rate of transition to continuum W_{fc} is smaller
in comparison with that of photoexcitation I/τ_{exc} and hence
the decay occurs for the most part after radiation the value
of absolute transfer to continuum is limited by the relative
population of the final excited level:

$$\eta \lesssim g_f / \sum_K g_K . \qquad (4)$$

_____

1) Effects of coherent interaction are disregarded in this
   case.

Yet if during photoexcitation the decay channel of the quantum system is open and its rate W_{fc} is rather high:

$$W_{fc} \gtrsim 1/\tau_{exc} \tag{5}$$

by such single irradiation it is possible to make the initial, all intermediate and final levels empty and hence to attain the maximum yield of photoprocess $\xi \simeq 1$.

It should be noted that the maximum yield of such a stimulated multiple-step photoprocess is limited by a relative fraction of the quantum systems being found in good resonance with multifrequency radiation. Owing to the original distribution of quantum systems over many initial states before radiation (for example, thermal distribution of atoms over the sublevels of a hyperfine structure or molecular distribution on rotational states) this value may be much less unity.

The multiple-step processes in atoms and molecules induced by laser radiation prove to be useful in many practically important applications. For each particular object and application we have to study very thoroughly the elementary process itself to choose an optimal sequence of induced quantum transitions. This affords, first, reasonable requirements for the length and energy fluence of laser pulse. And secondly, it enables us to obtain a wanted selectivity value of photoprocess with desired atoms and molecules. The both requirements can be fulfilled more easily through a multiple-step photoprocess rather than any single-step one. Really, a sequence of allowed single-step transitions enables an atom or a molecule to absorb high energy under rather moderate requirements on laser radiation intensity. Besides, a sequence of resonant transitions makes it possible to use excitation selectivity, in principle, at every step of excitation. This is of particular importance in the absence of excitation selectivity at the first step that is an ordinary linear absorption spectrum.

All these important features of multiple-step photoprocesses induced by laser radiation can be illustrated by specific photoprocesses with atoms and molecules.

2. Multiple-Step Photoionization of Atoms and Nuclear Physics.

The process of multiple-step photoionization (MSPI) of atoms has being studied intensively since 1970 [3], mainly from the standpoint of laser isotope separation. In this particular problem the required excitation selectivity of a chosen isotope is usually attained at the first step and the subsequent induced transitions are chosen from the condition of the most effective transition of selectivity excited atoms into continuum.

In the earlier works nonresonance photoionization of excited atoms to continuum was used, while now they use either quantum transitions to narrow autoionized resonances in the continuum or transitions to highly excited (Rydberg) states with their subsequent ionization by electric field pulses(see review [4]). Actually the scope of multiple-step atomic photoionization is much wider than the concrete problem of isotope separation.

First, during laser excitation of atoms, selection may be accomplshed not only among different isotopes but also by all other characteristics of atoms and nuclei showing up in the optical spectrum. This possibility is of particular value for laser selection of nuclei with different excitation energies (separation of nuclear isomers), with different velocities of motion or kinetic energies (monochromatization of proton beam [5]) or different spin orientation (polarization of proton beam [6]). Fig.2 shows in a simple way various schemes of MPSI of hydrogen atoms which may be used in laser control over velocity and polarization of proton beams.

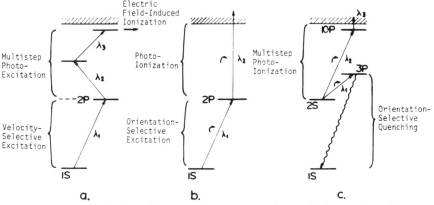

a.   b.   c.

Fig.2 Simplified schemes of quantum transitions for the control of the velocity (a) and polarization (b,c) of a proton beam, based on the multistep selective photoioniza-tion of hydrogen atoms.

Fig.3 presents corresponding simplified geometries of some experiments on interaction of a hydrogen atomic beam with laser radiation where either monochromatization or polarization of protons is under way.

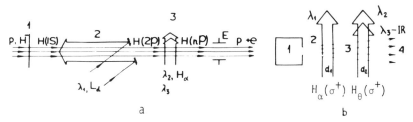

a   b

Fig.3 a) Principal scheme of laser monochromatization of relativistic proton beam: (1) conversion of protons or H^- ions on a thin target to neutral hydrogen atoms, (2) velo-city-selective excitation of hydrogen atoms, (3) stepwise photoionization of selectively excited hydrogen atoms. b) Principal scheme of laser polarizer of protons: geomet-ry for the irradiation of a wide-aperture beam of a meta-stable hydrogen atoms by a selectively quenching (λ_1) and photoionizing (λ_2, λ_3) laser beams (from works [5,6]).

Monochromatic laser radiation can be used to excite hydrogen atoms with a specified projection of their velocity on the chosen direction (Fig.2a). In the case of hydrogen atoms moving with relativistic velocities such excitation by a counterrunning laser beam can be done at the transition L_α even without laser radiation used in the vacuum ultraviolet (Fig.3a).

Each excited hydrogen atom can be ionized with a ultimate absolute yield by the method of MSPI through higher states. The evaluations show [5] that even at the present level of tunable lasers it is quite possible to monochromatize proton beams with energies between 200 and 600 MeV to values from 0.1 to 1.0 keV on real accelerators. By tuning and measuring the wavelength of laser radiation λ_1 at the first step of excitation it is possible to tune and measure the absolute energy value of relativistic protons.

Atoms can be excited by circular-polarized laser radiation to certain sublevels of a hyperfine structure and then ionized by the method of MSPI(Fig.2b). As a result of electron-nucleus interaction, the moment of absorbed photon is distributed between the electron and the nucleus, and certain nuclear orientation takes place in the ensemble of excited atoms and in that of photoions after all. The laser polarization of nuclei by such a method was considered in work [7]. The absence of proper laser sources at $\lambda (L_\alpha)$=1215 Å makes it still impossible to realize this scheme of proton polarization on accelerators. Yet even now we may apply the scheme of MSPI of oriented hydrogen atoms from the metastable $2S_{1/2}$ state. In this method the nuclei become oriented first through selective quenching of some sublevels of hyperfine structure by the action of circular-polarized laser radiation at the wavelength of H_α - transition (Fig.2c). Such a method can be used to obtain wide-aperture beams of polarized protons as schematically illustrated in Fig.3b.

The realization of multiple-step atomic photoionization with a high absolute yield (up to values approximating unity) automatically makes it possible to selectively detect single atoms, that is to realize the ultimate sensitivity of spectroscopy. In our earlier works as well as later in the lectures and reports of 1975-76 [8] particular emphasis was placed upon this interesting possibility of the MSPI method. Now the method is under intensive investigations in a number of laboratories, and a special report [1] is dedicated to it. Here we should just note that the MSPI method gives us a unique possibility at the same time to realize exceptionally high selectivity in detecting different nuclei of one and the same element that can not be attained by other physical methods[9]. This possibility is based on multiple use of the shift (isotopic, isomeric, etc.) of spectral lines at several successive quantum transitions as it is shown in Fig.4. Inevitable overlapping of the wings of adjacent spectral absorption lines confines the selectivity of one-step excitation of nuclear isotopes or isomers at the level $S_1 \simeq 10^4$-10^6. But in a

three-step process when the spectral line shift is used three times it is expected that the selectivity value of MSPI may reach $S=S_1S_2S_3 \simeq 10^{12}$ - 10^{18}. It is essential that in the ab-

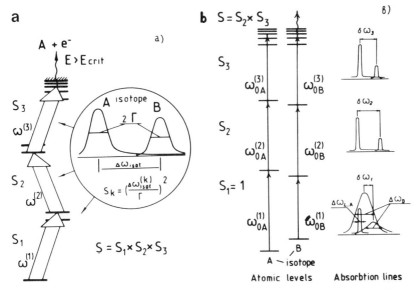

a) Multiplication of selectivity at subsequent selective stepwise excitation of atoms and (b) case of overlapping Doppler-broadened absorption lines without selectivity on the first step (from work [9]).

Fig.4 a) Multiplication of selectivity at subsequent selective stepwise excitation of atoms and (b) case of overlapping Doppler-broadened absorption lines without selectivity on the first step (from work [9]).

sence of selectivity at the first step of excitation (because of Doppler-broadened spectral lines in a gas, for example) we may realize considerable selectivity $S=S_2 \times S_3$ at the subsequent steps of excitation (Fig.4b).

I suppose that a combination of extremely high sensitivity and selectivity of detection of atoms and nuclei by the method of MSPI will be very beneficial in a search for artificial and natural exotic nuclei which are accessible in negligible concentrations on background of a huge amount of normal nuclei.

3. Multiple-Step Photodissociation of Molecules and Chemistry.

Since 1970 this process has being studied also for the purpose of isotope separation [10]. At present we have got a various methods for selective multiple-step photodissociation (MSPD) of molecules both through excited and ground electronic states. Fig.5 illustrates the main schemes for selective MSPD of molecules applied with success to separation of isotopes by our laboratory. In all these schemes selectivity is attained by excitation of vibrational levels through IR laser radiation, and for multiple-step excitation of vibrational levels of polyatomic molecules it is enough to use one- or two-frequency intense IR radiation. Indeed, a polyatomic molecule usually has a great number of similar-in-frequency vibrational-rotational transitions between its successive vibrational

509

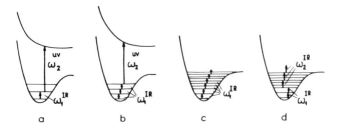

Fig.5 Schemes of selective multistep photodissociation of
molecules by laser radiation through an excited electronic
state (a,b) and ground electronic state (c,d): a) two-step
IR + UV photodissociation; b) multistep selective excitation of high vibrational levels and following photodissociation by UV radiation; c) multistep selective excitation
and dissociation by single frequency intense IR pulse;
d) multistep selective excitation of vibrational levels and
following their photodissociation by off-resonant intense
IR pulse.

levels. Besides, originally at room temperature the molecules
are distributed over many vibrational-rotational states. Since the anharmonicity of vibrations is usually less than the
width of vibrational-rotational band, several vibrational-rotational transitions turn out to be in resonance with rather
intense monochromatic IR radiation. In some cases the detuning of radiation frequency and transition frequencies is compensated by nonlinear effects in a strong field.

Of course, such simplification of multiple-step excitation
when using the method of "brute force" to excite a multilevel
system (Fig.6a) is achieved at the expense of inevitable reduction of excitation selectivity. This shows up in the fact
that, though the frequency dependence of the rate of multiple-
step molecular excitation by intense one-frequency radiation
is resonant by its nature, the resonance width ($\Delta\nu_{res}/\nu_{vib} \simeq 10^{-3}$)
is usually hundreds times larger than the width of individual
vibrational-rotational absorption lines. When the molecules
are cooled we can compress their original distribution over
many levels and thus increase the sharpness of resonance. This
is practically important for separation of heavy isotopes
(Fig.6b). Yet the necessity to use an intense field makes it
impossible to realize the maximum selectivity even in this
case. Only the use of multifrequency IR radiation with its
frequencies tuned in precise resonance to successive allowed
vibrational-rotational transitions (Fig.6c), as is done in the
case of multiple-step excitation of atoms, enables us to realize high selectivity of excitation in molecular gases at
room temperature and moderate requirements on laser radiation
intensity. Nevertheless, the method of multistep vibrational
excitation and dissociation of polyatomic molecules by one-
or two-frequency intense IR radiation, which is rather simple
and accessible for many laboratories, is under extensive study and coming into use in practice.

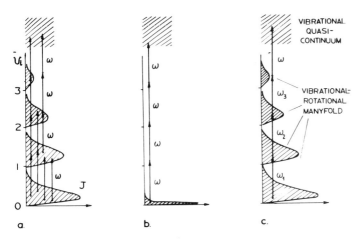

Fig.6 Various schemes of multistep selective excitation of vibrational rotational levels of polyatomic molecules: a) low selective excitation at room temperature by single frequency IR intense pulse; b) highly selective excitation at low temperature by single frequency IR intense pulse; c) highly selective excitation at room temperature by multifrequency IR laser pulses of moderate intensity.

The qualitative nature of MSPD by IR radiation is clear, and the studies conducted within two last years are aimed at obtaining experimental data on true characteristics of the process and their correlation with theoretical models. Certainly, if we could measure the vibrational distribution of molecules induced by multistep IR excitation it would be possible to check in detail various theoretical models. But until very recently, we were able to measure only the average energy absorbed per molecule in the region of interaction between IR field and molecular gas. These measurements are rather rough since a powerful IR radiation pulse excites just some fraction (q) of molecules while the rest (1-q) molecules remain practically unexcited (Fig.7a). Therefore, to measure the true characteristics of the process not averaged over two ensembles ("hot" and "cold") of molecules, it is necessary that we should know the value q of the relative fraction of excited molecules as well as its dependence on the intensity or energy fluence and the laser pulse frequency. In works [11] there were various methods applied to measure the value q and its dependence on energy fluence for OsO_4, SF_6 and CF_3I molecules. Fig.7b shows how the value q depends on the energy fluence of a laser pulse of standard length (100 ns), the results obtained in this work. It is very essential to know these dependences for comparison of theoretical models with experiment.

It was an important step we took in the last year when we elaborated some theoretical models which enable us to determine the value q by fitting theoretical data to experimental ones [12]. The next step is to calculate directly the value q starting from the spectroscopic information about vibrational

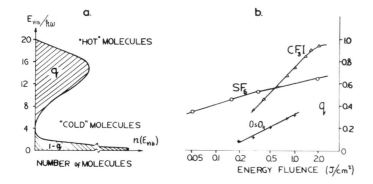

Fig.7 a) Generation of two molecular ensemble ("hot" and "cold" at multistep excitation by intense IR laser pulse; b) Relative fraction of highly-excited ("hot") molecules as a function of energy fluence of CO_2 laser pulse (data from works [11]).

-rotational molecular transitions between low-lying vibrational levels responsible for molecular "leakage" into the vibrational quasicontinuum.

It should be noted that the formation of two ensebles with greatly different vibrational temperatures under action of a laser pulse and observation of their temporal evolution to one ensemble due to V-V exhange between them with some medium temperature give us a convenient method to investigate vibrational exchange with highly excited molecules participating.

The ensemble of highly excited "hot" molecules has near-equilibrium distribution of vibrational energy. The seeming strong nonequilibrium (Fig.7a) is explained exclusively by strong degeneration of highly excited vibrational states. As an illustration, Fig.8 shows separately the distribution of

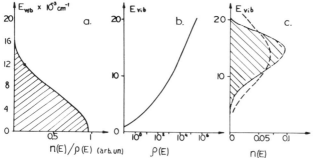

Fig.8 Vibrational energy distribution on quantum states (a), density of quantum states (b) and distribution of vibrational energy per unit of energy (c) for the ensemble of "hot" CF_3I molecules after the multistep excitation by intense CO_2 laser pulse. Dashed curve (c) shows the vibrational energy distribution for the thermal model (from work [12]).

vibrational energy per one quantum state (a), an increase in the density of vibrational quantum states with increasing excitation energy (b) and the resulting distribution of vibrational energy per unit energy interval (c). All these curves have been obtained by calculating the excitation of CF_3I molecules by a CO_2 laser pulse at the line R(14) (1074.65 cm^{-1}) with energy fluence of 1 J/cm^2 within the frameworks of the theoretical model [12] which provides the good agreement with the experimental data on absorbed energy and dissociation yield. Within this model the vibrational distribution of energy over all modes in the ensemble of highly excited molecules somewhat differs from the equilibrium thermal distribution where the absorbed energy is also distributed over all vibrational modes [13]. But this difference is comparatively small (dashed curve in Fig.8c). Such an effect of difference between equilibrium statistical and real kinetic distributions has been already discussed in works [14, 11].

A similar situation of simultaneous excitation of all vibrational modes must be typical of all polyatomic molecules comparatively small in size and laser pulses from 10^{-6} to 10^{-8}sec duration. Indeed, a powerful IR pulse generating vibrations of a group of atoms at a large vibration amplitude due to anharmonicity inevitably involves quickly all neighbouring atoms in motion. This may be distinctly observed in numerical computation of the atoms in SF_6 molecules within the classical model [15]. In other words, this corresponds to laser excitation of a great number of compound transitions in vibrational quasicontinuum with all the vibrational modes participating. It is clear that the excitation of many vibrational modes in quasi-continuum occurs during a laser pulse directly.

It is expected, however, that in large polyatomic molecules and specifically in long chain molecules the situation may differ materially. Really, let a CO_2 laser excite the vibration of a functional group, CF_3, for example, tied to one end of the molecule. Since this group is not directly connected with the motion of atoms at the opposite end of the molecule the vibrational excitation will reach this end in a characteristic time $\tau \simeq N/\delta\nu_{anh}$, where N is the number of links in the chain proportional to the number of atoms, $\delta\nu_{anh}$ is the characteristic anharmonicity constant governing the interaction of vibrations or the anharmonic bond of adjacent links in the molecule. When exciting the molecule even by a pulse with τ_p=1 nsec, when $\delta\nu_{anh} \simeq 1$cm^{-1} and $N \simeq 10^2$, a considerable difference from the predictions of the statistical model may be expected. Such a difference can be registered through observing the hot bands in the UV or IR spectra of another functional group at the opposite end of the molecule or directly by observing the dissociation of the molecule through chan - nels which differ from one with the lowest dissociation energy. Perhaps, just such an effect has been observed in work [16]. Experiments in this direction are to reveal the conditions which will make possible IR multistep photochemistry selective by intramolecular bonds (site-selective IR MSPD).

4. Multistep Photoreactions of Biomolecules.

The possibility of selective laser action on biomolecules using multistep excitation of certain molecular bonds or fragments has being discussed for several years [17,8c]. The first hopeful experimental results, however, have been obtained quite recently from the fragments of DNA and RNA, the most interesting biomolecules. The paper by Angelov et al. 2 is concerned with these results.

From the standpoint of obtaining high selectivity it is natural to use selective excitation of vibration and then to transfer the vibrationally excited molecules to those excited electron states which have a high quantum yield of photoreaction (schemes in Fig.5a,b). But the absorption of solvents in the IR range, an abundance of vibrational levels in biomolecules, a low energy limit of vibrational quasi-continuum and a very fast intra- and intermolecular vibrational relaxation make it very hard to realize this approach at the present-day level of picosecond tunable IR lasers. The first successful results [18,19] in selective excitation of nucleic acid fragments have been obrained by a less evident way, that is by two-step excitation through intermediate electron state. Despite the fact that the UV spectra of electron absorption are overlapped almost completely, i.e. there is no selectivity at all at the first step of excitation, the total photoreaction yield differs substantially for various bases of nucleic acids. As an example, Fig.9 gives normalized dependences of photoreaction yield on picosecond pulse intensity (pulse length of 20 ps) at the wavelength of 265 nm for two RNA bases (adenine and uracil). The photoreaction yield here is measured from the value of irreversible bleaching in the UV absorption bands of these bases' solution, with the value of total bleaching during many irradiating pulses normalized to the total energy fluence of irradiation pulses (J/cm^2), that is, to the value of UV radiation dose. With intensities below the saturation level of the first electron transition the photoreaction yield squarely depends on the intensity of pulses. This is a proof of two-step photoexcitation. With the electron transition saturated, the photoreaction yield becomes linearly dependent on intensity because the second electron transition from excited state results in a very fast photoreaction or, in other words, it is an unsaturable transition.

Ionization, dissociation or a chemical reaction of bases subsequent to absorption of two UV radiation quanta with their total energy of about 8 eV may be a potential mechanism of such photoreactions. Our studies are aimed at revealing this mechanism. The selectivity in such multiple-step photoreactions may arise due to the differences in the lifetimes of intermediate state, in the cross-sections of the subsequent (second) electron transition or in the yields of the last photochemical step of the process. The use of shorter picosecond pulses and independently tunable picosecond radiation at the second electron transition between excited states will enable us to gain some insight into the mechanism of selectivity origination.

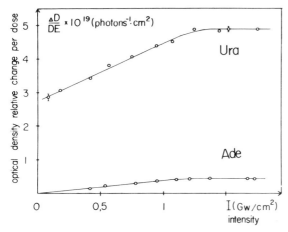

<u>Fig.9</u> The dependence of the optical density relative chan-
ge in the maximum of the difference spectrum per dose on
the irradiation intensity for uracil and adenine (from [19]).

The selectivity observed in two-step photoreactions of nucleic acids bases remains in more complex compounds as well
as in the fragments of nucleic acids with two different bases
[2]. This fact has every reason to hope that such a selective
step photoprocess may be successfully applied to real DNA and
RNA molecules. The advent of a laser photoion projector for
reading the sequence of DNA bases is, to my opinion, one of
the most promising applications of selective photoaction on
various nucleic acid bases [20]. To create it, we are to realize the selective-in-bases process of multistep photoionization of DNA gragments adsorbed on a surface. Now we are conducting our investigations in this direction.

References

1. G.I.Bekov, E.P.Vidolova-Angelova, V.S.Letokhov,V.I.Mishin,
   paper in present volume.
2. D.A.Angelov, P.G.Kryukov, V.S.Letokhov, D.N.Nikogosyan,
   A.A.Oraevsky, to be published.
3. V.S.Letokhov, USSR. Patent No.65743 (Appl.30.03.1970);
   R.V.Ambartzumian, V.P.Kalinin, V.S.Letokhov. Pis'ma ZhETF
   (Russian) 13, 305 (1971) /JETP Lett., 13, 217 (1971)/.
4. V.S.Letokhov, V.I.Mishin, A.A.Puretzky, in Progress in
   Quantum Electronics, ed. by J.Sanders and S.Stenholm (Pergamon Press, New York, 1977), vol.5, part 3, pp.139-204.
5. V.S.Letokhov, V.G.Minogin. Phys.Rev.Lett., 41, 775 (1978).
6. V.S.Letokhov, V.M.Lobashev, V.I.Minogin, V.I.Mishin. Pis'
   ma ZhETF (Russian) 27, 305 (1978).
7. N.Delone, B.A.Zon, M.V.Fedorov. Pis'ma Zh.Tekh.Fiz.(Russian) 4, 229 (1978); M.V.Fedorov. Optics Comm., 26, 183
   (1978).

8. V.S.Letokhov, a) in "Frontiers in Laser Spectroscopy",
   Proc. of Les-Houshes Summer School on Theoretical Physics,
   July 1975, vol.2, ed.by R.Balian, S.Haroche, S.Liberman
   (North-Holland, Amsterdam, 1977), p.771; b) Usp.Fiz.Nauk
   (Russian) 118, 199(1976); c) in "Tunbale Laser and Applications", Proc. of Loen Conf., Norway, June 1976, ed. by
   A.Mooradian, T.Jaeger, P.Stokseth (Springer-Verlag, Heidelberg, 1976), p.122.
9. V.S.Letokhov, V.I.Mishin. Optics Comm., (in press) 1979.
10. V.S.Letokhov. USSR Patent No.65744 (Appl.30.03.1970);
    R.V.Ambartzumian, V.S.Letokhov. Appl.Optics. 11, 354(1972).
11. a) R.V.Ambartzumian, G.N.Makarov, A.A.Puretzkii. Pis'ma
    ZhETF (Russian) 28, 246 (1978).
    b) V.N.Bagratashvili, V.S.Doljikov, V.S.Letokhov. Zh.Eksp.
    Teor.Fiz. (Russian) 76, 18 (1979);
    c) V.N.Bagratashvili, V.S.Doljikov, V.S.Letokhov,E.A.Rya-
    bov in "Laser-Induce Processes in Molecules", ed.by K.L.
    Kompa and S.D.Smith, Springer Series in Chemical Physics,
    vol.6 (Springer-Verlag, Heidelberg, 1979), p.179.
12. V.N.Bagratashvili, V.S.Doljikov, V.S.Letokhov, A.A.Maka-
    rov, E.A.Ryabov, V.V.Tiaht. Zh.Eksp.Teor.Fiz. (in press).
13. J.G.Black, E.Yablonovitch, N.Bloembergen, S.Mukamel.
    Phys.Rev.Lett., 38, 1131 (1977).
14. E.R.Grant, P.A.Schulz, Aa.S.Sudbo, Y.R.Shen, Y.T.Lee,
    Phys.Rev.Lett., 40, 115 (1978).
15. W.E.Lamb,Jr. (private communication).
16. A.Kaldor (in press).
17. V.S.Letokhov. Journal of Photochemistry, 4, 185 (1975).
18. P.G.Kryukov, V.S.Letokhov, D.N.Nikogosyan,A.V.Borodavkin,
    N.A.Simukova. Chem.Phys.Lett., 61, 375 (1979).
19. P.G.Krykov, V.S.Letokhov, D.N.Nikogosyan, D.A.Angelov.
    Nature (in press).
20. V.S.Letokhov. Kvantovaya Elektronika (Russian), 2, 930
    (1975); Phys.Lett., 51, 231 (1975).

Laser Spectroscopy of Short-Lived Isotopes in Fast Atomic Beams and Resonance Cells

H.-J. Kluge, R. Neugart, and E.-W. Otten

Institut für Physik, Universität Mainz
D-6500 Mainz, Fed. Rep. of Germany

1. Introduction

The introduction of laser techniques to optical spectroscopy of hyperfine structure (HFS) and isotope shift (IS) has put new life into this field at the intersection between atomic and nuclear physics which is now about fifty years old. Two severe limitations of classical spectroscopy could be reduced drastically, i.e., (i) the amount of atoms needed for optical spectroscopy and (ii) the Doppler width of optical lines. The increase in sensitivity went together with the development of powerful accelerators or reactors so that exotic nuclei with half lives down to 10 msec can now be produced and optically analysed. The increase in resolution allows the determination of nuclear spins, nuclear moments, and changes of charge radii between different isotopes even for very low Z nuclei. In the case of the measurement of spins and moments, optical spectroscopy has to compete with very elaborate atomic and nuclear spectroscopic methods as, e.g., the atomic beam magnetic resonance, optical rf techniques or γ spectroscopy. However, only optical spectroscopy gives access to the nuclear-volume dependent IS of short-lived isotopes which provides a very sensitive measure of the radial change when the neutron number is varied. It is this information that turned out to be very fruitful as will be shown for the case of Hg isotopes. This is because the radial change in an isotopic chain reflects collective as well as single particle effects, namely the gross behaviour of an expanding nucleus when neutrons are added as compared to the standard and generally used $A^{1/3}$ law (so called IS discrepancy), the effect of the addition of a single neutron or a neutron pair (odd-even staggering), the influence of the orbit of different unpaired neutrons (isomer shift), and changes in the nuclear shape (deformation effect).

If the isotopes to be investigated are short-lived ($T_{1/2}$ in the range of min or less), the experiment has to be performed on-line with the production. In addition, the unselective production process for nuclei far from stability requires a mass separation and some chemically selecting process. Two on-line laser experiments on mass-separated isotopes, both performed by groups of the University of Mainz, will be reported in this contribution. These are, firstly, Doppler-limited fluorescence spectroscopy on short-lived Hg isotopes in resonance

517

cells performed by F. Buchinger, P. Dabkiewicz, C. Duke (Grinell), H. Fischer, H.-J. Kluge, H. Kremmling, T. Kühl, A.C. Müller, E.-W. Otten, and H.A. Schuessler (Texas A&M) and, secondly, sub-Doppler spectroscopy on a fast atomic beam performed on neutron-rich Rb and Cs isotopes at the TRIGA reactor/Mainz by J. Bonn, S.L. Kaufman, W. Klempt, H. Lochmann, G. Moruzzi (Pisa), R. Neugart, E.-W. Otten, L.v. Reisky, B. Schinzler, K.P.C. Spath, J. Steinacher, and D. Weskott.

2. Nuclear Properties obtained from Optical Spectroscopy

The electromagnetic interaction between the nucleus and the shell electrons gives rise to the HFS splitting and the IS of spectral lines and hence access to a determination of nuclear properties. Because of lack of space, the reader is referred to the literature |1-4|.

3. Production of Short-Lived Isotopes

Three types of nuclear reactions are capable of producing unstable nuclei far from β stability: thermal neutron-induced fission, heavy-ion reactions, and reactions with high-energy protons. Fig.1 gives, as an example, the production yield for mass-separated Cs isotopes. The 600 MeV proton spallation of La (yielding neutron-deficient isotopes) and the 600 MeV proton induced fission of U (yielding neutron-rich isotopes) are clearly superior to the other production schemes. This is mainly due to the high penetrating power of the protons allowing thick targets of typically $100g/cm^2$ and the high beam intensities available from modern accelerators (600 MeV synchrocyclotron at CERN: up to $3x10^{13}$ protons/sec). For comparison: the 235U target used at the TRIGA reactor at Mainz has about $1g/cm^2$ in a flux density of $2x10^{11}$ neutrons/sec cm^2 and typical targets for heavy ion reactions have a thickness in the range of mg/cm^2, while the beam intensity is of the order of 10^{12} heavy ions/sec.

The broad distribution of isotopes (Fig.1) and elements produced by such reactions demands for a mass separation and some chemically selective process (target-ion source configuration). The on-line mass separator with the broadest spectrum of produced isotopes is the ISOLDE facility at CERN|6|. Since its construction in 1969, it has provided 20 different elements in a chemically clean form and an additional 20 as chemical mixtures. Since 20-30 isotopes are available for each element, several hundred nuclear species are accessible for investigation. In favorable cases such as the alkali or Hg isotopes, the yield reaches up to 10^{10} atoms per sec and per mass number (Fig.1).

4. Laser Fluorescence Spectroscopy of Hg Isotopes in a Resonance Cell

The first experiments on very neutron-deficient Hg isotopes were carried out by nuclear radiation detected optical pumping

(RADOP) |4|. In these experiments, the nuclear spin is polarized by optical pumping and the nuclear polarization is monitored by the asymmetry of the β decay. The results of these investigations (discussed in Section 7) demanded for a measurement of the even light Hg isotopes, which have zero nuclear spin and therefore cannot be studied by the RADOP technique. Hence purely optical methods had to be used.

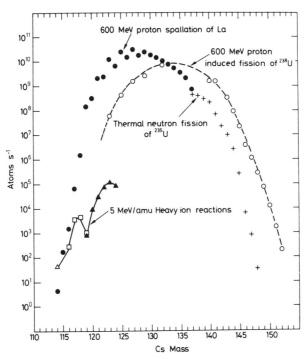

Fig.1 Production of mass-separated Cs isotopes by various techniques: (i) spallation of molten La with 600 MeV protons (ISOLDE/CERN), (ii) fission of 238U with 600 MeV protons (ISOLDE/CERN), (iii) heavy ion reaction (GSI, Unilac/Darmstadt) and (iv) thermal neutron-induced fission of 235U (TRIGA reactor/Mainz) |6|.

The output of a pulsed dye laser system pumped by a nitrogen laser is frequency doubled in an temperature-matched ADP crystal. The λ = 2537 Å laser beam for exciting the 6s6p $^3P_1$ state passes a resonance cell, which is periodically filled with the isotope under investigation |7|. The resonance radiation of the unstable atoms is observed with a photomultiplier for several atomic lifetimes (τ = 115 nsec) after laser excitation of the $^3P_1$ state. Scanning the λ = 2537 Å line of the isotope under investigation, the frequency is calibrated by the simultaneously recorded resonance light from the stable even-even isotopes in a magnetic field.

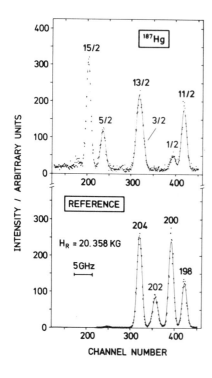

Fig.2 Intensity of the fluorescence light in the $6s^2\ ^1S_0$ - $6s6p\ ^3P_1$, λ = 2537 Å line of 187Hg (I = 3/2) and 187mHg (I = 13/2) and of the σ^+ Zeeman components of the even stable Hg isotopes in a magnetic field versus the frequency of the exciting laser light.
The points indicate the experimental values, the solid line is the best fit by a Gaussian

Fig.2 shows as an example the HFS pattern of the ground and isomeric states of 187Hg [9]. Whereas the spin of the ground state has already been measured by the RADOP experiments, the isomeric-state spin was determined as I = 13/2 by time differentially observed Hanle quantum beats [20].

HFS and IS data obtained by classical optical spectroscopy, level crossing, double resonance, RADOP and laser spectroscopy are now available for 206Hg down to 181Hg ($T_{1/2}$=3.6 sec). Although laser spectroscopy did not reach the ultimate sensivity of the RADOP technique (181Hg: 100 atoms/cm^3 [4]), it worked with a density of 10^6 atoms/cm^3 in the case of 185mHg. This limitation in sensitivity is due to the low duty cycle of a pulsed laser system (10^{-7}), adsorption of the atoms on the walls of the resonance vessel, and losses due to the time needed to transfer the ion beam into the cell in order to build up an atomic vapour.

5. Fast Beam Laser Spectroscopy

The drawbacks in respect to sensitivity as mentioned above can principally be avoided by (i) use of a cw laser, (ii) investigation of free atoms in an atomic beam and by (iii) directly using the output of the isotope separator. This led to the

concept of collinear spectroscopy on fast atomic (or ionic)
beams |10|.

The basic idea is very simple: Sufficiently large excitation rates are obtained, if the time of interaction between
the beam and the laser light is long enough. Therefore, the
laser beam has to be superposed collinearly to the output beam
of the separator. And, since resonance lines of ions are
usually not accessible to cw dye lasers, the ion beam has to
be converted into a fast atomic beam. This neutralization is
efficiently performed by charge transfer in an alkali-vapour
cell.

The striking advantage of this concept is the avoidance of
any loss of material between isotope separation and laser excitation. But this advantage is only profitable, if all (or a
sufficiently large number of) atoms interact with the light.
This is a question of the absorption Doppler width.

Since the spread of kinetic energy in the beam remains unchanged under electrostatic acceleration

$$\delta E = \delta(\frac{mv^2}{2}) = mv \, \delta v = \frac{mc^2}{v^2} \Delta v_D \, \delta v_D = \text{const},$$

the product of the average velocity v and the velocity spread
δv, or likewise the product of the Doppler shift Δv_D and the
Doppler width δv_D are constants of the motion. In other words:
The Doppler width is considerably reduced from its original
value $\delta v_D(0)$ in the ion source. For the ideal case of a
thermal distribution (which is realized in surface-ionization
sources)

$$\delta v_D = \frac{1}{2}(kT/eU)^{1/2} \, \delta v_D(0).$$

The reduction factor is about 400 for a source temperature of
1500 K and an acceleration voltage of 5 kV. It corresponds to
a Doppler width of 4 MHz for the blue Cs line (λ=455.5 nm),
compared to a natural linewidth of 1.2 MHz. Natural linewidths
of the strongest resonance lines are even larger by a factor
of 10. It may be concluded that essentially all atoms in the
beam are excited and that the resolution is comparable to
Doppler-free methods.

It is presumed, of course, that the charge-transfer process
doesn't change the velocity distribution. This is actually
the consequence of the large cross-section of about 10^{-14} cm^2
which exceeds the kinetic cross section by two orders of
magnitude. In the resonant case, e.g.

$$Cs^+ + Cs(6s) \rightarrow Cs(6s) + Cs^+$$

the beam energy remains unchanged, whereas in the non-resonant case, e.g.

$$Cs^+ + Cs(6s) \rightarrow Cs(6p) + Cs^+ - 1.4 \text{ eV}$$

or in collisions between different partners, the fast-beam kinetic energy is changed by almost exactly the amount of the energy defect ΔE. Conservation laws require that in forward scattering the kinetic energy transferred to the target atom is of the order $(\Delta E)^2/eU$, which is negligible for beam energies in the range of keV. In any case, the width of velocity distribution and the angular divergence is not affected by charge transfer. However, the Doppler-shifted sharp lines may be split by the energy-loss spectrum corresponding to different reaction channels.

Fig.3 shows the experimental set-up used with the mass-separated beam of alkali isotopes from nuclear fission (see Fig. 1). The ion beam is deflected to merge with the laser beam and neutralized in the heated charge-exchange cell containing alkali metal at a vapour pressure of 10^{-3} Torr. Laser light is absorbed by the atoms whenever the Doppler-shifted frequency is tuned to one of the resonance lines.

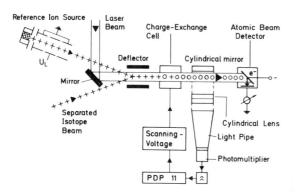

Fig.3 Experimental set-up for fast beam spectroscopy

This is detected by counting fluorescence photons, collected along a path length of 20 cm by a cylindrical lens and guided into the photomultiplier by a light pipe. Varying the Doppler shift at fixed laser frequency offers a convenient way of scanning the absorption spectra. For this purpose a programmable voltage is applied to the charge-exchange cell, post-accelerating the ions prior to neutralization. The measurement of the IS additionally requires the comparison to a reference isotope. For this purpose a beam of stable 133Cs from a separate ion source is focussed through the apparatus alternatively to the unstable isotope beam. Uncertainties of beam-energy calibration are eliminated by relating all IS to 137Cs obtained from the mass separator. Measurements were performed in the transitions 6s $^2S_{1/2}$ – 7p $^2P_{3/2}$ (λ=455.5 nm) of Cs and 5s $^2S_{1/2}$ – 6p $^2P_{3/2}$ (λ=420.2 nm) of Rb.

A spectrum of 139Cs may serve as a typical example (Fig.4).

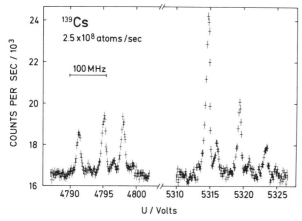

Fig.4 Spectrum of the 6s $^2S_{1/2}$-7 $^2P_{3/2}$ transitions in 139Cs (I=7/2). Transitions from the F=3 ground state HFS level are shown on the left, from the F=4 level on the right-hand side of the pattern

Both groups of hfs components are scanned at a fixed laser frequency, simultaneously with the 133Cs reference. The voltage scale is used for calculating relative Doppler shifts by

$$(\nu_2 - \nu_1)/\nu_0 = (2eU_1/M_1c^2)^{1/2} - (2eU_2/M_2c^2)^{1/2}$$

from which the hfs parameters and the IS can be evaluated. Atomic masses are known with sufficient accuracy from standard tables |11|.

The linewidth corresponds to 10 MHz in a frequency scale. By a computer fit the resolution of the excited-state hfs is better than 1 MHz, yielding the very small quadrupole interaction constants with an accuracy of about 20%. Errors in the large ground-state splittings and in isotope shifts are usually determined by the voltage calibration and amount to several MHz.

A characteristic of all spectra are the satellites of each line, separated from the main peak by 1.4 volt. They are due to the non-resonant branch of charge transfer to the first excited 6p state of Cs, giving rise to a loss of kinetic energy equal to the 6s - 6p energy difference. In this way the distribution of charge-transfer reaction products over final states is directly observed. We have shown that this distribution strongly depends on the reaction partners |12,13|. In fact, these high-resolution measurements can be used to evaluate branching ratios of charge-transfer reaction channels.

Up to now, fast-beam laser spectroscopy has been applied to short-lived isotopes with a yield exceeding 10^7 atoms/sec. Further effort may tackle the problem of sensitivity by improving the efficiency of light collection and suppressing the

background. Useful techniques may be the excitation and observation at different wavelength where it is possible, and,
for longer-lived states, the pulsed excitation and observation
of the subsequent decay. A non-optical detection scheme might
combine fast-beam laser spectroscopy with the RADOP technique
|14| by implantation of optically pumped atoms into a suitable
crystal, and observation of the asymmetry in the β-decay of
polarized nuclei.

6. Results of the Fast-Beam Spectroscopy

Measurements were performed at the TRIGA reactor/Mainz for the
neutron-rich isotopes 138-142Cs and 89-94Rb. Their half lives
extend from 32 min for 138Cs to 1.7 sec for 142Cs, and simi-
larly for the Rb isotopes between 89Rb and 94Rb |15-17|.

The information obtained includes spins, magnetic dipole
and electric quadrupole moments as well as differences of
nuclear charge radii. Most spins and magnetic moments of Rb
and Cs isotopes had previously been measured by C. Ekström et
al. |18| using atomic beam magnetic resonance on-line with
ISOLDE, and a few spins were measured by optical pumping
(RADOP) |4|. These quantities are accessible in the atomic
ground state.

During the last year, the Orsay group at ISOLDE/CERN
applied a different technique (optical pumping of slow atomic
beams with magnetic detection) also to neutron-rich Rb and Cs
isotopes and, in addition, to spallation-produced neutron-deficient ones. Due to the about 100 times higher yield of
ISOLDE as compared to that of the TRIGA mass separator (Fig.1)
they were able to extend the measurements up to 145Cs and
down to 118Cs and in the case of Rb to the masses 76<A<98.
These measurements which include our results will be discussed
by S.LIBERMANN |8| in a contribution to this session.

7. Results of Laser Spectroscopy in Resonance Cells

Laser spectroscopy in resonance cells has been applied to
short-lived Hg isotopes within the mass range $206 > A > 184$.
It would be beyond the scope of this contribution to discuss
all the results in detail. Instead, the discussion will concentrate on a brief outline of the most significant results of
the Hg measurement in order to demonstrate the richness of
nuclear information obtainable by optical spectroscopy. Putting
all data from the older spectroscopic and the recent laser
work together, one obtains the following properties of the Hg
chain:

7.1 Spin and Moments

Spins and nuclear moments are now known for 21 ground states
or isomers in the Hg chain. These are useful in assigning the
nuclear configuration. Especially interesting is the smooth
trend of the quadrupole moments of the eight $I = 13/2$ isomers
which can be explained in the frame of the rotation-aligned
coupling scheme |9,19|.

7.2 Isotope and Isomer Shift

Fig.5 shows the systematics of the Hg radii relative to 204Hg.

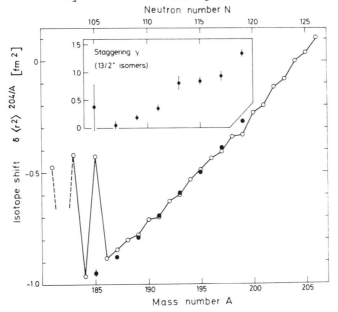

Fig.5 Changes of the charge radii of Hg isotopes relative to 204Hg. Open circles indicate ground states and full ones indicate isomers. Inset: odd-even staggering γ of the isomers

Isotopes shift discrepancy. Correcting the experimentally found $\delta<r^2>$ values for deformation effects and comparing these numbers with the $\delta<r^2>$ values of the liquid drop model, the ratio $\rho=0.74(10)$ is found from Fig.5 for the regular IS's from 206Hg down to 184Hg. This is much less than expected from the $A^{1/3}$ law.

Odd-even staggering of isomers. In the inset of Fig.5 the staggering parameter of the isomers (defined in $|5|$), is plotted. The striking regular trend is explained by the rotation-aligned coupling scheme $|21|$.

The anomalous IS of 181, 183, 185Hg. The jump in the IS of the odd light Hg ground states with A<185 is due to a sudden onset of strong prolate deformation. The change in deformation amounts to $\delta<\beta^2>^{185/186} = 0.053(5) |22|$. The reason is a competition between the proton and the neutron configuration. The protons with Z=80 near the magic value Z=82 favour a spherical shape while the neutrons with N midway between the magic numbers N=82 and N=124 favour a deformed shape ($\beta\approx0.3$).

The odd-even staggering in the light Hg isotopes. The IS of the even isotopes 186Hg and 184Hg follows the slope of the heavier isotopes. Here, the gain in pairing energy favours the oblate deformation of $\beta\approx-0.15$ and a huge staggering of $\gamma=13(1)$ is therefore observed $|22|$.

The isomer shift in 185Hg. The difference in the mean-square radii of the ground and isomeric state in 185Hg amounts to 0.52(2) fm^2 |9|. This is by far the largest isomer shift ever observed for any nucleus. It is caused by tiny energy differences of an $i_{13/2}$ and a $p_{3/2}$ neutron (|521 1/2|Nilsson level). Thus, a shape isomerism and a coexistence of two very different shapes are observed in 185Hg, which are almost degenerated in energy.

8. Conclusion

We have tried to show that optical spectroscopy has recovered its importance for nuclear research. The information obtained on nuclear ground and isomeric states far from β stability yields a number of basic nuclear properties for developing and testing nuclear models. Optical spectroscopy has a monopoly on measuring charge radii of short-lived isotopes which have turned out to be especially instructive. Present experiments using laser techniques have reached a grade of sensitivity and resolution for investigating systematically long isotopic chains which still provide key information about nuclear structure.

This work was supported by the Bundesministerium für Forschung und Technologie and the Deutsche Forschungsgemeinschaft.

References

1 H.Kopfermann: Nuclear Moments (Academic Press, New York 1958)
2 D.N.Stacey: Rep.Progr.Phys. 29, 171 (1966)
3 E.-W.Otten: Hyperfine Interactions 2, 127 (1976)
4 H.-J.Kluge: In Progress in Atomic Spectroscopy, Part B, ed. by W.Hanle, H.Kleinpoppen, (Plenum, New York, 1979) p. 727
5 W.J.Tomlinson III, H.H.Stroke: Nucl.Phys. 60, 614 (1964)
6 H.L.Ravn: Submitted to Physics Reports
7 C.Duke et al.: Phys.Lett. 60A, 303 (1977)
8 S.Liberman et al.: This volume
9 P.Dabkiewicz et al.: Phys.Lett. 82B, 199 (1979)
10 S.L.Kaufman: Opt.Comm. 17, 309 (1976)
11 A.H.Wapstra, K.Bos: Atomic Data and Nuclear Data Tables 19, 177 (1977)
12 K.R.Anton et al.: Phys.Rev.Lett. 40, 642 (1978)
13 R.Neugart et al.: In Laser Spectroscopy III, (Springer, Berlin 1977)p. 446
14 E.-W.Otten: In Atomic Physics V(Plenum,New York 1977)p.239
15 B.Schinzler et al.: Phys.Lett. 79B, 209 (1978)
16 J.Bonn et al.: Z.Phys. A278, 227 (1979)
17 W.Klempt et al.: Phys.Lett. 82B, 47 (1979)
18 C.Ekström et al.: Nucl.Phys. A292, 144 (1977) and private communication
19 F.S.Stephens: Rev.Mod.Phys. 47, 43 (1975)
20 H.Kremmling et al.: Submitted to Phys.Rev.Lett.
21 H.H.Stroke et al.: Phys.Lett. 82B, 204 (1979)
22 T.Kühl et al.: Phys.Rev.Lett. 39, 180 (1977)

Hyperfine Spectroscopy of Radioactive Alkali Isotopes

S. Liberman, J. Pinard, H.T. Duong, P. Juncar, J.L. Vialle, P. Pillet, and P. Jacquinot

Laboratoire Aimé Cotton, Centre National de la Recherche Scientifique
F-91405 Orsay, France

G. Huber[1], F. Touchard, S. Büttgenbach[2], C. Thibault, and R. Klapisch
Laboratoire René Bernas, Centre de Spectrométrie de Masse
F-91406 Orsay, France

A. Pesnelle[3]

The ISOLDE Collaboration, C.E.R.N., CH-1211 Geneva, Switzerland

Systematic experimental studies of long series of radioactive isotopes of the same element are now becoming a rapidly developing area. The results obtained in the case of Hg isotopes for instance [1], give a good idea of the attractiveness of such a type of investigation . Moreover, similar studies of radioactive Na isotopes of mass number 21 to 31 have already been performed using high resolution laser spectroscopy [2], and have shown for the first time for such a light element, an appreciable volume shift contribution to the total isotope shift. Obviously, hyperfine results, as well as nuclear spin determinations and precise isotope shift measurements give an ensemble of data that seem to be of crucial importance to a better understanding of nuclear properties and suggest new refinements in theoretical models. The further the isotopes are from magic numbers, the larger the nuclear deformations are expected to be. One has therefore to deal with essentially short-lived species, which requires one to work on-line behind the accelerators where they are produced. In the previously mentioned case of Na atoms for instance, the nuclei were obtained by fission reactions of uranium targets bombarded by 20 GeV protons delivered by the PS synchrotron of C.E.R.N. .

One way to pursue this kind of investigations is to systematically study alkali atoms, which all have an isotope with a magic number of neutrons. The most appropriate installation for providing such atoms is certainly the ISOLDE facility of the 600 MeV proton synchrocyclotron at C.E.R.N. . In fact, the atoms are delivered as singly ionized mass separated isotopes, accelerated to an energy of about 60 KeV . As we are mostly interested in the study of neutral atoms, the ions are converted into thermal atoms of an atomic beam after implantation on, and reevaporation from the surface of a tantalum target heated to 900°C . Then the atoms of the beam interact, through their D_1 or D_2 resonance line, with a single mode tunable C.W. dye laser beam, frequency controlled and step by step scanned using a stabilized sigmameter interferometer (see Figure 1) [3]. The precision in frequency measurements can reach 1 MHz in the best cases. The resulting optical pumping occuring in the ground state hyperfine sublevels provides changes in the effective magnetic moment which are analyzed by a six-pole magnet (whose effect consists in focusing $m_J = +1/2$ atoms and defocusing $m_J = -1/2$ atoms). After being ionized on a hot tantalum conical surface,

[1]Presently at GSI D-6100 Darmstadt 1, Fed. Rep. of Germany
[2]Presently at the University of Bonn, D-5300 Bonn, Fed. Rep. of Germany
[3]Presently at Service de Physique Atomique, CEN-Saclay, France

the atoms are selected by a mass spectrometer, and finally counted by an electron multiplier. The overall efficiency of the apparatus, defined as the number of counted ions versus the number of ISOLDE ions, lies between 10^{-6} and 10^{-5} .

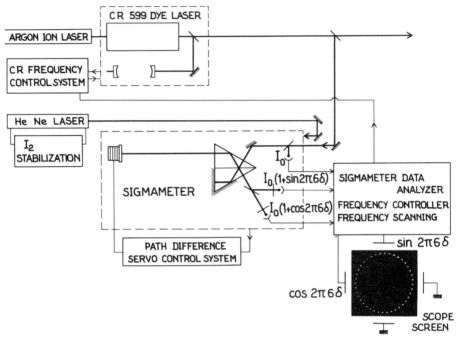

Figure 1

Rubidium isotopes.

Rubidium isotopes were produced using two types of reactions. For the heavier isotopes, fission reactions in uranium targets were utilized. For the lighter isotopes, spallation reactions in niobium targets were utilized, according to the mechanism : ^{93}Nb (p, 5p x n)$^{(89-x)}$Rb , such that only isotopes of mass number smaller than, or equal to 89 can be produced.

The hyperfine properties have been studied using the D_2 resonance line $(5s\ ^2S_{1/2} - 5p\ ^2P_{3/2})$ which is located at the wavelength $\lambda = 780,0$ nm. To oscillate at that frequency in a C.W. regime, the dye laser solution contained DEOTC excited by the strong red lines of a Kr^+ laser.

As in the previous experiments on Na atoms, the isotope shift measurements have been done using an auxiliary atomic beam of the stable isotopes, whose line components are taken as frequency references.

Complete structures and relative positions of all isotopes and isomers which have been studied are displayed in Figure 2. The different nuclei are

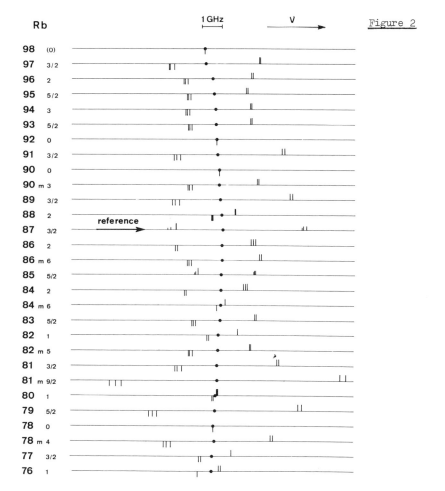

Figure 2

marked with their mass number (m indicates an isomer) and with their measured nuclear spin. The dots represent the centre of gravity of the line structures. Their position relative to each other gives the isotope shift. It is to be noticed that, due to the method of detection [4], low frequency hyperfine components appear as negative amplitudes, whereas high frequency hyperfine components appear as positive ones. For the stable isotopes however (mass number 85 and 87), because of the fluorescence detection of the laser excited atoms of the auxiliary atomic beam, all components appear as positive on a near zero background.

In all, 29 isotopes and isomers have been studied. Precise numerical results have been obtained from recordings, and are currently being studied theoretically. Complete understanding of the evolution of nuclear properties along that series of isotopes is yet far from being achieved.

Cesium isotopes.

Cesium atoms were basically studied the same way as rubidium atoms .
They were produced either by fission of uranium targets, or by spallation
reactions using lanthanum targets for isotopes lighter than 137, according
to the process :

$$^{139}La\ (p,\ 3p\ x\ n)^{(137-x)}Cs\ .$$

At first, because of the availability of dyes in the blue range (such
as coumarins and stilben), preliminary studies have been done on the second
D_1 line 6s $^2S_{1/2}$ – 7p $^2P_{1/2}$ at λ = 459.4 nm [5]. Later as dyes became
available in the near infrared region, new experiments were performed directly on the first D_2 line 6s $^2S_{1/2}$ – 6p $^2P_{3/2}$ at λ = 852.1 nm . In that
case, the proper dye to be used is HITC, and, as for DEOTC, excitation must
be provided by the red lines of a Kr^+ laser. The resulting recordings
are schematically displayed in Figure 3. In total, 36 different species
have been studied, composed of 28 isotopes and 8 isomers. Here again we
have obtained precise numerical results concerning hyperfine structure
constants as well as nuclear deformations. To illustrate the latter point,
Figure 4 shows a plot of the variation of the mean square charge radii $\delta\langle r^2\rangle$
for all the nuclei studied. Clearly this plot exhibits 3 zones of different
characteristic behaviour. The first zone, of neutron numbers higher than
82 (magic number corresponding to a closed nuclear shell), shows an almost
linear and rapid increase of $\delta\langle r^2\rangle$ which could be attributed, in a crude
way, to the behaviour of nuclei when adding neutrons in a shell which is far
from being closed : in a sense, the added neutrons could be considered as
less bound and would therefore lead to a rapidly growing volume of nuclei.
The second zone is located between neutron numbers 68 and 82 and it corresponds to a slowly increasing value of $\delta\langle r^2\rangle$, with evidence for an odd-even
staggering. Contrary to the preceding one, this second zone describes the
behaviour of the nuclear shape when one approaches the closed shell. The
third zone is located below neutron number 67 and it exhibits very large
fluctuations of $\delta\langle r^2\rangle$, as was observed for the first time in the Hg series
of isotopes and isomers [1]. Theoretical investigations are now in progress
to interpret these features. It could manifest the possible existence of
two quite different "pseudo stable" nuclear shapes of very nearly the same
energy : one corresponding to the ground state of the nucleus, the other
corresponding to the metastable excited state (isomer). This study comprises the longest series of isotopes and isomers that have ever been investigated by optical techniques.

Francium isotopes.

As has already been pointed out, these optical methods comprising the
magnetic detection and the counting of atoms may present highly sensitive
capabilities. It made it possible to set up a specific arrangement allowing
to search for the still unknown resonance lines of Fr . The D_2 line of
that element was then observed for the first time, and its wavelength roughly
measured in 1978 [6]. The experiment has afterwards been repeated in order
to increase the precision of the wavelength measurement, and to investigate
the high resolution spectrum of that line. The dye utilized in these experiments was an oxazine. The most accurate value which can be given at the
present time for the wavelength is $\lambda = (717.97 \pm 0.01)$ nm. The hyperfine

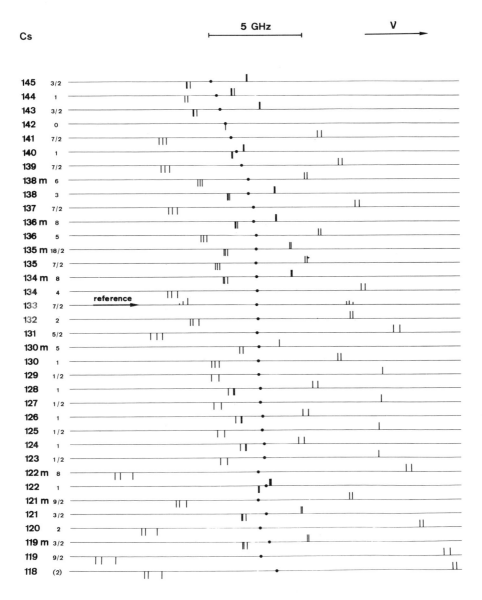

Figure 3

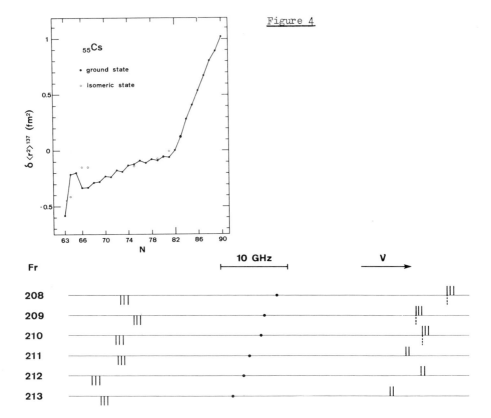

Figure 4

Figure 5

structure of that line has been analyzed for 6 isotopes of mass number 208
up to 213. The observed structures are displayed in Figure 5. For all 6
isotopes these structures exhibit remarkably large hyperfine splittings (up
to 50 GHz) of the atomic ground state sublevels. It is worth while to notice
that all isotopes of Fr being unstable it is not possible to use an auxiliary atomic beam of this element to provide a frequency reference. Basically, it is the sigmameter which serves as a frequency scale reference.
But the scanning procedure of the laser over more than 50 GHz cannot be done
continuously : it is actually necessary to manually operate a controlled
mode hoping. This procedure introduces a small indetermination in the frequency interval measurement corresponding to the hyperfine splitting of the
groundstate, which can be estimated to about 50 MHz.

References.

[1] E. W. Otten, Atomic Physics 5, p. 239. Editors : R. Marrus, M. Prior, H. Shugart, Plenum Press New York and London (1977).

[2] G. Huber, F. Touchard, S. Büttgenbach, C. Thibault, R. Klapisch, and H. T. Duong, S. Liberman, J. Pinard, J.-L. Vialle, P. Juncar, P. Jacquinot, Phys. Rev. C 18, 2342 (1978).

[3] P. Juncar and J. Pinard, Opt. Commun. 14, 438 (1975).

[4] H. T. Duong and J.-L. Vialle, Opt. Commun. 12, 71 (1974).

[5] G. Huber, F. Touchard, S. Büttgenbach, C. Thibault, R. Klapisch, and S. Liberman, J. Pinard, H. T. Duong, P. Juncar, J.-L. Vialle, P. Jacquinot and A. Pesnelle, Phys. Rev. Lett. 41, 459 (1978).

[6] S. Liberman, J. Pinard, H. T. Duong, P. Juncar, J.-L. Vialle, P. Jacquinot, G. Huber, F. Touchard, S. Büttgenbach, C. Thibault, R. Klapisch, A. Pesnelle, C. R. Acad. Sci. 286B, 253 (1978).

Atomic Beam Laser Spectroscopy of Radioactive Ba Atoms

G. Schatz

Kernforschungszentrum Karlsruhe GmbH, Institut für Angewandte Kernphysik
D-7500 Karlsruhe, Fed. Rep. of Germany

1. Introduction

The high sensitivity of laser spectroscopy offers the possibility to
perform spectroscopic measurements on very small amounts of material. It
has thus become possible to study the optical hyperfine structure of
radioactive atoms only a small number of which can be produced at one time.
In this way, one can obtain information on nuclear properties (especially
radii) which is inaccessible by other methods. Therefore laser spectroscopy
has become a powerful tool also in nuclear physics and has been applied to
the study of short lived isotopes |1|.

   This contribution describes a measurement of the hyperfine structure of
the resonance transition $6s^2$ $^1S_0$ - $6s6p$ $^1P_1$ (λ = 553.6 nm) in radioactive
Ba atoms with the aim to determine nuclear moments and radii. So far, 9 un-

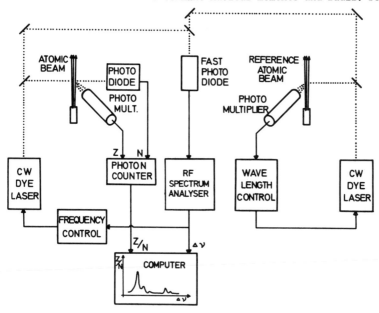

<u>Fig.1</u> Schematic diagram of the experimental set-up. The beam of radio-
active atoms is shown in the left part, the right part of the figure
shows the optical reference frequency system.

stable isotopes and isomers have been studied with half lives down to 11.9 min.

2. Experiment

In view of the small isotopic shifts in Ba (for two neighbouring isotopes on the average 1.5 times the natural line width of the transition studied) the measurements were carried out on a well collimated atomic beam. A schematic diagram of the experimental set-up is shown in Fig.1. The beam of radioactive atoms is shown in the left part of the figure. It is crossed at right angles by the beam of a single mode tunable dye laser the intensity of which is measured behind the apparatus for normalization purposes. The laser frequency is scanned step-wise over the resonance line, and the fluorescence quanta are counted by a photomultiplier in a direction perpendicular to both beams. The frequency is compared to that of a second identical dye laser which is stabilized to the same transition in a stable even Ba isotope by use of a second small atomic beam apparatus. The two laser beams are mixed on a fast diode, and the difference frequency is stabilized by usual rf methods. The electronics is interfaced to a small computer which registers the data and sets the laser frequency automatically once the scanning range and step width have been chosen.

The dye lasers are home-made and essentially copies of a design by HARTIG and WALTHER |2|. They have a frequency jitter of about 5 MHz each, which is negligible as compared to the natural line width of 19 MHz.

The quality of spectra which can be obtained is demonstrated in Fig.2. For this measurement approximately 50 ng of each ^{137}Ba and ^{130}Ba were implanted into the atomic beam oven. The line width obtained is typically 10 to 20 % larger than the natural line width at low laser power. This excess is due to the laser line width and some residual Doppler broadening.

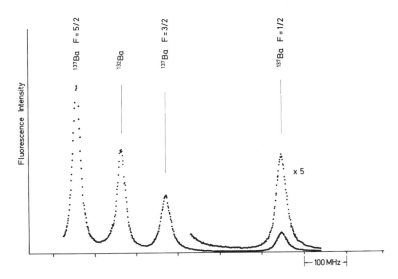

Fig.2 Fluorescence spectrum measured on a 50 ng sample of two separated stable Ba isotopes (A = 137 and 132).

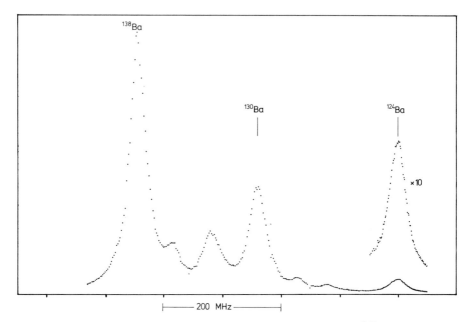

Fig.3 Fluorescence spectrum measured on a 10 pg sample of ^{124}Ba. The other lines with the exception of ^{130}Ba, which was added as a reference, are due to contamination by stable barium.

An experiment proceeds in the following way: In a first step the nuclides to be studied are produced by a nuclear reaction (all but one by means of the Karlsruhe Isochronous Cyclotron) from stable isotopes of Ba, Cs, or Xe. They are then passed through an electromagnetic mass separator. This step is necessary in order to reduce the amount of unwanted, especially stable, Ba isotopes. Due to the small isotopic shifts of Ba and due to the long Lorentzian tails of the lines, the abundant isotopes would otherwise completely mask the signals from the small amounts of radioactive atoms. The sample is then transferred into the atomic beam oven, and the measurement of the resonance fluorescence is started.

Fig.3 shows the result of a measurement on ^{124}Ba which has the shortest half-life (11.9 min) of all isotopes studied so far. The amount of isotope brought into the atomic beam oven was 10 pg in this case, as determined by measuring the γ-decay rate. The line assigned to ^{124}Ba is clearly visible on the background from the tails of lines due to stable Ba isotopes. It is identified by its radioactive decay.

3. Results

Altogether 9 unstable isotopes and isomers have been studied so far, in addition to the 7 stable ones. The results are summarized in Table 1. The implications of the results for nuclear physics have been discussed in several publications |3,4,5| and will not be considered here. In these papers, also details of the production reactions and of the amount of material available (ranging between 10 pg and 30 ng, depending on the isotope) are given.

Table 1 Measured isotope shifts and hyperfine structure constants of the
λ = 553.6 nm BaI resonance line

A	half-life	isotope shift $\lvert$MHz$\rvert$	A-factor $\lvert$MHz$\rvert$	B-factor $\lvert$MHz$\rvert$
138	stable	0	–	–
137	stable	215.0 (7)	-109.8 (5)	49.7 (5)
136	stable	128.9 (5)	–	–
135g	stable	260.9 (7)	- 98.3 (5)	32.5 (4)
135m	29 h	161.7 (6)	32.0 (1)	210.1 $(^{+33}_{-12})$
134	stable	143.0 (5)	–	–
133g	10.5 a	250.0 (9)	273.3 (8)	–
133m	39 h	202.0 (10)	29.1 (13)	196.0 (12)
132	stable	167.9 (5)	–	–
131	11.5 d	249.2 (21)	249.2 (20)	–
130	stable	207.3 (7)	–	–
129g	2.2 h	312.3 (20)	139.9 (21)	–
129m	2.1 h	362.7 (15)	- 46.7 (3)	351.2 (20)
128	2.4 d	271.1 (8)	–	–
126	97 min	355.8 (6)	–	–
124	11.9 min	447.0 (10)	–	–

4. Sensitivity - Achieved and Achievable

In this section we want to discuss the sensitivity obtained so far in this
type of experiment and its possible improvements. As a first step we should
like to present a detailed analysis of the measurement on ^{124}Ba (the results
of which are shown in Fig. 3) since this was the one with the smallest
sample available. The particle flux in the atomic beam can be estimated from
the measured counting rate, the fluorescence light detection efficiency
($\sim$ 2 %) and the measured laser beam intensity. The result is a flux of
500 atoms/sec crossing the laser beam. Each of the atoms emits $\sim$ 150 spontaneous quanta. The latter figure can be determined independently from the
measured power broadening of 8 MHz, and satisfactory agreement is obtained.
At the maximum of the resonance line in Fig.3 one thus obtains 5 counts per
atom crossing the atomic beam when the anisotropy of the fluorescence light
with respect to the laser light polarization is taken into account. In order
to reduce the Doppler width to a tolerable value (13 MHz in this experiment) the atomic beam has to be sharply collimated. The loss of atoms by
collimation is not well known because of lack of precise knowledge about
the angular distribution of the atoms leaving the canals of the oven. But
the loss factor is probably not less than 10^3, so the flux of atoms from
the atomic beam oven is larger than $5 \cdot 10^5$ sec^{-1} at the beginning of the measurement. The total experiment which comprises 4 complete scans of the
fluorescence intensity such as shown in Fig.3 lasted approximately 50 min.
The constancy of the atomic beam flux during this period can be ascertained
from the lines of the stable Ba atoms. During the experiment therefore
$5 \cdot 10^8$ atoms left the atomic beam oven if radioactive decay is taken into
account. If the latter were to be neglected this would increase to
$1.5 \cdot 10^9$. This figure is to be compared to the number of $5 \cdot 10^{10}$ atoms
(corresponding to 10 pg) which we know from gamma ray measurements to be
present in the oven at the beginning of the experiment. Much of the difference between these two figures is due to radioactive decay. This means
that for an isotope of longer half-life the time of measurement could have
been extended leading to an improved accuracy. We know, though, from other

experiments that only a fraction of the atoms brought into the atomic beam oven can be evaporated at the small amounts of material available in this type of experiment.

It should be pointed out that the example considered here is that of an even isotope. For odd isotopes, optical pumping occurs, and the atoms are redistributed among the Zeeman levels of the groundstate. This leads to a saturation of the fluorescence light at much lower levels of laser intensity for at least one of the three hyperfine components if the nuclear spin is 1 or larger. Therefore the sensitivity is roughly one order of magnitude smaller than described above for odd isotopes unless their spin is 1/2. It appears possible, though, to reduce this difference between even and odd isotopes by switching the laser polarization several times during the passage of the atoms through the laser beam.

Which are the probable limits of sensitivity of this method? The resonance line of ^{124}Ba in Fig.3 has a signal to background ratio of ∿5. In addition, some improvements of the statistics appear possible (measuring a larger number of samples, increasing laser intensity together with expanding the laser beam, opening the collimating slit etc.), though partly at the expense of background or resolution. It is therefore probably not optimistic to say that a reasonable measurement on an even Ba isotope should be possible with ∿ 1 pg (and with ∿ 10 pg of an odd isotope). The background is dominated by the tails of lines from stable isotopes. It appears very difficult to reduce this contamination further because barium seems to be everywhere on the ng level. In fact, these experiments have only become possible after hard and tedious work which reduced the contamination from stable Ba by several orders of magnitude. The situation is more favourable for elements which either have a larger isotopic shift or are not as omnipresent as barium. In such a case the background would be due to other sources (laser light scattering, thermal emission of the photocathode, response of the multiplier to gamma rays from the sample etc.) which amount to ∿ 100 counts/sec and therefore are unimportant in the present experiment. It is difficult to estimate the importance of these background contributions, though, without referring to a definite case.

Acknowledgements

The work described in this contribution was done by a group which is identical to the authors in refs. 3 to 5. In addition we should like to thank Mr. B. Feurer for his invaluable aid in preparing the mass separated samples and the cyclotron crew for their dedicated effort. The work would not have been possible without the advice and encouragement of many colleagues. Of these we should like to thank especially Professors H. Walther, Munich, and E.W. Otten, Mainz.

References

1. See, e.g., the preceding contributions to this volume by H.J. Kluge and S. Liberman.
2. W. Hartig and H. Walther, Appl. Phys. 1, 171 (1973).
3. G. Nowicki, K. Bekk, S. Göring, A. Hanser, H. Rebel, and G. Schatz, Phys. Rev. Lett. 39, 332 (1977).
4. G. Nowicki, K. Bekk, s. Göring, A. Hanser, H. Rebel, and G. Schatz, Phys. Rev. C 18, 2369 (1978).
5. K. Bekk, A. Andl, S. Göring, A. Hanser, G. Nowicki, H. Rebel, and G. Schatz, Z. Physik, in press.

Dye Laser Saturation Spectroscopy with Beams of Atoms and Ions[1]

J. Kowalski, R. Neumann, and F. Träger

Physikalisches Institut der Universität Heidelberg, Philosophenweg 12
D-6900 Heidelberg, Fed. Rep. of Germany

G. zu Putlitz

Gesellschaft für Schwerionenforschung, Planckstraße 1
D-6100 Darmstadt und
Physikalisches Institut der Universität Heidelberg, Philosophenweg 12
D-6900 Heidelberg, Fed. Rep. of Germany

Introduction

In this paper we report on two experiments, which have been carried out applying the method of dye laser saturation spectroscopy with observation of resonance fluorescence.

In the first experiment the calcium intercombination line was used for measurements of isotope shifts and hyperfine structure splittings (hfs) of stable and radioactive Ca-isotopes. New principles for the production and detection of saturation signals in very weak spectral lines had to be developed. Previously unknown results about nuclear charge radii and quadrupole moments of Ca-isotopes were obtained. The second experiment aims for a precise determination of the fine and hyperfine structure of the first excited ^3S and ^3P states of the helium-like ion Li^+. This experiment contributes new data to the quantummechanical problem of the two-electron systems.

Dye Laser Saturation Spectroscopy in the $4s^2 \, ^1S_0 - 4s4p \, ^3P_1$ Transition in Calcium

Saturated absorption experiments on the extremely weak calcium intercombination line have recently proved to be of great interest:

1) This line is supposed to be an appropriate and promising candidate for a future frequency or wavelength standard.

2) The high resolution associated with signals created by saturated absorption of this transition yields accurate spectroscopic information on isotope shifts and hyperfine structure splittings (hfs). They permit a study of the nuclear charge distribution of the Ca-isotopes. Incorporation of $f_{7/2}$-shell neutrons or neutron pairs into the double magic ^{40}Ca nucleus up to the closed configuration of double magic ^{48}Ca opens the way for a study of changes of the radii if the mass number is changed by 20%, the neutron number by even as much as 40%. In this respect the long chain of Ca-isotopes (6 stable isotopes) can be regarded as a unique testing ground for a study of the distribution of nuclear matter, in particular for light nuclei. Moreover, the radioactive isotope ^{41}Ca is of special

---

[1] The work described here is in part based on the publications in ref.(1-3, 11-13)

interest because it offers a chance to study the influence of a single neutron on the double magic ^{40}Ca core. ^{41}Ca and its mirror nucleus ^{41}Sc also play an important role in Coulomb displacement energy calculations.

3) The weak Ca intercombination transition, the long lifetime of the 4s4p $^3P_1$ state, and other spectral lines originating from this excited state are rather well suited to develop new schemes for the detection of laser in-duced transitions, application of rf-techniques in combination with laser excitation, optical Ramsey resonances (4) and for the study of various effects associated with resonance light pressure (5).

At first glance the Ca-intercombination line $4s^2\ ^1S_0$ - 4s4p $^3P_1$ with a natural linewidth of only 410 Hz corresponding to the very small oscillator strength of f = 3.7·10^{-5} does not seem to be favourable for spectroscopic studies in particular for saturation spectroscopy. However, by applying a technique developed for this transition (1,2) (see Fig. 1), strong signals can be produced. The beam of a frequency stabilized cw dye laser intersects an atomic beam of natural Ca at right angles where the different isotopes can be excited selectively to the 4s4p $^3P_1$ state.

Its lifetime being rather long (τ=0.39 ms), the beam emits photons along the path of flight of the atoms. Therefore resonance fluorescence radiation can be monitored some 12 cm downstream the beam, which leads to the total suppression of any disturbing instrumental background. Care must be taken to collect the reemitted radiation as efficiently as possible. This is accomplished by means of a lightpipe surrounding the atomic beam. Reflecting the laser beam back onto itself, Lamb-dips have been recorded (2) in the Doppler profiles of the transitions with laser powers lower than 2 mW. There is little chance to detect these saturation dips in absorption by the common technique with a weak probe beam, because the absorption of the order of 10^{-5} cm^{-1} is much too small. A complete description of the apparatus used in these experiments can be found in (1,2). Fig. 2 shows a more recently developed scheme of the experimental set-up, which is being used for high precision tuning of the dye laser by means of a rf-synthesizer. The laser system which is based on the frequency-offset-locking technique can also be applied for the stabilization of the dye laser to hfs components or any other signal.

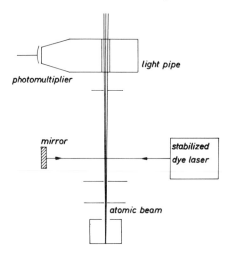

Fig. 1. Experimental set-up for the delayed observation of resonance fluorescence light from the metastable Ca-$^3P_1$-term after excitation with a dye laser.

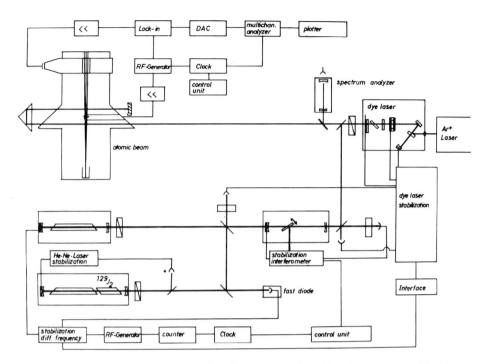

Fig. 2. Schematic arrangement of a large range tunable frequency -offset-locking spectrometer.

The apparatus provides rf-precision tuning of the dye laser over a range of more than 20 GHz. This is accomplished by repeatedly scanning the tunable He-Ne-laser for exactly the free spectral range of the transfer interfero-meter, and then jumping back with the laser frequency to a neighboring transmission maximum while keeping the length of the interferometer constant.

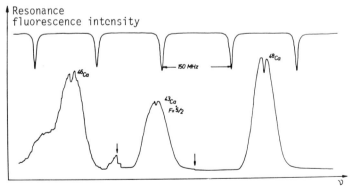

Fig. 3. Lamb-dip signals in the fluorescent light of rare abundant Ca-iso-topes. The arrows indicate a change in the amplification of the detection system

Fig. 3 demonstrates the high resolution associated with the Lamb-dips and their narrow linewidth of 900kHz (2), as well as the high sensitivity of the apparatus. Even the very rare ($3 \cdot 10^{-5}$ abundance) isotope ^{46}Ca (2) has been detected in the natural isotope composition including a Lamb-dip with good signal-to-noise ratio. The width of the dips is determined by the width of the laser line and by transit-time broadening which are slightly enhanced by saturation broadening.

This experimental technique of atomic beam saturation spectroscopy with delayed detection of resonance fluorescence developed for our experiments here is capable of producing lines with widths of a few kHz only, which might be used for frequency standard purposes. Very recently this method has been used by BARGER et al. (6) for the detection of optical Ramsey reconances in the Ca-intercombination transition.

Combined Laser Saturation rf-Spectroscopy of rare Isotopes

Saturation of the intercombination line of Ca can also be used for a novel scheme to detect the rf-transitions induced between sublevels of the excited $4s4p\ ^3P_1$ state. This technique is similar to the well-known classical double resonance method (7) and carries the resolution for hyperfine structure measurements even further (see Fig. 4): The atoms travel across a laser beam 1, which saturates the optical transition from the ground state to e.g. hfs level a. In the subsequent rf-coil the population of state a and the neighboring hfs level b is equalized. Therefore level a can be refilled from the ground state when the atoms cross a laser beam 2. This results in an increase of emitted resonance fluorescence radiation which is observed by a photomul-

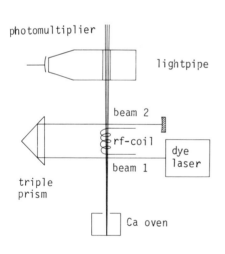

Fig. 4. Schematic arrangement for the detection of r-transitions between sublevels of the excited state in a saturated absorption experiment.

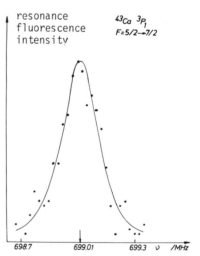

Fig. 5. rf-signal originating from an induced magnetic dipol transition between the fhs levels F=5/2, 7/2 in the $4s4p\ ^3P_1$ state of ^{43}Ca. The laser was stabilized to the $^1S_0$-$^3P_1$, F=5/2 transition. The linewidth is 220 kHz.

tiplier via the light pipe. This method, which is in some sense similar to experiments recently performed in atomic ground states (8,9) has been applied to ^{43}Ca. The width of the signals (Fig. 5) is determined by the inverse transit time, which the atoms take to cross the rf-coil ($\sim$ 50 kHz) and by Zeeman broadening due to the residual earth magnetic field of $\sim$ 50 mGauss. Signals as shown in Fig. 5 were obtained using the natural isotopic mixture with an abundance of only 0,14% of the isotope ^{43}Ca thus demonstrating the high sensitivity of this method. The technique introduced here can also be applied to short lived excited states. In this case the rf-field and the laser light field must interact simultaneously with the atoms thus coupling together the two excited state sublevels and the ground state.

From the rf experiment and the scan experiments as described above the quadrupole moment $Q(^{43}$Ca$) = -0.06(2)$ b was obtained, which coincides with a value from recent measurements in the metastable $4s4p\ ^3P_2$ term (10).

Nuclear Charge Radii of Ca Isotopes

For a measurement on the stable isotopes the natural isotopic mixture was used (11,12). For a study of radioactive ^{41}Ca (13) a quantity of only 20 μg with a concentration of one part in a thousand was available. This small amount was sufficient to produce signals including Lamb-dips using an atomic beam for one hour duration. A second beam of natural Ca served for adjusting the laser to the intercombination line and for an easy identification of the recorded signals.

From the field shift part of the total isotope shifts of the 6 stable Ca-isotopes and ^{41}Ca the changes of the mean square nuclear charge radii have been derived. The results are shown in Fig. 6. The radii of ^{40}Ca and ^{41}Ca are

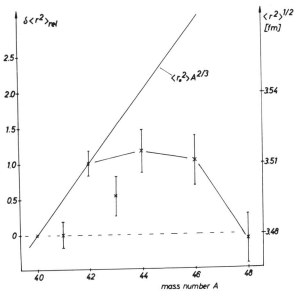

Fig. 6. Nuclear charge radii for even and odd Ca isotopes between ^{40}Ca and ^{48}Ca determined from optical isotop shifts in the Ca-intercombination line.

equal within the experimental errors, so the additional $f_{7/2}$-neutron in ^{41}Ca outside the closed proton- and neutron shells does not penetrate very much into the ^{40}Ca core and does not polarize it significantly either. However, two paired neutrons in ^{42}Ca lead to an increase of the charge radius of $\sim 1\%$, which stays about constant through ^{44}Ca to ^{46}Ca, and rapidly falls off at ^{48}Ca to the same charge radius as ^{40}Ca.

Surprisingly ^{43}Ca has a smaller radius than both neighbouring even iso-topes whereas the opposite behaviour has been found throughout the periodic system except ^{135}Ba.

The peculiar behaviour of the Ca charge radii has been suggested to be due to changes in the deformation of the Ca-nuclei (14). The deformation increases in the first half of the $f_{7/2}$-shell and decreases in the second half. Whether or not this explanation holds can be tested in the immediate future by measuring the hfs of ^{41}Ca and ^{45}Ca and deducing the nuclear quadru-pole moments and hence the deformations of these isotopes. Moreover, the staggering parameters of the odd isotopes also give information on this point.

Dye Laser Saturation Spectroscopy on the $2s\,^3S_1$ - $2p\,^3P$ Transition in ^{6,7}Li$^+$

The second experiment, which will be discussed here, represents a first at-tempt to investigate the $1s2s\,^3S_1$ and the $1s2p\,^3P$ levels of Li$^+$ (Fig. 7) with high precision by tuned dye laser spectroscopy. Li$^+$ is like helium a two electron system and belongs to the fundamental quantummechanical problems in the sense, that its spectroscopic properties - in particular fs and hfs - are accessible to very precise calculations using advanced methods of pertu-bation theory (15,16). Thus the He atom and its ^3P fine structure is the source of one of the most accurate values for the fs constant α to date (17). Numerous experimental as well as theoretical studies are devoted to the metastable $1s2s\,^3S_1$ state and the lowest members of the excited ^3P sequence of He. For two electron ions measurements have not reached a precision com-parable to the one in helium. Results more recently published for Li$^+$ have been obtained by the methods of interferometric analysis of a Li hollow catho-de lamp (18), beam foil spectroscopy (19), and Doppler tuned ion-beam laser spectroscopy (20). In this experiment a complete set of parameters determining the hfs of 2^3S_1 and $2^3P_{1,2}$ and the fs of the 2^3P in both ^{6,7}Li$^+$ is obtained with an accuracy better than in previous experiments. For both the hfs and

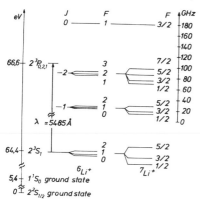

Fig.7. Level diagram with the first excited states of Li$^+$. The lifetime of the ^3P state is $\tau = 45$ ns. The nuclear moments of ^{6,7}Li are I=1 and I=3/2 re-spectively.

the fs precise calculations exist (15,16), whose results are compared with the experimental values. Furthermore a precise value for the isotope shift (I.S.) in the $^3S_1$ - ^3P transition is obtained. The data collected so far are not representative for the precision achievable, which can be increased by two or three orders of magnitude if a combination of laser and rf-techniques is employed.

Method and Experimental Arrangement

The experimental set-up is very similar to the one used in the Ca-experiment. In this experiment, however, one needs an intense beam of slow metastable ions in the 2^3S_1 level, which is excited by electron impact on the Li atoms leaving the aperture of the oven. A 3 mm diameter heated tungsten loop above the oven serves as cathode. Though the cross-section for combined ionization and excitation into the 2^3S_1 state is very low ($\sigma = 10^{-22 \pm 1} cm^2$ at threshold (21)) a sufficient part of ions can be produced in this state forming a beam of $\sim 10^9/mm^2 \cdot sec$. The ions are accelerated to 200 eV ($v = 7 \cdot 10^6 cm/s$) and focussed into a beam of low divergence. The Doppler profile perpendicular to the beam axis has a measured FWHM of about 1 GHz.

The ion-beam is crossed at right angles by a cw dye laser light beam which is reflected onto itself in order to produce saturation dips in the 2^3S_1 - 2^3P transitions. The fluorescent light is registered with a cooled photomultiplier looking directly into the excitation region of the laser. The freejet dye laser has a frequency bandwidth of about 1 MHz and can be scanned over more than 60 GHz. The frequency variation is calibrated with a stabilized confocal Fabry-Perot interferometer of a free spectral range (fsr) of about 75 MHz. This interferometer is calibrated by locking it to an iodine stabilized He-Ne laser and by stabilizing a dye laser with a frequency offset of about 8 GHz away from the He-Ne laser frequency on a different transmission maximum. The difference frequency of the lasers, which is a multiple of the fsr, is measured by means of an avalanche diode.

Measurements and Results

The hfs of the 2^3S_1 and $2^3P_{1,2}$ ion states of both stable isotopes $^{6,7}Li$, the fs of the 2^3P level and the isotope shift in the 2^3S_1 - 2^3P transition have been investigated in this work. Fig. 8 shows an example of a recorder trace of fluorescence signals from the $^3P_1$ term in $^6Li^+$. The hyperfine structure of the ground state and thereby its A-factor can be determined directly from the $^3S_1$ - $^3P_0$ transitions (Fig. 9), because the $^3P_0$ term has no hfs. From their distance to the corresponding $^3S_1$ - $^3P_0$ transitions of $^7Li^+$ the isotope shift of this transition can be determined. For the determination of the fs splitting of one isotope, fs terms of the other isotope lying inbetween are used as auxiliary reference points for the laser scans. An example of such a measurement is shown in Fig. 10.

For most of the measurements the laser beam was expanded by a factor of 6 to reduce the width of the saturation dips, which was predominantly due to time of flight and power broadening. Typically widths of 30 MHz were obtained at laser powers of 25 mW.

In table 1 the measured magnetic hfs constant of the 2^3S_1 term and the hfs constants and fs splittings of the 2^3P term are summarized for both isotopes $^{6,7}Li^+$ together with theoretical data. For the hfs of 2^3S_1 term the agreement between experimental and theoretical results is very good.

The situation is more complex in the $1s2p\,^3P$ multiplet: Fs and hfs of the 2^3P state are of comparable magnitude in both Li isotopes, more pronounced in ^{7}Li. As a consequence states with equal hyperfine structure quantum number F in different fs levels disturb each other considerably and distort the term pattern as a whole. This leads to an energy matrix diagonal with respect to the operator $\vec{F}$ but not to $\vec{J}$, the operator of the total angular momentum of the electron. Therefore undisturbed fs intervals and the hfs constants cannot be calculated directly from the measured splittings. They are deduced by diagonalization of the energy matrix and by identification of the energy eigenvalues with the measured term distances. Five parameters are taken into consideration for the theoretical description of the 2^3P fine and hyperfine multiplet: the undisturbed fs distances $\Delta\nu_{01}$ and $\Delta\nu_{02}$ of J=1 and J=2 relative to J=0, and the contact, dipolar and orbital hfs constants A_c, A_d, and A_o, where A_o and A_d are related by $A_o = -4.925\,A_d$. The hfs constant B which represents the small influence of the electric nuclear quadrupole moment can be neglected within the present accuracy.

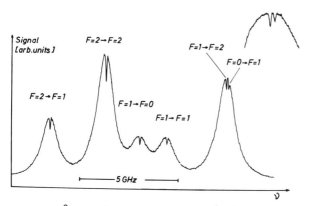

Fig. 8. $2^3S_1 - 2^3P_1$ transitions in ^{6}Li$^+$ for the determination of the $^3P_1$-hfs. The enlarged section shows two close lying signals well resolved.

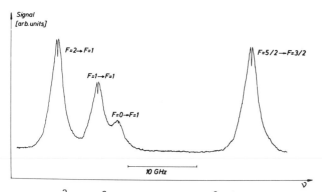

Fig. 9. $2^3S_1 - 2^3P_0$ transitions in ^{6}Li$^+$ (left part) and one of the corresponding isotope shifted transitions in ^{7}Li$^+$ (right part). The three transitions on the left reflect the $^3S_1$-hfs.

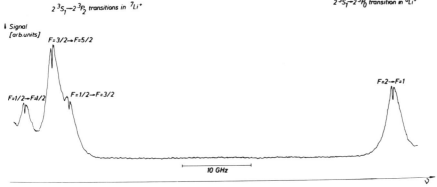

Fig. 10. Laser scan over 60 GHz for the determination of the $^3P_2$ - $^3P_0$ fs splitting of $^7Li^+$. The transition in $^6Li^+$ is used as reference line. An enclosing scan of 26 GHz leads to the $^3S_1$ - $^3P_0$ transition in $^7Li^+$ as shown in fig. 9.

The parameters were obtained by means of a least square fit to the measured energy splittings using the analytically known relations between these parameters and the energy splittings from the eigenvalue equations.

The measured values of the $2^3P_{1,2}$ hfs constants for both isotopes are generally in good agreement with theory, more pronounced for $^6Li^+$. Regarding the two fs separations $\Delta \nu_{01}$ and $\Delta \nu_{02}$ one gets consistent values for $^6Li^+$ and $^7Li^+$, which represent two independent measurements. However both values differ from the theoretical ones, markedly for the smaller fs splitting $\Delta \nu_{02}$ which differs by about 50 MHz from the theoretical value. No explanation can be given at the moment for this surprisingly large deviation. The theoretical values are believed to be of a precision of about 0.3 MHz (16). The results

Tab. 1. Hfs constants and fs splittings of $^{6,7}Li^+$, all values in MHz.

		$^6Li^+$		$^7Li^+$	
		This work	N.A.Jette et al. *B.Schiff et al.	This work	N.A.Jette et al. *B.Schiff et al.
2^3S_1	A	2998.3(3.0)	2998.23	7918.9(4.0)	7918.34
2^3P	A_c	1390(6)	1392.8	3669(6)	3678.4
	A_o	20.8(4.0)	18.4	57(5)	48.6
	A_d	−4.2(1.0)	−3.7	−11.5(1.0)	−9.86
	$\Delta \nu_{01}$	−155698(20)	−155725.1*	−155694(24)	−155725.1*
	$\Delta \nu_{02}$	− 93023(9)	− 93072.1*	− 93019(7)	− 93072.1*

quoted above make new calculations as well as even more precise measurements on this fundamental three-body problem desirable.

An additional result is the isotopic shift (IS) of the 2^3S_1 - 2^3P transition. It was extracted from the 2^3S_1 - 2^3P_0 transition. One obtains a value of 36233 (9) MHz, which is composed of the normal mass shift of 7057 MHz and the sum of the specific and the very small volume shift of 29176 (9) MHz. No theoretical value for the specific mass shift is available up to now, although, in contrast to many-electron systems, Li$^+$ with two electrons only should be accessible to theoretical treatments of the specific mass shift.

This work was sponsored by the Deutsche Forschungsgemeinschaft.

References

1 F. Träger, R. Neumann, J. Kowalski, G. zu Putlitz: Appl.Phys. 12, 19 (1976)
2 U. Klingbeil, J. Kowalski, F. Träger, H.-B. Wiegemann, G. zu Putlitz: Appl.Phys. 17, 199 (1978)
3 U. Klingbeil, J. Kowalski, F. Träger, H.-B. Wiegemann, G. zu Putlitz: Kvantovaya Electronica 5, 1716 (1978)
4 Ye. V. Baklanov, B.Ya. Dubetsky, V.P. Chebotayev: Appl.Phys. 9, 171 (1976)
5 A.P. Kazantsev: Sov.Phys. Usp. 21, 58 (1978)
6 R.L. Barger, J.C. Bergquist, T.C. English, D.J. Glaze: to be published
7 J. Brossel, F. Bitter: Phys.Rev. 86, 308 (1952)
8 S.D. Rosner, R.A. Holt, T.D. Gaily: Phys.Rev.Lett. 35, 785 (1975)
9 W. Ertmer, B. Hofer : Z.Physik A 276, 9 (1976)
10 P. Grundevik, M. Gustavsson, I. Lindgren, G. Olsson, L. Robertsson, A. Rosén, S. Svanberg: Phys.Rev.Lett. 42, 1528 (1979)
11 R. Neumann, F. Träger, J. Kowalski, G. zu Putlitz: Z.Physik A 279, 249 (1976)
12 U. Klingbeil, J. Kowalski, F. Träger, H.-B. Wiegemann, G. zu Putlitz: Z.Physik A 290, 143 (1979)
13 J. Kowalski, F. Träger, S. Weißhaar, H.-B. Wiegemann, G. zu Putlitz: Z.Physik A 290, 345 (1979)
14 H.D. Wohlfahrt: in Proceedings of the Conference "What do we know about the radial shape of nuclei in the Ca-region?", Karlsruhe, West-Germany, May 2-4, 1979
15 B.Schiff, Y. Accad, C.L. Pekeris: Phys.Rev. A 1, 1837 (1970)
16 N.A. Jette, T. Lee, T.P. Das: Phys.Rev. A9, 2337 (1974)
17 A. Kponou, V.W. Hughes, C.E. Johnson, S.A. Lewis, F.M.J. Pichanick: Phys.Rev.Lett. 26, 1613 (1971)
18 R. Bacis, H.G. Berry: Phys.Rev. A 10, 446 (1974)
19 W. Wittmann: Thesis, Berlin 1977
20 B. Fan, A.Lurio, D. Grischkowsky: Phys.Rev.Lett. 41, 1460 (1978)
21 A. Adler, W. Kahan, R. Novick, T. Lucatorto: Phys.Rev. A 7, 967 (1973)

Laser Induced Nuclear Orientation

M.S. Feld[1]

Spectroscopy Laboratory and Physics Department,
Massachusetts Institute of Technology, Cambridge, MA 02139, USA

and

D.E. Murnick

Bell Laboratories, Murray Hill, NJ 07974, USA

Introduction

There is a growing awareness that the techniques of laser spectroscopy can be brought to bear on nuclear physics problems in
a variety of ways.[1]   Laser induced nuclear orientation [2,3]
is one such example of the intersection of laser spectroscopy and
nuclear spectroscopy. In this technique the angular momentum
transferred to a resonantly absorbing atomic (or molecular) transition by polarized laser radiation can orient the atomic nuclei
along a given axis. If the nuclei are unstable,the angular distribution of the subsequent nuclear radiation (gamma, beta, fission fragments, etc.) will be anisotropic. This spatial anisotropy can be a sensitive probe for resolving hyperfine splittings
and other nuclear structure effects, especially when coupled with
laser saturation or atomic-molecular beam techniques. Another
important aspect is the ability to use laser photons to control
gamma or other nuclear radiation. The new physics to be studied
includes precision measurements of isomer shifts and excited
state nuclear moments. The technique can also be used to produce
polarized targets and beams, for nuclear isomer separation, weak
interaction studies, and possibly to produce narrow band gamma
radiation. Some of these potential applications are described
elsewhere.[2,3] The present discussion will be restricted to
three topics:

- describing the technique of laser induced nuclear orientation
- presenting preliminary observations of the effect in ^{24m}Na
- proposing a technique for producing tunable, narrowband gamma
  radiation

    Important previous studies of nuclear orientation using con-
ventional light sources have been performed by the Mainz group[4]
and by CAPPELLER and MAZURKEWITZ[5]. A different approach to
producing tunable gamma radiation has been proposed by LETOKHOV
and his collaborators[6].

Concept of Laser Induced Nuclear Orientation

Consider a vapor containing isomeric atoms, i.e. atoms whose
nuclei are in an excited state (Fig.1). Since the concentration
of such atoms is always very low, the optical spectrum either in
absorption or spontaneous emission is exceedingly weak and diffi-

---

[1]Work supported by the U.S. Department of Energy

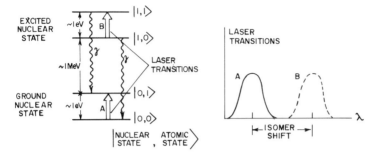

Fig.1 Schematic representation of coupled atomic and nuclear transitions

cult to resolve. An enormous enhancement in sensitivity can be obtained by studying the nuclear radiation rather than optical photons. Although this radiation is ordinarily isotropic, its angular distribution may be altered by a circularly (or linearly) polarized laser field which is tuned into coincidence with one of the hyperfine transitions of the isomeric atoms. Circularly polarized laser photons exert a torque on the atomic electrons which tends to align their angular momenta along the propagation axis of the laser beam. Since the electrons and nuclei are coupled by the hyperfine interaction, the atomic nuclei can also become aligned, and hence radiate anisotropically (Fig.2). In this way, as the laser is tuned the hyperfine structure of the isomeric atoms will manifest itself as a series of resonant changes in the angular distribution of the nuclear radiation.

The degree of nuclear orientation can be greatly enhanced by taking advantage of optical pumping techniques. The optical pumping effect [7] gives rise to a net transfer of atoms to higher or lower orientation quantum numbers, depending on the laser polarization. If the laser is right circularly polarized, for example, the selection rule for absorption is $\Delta M_F = +1$, whereas for spontaneous re-emission $\Delta M_F = 0, \pm 1$ is generally allowed. Hence the pumping cycle tends to shift the population to higher M_F levels, leading to a nuclear polarization in ground and excited atomic states. The degree of polarization depends on the laser intensity and the oscillator strength of the atomic transition, as well as on the extent of relaxation effects, such as collisional disorientation.

Line Narrowing Using Laser Saturation Effects. Under suitable conditions (relatively low pressure, so that the transition is Doppler broadened) laser optical pumping of the isomeric transition can lead to a high degree of nuclear orientation over a narrow range of velocities. Discussions at this conference and elsewhere[8-10] have described the qualitatively different features of the laser saturation process when optical pumping is present. The large (up to several orders of magnitude) reduction of the saturation threshold and other related behavior leads to greatly enhanced nuclear orientation efficiency at reduced laser intensity levels. Accordingly, in combination with the

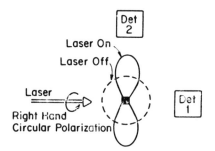

Fig.2 Schematic experimental set-up. Det 1 and 2 are gamma ray detectors and the solid and dashed lines represent gamma ray angular distributions.

laser saturation techniques, resolution at the natural width level can be achieved. Laser induced nuclear orientation can also be combined with rf double resonance techniques of the type used by CAPELLER and MAZURKEWITZ[5].

Applications Laser induced nuclear orientation should be applicable to unstable nuclei with lifetimes as short as ~1 nsec, the lower limit being determined by the minimum optical pumping rate and the uncertainty broadening introduced into the atomic resonance line. In optimal cases the single particle sensitivity of nuclear counting techniques allows experiments to be performed on equilibrium concentrations as low as ~ 1 atom/cm^3.

Laser induced orientation and polarization of nuclei, both stable and unstable, may have other interesting applications. Possibilities include production of polarized beams and targets [11,12] and, in certain unstable systems, polarized beta and gamma radiation. If the polarized nuclei are beta unstable, beta-gamma coincidence experiments are also feasible. As in low temperature nuclear orientation experiments, the hyperfine inter-action in the beta unstable ground state, as well as in gamma emitting intermediate states, can be studied. The application of this technique to produce narrowband gamma radiation is discussed below.

Laser Induced Nuclear Orientation Experiments

Three experiments to observe laser induced gamma anisotropies are in progress at out laboratories. The long lived isomer ^{137m}Ba is being studied in a cell configuration. ^{22}Na is being studied, also in a cell, as a prototype example of a system in which both beta and gamma anisotropies can be observed, and an on-line study of ^{24m}Na, a short lived state produced by a d,p reaction, is in progress[13]. Preliminary results of the latter study are presented here.

The short lived state (τ=20 msec) under study is the first excited level of the ^{24}Na nucleus (Fig.3). According to a shell model description, it is formed by three $d_{5/2}$ protons with total angular momentum 3/2 coupling with a $d_{5/2}$ neutron hole. The resulting excited state nuclear level, an I=1$^+$ state, decays to the I=4$^+$ ground state by a spin flip of the total proton angular

551

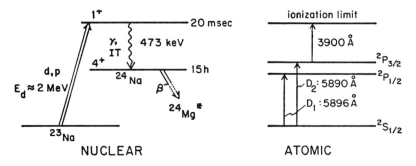

NUCLEAR   **ATOMIC**

<u>Fig.3</u> Relevant nuclear and atomic levels and transitions for
^{24}Na.

momentum, emitting a 473 keV gamma ray. A modified shell model
theory[14] predicts nuclear g factors of 0.385 and -1.86 for the
ground and excited states, respectively, the magnitudes of
which agree reasonably with the corresponding experimental values
of 0.42 [15] and |1.928|[16]. The hyperfine A constant of the
^{24}Na ground state is 253.2 MHz[15]. From the ratio of the measured g factors we therefore obtain an estimate of -1163 MHz
for the ^{24m}Na A constant. The ^{24}Na nucleus is nearly spherical
with a small quadrupole moment;[17] the nuclear volume change
associated with the isomeric level is expected to be small.
Thus, according to the above considerations we expect i)a hyperfine structure splitting of

the same order as that of ^{23}Na, dominated by the magnetic moment
contribution; ii)inverted hyperfine multiplet structure in the
isomeric atom (largest F state lying lowest), due to the negative value of g; and iii)a small (<100 MHz) isomer shift relative to the ^{24}Na ground state.

The ^{24m}Na isomer is produced on line at the Bell Laboratories
4MV van de Graaff accelerator. The experimental arrangement is
shown in Fig.4. The nuclear isomer is populated by the reaction

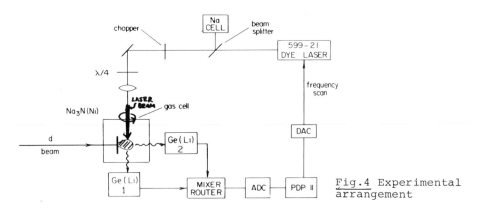

<u>Fig.4</u> Experimental
arrangement

^{23}Na (d,p) ^{24m}Na at deuteron energies of about 2 MeV. The target
is a thin layer of NaCl, deposited on a nickel foil. As the excited state nuclei are driven through the foil they lose kinetic
energy. The foil thickness is adjusted so as to reduce the energy of the ^{24m}Na nuclei to a few keV. The target is placed in
25-150 torr of argon buffer gas, which serves to stop the emitted
nuclei in a few mm and thus confine them before they decay (20
msec). The buffer (as well as the plasma formed by the emerging
beam) also helps to rapidly neutralize the isomeric atoms. ^{24m}Na
production rates are typically 10^3-10^4/sec. Thus, typically 10
-100 isomeric atoms are in the observation volume at a given time.
Optical pumping is achieved by means of dye laser radiation at
589 nm. The anisotropy is observed by monitoring 473 keV gamma
radiation with a pair of Ge(Li) detectors placed at right angles
as shown. Changes in quantization due to stray fields are prevented by means of a "keeper field" of a few gauss. Saturation
resonances from a ^{23}Na calibration cell are used as frequency
markers.

   In preliminary results on the D1 line (Fig.5) we have obtained a statistically significant frequency dependent anisotropy of 0.9%, about 10% of the expected maximum. As can be
seen, the peak of the anisotropy spectrum is shifted several
hundred MHz above the center of gravity of the ^{23}Na D1 line.
In analyzing the data several correction factors must be taken
into account: i) the measured [17] value of the ^{23}Na - ^{24}Na isotope shift (706) MHz); ii) the pressure shift induced by the Ar
buffer (-9 MHz/torr) [18]; and iii) the fact that the gamma anisotropy should be centered on the atomic ground state F=3/2 ^{24m}Na
(I=1) hyperfine doublet. This occurs because the ground state

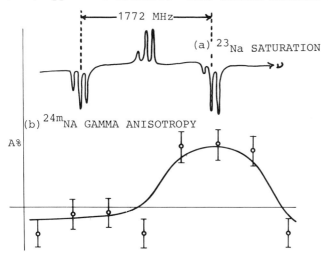

Fig.5 Typical observed gamma ray anisotropy vs. laser frequency.
a) Lamb dip calibration spectrum, ^{23}Na D1 line. b) Observed gamma
anisotropy A.

of the other doublet is F=1/2, and therefore cannot give rise to
nuclear orientation in the optical pumping process. With these
corrections, we obtain a preliminary value for the ^{24m}Na - ^{24}Na
isomer shift of 0±200 MHz, assuming the nuclear model prediction
of a negative g factor to be correct. The assumption of a positive value of g gives an unphysically large (>1000 MHz) value
for the isomer shift.

Thus, our preliminary results confirm the negative g factor
and small isomer shift predictions of the nuclear model. Work
is in progress to improve the gamma anisotropy signal-to-noise
ratio. Once this is accomplished, we hope to use standing-wave
techniques at reduced buffer gas pressures to observe narrow
saturation resonances manifested on the gamma anisotropy.

Production of Narrowband Tunable Polarized Gamma Radiation

Gamma rays emitted from a gaseous sample of isomeric atoms are
Doppler broadened, and thus distributed over a spectral range
$\sim \omega_\gamma u/c$ with $\hbar\omega_\gamma$ the energy separation between initial and final
nuclear states (Fig.1) and u the thermal velocity. Since at room
temperature $u/c \sim 10^{-6}$, the Doppler widths of gamma radiation in
the 100 keV- 1 MeV range are ~0.1-1 eV. The gamma ray absorption
profile of the gaseous sample is similarly broadened. Furthermore, because of the momentum transfer between the gamma rays
and the recoiling nuclei, absorption and emission frequencies do
not coincide. The center frequency of the emission line is down-
shifted to

$$\omega^\circ_{em} = \omega_\gamma (1-R) ,$$

and the absorption center frequency is upshifted to

$$\omega^\circ_{abs} = \omega_\gamma (1+R) ,$$

with

$$R = \frac{\hbar\omega_\gamma}{2Mc^2} ,$$

and M the nuclear mass. The fractional shift 2R between emis-
sion and the absorption resonances is typically $\sim 10^{-5}$-10^{-6} ($2R\omega_\gamma$
~0.1-10 eV).

It should be possible to use laser saturation techniques to
enhance or suppress a narrow portion of the gamma ray Doppler
profile, and thus produce sharp gamma emission and absorption
resonances which can be tuned over the entire Doppler profile,
a tuning range which exceeds that of the Mössbauer effect by 1-2
orders of magnitude. As is well known, an intense monochromatic
laser field propagating in the $\hat{x}$ direction selectively saturates
a Doppler broadened atomic resonance over a narrow section Δv
of the $\hat{x}$-component of the velocity profile centered at v_0:

$$\Delta v = \gamma/k ,$$
$$v_0 = \Delta/k ,$$

with γ the radiative decay rate of the atomic transition, $k=\omega/c$
its wave number, and Δ the detuning of the laser from the atomic

554

center frequency ω. Consider now the gamma radiation emitted from such a system. The relevant energy levels and transitions are shown in Fig.1, in which $|0,0>$ and $|0,1>$ denote ground and excited electronic states of the atomic-nuclear system with the nucleus in its ground state, and $|1,0>$ and $|1,1>$ the corresponding electronic states with the nucleus excited. Since the coupling between nucleus and electrons is very weak, in a transition between states the atomic and nuclear quantum numbers cannot both change, and so there will be two atomic transitions and two gamma transitions, as indicated in the figure. If the laser field selectively interacts with the isomeric atomic transition a build-up of population in the $|1,1>$ state will occur over a narrow range of velocities, with a corresponding depletion in the $|1,0>$ velocity profile. Thus, viewed along the $\hat{x}$-axis the gamma rays emitted from the $|1,1>$ state will be centered at frequency

$$\omega_{em} = \omega_\gamma (1-R + v_0/c) = \omega_{em}^o + \frac{\omega_\gamma}{\omega} \Delta \qquad (1)$$

with a spectral linewidth

$$\Delta\omega \simeq \omega_\gamma \frac{\Delta v}{c} \simeq \frac{\omega_\gamma}{\omega}\gamma \quad . \qquad (2)$$

Narrow gamma absorption resonances can be similarly obtained if the laser interacts with the $|0,0> \rightarrow |0,1>$ transition. For typical allowed transitions $\gamma \sim 25$ MHz $\simeq 10^{-7}$ eV, giving $\Delta\omega \simeq 10^{-2}$ -10^{-3} eV (80-800 cm^{-1}); weaker transitions can be narrower by a factor of 1,000, giving correspondingly smaller values of $\Delta\omega$. Similarly, a narrow depletion will occur in the gamma-emission arising from the $|1,0>$ state.

Unfortunately, these narrow features are not ordinarily observable, since splitting between ground and isomeric atomic transitions is small ($\sim 10^{-5}-10^{-6}$ eV), and so in the net gamma emission from the isomeric atom the narrow gamma peak from the $|1,1>$ level will be compensated for by the depletion in the $|1,0>$ emission profile. To observe the narrow features requires changing the velocity distribution of one state, for example, by adding a buffer gas which preferentially induces velocity changing collisions in one state. Another approach would be to selectively photoionize and electrostatically (or otherwise) remove the $|1,1>$ excited state atoms. However, the laser induced nuclear orientation technique probably is of greater feasibility and interest. As discussed earlier, under suitable conditions laser optical pumping can lead to a high degree of orientation over a narrow range of velocities. Depending on the polarization state of the laser field (circular or linear), and the F values of the transition being optically pumped ($\Delta F = +1$, 0 or -1), population will be transferred either to $M_F = \pm F$ or $M_F \simeq 0$ ground state levels. Depending on the nuclear spin change in the subsequent gamma emission, the gamma ray angular distribution will be sharply peaked, either along the laser propagation direction or at right angles to it.

The combination of population changes in the M_F-states induced over a narrow velocity range with sharply peaked an-

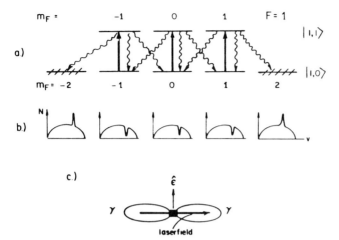

Fig.6 Anisotropic tunable gamma radiation induced by optical pumping. a) Laser pumping and radiative decay channels using linearly polarized light to pump an F=2→1 atomic transition. b) Velocity distributions of population in different M_F states of F=2 (lower) level. c) Schematic angular distribution of gamma radiation from the oriented atoms.

gular distribution patterns overcomes the overlap (compensation) problem and makes possible narrow tunable gamma resonances. For example, if the atomic transition is a $\Delta F = -1$ transition pumped by linearly polarized laser light (Fig.6), then within the narrow velocity band selected, population will be transferred to the $M_F=\pm F$ states from all the other M_F states. The gamma radiation

will be sharply peaked along the laser propagation direction. The center frequency and linewidth of this radiation are given by (1) and (2), respectively. The gamma radiation emitted from velocities outside this frequency band will be isotropic, since the M_F levels are equally populated at all other velocities. Hence the net gamma radiation spectrum in the forward direction will consist of a large narrow-frequency peak superimposed on a broad Doppler background.

With other laser polarizations, ΔF transitions, etc. one can similarly obtain a broad emission band with a narrow <u>depletion</u> occuring at a frequency corresponding to the resonant velocity group. Similarly, by optically pumping the ground state ($|0,0\rangle$ - $|0,1\rangle$) atomic transition, narrow gamma absorption resonances can also be induced.

<u>Observation of Narrow Gamma Resonances</u> Observation of narrow gamma resonances using conventional gamma ray techniques is likely to be difficult because of the absence of detectors of sufficient frequency resolution. A Mössbauer absorber of the appropriate frequency would be a suitable detector, but finding a frequency coincidence between a source of interest and a Mössbauer absorber may prove to be difficult. An absorbing sample of the same type as the source can be used if the recoil shift

is smaller than the Doppler width, so that source and absorber can be tuned to the same frequency. It should be noted, however, that direct detection of absorption is not feasible, because nuclear absorption cross sections are small and the gas density is low. Observation of conversion electrons or reradiated absorbed gammas emitted from the sides of the absorber may prove to be a useful means of detection, though the Compton effect and Rayleigh scattering will produce a large unavoidable background.

Discussions with Martin Deutsch and Thomas Kuhl are gratefully acknowledged. The ^{24m}Na experiments reported here are being done in collaboration with Hyatt Gibbs and Obert Wood of Bell Laboratories.

[1] D. E. Murnick and M. S. Feld, "Applications of Lasers to Nuclear Physics," Ann.Rev.Nucl.and Particle Sci. *29*(1979).

[2] M. Burns, P. Pappas, M.S. Feld, and D.E. Murnick, Nucl.Instruments and Methods *141*,429(1977).

[3] M. Burns, P. Pappas, M.S. Feld, and D.E. Murnick, in *Hyperfine Interactions IV*, edited by R.S. Raghaven and D.E. Murnick (North Holland, Amsterdam, 1978), p. 50.

[4] H.J.Kluge, in *Progress in Atomic Spectroscopy*, W. Hanle and H. Kleinpopen eds. (Plenum, New York, 1979), p. 727.

[5] U. Capeller and W. Mazurkewitz, J.Mag.Resonance *10*,15(1973).

[6] V.S. Letokhov, in *Frontiers in Laser Spectroscopy*, R. Balian, S. Haroche and S. Leiberman, eds. (N. Holland, Amsterdam, 1977), p. 713.

[7] R.Bernheim, *Optical Pumping* (Benjamin, New York, 1965); W.Happer, Revs.Mod.Phys. *44*,169(1972).

[8] D.E. Murnick, M.M. Burns, T.U. Kuhl, and P.G. Pappas, in *Laser Spectroscopy IV*, Proc.4th.Intern.Conf.,H. Walther and K.W. Rothe, eds. (Springer, New York, 1979); and references therein.

[9] G.W. Series and W. Gawlik, in *Laser Spectroscopy IV*, Proc.4th. Intern.Conf., H. Walther and K.W. Rothe, eds. (Springer, New York, 1979); and references therein.

[10] M.Pinard, C.G. Aminoff and F. Laloe, Phys.Rev.A, to be published.

[11] L.W. Anderson and G.A. Nimmo, Phys.Rev.*42*,1520(1979).

[12] G. Baum, C.G. Caldwell and W. Schroder, submitted to *Optics Communications* (1979).

[13] M.M.Burns, M. Deutsch, M.S. Feld, H.M. Gibbs, T.U. Kuhl, D.E. Murnick, P.G. Pappas, and O.R. Wood, to be published.

[14] L. Zamick, M.M. Burns, M.S. Feld, H.M. Gibbs, T.U. Kuhl, D.E. Murnick, P.G. Pappas, and O.R. Wood, to be published.

[15] Y.W. Chan, V.W. Cohen and M. Lipsicas, Phys.Rev.150,933(1966); V.W. Cohen, Bull.Am.Phys.Soc.18,727(1973).

[16] W. Buttler, et.al.D.P.G.(6)14,496(1979)

[17] G. Huber, F. Touchard, S. Buttgenback, C. Thibault, R. Klapisch, H.T. Duong, S. Liberman, J. Pinard, J.L. Vialle, P. Juncar and P. Jacquinot, Phys.Rev.C. 18,2342(1978).

[18] D.G. McCartan and J.M.Farr, J. Phys.B 9, 985(1976).

Information on Nuclear Quadrupole Moments from Laser-rf Double-Resonance Spectroscopy

S. Penselin

Institut für Angewandte Physik der Universität Bonn
D-5300 Bonn 1, Fed. Rep. of Germany

1. Introduction

The determination of nuclear quadrupole moments of stable nuclei
from the hyperfine structure (hfs) of atomic spectra is still a
challenge to atomic physics mainly for three reasons: 1) The
nuclear quadrupole moment Q cannot be determined directly from
hyperfine structure measurements, but only the quadrupole interaction constant B, which contains relativistic and configuration
interaction effects including the Sternheimer quadrupole shielding and antishielding. This limits in many cases the accuracy,
with which Q can be determined from such measurements, the uncertainty for Q beeing typically of the order of 10...20% or
even bigger. 2) There are some very rare isotopes, for which the
quadrupole interaction could not be measured in the past, because the resonance signals of all methods used so far were too
low for detection. 3) For some isotopes the hyperfine structure
could be measured only in atomic ground states, which have no
gradient of the electric field and thus have no quadrupole interaction. - For all three cases the laser-rf double-resonance spectroscopy with the ABMR-LIRF method (atomic beam magnetic resonance, detected by laser induced resonance fluorescence) has
brought considerable progress recently, which will be reported
here with three examples. The method itself, which was developed in our laboratory in Bonn by ERTMER and HOFER [1] and at
the same time independently for molecular beam magnetic resonance at London (Ontario) in Canada by ROSNER, HOLT and GAILY
[2], is described in detail in the next paper of this volume by
CHILDS, GOODMAN and POULSEN, and the description will therefore
be omitted here.

2. Sternheimer effect. The hfs of $3d^6 4s \; ^6D_{1/2}...9/2$ of ^{55}Mn and the nuclear quadrupole moment of ^{55}Mn [3]

The modern theory of hfs has developed, by means of the technique
of effective operators, possibilities to take into account the
effects of configuration interactions and relativity on hfs (see
e.g. [4]). The hamiltonian of the quadrupole interaction can
then be written [4] for electronic states of an atom with an
open shell of N electrons with angular momentum l as

$$\mathcal{H}(E2) = \frac{T_n^{(2)}}{eQ} \cdot \sum_{i=1}^{N} \quad b_1^{02} \sqrt{\frac{21(1+1)(21+1)}{(21-1)(21+3)}} \ U_i^{(02)2} \ + \tag{1}$$

$$+ \sqrt{3/10} \ b_1^{13} \ U_i^{(13)2} + \sqrt{3/10} \ b_1^{11} \ U_i^{(11)2}$$

where b_1^{02}, b_1^{13} and b_1^{11} are to be treated as free parameters. They are proportional to Q and $\langle r^{-3}\rangle^{k_sk_l}$. SANDARS and BECK have shown [5] that all three interactions contained in (1) are necessary to account for relativistic effects, whereas it can be seen from WYBOURNE [6] that in lowest order the presence of configuration interaction does not require the operators $U_i^{(13)2}$ or $U_i^{(11)2}$ but simply multiplies the matrix element of the first operator $U_i^{(02)2}$ by a constant which is independent of the state of the configuration and which is equivalent to Sternheimer's anti-shielding or shielding corrections. This offers the opportunity of determining the quadrupole moment Q from the two purely relativistic parameters b_1^{13} and b_1^{11} independently of the Sternheimer correction, if these two parameters can be determined experimentally with high enough accuracy and if the theoretical scheme is realistic.

In order to test these assumptions, DEMBCZYŃSKI, ERTMER, JOHANN, PENSELIN and STINNER measured the hfs of the metastable $3d^6 4s$ $^6D_{1/2}...9/2$ fine structure multiplet (about 17300 cm^{-1} above the ground state) of ^{55}Mn using the ABMR-LIRF method. The population of these fairly high lying metastable states was achieved mainly by electron impact from the electron beam which heated the atomic beam oven. The beam oven had its exit hole of 3 mm ∅ in the top and the electron beam was focussed through this hole from above. The hfs interaction constants A_J and B_J of all 5 fine structure states were measured. Adequate eigenfunctions for the states of the multiplet were derived after a very careful analysis of the available fine structure data. All 49 known states of the $3d^5 4s^2$, $3d^6 4s$ and $3d^7$ configurations were considered simultaneously and the varied parameters included those of spin-other-orbit interactions, electrostatically correlated spin-orbit interactions and electrostatic interactions with higher configurations and effective three body interactions. Using these eigenfunctions the following values for the parameters $b_1^{k_sk_l}$ could be derived from B_J's (in MHz):

$$b_{3d}^{02} = 229.883(28) \quad b_{3d}^{13} = 9.877(200) \quad b_{3d}^{11} = -4.375(100) \tag{2}$$

According to ARMSTRONG [7] and BAUCHE-ARNOULT [8] the effective radial parameters $\langle r^{-3}\rangle_{3d}^{k_sk_l}$ ($b_{3d}^{k_sk_l} = e^2/(h4\pi\varepsilon_o) \cdot Q \cdot \langle r^{-3}\rangle_{3d}^{k_sk_l}$ can be expressed as follows:

$$\langle r^{-3}\rangle_{3d}^{02} = \langle r^{-3}\rangle_{3d} \ R^{02} (1+\Delta_1^{02}+\Delta_2^{02}) \tag{3}$$

$$\langle r^{-3}\rangle_{3d}^{13} = \langle r^{-3}\rangle_{3d} \ R^{13} \tag{4}$$

$$\langle r^{-3}\rangle_{3d}^{11} = \langle r^{-3}\rangle_{3d} \ R^{11} \tag{5}$$

where R^{02}, R^{13} and R^{11} are relativistic corrections and Δ_1^{02} and Δ_2^{02} are corrections taking into account configuration interactions. Thus $(1+\Delta_1^{02}+\Delta_2^{02})$ is equivalent to the Sternheimer correction factor $(1-R)$. The parameter $<r^{-3}>_{3d}$ should be the same in all three equations. It was possible to deduct the value $<r^{-3}>_{3d} = 2.999(3)$ a_0^{-3} for this parameter from a similar analysis of the parameters $a_J^{k_s k_l}$ as determined from the experimental interaction constants A_J^{3d}. Extensive calculations of the relativistic correction factors (RCF) R^{02}, R^{13} and R^{11} have been made for many elements by LINDGREN and ROSÉN [9] using restricted relativistic Hartree-Fock wave functions, but unfortunately not for Mn. Therefore values for the RCF of the $3d^64s$ configuration of ^{55}Mn were obtained by interpolating between values calculated for the $3d^44s$ and $3d^84s$ configurations of the atoms ^{51}V and ^{59}Co respectively. As can be seen from Table 1, the ratio of the purely relativistic parameters b^{11} and b^{13} agrees very well with the ratio of the correction factors R^{11} and R^{13} within the limits of error, which is a good argument in favour of the relativistic corrections and the assumptions made. Table 1 shows also the same comparison for the two other possible ratios b^{11}/b^{02} and b^{13}/b^{02} with the corresponding R^{11}/R^{02} and R^{13}/R^{02},

Table 1 Experimental and theoretical ratios of the effective hyperfine parameters. The theoretical ratios are the ratios of the corresponding relativistic correction factors as calculated with restricted relativistic Hartree-Fock wave functions [9]

	b^{11}/b^{13}	b^{13}/b^{02}	b^{11}/b^{02}
experimental	-0.443(16)	0.0430(8)	-0.0190(40)
theoretical	-0.433	0.0437	-0.0189

and the agreement is again whithin the limits of error. This means, that in this case two corrections Δ_1^{02} and Δ_2^{02} cancel each other, resulting in $R = 0.00(2)$. Using the above mentioned value for $<r^{-3}>_{3d}$, one can calculated now the quadrupole moment Q from all three b-parameters with the results:

from b_{11}^{13}: $Q = 0.324(14)$ barns
from b_{02}: $Q = 0.332(8)$ barns
from b^{02}: $Q = 0.329(8)$ barns

Because all three values agree very well within the limits of error, the final result can be given as

$Q(^{55}Mn) = 0.33(1)$ barns.

In conclusion the results show, that the very detailed analysis of the hyperfine structure data with the effective operator theory is in very nice agreement with the experimental data, the most remarkable result being the determination of the nuclear quadrupole moment from the experimental purely relativistic parameters b^{13} and b^{11}, which are not affected by Sternheimer corrections.

3. Rare isotopes. The quadrupole moment of ^{50}V [10]

The isotope ^{50}V (I=6) is one of the very few rare stable odd-odd nuclei with natural abundance of only 0.25%. The only information on the ^{50}V nuclear ground state quadrupole moment obtained so far was the absolute value of the ratio $Q(^{50}V)/Q(^{51}V)$. It was derived from line widths of NMR signals to be about 1.3 [11]. Such NMR results are sometimes quite unreliable unless they are performed very carefully. Therefore ERTMER, JOHANN and MEISEL in our laboratory [10] made use of the very high sensitivity of the ABMR-LIRF method in order to detect the very weak rf signals of hfs transitions in the electronic ground state $3d^34s^2$ $^4F_{3/2}$ of the rare isotope ^{50}V. They were able to detect and measure these rf transitions by tuning the cw dye laser to the spin forbidden $3d^34s^2$ $^4F_{3/2}$ to $3d^34s4p$ $^6D^o_{7/2}$ transition at 5527.64 Å. Despite of this very weak transition a depopulation to about 40% could be achieved for the single hfs levels in the ground state. The resonance fluorescence was observed by the strong decays from $3d^34s4p$ $^6D^o_{7/2}$ to $3d^44s$ $^6D_{1/2}$ at 6258.57 Å and to $3d^44s$ $^6D_{3/2}$ at 6274.65 Å using appropriate filtering in front of the detector multiplier for discriminating against scattered laser light. Very careful shielding was necessary to suppress the oven stray light to a level that allowed observation of the very weak ^{50}V rf transitions. The measured hyperfine intervals at zero magnetic field after applying suitable corrections for the influences of hfs matrix elements non-diagonal in J and the corrected A- and B-factors of the state are the following:

$$\Delta\nu_{corr}(F = 9/2 \leftrightarrow F = 11/2) = 1177.669(20) \text{ MHz}$$

$$\Delta\nu_{corr}(F = 11/2 \leftrightarrow F = 13/2) = 1382.315(20) \text{ MHz}$$

$$\Delta\nu_{corr}(F = 13/2 \leftrightarrow F = 15/2) = 1582.223(20) \text{ MHz}$$

$$A_{corr} = 212.299(2) \text{ MHz} \qquad B_{corr} = -16.04(2) \text{ MHz}$$

Because similar measurements with the ABMR-LIRF method were performed recently by CHILDS, POULSEN, GOODMAN and CROSSWHITE on the 99.75% isotope ^{51}V [12], the corrected B-factor of +3.982(24) MHz for ^{51}V could be taken from their work in order to obtain the quadrupole moment ratio

$$Q(^{50}V)/Q(^{51}V) = B_{corr}(^{50}V)/B_{corr}(^{51}V) = -4.03(3).$$

This shows that the ratio obtained from NMR line widths was in error. From the quadrupole moment of ^{51}V $Q(^{51}V) = -0.052(10)$ barns as given in [12] one gets the following value for the quadrupole moment of the ^{50}V nucleus:

$$Q(^{50}V) = +0.21(4) \text{ barns.}$$

Since the $Q(^{51}V)$ reference value is uncorrected for the Sternheimer effect, the same applies to the $Q(^{50}V)$ value; therefore as usual a 20% error is attributed to the result to account for this fact.

4. Atoms with no quadrupole interaction in the ground state. The quadrupole moments of ^{95}Mo and ^{97}Mo [13]

The ground state of the Mo atom is $4d^5 5s\ ^7S_3$ and thus nominally spherically symmetric. Therefore it should have no quadrupole interaction. Despite of this fact, small but nevertheless very accurate B-factors have been measured for the two stable isotopes ^{95}Mo (I=5/2, 15.7%) and ^{97}Mo (I=5/2, 9.5%) with the atomic beam magnetic resonance (ABMR) method [14,15]. These B-factors are completely due to relativistic effects and deviations from LS coupling. Thus the results derived from these B-factors for the nuclear quadrupole moments and even for their ratio depend strongly on calculations of these effects and, in particular, on estimates of their dependability. Further results for the quadrupole moment ratio were found by KAUFMANN et al. with the NMR method [16,17] which can give only the absolute value since it is derived from relaxation time observations. The most recent NMR Q-ratios [17] were obtained from a very careful measurement of the spin-lattice relaxation times T_1 and assuming that quadrupole interaction is the main cause of this relaxation, in which case the Bloembergen theory can be applied [18]. As can be seen from Table 2, the results of the several measurements are not at all in agreement. Therefore DUBKE, JITSCHIN, MEISEL

Table 2   Results for the quadrupole moment ratio $Q(^{97}$Mo$)/Q(^{95}$Mo$)$ from different experiments

method	$Q(^{97}$Mo$)/Q(^{95}$Mo$)$	Ref.
ABMR	−2.0(4)	[14]
ABMR	−5.3(40)	[15]
NMR	± 9.2(9)	[16]
NMR	± 11.4(3)	[17]
ABMR-LIRF	− 11.50(3)	[13]

and CHILDS in our laboratory used the ABMR-LIRF method for a measurement of the hyperfine structure of the two isotopes in the metastable excited states $4d^4 5s^2\ ^5D_1$ and $^5D_2$ which are not spherically symmetric and should thus have a regular quadrupole interaction [13]. Despite of the fairly high excitation energy of about 11,3oo cm^{-1} the two states were so well populated by electron impact from the electrons heating the atomic beam source, that with the very high sensitivity of the ABMR-LIRF method the rf resonances could be detected and measured. The ABMR-LIRF method was simplified in this case in such a way that all three interaction regions of the atomic beam (two with the laser beam and one with the rf) coincided spatially. The results are shown in Table 3. The ratios of the quadrupole moments as determined from the two different fine structure states agree well within the limits of error, the average result for the ratio and for

Table 3 Results of the ABMR-LIRF measurements of the hfs of the states $4d^45s^2\ ^5D_{1,2}$ of ^{95}Mo and ^{97}Mo. The A- and B-factors are corrected for second order effects

	^{95}Mo	^{97}Mo
J=1:		
A_{corr}	-4o.812 (3) MHz	-41.671 (2) MHz
B_{corr}	- 2.960 (8) MHz	+34.o16 (6) MHz
$B^{97}_{corr}/B^{95}_{corr}\ =\ Q(^{97}\text{Mo})/Q(^{95}\text{Mo})\ =\ -11.49(3)$		
J=2:		
A_{corr}	-5o.6399(1o) MHz	-51.6991(16) MHz
B_{corr}	- 2.113 (7) MHz	+24.329 (11) MHz
$B^{97}_{corr}/B^{95}_{corr}\ =\ Q(^{97}\text{Mo})/Q(^{95}\text{Mo})\ =\ -11.51(4)$		

the quadrupole moments being:

$$Q(^{97}\text{Mo})/Q(^{95}\text{Mo})\ =\ -11.5o(3)$$
$$Q(^{95}\text{Mo})\quad\quad\ =\ -\ o.o15(4)\ \text{barns}$$
$$Q(^{97}\text{Mo})\quad\quad\ =\ +\ o.17\ (4)\ \text{barns}.$$

For the calculation of the quadrupole moments the necessary values for $<r^{-3}>$ have been taken from the analysis of the magnetic dipole interaction. The errors take into account the uncertainty of this analysis and the fact that a Sternheimer correction was not calculated and not taken into account.

5. Conclusion

All three examples show that the ABMR-LIRF method is a very powerfull tool for getting information on nuclear deformations. It makes possible hfs studies that were impossible in the past for experimental reasons.

The work was supported by the Deutsche Forschungsgemeinschaft.

6. References

1 W.Ertmer, B.Hofer: Z.Phys. A 276, 9 (1976)
2 S.D.Rosner, R.A.Holt, T.D.Gaily: Phys.Rev.Lett. 35, 785 (1975)
3 J.Dembcziński, W.Ertmer, U.Johann, S.Penselin, P.Stinner: Z.Physik, accepted for publication
4 W.J.Childs: Case Stud.Atom.Phys. 3, 215 (1973)

5   P.G.H.Sandars, J.Beck: Proc.Roy.Soc.(London) A 289, 97(1965)

6   B.G.Wybourne: "Spectroscopic Properties of Rare Earths", Interscience Publishers Inc., New York 1965

7   L.Armstrong jr.: "Theory of the Hyperfine Structure of Free Atoms", Wiley Interscience, New York 1971

8   C.Bauche-Arnoult: Journal de Physique 34, 3o1 (1973)

9   I.Lindgren, A.Rosén: Case Stud.Atom.Phys. 4, 25o (1974)

1o W.Ertmer, U.Johann, G.Meisel: Physics Letters B, accepted for publication

11 O.W.Howarth, R.E.Richards: Proc.Phys.Soc.(London) 84, 326 (1964)

12 W.J.Childs, O.Poulsen, L.S.Goodman, H.Crosswhite: Phys.Rev. A 19, 168 (1979)

13 M.Dubke, W.Jitschin, G.Meisel, W.J.Childs: Physics Letters A 65, 1o9 (1978)

14 J.M.Pendlebury, D.B.Ring: J.Phys. B 5, 386 (1972)

15 S.Büttgenbach, M.Herschel, G.Meisel, E.Schrödl, W.Witte, W.J.Childs: Z.Phys. 266, 271 (1974)

16 J.Kaufmann: Z.Physik 182, 217 (1964)

17 J.Kaufmann, J.Kronenbitter, A.Schwenk: Z.Physik A 274, 87 (1975)

18 N.Bloembergen: "Nuclear Magnetic Relaxation", Benjamin Inc., New York 1961

Laser-rf Double-Resonance Spectroscopy

W.J. Childs, L.S. Goodman, and O. Poulsen[1]

Argonne National Laboratory, Argonne, IL 60439, USA

1. Introduction

Although optical-microwave and optical-rf double-resonance experiments have
been performed for many years now, perhaps the first such experiment to
use a single-frequency, tunable dye laser as the source of the optical
photons was that of R. W. Field, et al. [1], who examined the molecular
structure of BaO vapor in 1973. Double resonance was detected as
microwave-dependent changes in the laser-induced fluorescence of the
vapor. The idea of substituting laser fields for the A and B magnets of
a conventional Rabi atomic-beam magnetic-resonance apparatus had occured
early in the laser era to many people, and its feasibility was demonstrated
by Schieder and Walther [2] in 1974. The first actual observation with
this arrangement of the laser-rf double-resonance technique was carried
out by Rosner, Holt and Gaily [3] in 1975 on a molecular beam of Na_2. The
technique was applied independently by Ertmer and Hofer [4], the same
year, to investigate the hyperfine structure (hfs) of metastable levels
of ^{45}Sc in a collimated atomic beam.

2. General Principles

The general features of the technique are easily understood from Fig. 1.
In the arrangement shown, a collimated, vertical atomic beam is produced
by electron-bombardment heating of an oven containing the source material.
If desired, a discharge can be maintained at the oven orifice to enhance
the population of metastable states too high to be populated thermally.
The light from a tunable, single-frequency dye laser ("pump" beam) is
arranged to intersect the atomic beam orthogonally. Because of the
virtual removal of the first-order Doppler effect, the tight collimation
of the atomic beam, and the extreme monochromaticity of the laser
(typically 1– 10 MHz), the light is resonant for only one hyperfine
component of the optical transition. In Fig. 1 (b), this is shown as
the transition $F \rightarrow F'$. Since the optical transition is allowed, the
upper state will decay immediately, and will normally drop back to a
different lower level than that from which it started. If it does
return to the original level, however, it will immediately absorb
another photon, and can in fact absorb and emit many times before it
crosses the laser beam. Thus after crossing the first ("pump") laser
beam, the number of atoms of the beam still in the original substate F

---

[1]Permanent address: Institute of Physics, University of Aarhus, DK-8000
Aarhus C, Denmark.

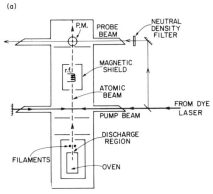

(a)

Fig. 1 (a) Schematic view of laser-rf double-resonance apparatus, and (b) energy-level diagram showing the levels and transitions involved. Note that in (b), the optical energy difference, $h\nu$, is about 10^6 larger than the hfs energy differences in either level alone

(b)

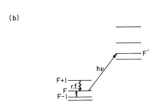

is normally very small. When the atomic beam is crossed by a second ("probe") laser beam of exactly the same wavelength as the pump beam, there are therefore very few atoms that can absorb the light, and consequently the fluorescence monitored from the "probe" region is very small. If, however, the depleted state can be repopulated in the region between the two laser interaction regions, strong fluorescence will be observed at the probe region. This repopulation can be achieved, as shown in Fig. 1 (b), by inducing radiofrequency transitions between adjacent hfs levels of the lower optical state. The transition probability for such rf transitions is normally very large, and an appreciable fraction of the atoms can be made to undergo resonance. Because the rf transition is essentially independent of whatever is going on in the two laser-atomic-beam interaction regions, the linewidth is not affected by the laser linewidth or the lifetime of the upper state of the optical transition. In addition, any Doppler effects are much smaller than for the optical line because the rf frequency is relatively so small, and the principal remaining sources of linewidth are Zeeman broadening and the uncertainty width caused by the transit time of the atoms through the rf region. The former effect can be reduced almost arbitrarily by careful magnetic shielding, and in any case at small field the broadening is symmetric about the line center. The transit-time linewidth can be made extremely small by using the well-known Ramsey [5] double-loop technique, with a very long loop separation if desired. Even with a single, conventional, 3-cm long rf region, linewidths of 10– 20 kHz are routine for rf frequencies up to 2 GHz. With the present geometry, however, it has been noted that the observed linewidth increases in proportion to the rf frequency, and this is consistent with a Doppler broadening. The observed width for rare-earth atoms at about 1600°C increases from about 15 kHz at 0– 2 GHz to about 70 kHz at 10 GHz. The rf geometry could certainly be altered to reduce this effect significantly.

It should perhaps be noted that although the double-resonance technique is very well suited for precise measurement of hfs splitting in a line, it cannot be used to measure isotope shifts.

3. Double-Resonance in Rare-Earth Atoms

Because of their extremely rich spectra, particularly in the Rh6G region, the rare-earth elements are a very attractive area for laser spectroscopy. The price to be paid for this richness, however, is the great difficulty in interpretation of results and of comparison with ab initio theory. A lot of atomic structure information can be obtained, however, particularly with regard to the effects of relativity and various types of configuration interaction such as core polarization and Sternheimer shielding on the hfs.

Our fist effort in the rare earths was on samarium, for which every optical line can be expected to exhibit complex hfs from both odd isotopes, and additional lines due to the several even isotopes. Figure 2 shows the fluorescence observed as the laser is scanned through the line λ5874.2 (the lower laser beam and the rf of Fig. 1 (a) are not used for this). The interpretation of the line, in terms of the initial and final values of F for each component, is given at the top of the figure. The weaker, unlabeled lines were identified as due to absorption from a highly excited state at 11,406 cm^{-1}, the interpretation for which is given in Fig. 3. The frequency markers are from a Fabry-Perot interferometer whose free spectral range is 300 MHz. The laser-rf double-resonance technique was then used to measure [6] the separations between the hfs components in these two lines, and for several others, to ±1 kHz, or about 2 ppm. The rf signals are large even for highly excited levels, and can easily be seen on a count-rate meter without any averaging. It is interesting to note that for the two lines of Fig. 2, the isotope shifts have opposite signs (mass numbers for the strong line are in large type, and those for the weaker line in smaller type).

We have also done double-resonance studies in Tb [7] and Pr, for which the interpretation of fluorescence patterns is much simplified by the fact that only one isotope occurs naturally. The number of fine structure levels is still immense, however, and since one often observes several lines per Angstrom, an interferometric lambda meter is almost a necessity to ensure that the line found is the one sought. As in the Sm, double-resonance was easy to observe in both Tb and Pr, and the lower half of Fig. 4 shows an example. The line λ6118.02 was known [8] to connect the 8250.22 cm^{-1}, 9/2- level with the 24590.84 cm^{-1}, 11/2+ level, and analysis of several laser scans through the optical line showed that the largest hfs interval (F = 7 ↔ 6) of the lower optical level was about 1494 ± 6 MHz. The laser was then set to be resonant with one of the optical absorption transitions from the F = 7 hfs level of the lower state (F = 6 could have been used instead), and the fluorescence was monitored as a function of the rf frequency applied, as shown in Fig. 1 (a). The resulting spectrum is shown in Fig. 4 (a). The resonant frequency is easily determined to ±1 kHz (better than 1 ppm), and curve fitting and longer

Fig. 2 Plot of fluorescence intensity observed as the exciting laser is swept through the line λ5874.2 of samarium. Two atomic lines are actually involved, as discussed in the text. The F-values that label the hfs components of the stronger line are shown above the lines. Mass numbers for the stronger line are shown in large type, and those for the weaker line in smaller type

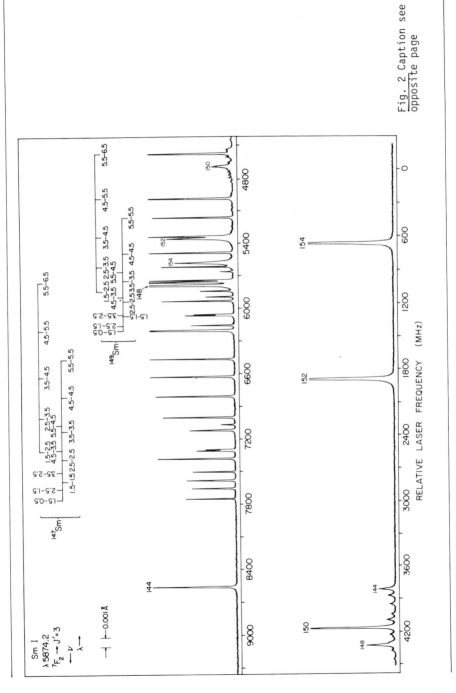

Fig. 2 Caption see opposite page

569

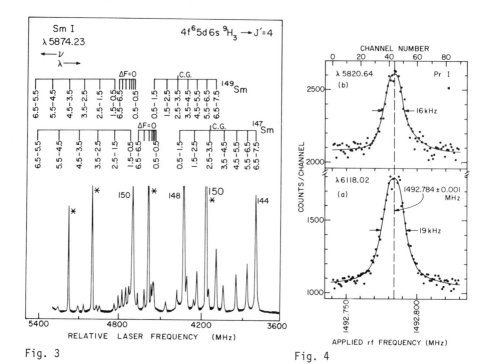

Fig. 3

Fig. 4

Fig. 3 Analysis of the weaker line of Fig. 2; it arises from absorption by a level at 11,406 cm^{-1}. The asterisks denote lines that are part of the strong line of Fig. 2. After obtaining rough values for the hfs splittings from fluorescence spectra such as Figs. 2 and 3, the laser-rf double-resonance technique was used to measure the separations to a few ppm

Fig. 4 Two measurements of the F = 7 ↔ 6 hfs intervals in the 8250.22 cm^{-1} metastable level of ^{141}Pr, obtained by laser-rf double resonance. Curve (a) was obtained using λ6118.02, and (b) with λ5820.64. The agreement of the measured splittings to better than 1 ppm demonstrates the great power of the double-resonance technique for identifying the lower levels of unassigned spectral lines

data collection could easily improve the precision if desired. Similar data have been taken on many metastable levels of Pr, and a systematic comparison of the results with the ab initio theory will be published when the work is complete.

Figure 4 (b) illustrates how the double-resonance technique, through high-precision measurement of hfs, can be used to identify the lower level of a transition uniquely. In this case, an optical line discovered accidentally while searching for another, was analyzed and found to originate from a lower level with the same J and very nearly the same hfs constants as for the lower level of λ6118.02, shown in Fig. 4 (a). To determine if the lower levels for the two transitions were the same. The F = 7 ↔ 6 hfs interval was measured for the lower level of the new

570

line, as shown in Fig. 4 (b). That the zero-field hfs intervals are identical to within 1 ppm is convincing evidence of the identity of the two lower levels. This technique has been found to be fast, certain, and widely applicable for identifying levels.

4. Double-Resonance in ^{235}U

The continuing, world-wide interest in accurately measuring the hyperfine intervals of ^{235}U has been stimulated recently by the concept of laser isotope separation. The double-resonance technique has been applied to the problem, and the results, [9,10] shown in Table I, are little more difficult to obtain than the fluorescence values, which are also shown.

The experimental values were analyzed with eigenvectors from H. Crosswhite [11] and relativistic Dirac-Slater radial expectation values from Lewis [12] to give values for the nuclear dipole and quadrupole moments. Although a great deal more experimental work could be done, the limitation now appears to be in the eigenvectors available.

Table 1 Hfs intervals measured by laser fluorescence and by laser-rf double-resonance in the ground and first excited levels of ^{235}U. The very high precision shown for the double-resonance technique is simple to achieve, and is by no means a limit

Atomic State (cm^{-1})	Hyperfine Interval F --- F'	Measured Values	
		By Laser Fluorescence (MHz)	By Laser-rf Double Resonance (MHz)
0.000	9.5 --- 8.5	816 ± 2	817.2177 ± 0.001
	8.5 --- 7.5	53 ± 2	51.536 ± 0.001
	7.5 --- 6.5	-486 ± 2	-487.5303 ± 0.001
	6.5 --- 5.5	-827 ± 2	-826.7115 ± 0.001
	5.5 --- 4.5	-992 ± 2	
	4.5 --- 3.5	-1009 ± 2	
	3.5 --- 2.5	-910 ± 2	
620.323	8.5 --- 7.5	-567 ± 2	-566.335 ± 0.001
	7.5 --- 6.5	-504 ± 2	-507.329 ± 0.001
	6.5 --- 5.5	-443 ± 2	-445.489 ± 0.001
	5.5 --- 4.5	-380 ± 2	-381.159 ± 0.001
	4.5 --- 3.5	-312 ± 2	-314.715 ± 0.001
	3.5 --- 2.5	-244 ± 2	
	2.5 --- 1.5	-178 ± 2	

Acknowledgements

One of the authors (O.P.) acknowledges partial support from the Danish
Natural Science Research Council. Work performed under the auspices of
the Division of Basic Energy Sciences of the U. S. Department of Energy.

References

1. R. W. Field, A. D. English, T. Tanaka, D. O. Harris, D. A. Jennings:
    J. Chem. Phys. 59, 2191 (1973).
2. R. Schieder, H. Walther: Z. Physik 270, 55 (1974).
3. S. D. Rosner, R. A. Holt, T. D. Gaily: Phys. Rev. Lett. 35, 785 (1975).
4. W. Ertmer, B. Hofer: Z. Physik A276, 9 (1976).
5. N. F. Ramsey: Phys. Rev. 76, 996 (1949).
6. W. J. Childs, O. Poulsen, L. S. Goodman: Phys. Rev. A19, 160 (1979).
7. W. J. Childs, L. S. Goodman: J. Opt. Soc. Am. (to be published).
8. W. F. Meggers, C. H. Corliss, B. F. Scribner: Tables of Spectral Line
    Intensities, Part 1, Natl. Bur. Stand. Monograph 145 (U.S. GPO,
    Washington, DC, 1975), p. 213.
9. W. J. Childs, O. Poulsen, L. S. Goodman: Opt. Lett. 4, 35 (1979).
10. W. J. Childs. O. Poulsen, L. S. Goodman: Opt. Lett. 4, 63 (1979).
11. H. Crosswhite, private communication, 1978.
12. B. Lewis, private communication, 1978.

cw VUV Generation and the Diagnostics of the Hydrogen Plasma

G.C. Bjorklund[1]

IBM Research Laboratory, San Jose, CA 95193, USA
and
R.R. Freeman

Bell Telephone Laboratories, Holmdel, NY 07733, USA

1. Introduction

For the last few years, one of the most interesting frontiers of nonlinear optics has been the vacuum ultraviolet spectral region. In this paper, we shall describe several nonlinear optics experiments concerning the generation of coherent VUV radiation or the excitation of VUV energy levels. These include the production of cw VUV radiation by four-wave mixing, hydrogen plasma diagnostics using resonant multiphoton ionization, and excitation of hydrogen Rydberg levels by three photon absorption.

2. Generation of CW VUV Coherent Radiation

We have utilized four-wave sum frequency mixing in Sr vapor to produce continuous wave vacuum ultraviolet radiation at several wavelengths near 1700Å.[1] By utilizing a 55 kG magnetic field to enhance the nonlinear susceptibility at the output frequency, approximately 10^7 photons/sec in a bandwidth of approximately 6 GHz were generated with cw fundamental beams of less than 1W power.

For four-wave mixing processes, the power P_4 generated at the sum frequency $\omega_4 = \omega_1 + \omega_2 + \omega_3$ obeys the proportionality relation[2]

$$P_4 \propto |\chi^{(3)}|^2 \, P_1 P_2 P_3 N^2 F(b\Delta k)$$

where P_1 is the power at frequency ω_i, N is the atomic number density, $\chi^{(3)}$ is the third order nonlinear susceptibility, and $F(b\Delta k)$ is a numerical function which describes focusing and phase matching effects. Our success in producing cw VUV was a result of adjusting conditions to produce the highest possible value for $\chi^{(3)}$. Phase matching was not utilized.

For nearly resonant sum frequency mixing, $\chi^{(3)}$ obeys the relation[3]

$$\chi^{(3)}(\omega_4) \propto \sum_{ijk} \frac{\mu_{gi} \, \mu_{ij} \, \mu_{jk} \, \mu_{kg}}{(\Omega_{ig} - \omega_1 - i\Gamma_i)(\Omega_{jg} - \omega_1 - \omega_2 - i\Gamma_j)(\Omega_{kg} - \omega_4 - i\Gamma_k)}$$

---

[1]The experimental work was performed when G.C. Bjorklund was at Bell Laboratories.

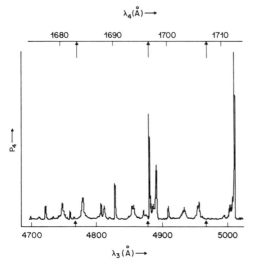

Fig. 1 Pulsed laser measurement of the susceptibility of Sr for the process $\omega_1+\omega_2+\omega_3\to\omega_4$ with $\omega_1+\omega_2$ constrained to the $5s^2$-$5s7s$ two photon resonant channel.

where μ is the dipole matrix element and i,j and k represent excited energy eigenstates of the atom with energies above the ground state g given by $\hbar\Omega$, and widths given by Γ. The value of $\chi^{(3)}$ was optimized by tuning the input lasers to utilize the $5s2(^1S_0)\to5s7s(^1S_0)$ two photon resonant channel. In previous work,[4] this channel had been shown to provide the strongest coupling (largest value of μ_{jk}) to the Sr autoionizing levels near 1700Å. Additional resonant enhancement of $\chi(3)$ was provided by utilizing an external magnetic field to shift the positions of the autoionizing levels and thus tune Ω_{kg} nearly into resonance with ω_4.[5]

The input fundamental beams were provided by an Ar^+ cw laser operating multiline and a cw Rd6G dye laser. The ω_1 beam was the fixed frequency Ar^+ laser line at 4764Å, while the ω_2 beam was provided by the dye laser and set at 5724Å so that $\omega_1+\omega_2$ was resonant with the 38444.1 cm^{-1} $5s^2$-$5s7s$ two photon transition. The ω_3 beam was one of several fixed-frequency Ar^+ laser lines and thus several values of ω_4 were generated. Each of the Ar^+ lines had a bandwidth of approximately 6 GHz and the dye laser was operated multimode with a 30 GHz bandwidth. An overall scan of $|\chi(3)|^2$ for Sr near λ_4=1700Å using the 5s7s two-photon channel is shown in Fig. 1. This data was experimentally determined using pulsed lasers.[4] The peaks in this scan are due to resonant enhancements of the susceptibility due to autoionizing levels.[6] The arrows mark the positions of the three fixed wavelengths λ_4 generated using the Ar^+ laser lines at 4764, 4880 and 4964Å to provide λ_3. In all cases, these fixed frequencies are not coincident with any of the large resonances in $\chi(3)$, although the 4880Å Ar^+ line produces a λ_4 which is quite close to the large resonance at λ_4=1697.03Å.

Figure 2 shows a high-resolution scan of the susceptibility near 1697Å. The arrow shows the position of the 4880Å laser line; clearly the susceptibility could be increased considerably if the autoionizing resonance could be shifted to be coincident with the laser line. Because the ground $5s^2$ and intermediate 5s7s states have J=0, all of the autoionizing states in Figs. 1 and 2 have J=1, and thus have three magnetic sublevels, M_J=0,±1. The paramagnetic energy shift of an M_J level in a magnetic field of magnitude H_0 is given by $g_J\mu_0M H_0$, where the magnetic

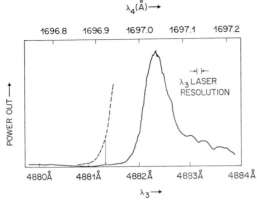

Fig. 2 High-resolution pulsed laser scan of the susceptibility between λ_3=4880 and 4884Å. The dotted line shows the calculated position of the M=+1 component in a 55 kG magnetic field.

moment in state J, $g_J\mu$ is equal to 1.4 MHz/G. The dotted curve on Fig. 2 shows the calculated position of the contribution to the susceptibility from the M_J=+1 sublevel for H_0=55 kG.

The experimental apparatus consisted of a Sr vapor cell contained in the room-temperature bore of a superconducting solenoid which was capable of producing 55±0.2 kG over the generation volume. The Sr cell had a 6 cm hot zone and provided Sr vapor number densities on the order of 5×10^{15} cm^{-3}. The cw input beams were linearly polarized, collinearly combined, and focused in the center of the cell with a confocal parameter of 0.09 cm. The generated output radiation was separated from the fundamental beams with a VUV bandpass filter and then passed through a VUV spectrometer to a solar blind PMT.

Figure 3 shows a spectrometer sweep of the generated cw radiation. The peaks all have a width given by the spectrometer resolution. In Fig. 3(b), the magnetic field is zero and the heights of the peaks reflect the relative strengths of the susceptibilities and laser powers at $\lambda_3(\lambda_4)$ of 4764 (1683Å), 4880 (1697Å), and 4964 (1707Å). Figure 3(a) shows an

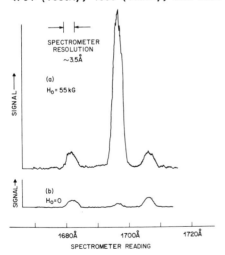

Fig. 3 VUV spectrometer scans of the generated output. (a) H_0=55 kG. (b) H_0=0G.

identical sweep, except that the magnetic field is 55 kG. Although the outputs at 1683 and 1707Å increased only slightly, the output at 1697Å increased by a factor of almost 20. Under these conditions, an output at 1697Å of 5×10^{-11}W or 4×10^{7} photons/sec was achieved with input powers of 0.85W at 4880Å, 0.425W at 4764Å, and 0.66W at 5724Å. Because of the narrow two-photon resonance, the bandwidth of the generated VUV is that of the λ_3 laser or about 6 GHz.

These results could be substantially improved by utilizing single axial mode fundamental lasers, a stronger magnetic field, and phase matching. A cw VUV bandwidth of less than 1 MHz could be achieved using narrow bandwidth input lasers. From Fig. 2, it can be seen that a factor of 5 enhancement in the generated power could be gained with a stronger magnetic field. Based upon demonstrated results in pulsed laser experiments,[4] an additional factor of 100 enhancement of the generated power would be expected if phase matching techniques were utilized.

3. Hydrogen Detection and Hydrogen Plasma Diagnostics

About 1 year ago, we demonstrated and reported in the literature the first application of resonantly enhanced multiphoton ionization spectroscopy to the detection of ground state hydrogen and deuterium.[7] In addition to providing a sensitive means of detection with good spatial and temporal resolution, this ionization technique provided a means of selective, laser induced ionization of the total H or D population contained within focal volumes as large as 2×10^{-4} cm^3. We then embarked upon a series of experiments which exploited this ionization capability to probe and perturb hydrogen discharge plasmas. Strong multiphoton optogalvanic[8] and laser-induced striation[9] effects in these plasmas were observed.

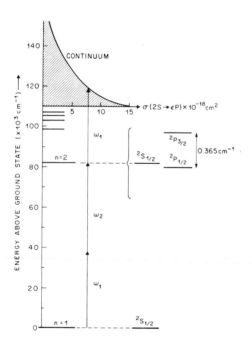

Fig. 4 Level diagram for H atom. Vertical arrows indicate photons absorbed in the ionization process. The cross section for transitions out of the $2^2S_{1/2}$ level to the ϵP continuum is also plotted (horizontal axis).

A schematic of the ionization process and the relevant energy levels is shown in Fig. 4. Ionization is produced by absorption of three photons, two at frequency ω_1 and one at frequency ω_2. The ω_1 radiation is always at a fixed wavelength of 2660Å while the ω_2 radiation is tunable about 2240Å. Resonant enhancement is achieved by tuning ω_2 such that $\omega_1 + \omega_2$ is resonant with the $1^2S_{1/2} - 2^2S_{1/2}$ two-photon transition at 82259 cm^{-1} in H or 82281 cm^{-1} in D. Since neither $2\omega_1$ or $2\omega_2$ is resonant with an intermediate level, significant ionization can occur only in the simultaneous presence of both beams.

In all of these experiments, the 2660Å ω_1 radiation was produced as the fourth harmonic of a commercial Q-switched 10-pps Nd:YAG laser and had a 6 nsec pulse duration, up to 2 MW peak power and a 0.3 cm^{-1} bandwidth. The tunable 2240Å ω_2 radiation was produced by pumping a dye laser amplifier with the Nd:YAG second harmonic to yield 300 kW of tunable 5660Å radiation, frequency doubling this radiation in angle tuned KDP to yield 50 kW of tunable 2830Å radiation, and then summing the 2830Å radiation with the 1.064 μm Nd:YAG fundamental in a second angle turned KDP crystal. The ω_2 radiation had a 4-nsec pulse duration, up to 20 kW peak power, and a 0.3 cm^{-1} bandwidth. It should be noted that with state of the art dye laser systems, an order of magnitude greater 2240Å power can be produced.

The multiphoton optogalvanic effects were observed using the apparatus shown in Fig. 5. A 15 cm long, 5 mm-i.d. Pyrex dc discharge tube containing a flowing mixture of He and H$_2$ or D$_2$ was utilized. The two quartz windows were attached to the sides of the tube to allow orthogonal illumination of the discharge. The discharge current varied between 30 and 100 mA and the voltage drop across the electrodes was about 500V. The He partial pressure was about 5 Torr and the H$_2$ or D$_2$ partial pressures varied between 0.1-2 Torr. Under these conditions, total dissociation of the H$_2$ or D$_2$ in the discharge was always achieved.

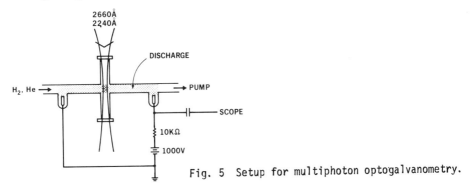

Fig. 5 Setup for multiphoton optogalvanometry.

The copropagating 2660Å and 2240Å laser beams were focused by a 25 cm lens to a point in the positive column of the discharge halfway between the electrodes. The transient optogalvanic signal due to the ~5 nsec laser pulses was detected through a dc blocking capacitor across the 10-kΩ discharge ballast resistor and was displayed directly on an oscilloscope with a differential amplifier plug-in. The signal had a broad asymmetric shape of ~5 μsec duration. At full laser power in both beams (2 MW at 2660Å and 20 kW at 2240Å tuned to the line center of the 1S to 2S transition) coincident on a focal volume of 10^{-3} cm^3 in the positive column, the galvanic signal due to three-photon ionization from the ground state was 20V for an H$_2$ or D$_2$ partial pressure of 2 Torr. Although the focal volume was small compared to the 12 cm^3 volume of the discharge, the large signal observed is consistent with the estimation that $>10^{12}$ electron-ion pairs per laser pulse were produced in a plasma with an electron density of $\sim10^9$ cm^{-3}.

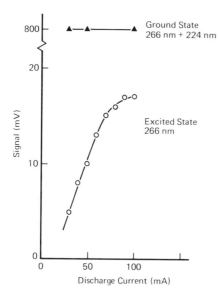

Fig. 6 Ionization signal vs. discharge current for illumination with both 2660Å and 2240Å beams and with the 2660Å beam only.

The optogalvanic signal was measured as a function of discharge current under conditions of illumination with both 2660Å and 2240Å beams and with the 2660Å beam alone. The results are shown in Fig. 6. The large signal obtained when both beams are present results from the three-photon ionization of the ground state H or D atoms. Since the fractional excited state and ion population produced in the discharge was small and since the dissociation of the H$_2$ or D$_2$ molecules was always complete, the ground state H or D population and hence the signal was constant as the discharge current was varied. The smaller signal obtained with just the 2660Å beam resulted from single photon ionization of the excited state atoms in the discharge. This excited state population and hence the signal is proportional to the discharge current.

The laser-induced striations were observed using the apparatus shown in Fig. 7. The discharge parameters were similar to those employed in the optogalvanic experiments and a 25 cm lens was again used to focus the copropagating 2660Å and 2240Å beams to a point in the positive column halfway between the electrodes. The spatial distribution of sidelight fluorescence along the path of the discharge was monitored by scanning the image of the positive column across the 150 μm entrance slit of a spectrometer set at one of the Balmer-line transitions. By utilizing time-gated detection, the transient sidelight fluorescence due to the laser excitation was easily observed against the cw background light of

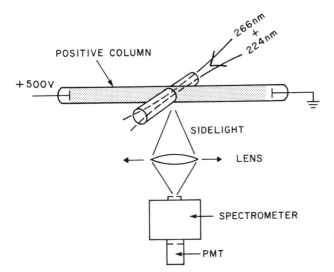

Fig. 7 Setup for observing laser-induced striations.

the discharge. This change in sidelight fluorescence is due to the recombination and de-excitation of the atomic hydrogen following the laser-pulse.

Figure 8 shows typical data taken by scanning along the length of the positive column and maintaining a constant 7 µsec delay from the laser pulse. The periodic structure observed was a standing-wave pattern that decayed with a 15 µsec lifetime. The location of the largest peak in the figure corresponds to the position of the region of intersection of the 0.4 mm wide focused laser beams. The effect of the laser photoionization

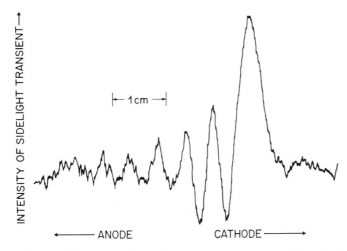

Fig. 8 Intensity of the transient change in sidelight fluorescence vs. position along the length of the positive column. The region of intersection with the laser beams is at the center of the large peak.

579

is strongly evident several centimeters along the discharge toward the anode and the signal shows the spatially periodic structure indicative of plasma striations.

In conclusion, resonantly enhanced three-photon ionization provides a convenient means of probing or perturbing hydrogen plasmas. The laser wavelengths employed pass easily through air, quartz windows, and dense hydrogen plasmas. By utilizing intersecting 2660Å and 2240Å beams, the laser induced ionization can be produced in a well defined region of three-dimensional space at an arbitary location within the plasma. Since the laser induced ionization takes place in 5 nsec, it should be possible to study the effects of these perturbations on a very fast time scale.

4. Selective Excitation of Rydberg Levels in Atomic Hydrogen by Three Photon Absorption

We have recently applied multiphoton spectroscopy techniques to the selective excitation of Rydberg levels in atomic hydrogen with values of n, the principal quantum number, as high as 35.[10] The atomic H was contained in a cell and the excitation was achieved directly from the ground state by absorption of three equal energy laser photons. An important feature of this excitation scheme is that as n increases, the cross section for ionization of the Rydberg atom in state n by the incident laser radiation falls off rapidly, allowing the chosen energy level to remain populated after the end of the laser pulse. This behavior is in marked contrast to the case of the previous section, where multiphoton excitation of a lower lying energy level in H led to complete ionization.

Figure 9 shows the energy level scheme of atomic H and indicates the multiphoton process. Each H Rydberg level is n-fold degenerate with respect to ℓ, the angular momentum quantum number, and the selection rules indicate that only the $\ell=1$ and $\ell=3$ quantum states are excited. Coverage of the range between n=10 and n=∞ requires a laser tunable between 2763Å and 2736Å. The Rydberg atom is subsequently collisionally ionized and the resulting ionization provides a convenient indicator of the excitation of a Rydberg level. Our assertion that the ionization of the Rydberg atoms by the UV laser is negligible is based on cross-sections calculated from the tables of Peach.[11] For n>10, the cross section for ionization of the Rydberg atom is less than 10^{-20} cm^2, which yields an ionization probability of less than 0.075 for typical optical power densities of 10^{10} W/cm^2 and 5 nsec laser pulses.

The tunable UV radiation was produced by frequency summing the output of a tunable dye laser with the second harmonic of a Nd:YAG laser. Over 2 mJ of linearly polarized UV radiation tunable between 2730Å and 2760Å with 5 nsec pulse duration and 0.3 cm^{-1} bandwidth was obtained. In order to cover the entire UV wavelength range in a single scan, the angular orientation of the frequency summing KDP crystal was automatically servo-controlled to maintain optimum phase matching as the dye laser was tuned.[12] The ground state atomic H was produced by dissociation of 2 Torr of H_2 gas flowing through a cw microwave discharge and the interaction with the focused laser beam took place in the immediate afterglow region of the discharge. The ionization current was collected by two planar metal electrodes separated by 2.5 cm and positioned to bracket the interaction region. The dc potential between the electrodes was adjusted

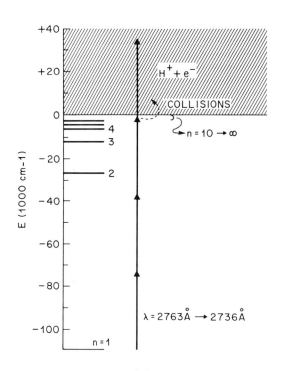

Fig. 9 Energy levels of
atomic hydrogen. The
solid arrows indicate
photons absorbed in the
three photon transition.
The Rydberg atoms are
predominantly ionized by
collisional mechanisms.

Fig. 10 Experimental
ionization signal as a function of UV laser wavelength.

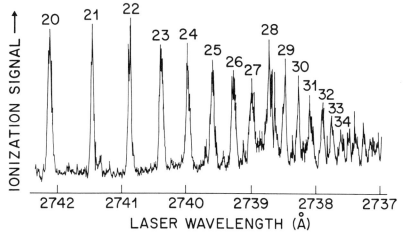

to values around 250V, near the breakdown limit, in order to produce gain
and the ionization signal was detected as a transient voltage drop across
a 200 kΩ resistor connected in series with the voltage source.

A typical high resolution spectrum between n=16 and n=35 is shown in
Fig. 10. The focused power density was 5×10^8 watts/cm^2. The resonances
in the figure are labeled by the value of n, the principal quantum number.
The full width at half maximum of each of the resonances is approximately

0.057Å of tuning of the laser, which corresponds to an equivalent resolution of 2.27 cm^{-1} for mapping out the H Rydberg states. This value is close to the resolution limit corresponding to the convolution of three laser linewidths of 0.3 cm^{-1} each with the 1.36 cm^{-1} Doppler width at room temperature. Reduction of the focused laser intensity did not improve this resolution, although a marked broadening of the Rydberg resonances, caused by level shifts due to the AC Stark effect, was observed if the laser intensity was increased. From the figure, it can be seen that for n between 20 and 25, the Rydberg levels are very well resolved, and that the limit of resolution corresponds to n=37. We attribute the spurious structures to resonances arising from multiphoton excitation or ionization of H_2 molecules.

Thus, Rydberg levels in atomic H can be selectively excited by three photon transitions from the ground state. Laser power densities on the order of 10^{10} W/cm^2, which are sufficient to produce easily detectable excitation signals and to cause large level shifts, will not destroy the Rydberg atoms by photo-ionization. Interesting future experiments would be to explicitly demonstrate the production of long-lived Rydberg atoms in an atomic beam environment and to study the Stark broadening phenomena with a single axial mode UV laser.

REFERENCES

1. R. R. Freeman, G. C. Bjorklund, N. P. Economou, P. F. Liao and J. E. Bjorkholm, Appl. Phys. Lett. 33, 739 (1978).

2. G. C. Bjorklund, IEEE J. Quantum Electron. QE-11, 287 (1975).

3. J. A. Armstrong, N. Bloembergen, J. Ducuing and P. S. Pershan, Phys. Rev. 127, 1918 (1962).

4. G. C. Bjorklund, J. E. Bjorkholm, R. R. Freeman and P. F. Liao, Appl. Phys. Lett. 31, 330 (1977).

5. N. P. Economou, R. R. Freeman and G. C. Bjorklund, Optics Letters 3, 209 (1978).

6. P. P. Sorokin, J. J. Wynne, J. A. Armstrong and R. T. Hodgson, Ann. New York Acad. Sci. 267, 30 (1976).

7. G. C. Bjorklund, C. P. Ausschnitt, R. R. Freeman and R. H. Storz, Appl. Phys. Lett. 33, 54 (1978).

8. C. P. Ausschnitt, G. C. Bjorklund and R. R. Freeman, Appl. Phys. Lett. 33, 851 (1978).

9. C. P. Ausschnitt and G. C. Bjorklund, Optics Lett. 4, 4 (1979).

10. G. C. Bjorklund, R. R. Freeman and R. H. Storz, to be published.

11. G. Peach, Mem. R. Astron. Soc. 71, 13 (1967).

12. G. C. Bjorklund and R. H. Storz, IEEE Journ. Quant. Elect. QE-15, 228 (1979).

Part IX

Laser Sources

Tunable Transition-Metal-Doped Solid State Lasers[1]

P.F. Moulton and A. Mooradian

Lincoln Laboratory, Massachusetts Institute of Technology
Lexington, MA 02173, USA

Transition-metal-doped solid state lasers appear to be quite useful as broadly tunable and powerful sources of coherent infrared radiation. The electronic energy levels of transition ions such as Ni, Co, V, and Fe in such hosts as MgF_2, MgO, ZnF_2, MnF_2 and $KMnF_3$ are strongly coupled to the lattice and have broad vibronic sidebands in both the absorption and emission spectra. The first demonstration of laser action on the vibronic sidebands of the electronic levels of divalent Ni impurity ions in MgF_2 was by L. F. JOHNSON [1] and coworkers in 1963. In the initial studies [2] of the Ni:MgF_2 laser, tuning over discontinuous segments in the range 1.62 to 1.84 μm was demonstrated using a lamp-pumped crystal cooled by flowing liquid oxygen. The output of this laser consisted of short spikes, the result of undamped relaxation oscillations.

This paper describes the demonstration of cw Ni:MgF_2 and Co:MgF_2 lasers which are continuously tunable from 1.61 to 1.74 μm and 1.63 to 2.08 μm, respectively, and have true cw, spike-free outputs. The use of a cw Nd:YAG laser as a pump source has considerably simplified cooling of the laser crystals. In addition the Ni:MgF_2 laser has been Q-switched to demonstrate large power enhancements, indicating the potential of this laser for use with nonlinear optical methods to efficiently generate a variety of infrared wavelengths. Such nonlinear techniques will be discussed and it will be shown how tunable coherent sources covering the entire near- and mid-infrared wavelength region can be constructed based on the use of transition-metal lasers. Finally, the operation of a 6 W cw Ni:MgO laser, another tunable infrared source, will be discussed briefly.

Figure 1 shows the absorption and fluorescence spectrum at 77 K resulting from the $^3A_{2g} \rightarrow \, ^3T_{2g}$ transition of Ni^{+2} in MgF_2. The purely electronic transitions, or zero-phonon lines, common to both absorption and emission, are evident in the 1.5 μm region. Vibronic, or phonon-assisted transitions appear also with a Stokes shift between absorption and fluorescence characteristic of phonon-assisted processes. The spectra for Co:MgF_2 are similar in nature. Because of this shift, four-level laser operation is possible in the phonon-assisted fluorescence wavelength region. Since the transitions are magnetic dipole in nature, the fluorescence lifetimes are long, 12.8 ms for Ni:MgF_2 and 1.3 ms for Co:MgF_2 at 77 K, which permits large enhancements in peak power in the cw-pumped, Q-switched mode of operation.

In this experimental work a cw Nd:YAG laser operating at 1.32 μm has been used to optically pump the Ni:MgF_2 and Co:MgF_2 lasers. Such an arrangement allows simple conduction cooling of the laser crystal on a liquid nitrogen cooled copper heat sink. The use of laser excitation at 1.32 μm also allows precise control of the optically pumped volume and minimizes heating in the crystal. Operation with

---

[1]This work was sponsored by the Department of the Air Force.

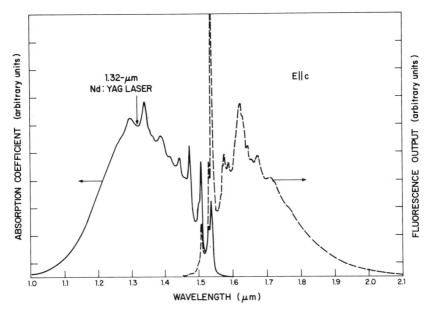

Fig. 1   Absorption and fluorescence spectra for Ni:MgF$_2$ at 77 K.

a 1.9 cm-long laser crystal of Ni:MgF$_2$ with laser mirrors formed directly on the end faces of the crystal has been reported [3]. Such a device generated cw power outputs as high as 1.7 W with a maximum conversion of absorbed pump power to laser output power of 37%.

The Ni:MgF$_2$ and Co:MgF$_2$ lasers described here are external cavity devices employing 1.2 cm-long, Brewster-angle crystals in a 3-mirror cavity shown in Fig. 2. The laser crystals were grown by a vertical-gradient-freeze technique described previously [4]. Two of the three mirrors of the laser cavity are inside the vacuum section of a liquid-nitrogen Dewar. The angle between the two arms of the cavity was chosen such that the astigmatism introduced by the laser crystal is cancelled by that of the spherical mirror [5]. The birefringent tuning element shown is a single plate of crystalline quartz, 1.4 mm thick, oriented with the "c" axis 57$^{\rm o}$ away from the surface normal. The tuning element was replaced in Q-switching experiments with a 2.5 cm long, Brewster-angle-cut crystal of LiNbO$_3$ with electrodes transverse to the direction of optical propagation.

The TEM$_{00}$ output powers vs wavelength of the three-mirror cavity lasers are shown in Figs. 3 and 4 for the Ni:MgF$_2$ and Co:MgF$_2$ lasers, respectively. The shapes of the tuning curves reflect structure in the vibronic emission spectra, possibly the effects of excited-state absorption and in the case of the Co:MgF$_2$ laser the presence of water vapor absorption in the laser cavity. The temporal output of the Ni:MgF$_2$ laser at 1.67 μm in response to a chopped pump laser input is shown in Fig. 5. The initial relaxation oscillations caused by the sudden onset of pumping damp out completely so that true cw output is obtained. The absence of continuous spiking in the output is the result of the stable optical pumping obtained with a laser excitation source and also the stable temperature conditions realized with conduction cooling.

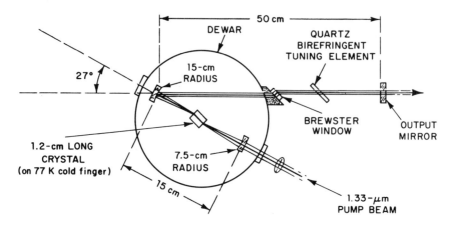

Fig. 2 Schematic diagram of three-mirror cavity used for laser experiments.

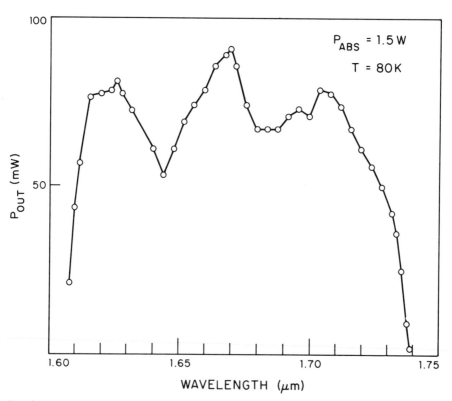

Fig. 3 Tuning curve from Ni:MgF$_2$ laser for 1.5 W of absorbed pump power.

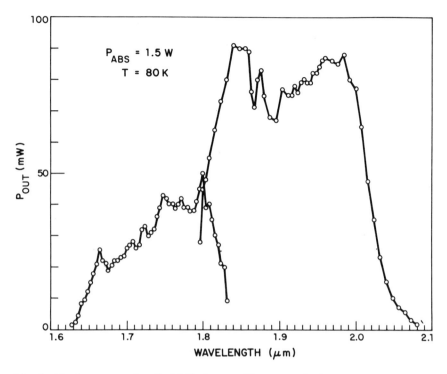

Fig. 4 Tuning curves for Co:MgF$_2$ laser. Two sets of mirrors were
required to cover the full tuning range. The output coupling for the short
wavelength set was less optimal than for the other set, hence the generally
lower output power at short wavelengths.

In Q-switching of the Ni:MgF$_2$ laser peak power outputs of 140 W were observed
in a 480 ns pulse at repetition rates of up to 100 Hz. Above that rate the peak power
dropped, decreasing to 40 W at 1 kHz. The 70 μJ energy of the pulses at low repetition rates corresponds to an output energy extraction from the laser medium of
about 1 J/cm^3. Larger excited volumes in larger laser crystals could reasonably be
expected to generate Q-switched pulses with energies in the 10-100 mJ range.

The longitudinal mode spectrum of the Ni:MgF$_2$ laser was observed using a
scanning confocal interferometer. Operation was typically on 3-5 longitudinal modes
contained in a 1.5 GHz (0.05 cm^{-1}) region. A single-frequency output of 20 mW was
obtained at 1.67 μm by placing two uncoated, tilted quartz etalons in the cavity. The
ease in obtaining single frequency operation and the few number of modes observed
without etalons both indicate that the Ni:MgF$_2$ laser transition is homogeneously
broadened.

There are a number of other transition-metal-doped crystals which have been or
could be used as laser media. Crystals in which laser operation was obtained in
early research [2] include Ni:MnF$_2$ and Co:ZnF$_2$, for which oscillation was observed
in the 1.9 and 2.15 μm wavelength regions, respectively. In addition, it is likely
that a Ni:ZnF$_2$ laser could be made to operate at around 1.7 μm.

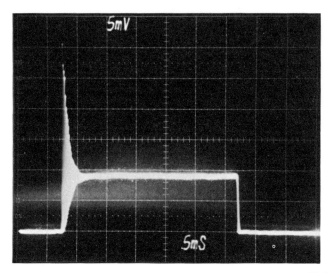

Fig. 5 Fully damped relaxation oscillations in output of Ni:MgF$_2$ laser, in response to a mechanically chopped pump laser input.

It is interesting to consider the use of transition-metal-doped lasers as pump sources for nonlinear processes. Figure 6 shows the infrared wavelengths generated with sources tuning in the 1.6 to 2.2 μm region. The two top lines indicate the spans of the first Stokes wavelengths from stimulated Raman scattering in H$_2$ and D$_2$ gas. A calculation of the threshold for stimulated scattering in H$_2$ gas at 20 atm pressure in a multipass cell [6] yields a value of 0.2 MW at pump and Stokes wavelengths of 2.0 and 11.8 μm, respectively. Though operation in H$_2$ gas out to wavelengths as long as 26 μm is possible in principle, thresholds would be high because of both the longer wavelengths and the pressure induced rotational absorption in H$_2$ gas [7]. The next three lines show the wavelengths generated by mixing in nonlinear crystals with common Nd:YAG laser lines. It is evident from Fig. 6 that the entire near- and middle-infrared wavelength region could be covered using several transition-metal lasers combined with nonlinear optical techniques.

Cw laser operation from Ni:MgO has recently been obtained [8]. The preliminary results are reported here to indicate the power output capabilities of transition-metal lasers. 6 W of output was observed at 1.32 μm from a 1.85 cm-long crystal, conduction-cooled by liquid nitrogen and optically pumped by the 1.06 μm line from a Nd:YAG laser. 2.5 W of power was generated in the TEM$_{00}$ mode and temperature

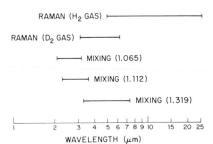

Fig. 6 Wavelengths generated by stimulated Raman scattering and mixing with common Nd:YAG laser lines, for lasers operating in the 1.6 to 2.2 μm wavelength region.

tuning produced operation in discontinuous segments from 1.316 to 1.409 μm. It is believed that equivalent power output performance should be possible from the other transition-metal-doped lasers mentioned in this paper.

References

1. L. F. Johnson, R. E. Dietz and H. J. Guggenheim, Phys. Rev. Lett. 11, 318 (1963).
2. L. F. Johnson, H. J. Guggenheim and R. A. Thomas, Phys. Rev. 149, 179 (1966).
3. P. F. Moulton, A. Mooradian and T. B. Reed, Opt. Lett. 3, 164 (1978).
4. T. B. Reed, R. E. Fahey and P. F. Moulton, J. Cryst. Growth 42, 569 (1977).
5. H. Kogelnik, E. P. Ippen, A. Dienes and C. V. Shank, IEEE J. Quantum Electron. QE-8, 373 (1972).
6. R. L. Byer and W. R. Trutna, Opt. Lett. 3, 144 (1978).
7. D. M. Bosomworth and H. P. Gush, Can. J. Phys. 43, 751 (1965).
8. P. F. Moulton, A. Mooradian, Y. Chen and M. M. Abraham (to be published).

New Methods for the Generation of Ultrashort Dye Laser Pulses

F.P. Schäfer

Max-Planck-Institut für biophysikalische Chemie, Abteilung Laserphysik
D-3400 Göttingen, Fed. Rep. of Germany

Two novel methods for the generation of tunable ultrashort dye laser pulses were developed recently in our laboratory. Both methods are somewhat unconventional and universally applicable and seem capable of considerable further development, although at the present state only pulsewidths around 50 ps have been obtained consistently.

The first method was developed by Dr. Szolt BOR of the Hungarian Academy of Sciences during his stay as a guest at our laboratory and exploits a self-Q-switching mechanism in distributed-feedback dye lasers [1-3]. The experimental arrangement is shown in Fig. 1.

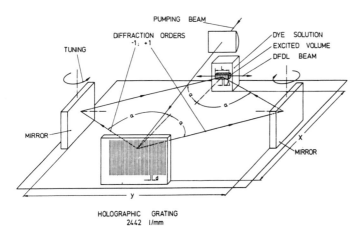

Fig. 1. Pumping arrangement of the distributed-feedback dye laser (from Ref. 3)

The pumping beam of a nitrogen laser is directed onto a holographic grating with 2442 lines/mm at very nearly normal incidence. This beam is then diffracted into the +1 and −1 orders at an angle α determined by the grating equation $\sin\alpha = \lambda_p/d$, where d is the grating constant. The diffracted beams are reflected from two mirrors that are aligned perpendicular to the grating surface and parallel to each other and to the grating

grooves. The focal line of the cylindrical lens which is used
to focus the beam on the dye cell is perpendicular to the grooves
of the grating. The dye cell itself is slightly tilted to avoid
feedback from the end windows. The two beams interfere at the
dye cell at a half angle α, resulting in a fringe spacing
$\Lambda = \lambda_p/2\sin\alpha$, which together with the grating equation gives
$\Lambda = d/2$. This means that the fringe spacing is independent of
the pump wavelength so that each spectral component of a braod-
band laser, for example a nitrogen laser, creates the same interference pattern. Moreover, one can show that even the low spatial
coherence of such a superradiant laser does not affect the visibility of the interference fringes if the distance between the
mirrors and the distance between grating and dye cell fulfill
a certain geometrical relation so that for each point of the
dye cell the two interfering beams have been diffracted from the
same point of the grating.

We now have a 100 % modulation of the inversion in the interference pattern and realize that the feedback increases with the
inversion and it goes down again as the inversion is depleted
after the start of the laser oscillations. By this self-Q-switching action the output pulses of the DFB-dye laser can become
very short as shown by the computer solution of the corresponding
rate equations for a typical case of pumping with a 4 ns nitrogen
laser pulse high above threshold. Several output pulses with
halfwidths between 33 and 90 ps are being generated in the case
of a $6.6 \cdot 10^{-3}$ M solution of rhodamine 6G in ethanol. This prediction was experimentally tested with a streak camera by
Prof. MÜLLER in our lab and Fig. 2 shows a typical streak camera
record. One observes good agreement between experiment and the
computer solution of the rate equation. If one now decreases
the pump power to about 20 % above threshold only one single
pulse is generated. When pump power, length of the pumped region
and concentration are optimized, single pulses of about 40 ps
halfwidth and 2 kW peak power are consistently obtained.

The spectral linewidth of the DFB-dye laser under conditions
when single pulses of 80 - 100 ps were generated was typically
0.05 to 0.06 Å, so that the time-bandwidth product was less
than 0.6, meaning that nearly transform-limited pulses were
observed.

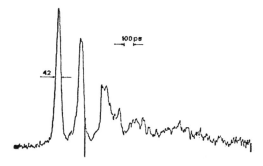

Fig. 2. Streak camera trace
of the distributed-feedback
dye laser pulse using a
$6.6 \cdot 10^{-3}$ M rhodamine 6G
solution in ethanol

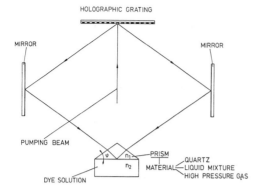

Fig. 3. Tuning of the distributed-feedback dye laser with a prism attached to the dye cell. The prism alters the angle of interference and the relative refractive index of the solvent (from Ref. 3)

When tuning the laser by any of the many possibilities offering themselves no mode hopping is observed but rather a smooth continuous tuning. The reasons for this could be that, firstly, the pulse length is too short for mode build-up and, secondly, there is a gradual transition between pumped and unpumped region instead of the sharp boundary conditions that are generally assumed in theory.

One well-known possibility for tuning a DFB-dye laser is the change of the refractive index of the solvent. Since the refractive index is pressure-dependent, one can easily fine-tune over a 10 - 20 Å range by applying pressures of up to 100 bar. This method is very convenient and without safety problems, since only the small volume of the dye cell is pressurized.

Another method that introduces an only slight chromatism tilts the two mirrors by an angle δ in opposite directions about a vertical axis, so that the angle of interference is changed. Linear tuning over 20 Å can easily be achieved by piezo-electric rotators.

Similarly, one can tune this laser by translating the grating or the mirrors, which is especially useful for fine-tuning.

Still another method, shown in Fig. 3, makes use of an additional prism attached to the dye cell to alter the angle of interference. Laser emission at 452 nm was observed with a quartz-prism with angle $\phi = 20°$ and a $3 \cdot 10^{-3}$ M solution of 7-diethyl-amino-4-methylcoumarin in ethanol. If the prism is constructed in the form of a liquid or high pressure gas cell, then continuous tuning is possible. There are several additional possibilities for stepwise tuning over a very wide range, which will not be discussed here.

At present we are trying to reduce the pulse length by using an atmospheric pressure nitrogen laser of 100 ps pulsewidth that should result in a DFB-laser pulse of less than 10 ps. We will then use an amplifier cell pumped by the same nitrogen laser to amplify the single, tunable ultrashort pulse to MW-peak powers at repetition rates of up to 100 Hz. This will then be a very versatile and economic tool for many problems in ps spectroscopy.

The second method to be discussed here was developed by
Jan JASNY of the Polish Academy of Sciences while he was a guest
in our lab, together with Jay JETHWA and the author. It relies
entirely on a novel type of Michelson interferometer that JASNY
had developed for a fast Fourier spectrometer [4,5].

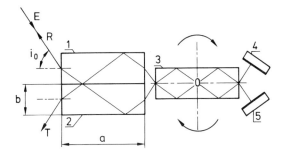

Fig. 4. Optical system
of the Jasny inter-
ferometer. 1,2,3 quartz
rectangular parallelepi-
peds, 4,5 mirrors (from
Ref. 4)

A schematic diagram of this interferometer is shown in Fig. 4.
It consists of three identical blocks of quartzglass with four
polished sides and two mirrors. The two blocks no. 1 and no. 2
are put together, but kept apart by a small ring of spacers of
108 nm thickness evaporated along the edges of the face. Block
no. 3 is mounted on the shaft of a motor whose axis of rotation
passes through the point marked O and is normal to the plane of
the screen. A light beam E entering block no. 1 under the
Brewster angle is split into two coherent beams of very nearly
equal intensity by frustrated total reflection at the air layer
between the two blocks. The two beams travel through the three
blocks as indicated, are reflected at mirrors 4 and 5 and, after
retracing their paths back to the beam splitting layer, combine
to form a reflected beam R and a transmitted beam T. When block 3
is rotated, one of the light paths inside it is shortened, the
other one lengthened by the same amount. Accordingly, at a constant speed of rotation the intensities of the transmitted and
reflected beams are sinusoidally modulated so that a minimum of
intensity in one beam coincides with a maximum in the other. The
modulation rate depends on the mechanical rate of rotation and
on the wavelength. For e. g. at a wavelength of 600 nm a mechanical rotation rate of 6000 rpm results in a modulation frequency
of 207 MHz. The modulation depth is 100 % at 580 nm and decreases
only slightly towards shorter and longer wavelengths due to the
slight chromaticity of the beam splitter.

The actual construction of the interferometer as it was built
in our optical workshop is shown in Fig. 5. The orientation of
the rotating block is uncritical because one can easily see that
all alignment errors cancel out because of the double pass of
each light beam through this block. Because of this we could
mount the block elastically on the shaft of a small servo-controlled DC-motor that could be run up to 10.000 rpm. However,
all the other parts of the interferometer must be kept in rigid
alignment, and so all parts were put together by optical con-
tacting, using a base plate 7, a spacer 6 and the mirrors 4

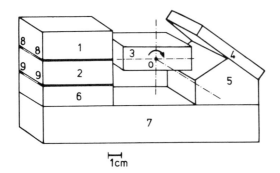

Fig. 5. Details of the construction of the Jasny interferometer. 1 through 5 as in Fig. 4, 6 spacer, 7 base plate. Parts 4 through 7 are made from Zerodur®. O and arrow indicate axis and sense of rotation, resp.

and 5 made of Zerodur®. An additional spacer of 1 μm evaporated along the edges of block 6 keeps the necessary distance to block 2 so that total internal reflection can occur in block 2. The whole interferometer is once aligned during assembly and then needs no further attention by the user.

In order to exploit the high light modulation frequencies possible with this interferometer for the generation of ultra-short tunable pulses, we used it instead of the totally reflect-ing mirror in a dye laser, as shown in Fig. 6. We used a small flashlamp-pumped dye laser and an outcoupling mirror of 94 % reflectivity on one end and the interferometer on the other end of the resonator. The length of the resonator resulted in a round trip time of 4.4 ns.

Trigger pulses for the flashlamps were derived from a photo-diode behind a stationary slit that was illuminated by an in-candescent lamp via a small concave mirror mounted on the motor shaft together with the rotating block.

The operation of this laser can be explained most easily in the following way. Let us assume that there exists a steady state with an ultrashort pulse travelling back and forth in the reso-nator. If the pulse arrives at the interferometer end when the rotating quartz block is in a position to give maximum reflecti-vity at the wavelength of the pulse, then it will be reflected back without significant losses since all glass-air interfaces are entered near the Brewster angle. If, after one round trip the pulse returns to the interferometer and that has in the mean-

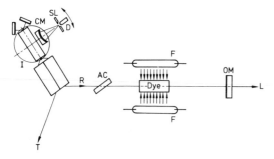

Fig. 6. Dye laser with Jasny interferometer re-placing the totally re-flecting mirror. CM con-cave mirror, SL slit, D diode, AC absorber cell (optional), F flashlamps, OM output mirror, R reflected beam, T trans-mitted beam (from Ref.4)

time moved to the next position of maximum reflectivity, then it is again reflected without losses. It is immediately clear that in this way for a definite wavelength and rotation rate a stable pulse train is generated at the output end. If some pulse would come before or after the maximum reflectivity positions of the interferometer it would be attenuated at every reflection and soon die out. If it had a somewhat different wavelength, it would also soon die out, since most of the reflections would occur at low reflectivity positions. One sees that for a given rotation rate spectral and temporal narrowing occurs that quickly leads to a train of ultrashort pulses of a wavelength determined only by the mechanical rotation rate.

It is obvious that the pulsewidths are shortest near threshold and that high above threshold multiple pulsing could occur.

50 ps halfwidth pulses were obtained most of the time near threshold and often pulses limited by the resolution of the streak camera of 5 ps, especially when firing the laser head single shot with long intervals between shots. Multiple pulsing high above threshold could be suppressed by the insertion of a cuvette with saturable absorber solution in the resonator. But it must be stressed that this absorber is not needed as long as one is only operating the laser near threshold.

In a $2 \cdot 10^{-4}$ M solution of rhodamine 6G in ethanol the laser wavelength could be tuned from 585 nm to 616 nm by changing the period of rotation from 9300 μs to 8800 μs. The bandwidth in this range was determined with a Fabry-Perot and found to be about 1 $\overset{\circ}{A}$ averaged over the whole pulse train. We have not yet switched out single pulses to see whether there is a frequency sweep during the development of the pulse train. The bandwidth was slightly increased when a saturable absorber was introduced in the resonator. Some experiments were also made with a ring-laser set-up, using the same laser head and two dielectric mirrors of 100 % and 96 % reflectivity, respectively, at two corners of a triangular resonator, whilst the interferometer was used in the transmission mode instead of the third mirror. The tuning characteristics and pulse widths were similar to the case of the linear resonator. In addition, one could easily weaken the pulses travelling in one direction while at the same time strengthen the pulses travelling in the opposite direction by judicious positioning of the laser head in the ring resonator. We have recently repeated most of these experiments in the blue region with coumarin 2 and obtained essentially similar results. Experiments with a cw dye laser which would avoid the thermal problems of flashlamp pumped dye lasers and should result in the reproducible generation of pulses of only a few ps halfwidth when operated near threshold are in progress.

A short remark is indicated here on another application of the Jasny interferometer that has nothing to do with the ultra-short pulses but should be of general interest. Since the Jasny interferometer is essentially equivalent to a linear Michelson interferometer with two moving mirrors applied recently by several groups together with a reference laser as direct reading wavelength meter, one can easily see that the Jasny interferometer could also be used in such an application with a number of advantages over the conventional Michelson. The useful angular range of 18° allows the measurement of more than 50 000 fringes at 600 nm resulting in a theoretical limit (with the usual 1/100 fringe detection) of better than $2 \cdot 10^{-7}$ accuracy which is sufficient for many applications provided the dispersion of the quartz-glass used is known with sufficient precision. The interferometer is very compact and rugged and insensitive to temperature changes. In addition, it allows very rapid repetitive measurements. We are presently building in our electronics workshop phase-locked loops for 3.4 MHz and 340 MHz which should result in a frequency measurement time of about 15 ms and six independent measurements per second at only 80 rpm. The present speed limit is determined by the electronic circuitry, since the intrinsic frequency limit of the mechanical and optical parts is at least a factor of 100 higher than this.

References

1. Zs. Bor: IEEE J. Quant. Electron. (accepted for publication)
2. Zs. Bor: Appl. Phys. 19, 39 (1979)
3. Zs. Bor: Opt. Commun. 29, 103 (1979)
4. J. Jasny, J. Jethwa and F. P. Schäfer: Opt. Commun. 27, 426 (1978)
5. J. Jasny, Polish Patent Appl. No. P-204729 from 17 Feb. 1978

Optically Pumped Molecular Lasers

B. Wellegehausen and H. Welling

Institut für Angewandte Physik, Universität Hannover
D-3000 Hannover, Fed. Rep. of Germany

Introduction

The vapors of many elements and mixtures of elements partly
consist of molecules, mostly diatomic molecules such as Na_2,
NaK, Te_2, I_2. These simple molecules, which exist in a stable
equilibrium with the atoms, are interesting candidates for the
realization of new laser systems in the ultraviolet or visible spectral region. Considering bound -bound electronic transitions, multiline step-tunable lasers can be obtained, and
on bound-free transitions also continuously tunable, excimer-
like laser systems appear feasible.

To explore the laser potentials of these materials, optical
pumping by means of suitable pump lasers is a very powerful
method prior to the application of the more complex discharge
or e-beam excitation. In this way, in the past two years laser pumped pulsed or cw oscillation has been achieved for a
number of homonuclear diatomic (dimer) molecules such as Li_2,
Na_2, K_2, Bi_2, S_2, Te_2, I_2. In this paper we will briefly review some basic facts and data of these dimer lasers and outline new developments and future possibilities. More detailed descriptions of individual dimer lasers and applications
performed with dimer systems are given in the Refs. 1-6 and
especially in a forthcoming review paper [7].

Principles of dimer lasers

A dimer laser consists of a suitable molecule cell in a cavity,
which basically is very similar to that used for cw dye lasers.
Fig. 1 shows a set-up with a linear resonator and a Brewster
prism to allow laser oscillation on individual lines. The
molecule cell may either be a sealed off glass or quartz
cell or conveniently a heatpipe, especially for the metal
vapors. A typical excitation – emission cycle between bound
electronic states of a diatomic molecule is shown schematically in Fig. 2. A pump laser with frequency ω_p excites
molecules from a certain rot.-vibr. level 1 of the electronic ground state X into a specific rot.-vibr. level 3 of an
excited electronic state A. Under suitable conditions population inversion can be achieved between this level 3 and
rot.-vibr. levels 2, 2', 2'' .. of the electronic ground state.
Laser oscillation is therefore possible on many lines all

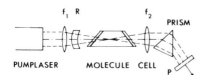

Fig. 1 Dimer laser set-up

PUMPLASER   MOLECULE CELL

belonging to the same fluorescence series. The laser cycles are finally closed by collisional relaxation processes (dashed arrows) as radiative transitions between rot.-vibr. levels of the same electronic state are not allowed in homonuclear diatomic molecules.

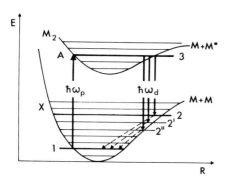

Fig. 2 Energy level diagram of a dimer molecule M_2, including dimer laser excitation-emission cycle

The most important features of these lasers are:
1. Multiline emission over broad spectral ranges, depending on the structure of the potential curves.

2. High amplification and low thresholds, due to strong transition moments and in most cases a negligible population density in the lower laser levels.

3. Good power conversion capabilities, as the energy of the pump and laser photons is nearly equal. To achieve high output powers, a lot of parameters such as length of vapor zone, temperature, buffer gas pressure have to be optimized.

4. Three level laser system with coupled Doppler broadened transitions. Due to the coupling of the pump and laser transitions by the common level, resonant two photon or Raman type processes contribute to the amplification. This results in strong forward - backward amplification and lineshape asymmetries. As a consequence, in a ring laser system pumped by a single frequency pump laser, unidirectional single frequency dimer laser oscillation is obtained without any unidirectional device or frequency selecting element. Furtheron, the dimer laser frequency simply can be tuned within the Doppler profile by tuning the pump laser frequency [6].

General dimer laser data

Fig. 3 gives a survey of the spectral ranges of cw dimer laser lines obtained from dimer systems. The ranges for the various systems are more or less densely covered with lines. In case of Li_2 and Na_2 laser emission from two molecular bands is obtained using argon and krypton lasers as a pump source. The K_2 system is pumped with the 647 nm line of the krypton laser and in all other cases argon laser lines are used for the excitation.

Typical operation conditions and some important data of the individual dimer laser systems are summarized in Table 1. Only the Na_2 (B-band), Bi_2 and I_2 systems have been investigated and optimized in more detail and these systems are well operating, powerful lasers. Also for Li_2, S_2 and Te_2 much better power data and efficiencies should be possible, whereas for K_2, at least for the 647 nm excitation wavelength, strong improvements are not expected due to intrinsic losses [7]. Except for the 488 nm excitation of Na_2, all data given in Table 1 were determined from lasers with linear resonators. Due to the mentioned asymmetry effects, in general ring resonators are better suited, and therefore further improvements of the output power data, especially for laser oscillation on individual lines, are expected with systematic ring laser investigations. With the presently operating dimer systems almost the whole visible and very near infrared spectral range can be covered with lines just by excitation with some argon laser lines. As the laser materials are stable and inexpensive and the laser systems easy to operate, dimer lasers may replace dye lasers in all cases where only

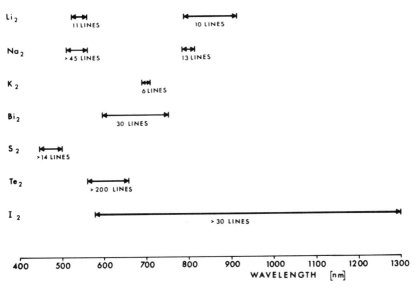

Fig. 3 Spectral ranges of cw dimer lasers excited by argon and krypton laser lines

Table 1 Data of cw dimer lasers

Molecule	Pump		Output Power [mW] Multiline	Observed Gain [b) [cm^{-1}]	Typical Threshold [mW]	Operation Conditions Remarks
	Line [nm]	Power [W]				
Li$_2$	476.5	1	15	0.05	< 100	Heatpipe ~1300 K, vapor zone 10 cm
	647	1	30			
Na$_2$	472.7	1	70			Heatpipe 650 - 850 K vapor zones 3 - 10 cm lowest thresholds < 1 mW single line-single frequency powers up to 200 mW buffer gas pressures up to 150 mbar
	488 [a)]	2.5	250	0.2	< 20	
	647	1	20			
K$_2$	647	1	3	-	< 200	Heatpipe ~ 600 K, vapor zone 20 cm
Bi$_2$	514.5[a)]	3.5	350	0.01	< 200	Heatpipe ~1200-1300 K, vapor zone 20 cm buffer gas pressures up to 150 mbar
S$_2$	363.8	2	3	0.005	< 600	Cell temperature 1100-1400 K reservoir temperature 400-500 K vapor zone 18 cm
Te$_2$	476.5[a)]	1	20	0.008	< 20	Cell 800-900 K, vapor zone 18 cm high density of lines >300(560-660 nm)
I$_2$	514.5[a)]	3.5	420	0.002	<100	Cell ~300 K, vapor zone 50-100 cm strong quenching by buffer gases (< 2 mbar) broad emission range 580-1340 nm

a) Single mode
b) determined from maximum output coupling

lines in certain spectral regions are needed, such as for frequency chains, wavelength standards, calibration purposes. Furthermore, these dimer systems are especially suitable for special spectroscopic investigations on the dimer molecules themselves [4,5].

New developments and future possibilities

Optically pumped lasers with homonuclear diatomic molecules seem to be sufficiently understood, although some open questions remain, especially regarding splitting effects of the forward-backward amplification profiles under the influence of strong pump and dimer laser fields [8].

To increase the variety of laser materials, also heteronuclear dimers or even molecules with more than two atoms can be considered. A laser cycle in these molecules should be similar to the one described in Fig. 2, however, a different relaxation behaviour in the lower laser level and additional relaxation losses in the upper laser level are expected, as radiative transitions between adjacent rot.-vibr. levels are now possible. As a tendency, laser oscillation in these molecules will be more difficult to achieve, also due to a greater complexity of the molecular spectra. Of special interest for laser purposes will obviously be the consideration of bound-free transitions with continuous emission spectra. Such spectra can be observed by excitation of stable molecules, for example Na$_2$, NaK [9-11], as well as by excitation of the excimer molecules such as Mg$_2$, Ca$_2$, Sr$_2$, which only have a weakly bound electronic ground state [12,13].

Experimental results

1. Among the heteronuclear dimers, the interhalogenes are of some interest. In first experiments with ICl vapor, cw laser oscillation was obtained on some lines in the range of 570 nm to 630 nm by excitation with the 514.5 nm line of the argon laser. However, as this excitation energy is slightly above the dissociation limit of the $B^3\pi$ state, there is presently some doubt whether this emission really results from ICl. Regarding the optimum temperature for laser oscillation of about $-20^\circ C$ and the emission wavelengths, oscillation of I_2, however, can be excluded. Further experiments with longer excitation wavelengths are in preparation.

2. Due to recent spectroscopic investigations [12], calcium seems to be a good candidate for laser investigations on bound-free transitions. Surprisingly, strong superfluorescent laser emission was found on very many lines in the spectral range of 480 - 674 nm, upon excitation of Ca vapor with a pulsed dye laser at some discrete wavelengths in the range of 450 - 480 nm. However, with the avaible spectroscopic data of Ca, Ca_2, the origin of these lines presently cannot be explained.

3. Intense continuous fluorescence bands have recently been observed following uv excitation of Na_2 molecules into higher electronic states [9,10]. We found similar diffuse bands in the emission spectra of Li_2 and K_2 and there seem to be good chances to realize laser oscillation in some of these bands. By excitation with a strong pulsed excimer laser pumped uv dye laser in the spectral range around 334 nm, we observed weak superfluorescent laser emission from sodium vapor in a narrow band around 460 nm and in addition and unexpectedly, strong broadband (5 nm) laser emission in the red wings of the sodium resonance lines around 591 nm [14]. While the blue emission is roughly within the range of the diffuse bands and may belong to a bound-free transition, the mechanism for the broadband yellow emission is not quite understood. The sodium resonance levels are strongly populated by dissociative transitions and the yellow emission may result from weakly bound molecular states of the B-band. This emission is presently also investigated towards the possibility to realize alkali-noble gas excimer systems.

References

1. H. Welling and B. Wellegehausen, Laser Spectroscopy III, Springer Series in Optical Sciences 7, (Springer Verlag, Berlin, 1977), p. 365
2. B. Wellegehausen , K.H. Stephan, D. Friede and H. Welling, Opt. Comm. 23, 157 (1977)
3. B. Wellegehausen, D. Friede and G. Steger, Opt. Comm. 26, 391 (1978)
4. J.B. Koffend, S. Goldstein, R. Bacis, R.W. Field and S. Ezekiel, Phys. Rev. Lett. 41, 1040 (1978)
5. J.B. Koffend, R. Bacis and R.W. Field, J. Mol. Spectrosc., to be published.
6. B. Wellegehausen and H.H. Heitmann, Appl. Phys. Lett. 34, 44 (1979)
7. B. Wellegehausen, IEEE J. Quantum Electron., to be published Oct. 1979

8. S. Ezekiel, private communication
9. J.P. Woerdman, Opt. Comm. <u>26</u>, 216 (1978)
10. M. Allegrini, G. Alzetta, <u>A</u>. Kopystynska, L. Moi and
    G. Orriols, Opt. Comm. <u>22</u>, 329 (1977)
11. E.J. Breford and F. Engelke, Chem. Phys. Lett. <u>53</u>, 282 (1978)
12. K. Sakurai and H.P. Broida, J. Chem. Phys. <u>65</u>, 1138 (1976)
13. C.R. Vidal and H. Scheingraber, J.Mol. Spectr. <u>65</u>, 46 (1977)
14. D. Friede, R. Bhatnagar, B. Wellegehausen, to be published.

Nd: YAG Pumped Tunable Sources and Applications to $\chi^{(3)}$ Spectroscopy

R.B. Byer

Applied Physics Department, Stanford University
Stanford, CA 94305, USA

1. Introduction

The successful application of the unstable resonator concept [1] to the
Nd:YAG laser [2,3] has led to significant improvements in Nd:YAG laser
pumped tunable sources. The properties of the unstable resonator Nd:YAG
source, important for pumping high peak power tunable sources, are briefly
reviewed. The Nd:YAG source advantages of high peak power and reliability
are illustrated by its application to stimulated Raman and four wave mixing
studies and to polarization CARS spectroscopy.

Nd:YAG Performance

The unstable resonator Nd:YAG laser source was first described by HERBST
et.al. [2] A primary goal of the design was to obtain a high power low
divergence uniphase output beam that could be efficiently converted in angle
phasematched KD*P crystals or by Stimulated Raman Scattering (SRS). The
positive branch confocal unstable resonator cavity, with appropriate compensation for the Nd:YAG rod thermal focussing, achieved the desired goal.
Table I lists typical output pulse energies at the fundamental and Nd:YAG
harmonics for the unstable resonator oscillator/amplifier system.

TABLE I Unstable resonator Nd:YAG source

Repetition Rate 10 Hz	Lamp Energy (J)	Output Energy (J)	Peak Power (MW)	Aver. Power (W)
Oscillator (6 mm dia)	70	0.25	31	2.50
Amplifier (9 mm dia)	80	0.75	93	7.50
Second Harmonic KD*P		0.25	35	2.5
Third Harmonic KD*P		.100	16	1.0
Fourth Harmonic KD*P		.06	8	.6
Fifth Harmonic KDP		.01	1.3	.1

The collimated output of the Nd:YAG source is efficiently doubled in angle
phasematched KD*P crystals. Figure 1 shows typical harmonic output energies
for both Type I and Type II phasematching. The recent discovery of 90°
phasematching for 1.064 μm plus .2660 μm in KDP at −35°C also allows more
than 10 mJ of .2120 μm to be obtained. [4]

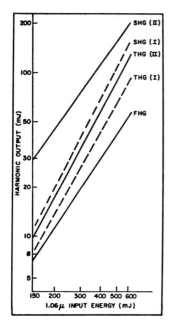

Fig. 1 Second (0.5320 μm),
third (0.3547 μm) and fourth
(0.266 μm) harmonic output
energies vs input 1.06 μm energy
for harmonic conversion in angle
phasematched KD*P crystals.
(Courtesy Quanta Ray Inc.,).

Under normal conditions the Nd:YAG oscillator operates in many axial
modes over a 0.4 cm⁻¹ wide linewidth. A single tilted etalon reduces the
linewidth to less than 0.1 cm⁻¹. Finally, proper programming of the Q-switch
voltage leads to a longer cavity build up time with resultant single axial
mode operation. [5,6] Single mode operation eliminates axial mode beating
modulation and reduces peak-to-peak pulse energy fluctuations to less than
1% as a direct result of the electronically controlled inversion density. [7]

At a fixed output coupling of 0.85, the Q-switched pulse length is
8 nsec for a 60 cm and 15 nsec for a 120 cm long cavity under strong
pumping conditions.

Nd:YAG Pumped Tunable Sources

The diffraction limited, high peak power output of the Nd:YAG source at the
fundamental and harmonics is ideal for pumping dye oscillator/amplifier or
LiNbO₃ OPO/OPA tunable sources.

The Nd:YAG pumped dye laser is a significant improvement over the
nitrogen pumped dye laser source in both peak power and spatial mode quality.
The near diffraction limited output beam profile of the dye laser source can
be efficiently doubled or summed with Nd:YAG wavelength in KD*P crystals to
provide ultraviolet tuning to 0.220 μm. Frequency mixing in LiNbO₃ of the
tunable dye output with .532 or 1.064 μm provides output to the 4.0 μm
wavelength range. The ability to provide frequency extension by nonlinear
interactions is an important advantage of the higher pulse energy Nd:YAG
pumped dye laser source.

Figure 2 shows typical dye laser output energies vs wavelength for 200 mJ
of input 0.532 μm. The conversion efficiency from input pump to dye output

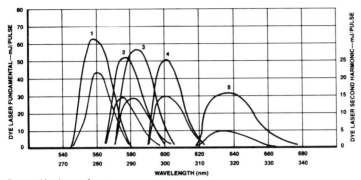

Dyes used for above performance
1. Exciton R590
2. Exciton R610
3. Exciton Kiton Red
4. Exciton 640
5. Exciton Cresyl Violet

Fig. 2 Dye laser output for 200 mJ of input 0.532 μm pump and second harmonic output (lower curves-right scale) for dye laser doubling in KD*P.

is typically 25% for visible dyes to 20% for red dyes. The .3547 μm pumped dyes tune over the .400 - .56 μm range and operate at 20% conversion efficiency.

The dye oscillator is a side pumped flowing cell with a telescopic prism beam expander [7,8] and higher order diffraction grating. Output linewidths are less than 0.5 cm⁻¹. The high efficiency of the dye oscillator,even at the expense of some increase in linewidth, is important if a high coherent output to superfluorescence background contrast ratio is to be maintained after the dye amplifier stages.

The high peak power and collimated dye output beam allow efficient harmonic generation and mixing to be used to extend the dye laser tuning range. Second harmonic conversion efficiencies are shown in Fig. 2 in the lower curves and the right hand scale. Efficient conversion to 0.22 μm by sum generation in KD*P and to the infrared by mixing in LiNbO3 has also been demonstrated. [3]

The high peak power output of the dye source allows SRS with conversion efficiencies of 20% to the first and second Stokes and 10% to the third Stokes wave in H2 gas. [10] Reasonable conversion efficiencies to the higher order anti-Stokes is also possible as shown in Fig. 3. The simplicity of SRS process make it a very practical technique for both infrared and ultra-violet conversion of the dye laser source.

The Nd:YAG pumped LiNbO3 OPO/OPA has been described previously. [11] Recent detailed studies of the OPO have led to improvements in the source reliability and linewidth control, [12] and to operation in a single axial mode. With the addition of a parametric amplifier (OPA) the output energy has been increased to the 20-50 mJ level. [13] Application of this tunable source for nonlinear spectroscopy of liquid water is described below.

The LiNbO3 OPO/OPA source can be frequency extended by mixing in nonlinear crystals and by four wave mixing in H2 gas. Recent measurements show that the

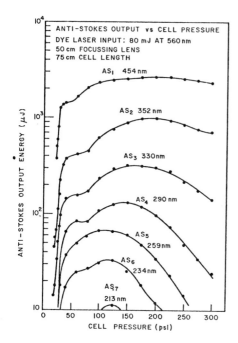

Fig.3 Anti-Stokes output energy vs H2 pressure for 80 mJ, 560 nm input from a dye laser source.

(Courtesy, S.E. Brosnan)

four wave mixing process is in good quantitative agreement with theory for conversion to both the Stokes and anti-Stokes outputs. [14]

Stimulated Raman and Four Wave Mixing Studies

The availability of high peak power Nd:YAG laser source has led to a renaissance in stimulated Raman scattering as a useful frequency conversion process and spectroscopic tool. Using the unstable resonator Nd:YAG source, we have investigated in detail the close coupling between stimulated Raman scattering and four wave mixing. [15]

The steady state coupled pump and Stokes equations for a multi-axial mode laser have the general form

$$\frac{\partial \tilde{E}_{sq}}{\partial z} = \frac{\omega_s}{2n_s c} \, \chi_R^{''} \, \tilde{E}_{pq} \, e^{i\Delta k_q z} \sum_n \tilde{E}_{pn}^* \, \tilde{E}_{sn} \, e^{-i\Delta k_n z} \tag{1a}$$

$$\frac{\partial \tilde{E}_{pq}}{\partial z} = \frac{-\omega_p}{2n_p c} \, \chi_R^{''} \, \tilde{E}_{sq} \, e^{-i\Delta k_q z} \sum_n \tilde{E}_{pn} \, \tilde{E}_{sn}^* \, e^{i\Delta k_n z} \tag{1b}$$

where $\chi_R^{''}$ is the Raman susceptibility, E_{pn} and E_{sn} are the pump and Stokes fields for the nth axial mode and Δk_n is the phase mismatch factor given by

$$\Delta k_n = \left[\frac{\partial k}{\partial \omega} \bigg|_{\omega_p} - \frac{\partial k}{\partial \omega} \bigg|_{\omega_s} \right] \Delta \omega_n \qquad (2)$$

where $\Delta \omega_n \equiv \omega_{pn} - \omega_{po} = \omega_{sn} - \omega_{so}$ is the detuning of the nth axial mode.

In the steady state single mode case the coupled equations reduce to

$$\frac{\partial \tilde{E}_s}{\partial z} = \frac{\omega_s}{2n_s c} \chi_R'' |E_p|^2 \tilde{E}_s$$

where pump depletion has been neglected.

The Stokes wave grows exponentially with a gain coefficient

$$g_s = \frac{\omega_s}{n_s c} \chi_R'' |E_p|^2 = \frac{2\omega_s}{n_s c} \chi_R'' \eta_p I_p \qquad (3)$$

where η_p is the impedance.

If two modes are present, the coupled equations reduced to the familiar four wave mixing equations [11] which have a solution first derived by GIORDMAINE and KAISER. [16]

Of interest is the case of n axial modes without dispersion. In this case, $\Delta k_n = 0$, and the Stokes wave grows as

$$\frac{\partial \tilde{E}_{sq}}{\partial z} = \frac{\omega_s}{2n_s c} \chi_R'' \tilde{E}_{pq} \sum_n |\tilde{E}_{pn}^*| |\tilde{E}_{sn}| e^{i(\phi_{pn} - \phi_{sn})}$$

Without dispersion present $\phi_{pn} \quad \phi_{sn} = \phi_o$ holds for all axial modes so that the Stokes intensity grows as

$$2n_s \frac{d I_s}{dz} = \frac{\omega_s}{n_s c} \chi_R'' \left\{ \frac{\left[\sum_n |\tilde{E}_{sn}| |\tilde{E}_{pn}| \right]^2}{\sum_n |\tilde{E}_{pn}|^2 \times \sum_n |\tilde{E}_{sn}|^2} \right\} \cdot \sum_n |\tilde{E}_{pn}|^2 \cdot \sum |\tilde{E}_{sn}|^2 . \quad (4)$$

The exponential Raman gain coefficient predicted from Eq.(4) is

$$\frac{d I_s}{dz} = \frac{2\omega_s}{n_s c} \chi_R'' \eta_p \beta \overline{I}_p \overline{I}_s = g_s \beta \overline{I}_s$$

where g_s is the previously derived plane wave Raman gain coefficient and β is the factor in brackets in Eq.(4). Consideration of Eq.(4) shows that $\beta = 1$ and the Raman gain is maximum if the pump and Stokes axial mode amplitudes are identical. This condition is precisely stated by the requirement that

607

$$\frac{\tilde{E}_{pn}}{\tilde{E}_{sn}} = \tilde{c}(z)$$

where $c(z)$ is a constant independent of the axial mode n. This requirement is identical to the four wave mixing conversion efficiency expression
derived by GIORDMAINE and KAISER and leads to the result that four wave mixing
is an intimate part of the Raman scattering process and drives the Stokes
spectral distribution to exactly reproduce that of the pump. This conclusion
is valid even in the limit of an incoherent pump source. [15]

When dispersion is included, the above conclusion is valid only over a
pump spectral interval which satisfies the phasematching relation
$\Delta k_{max} \ell_{coh} = \pi$ where Δk_{max} is the largest Δk_n in the pump Stokes spectrum
This limitation quickly leads to a finite pump acceptance bandwidth for SRS
given by

$$\Delta \nu_{max} = \frac{c}{4(n_p - n_s)L} \tag{6}$$

where L is the interaction length.

For example, for H_2 gas the pump acceptance bandwidth for forward scattering at 1.06 μm for Q(1) transition for a 1 m cell with 20 atm gas pressure
is 195 cm^{-1}. This is orders of magnitude larger than the Raman linewidth of
approximately 0.1 cm^{-1}.

The above result that SRS is independent of pump laser bandwidth had
been predicted and tested earlier by AKHMANOV et.al. [17] in liquid N_2 but
was not widely recognized.

Using the unstable resonator Nd:YAG source operating both in a single
axial mode and in a multi-axial mode we verified the bandwidth independence
of SRS for H_2 gas for both the vibrational and rotational scattering case. [15]
Figure 4 shows the SRS threshold in a 25 pass multipass cell [17] for single
mode and multi-axial mode excitation. The clear verification of the wide
pump acceptance bandwidth for SRS in H_2 gas led to an immediate simplification
and successful demonstration of the CO_2 laser pumped stimulated rotational
Raman scattering experiment in para-hydrogen. [18] The SRS process is also
useful as an efficient conversion process for wide band as well as narrow
band input pump sources.

Polarization CARS Spectroscopy of Liquid H_2O

The four wave mixing interaction provides an equally useful tool for spectroscopic studies. AKHMANOV et.al. [19] have shown that the coherent Raman
ellipsometry method (CREM) may be used to resolve the sub-structure of wide
Raman bands via an internal interference of adjacent modes in the band. We
chose to use the computer controlled $LiNbO_3$ OPO/OPA source and the polarization CARS technique to study the broad liquid water Raman band centered
at 3200-3600 cm^{-1}. [20] This band has been previously studied by spontaneous
Raman scattering and models have been proposed to explain the unresolved but
slightly assymetric shape. [21]

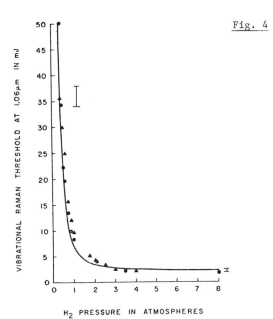

Fig. 4 Vibrational SRS thres-
in H_2 gas vs pressure
for 1.064 μm pumping
in a 25 pass multiple
pass cell. The ▲
corresponds to single
axial mode pumping and
the ● to multiple axial
mode pumping (0.4 cm^{-1}
linewidth) in the steady
state limit.

H_2 PRESSURE IN ATMOSPHERES

Figure 5 shows a schematic of the experiment. A key element in the success of the measurement was the PDP11 minicomputer which actively tuned the $LiNbO_3$ OPO, collected and normalized the data and tuned the spectrometer. The computer also recorded the spectra, calculated the rms deviation and normalized the signal for detector and optical response variations with wavelength over the wide tuning range.

Figure 6 shows the polarization CARS spectra of H_2O for various analyzer setting angles with and without a quarter wave bias. The spectra are a result of four independent scans at each analyzer setting with fifty averaged pulses per point. Each scan took twenty minutes to complete. The spectra could not have been obtained without the long term reliable operation of the Nd:YAG pumped $LiNbO_3$ tunable source and computer data processing capabilities.

The observed spectra shows that the broad Raman band of H_2O does indeed have sub-structure. By varying the polarizing angle of the analyzer the experimentalist is actively controlling the local interference of the susceptibility components of the individual spectral components that compose the liquid water band. In effect, the spectroscopist is observing a four dimensional spectra from selected angles. At each observation angle interference between the real and imaginary susceptibility components of each spectral component is occurring. In theory, the spectral observations can be inverted to obtain the real and imaginary parts of the resonant susceptibilities of the individual modes of the liquid water band.

This spectral study, which resolved for the first time structure in the liquid H_2O band, clearly demonstrates the power of the coherent Raman techniques.

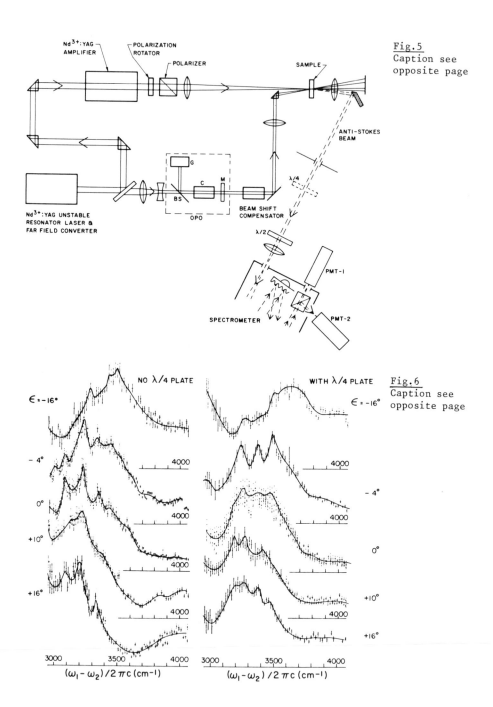

Fig.5
Caption see
opposite page

Fig.6
Caption see
opposite page

Conclusion

In conclusion it is clear that interactions involving the third order susceptibility are useful for both efficient frequency conversion and spectroscopic analysis. The increased power spectral brightness brought about by the unstable resonator Nd:YAG source and Nd:YAG pumped tunable sources is essential if full advantage is to be taken of $\chi^{(3)}$ spectroscopy. The wide tuning range and very reliable operation of the Nd:YAG pumped tunable sources mean considerably more effort will be spent on spectroscopy than on the laser source.

References

1. A.E. Siegman, Proc. IEEE 53, p.277 (1965); Applied Optics 13, p.353 (1974); IEEE Journ. Quant. Electr. QE-12, p.35 (1976).

2. R.L. Herbst, H. Komine and R.L. Byer, Opt. Commun. 21, p.5 (1977).

3. R.L. Byer and R.L. Herbst, Laser Focus 14, p.48 (1978).

4. G. Massey, M.D. Jones and J.C. Johnson, IEEE Journ. Quant. Electr. QE-14, p.527 (1978).

5. D.C. Hanna, B. Luther Davies, H.N. Rutt and R.C. Smith, Opto. Electr. 3, p.163 (1971).

6. W.R. Trutna, Y.K. Park and R.L. Byer, "Stimulated Raman Threshold Measurements in H_2 Gas Using a Single Axial Mode Q-switched Nd:YAG Laser", Paper W110, presented at the Optical Society of America Meeting, San Francisco, California, October 1978.

7. R.L. Herbst, Quanta Ray Inc., (private communication).

8. M.A. Novikov and A.D. Tertyshnik, Soviet Journ. Quant. Electr. 5, p.848 (1975).

9. V. Wilke and W. Schmidt, Applied Physics, 16, p.151 (1978); Applied Physics 17, p.177 (1979).

Fig. 5 Schematic of the polarization CARS experiment of liquid water. The broadband half wave plate allowed polarization rotation of the anti-Stokes beam. The polarization ratios were detected by photo-multiplier tubes 1 and 2 following a scheme proposed by OUDAR et.al., [22] and recorded by the computer for processing.

Fig. 6 Polarization CARS spectra of liquid H_2O for various angles between the analyzer and the normal to the background polarization. The spectra on left show the dispersion ratio of the signal and reference photo-multiplier signals averaged over 50 pulses per data point with the rms deviation. The spectra on the right show the same, but with a quarter wave plate present in the anti-Stokes beam with its principal axis angle at -3^0 to the plane of the non-resonant susceptibility polarization plane.

10. J.A. Paisner and R.S. Hargrove, "Tunable, Efficient VUV and IR Generation Using High Order SRS in H_2", Post Deadline Paper, C.L.E.A. Conference, Washington D.C. June 1979.

11. R.L. Byer and R.L. Herbst, "Parametric Oscillation and Mixing", in Topics in Applied Physics, vol. 16, <u>Nonlinear Infrared Generation</u>, ed. Y.R. Shen, Springer-Verlag, Berlin 1977.

12. S.J. Brosnan and R.L. Byer, "Optical Parametric Oscillator Threshold and Linewidth Studies", to be published IEEE Journ. Quant. Electr. June 1979.

13. R.A. Baumgartner and R.L. Byer, "Optical Parametric Amplification", to be published IEEE Journ. Quant. Electr. June 1979.

14. S.J. Brosnan, "Tunable Infrared Generation Using Parametric and Raman Processes", Ph.D. Thesis, Stanford University, March 1979.

15. W.R. Trutna, Y.K. Park and R.L. Byer, "The Dependence of Raman Gain on Pump Laser Bandwidth", to be published IEEE Journ. Quant. Electr. July 1979.

16. J.A. Giordmaine and W. Kaiser, PHys. Rev. <u>144</u>, p.676 (1966).

17. S.A. Akhmanov, Yu. E. D'Uakov and L.I. Pavlov, Sov. Phys. JETP, 39, p.249 (1974).

18. R.L. Byer and W.R. Trutna, Optics Letters, <u>3</u>, p.144 (1978).

19. S.A. Akhmanov, A.F. Bunkin, S.G. Ivanov and N.I. Koroteev, JETP, <u>47</u>, p.667 (1978).

20. N.I. Koroteev, M. Endemann and R.L. Byer, "Resolved Structure Within the Broadband Vibrational Raman Line of Liquid H_2O Using Polarization CARS", (to be published).

21. G.E. Walrafen in "Water, A Comprehensive Treatise", ed. by F. Franks, vol. 1, Plenum Press, New York, N.Y. Chapter 5, 1972.

22. J.L. Oudar, R.W. Smith and Y.R. Shen, Appl. Phys. Letts. (in press).

Acknowledgement

I want to acknowledge the contributions to this work by S.J. Brosnan, Quanta Ray Inc., M. Endeman; N.I. Koroteev, Moscow State University; Y.K. Park, and W.R. Trutna, Max Plack Institute. I want to thank Mary Farley for the preparation of this manuscript.

Part X

Postdeadline Papers

Atomic-Beam and Resonance Spectroscopy of Alkali and Alkaline-Earth Atoms

K. Fredriksson, P. Grundevik, M. Gustavsson, I. Lindgren, H. Lundberg,
L. Nilsson, G. Olsson, L. Robertsson, A. Rosén, and S. Svanberg

Department of Physics, Chalmers University of Technology
S-412 96 Göteborg, Sweden

During the last few years an extensive program of laser-spectroscopy investigations for sequences of alkali atomic states has been pursued in our laboratory. Very recently two fine-structure investigations were completed. The interesting Cs F-sequence was studied for $n = 10$-17 in experiments on a collimated beam [1]. With a CW dye laser the $7\ ^2P_{3/2}$ state was excited, from which about 10 per cent of the atoms cascade to the $5\ ^2D_{5/2}$ state. A single-mode, electronically tunable laser was then used to scan the $5\ ^2D_{5/2}$ - $n\ ^2F_{7/2,5/2}$ transitions. The fine-structure, that is inverted for all the studied F states, can be expressed as an expansion in odd powers of the effective principal quantum number. In addition to collimated-beam measurements, precision level-crossing determinations of the fine structure for the 5 and $6\ ^2D$ K states were performed in a strong external magnetic field [2]. Also using precision values for high-n states, obtained by Gallagher and Cooke [3], an accurate expansion describing the K fine-structure splittings was determined.

The interest of our research group has gradually moved to the alkaline-earth atoms. Several groups have recently been involved in hyperfine structure and isotope shift measurements for Ca. We have performed precision hyperfine-structure measurements in the $4s4p\ ^3P_2$ metastable state of ^{43}Ca using the atomic-beam magnetic-resonance (ABMR) method combined with a single-mode dye-laser for the detection. The electric quadrupole moment for the particularly interesting ^{43}Ca nucleus was accurately determined: $Q(^{43}Ca) = -0.065(20)$ b [4]. In addition, isotope shifts and hyperfine structure in the transition $4s4p\ ^3P_2 \leftrightarrow 4s5s\ ^3S_1$ were obtained using high-resolution spectroscopy.

In our experiments on the ^{43}Ca isotope, which has a very low natural abundance (0.145%), two separate atomic-beam set-ups were used in connection with a narrow-band single-mode dye laser spectrometer. In a collimated-atomic-beam apparatus, high-resolution optical spectroscopy studies were performed. A recording obtained in a laser scan of the 6162 Å line is shown in Fig. 1. Radio-frequency transitions between hyperfine levels were induced in an ABMR apparatus and the single-mode laser was used for a selective detection.

We have also performed a detailed hyperfine-structure study for 7 states belonging to the 5d6p configuration of Ba. The states were excited from the metastable states belonging to the 6s5d configuration and isotope shifts of a large number of lines were determined. Theoretical calculations of the hyperfine interactions are in progress [5].

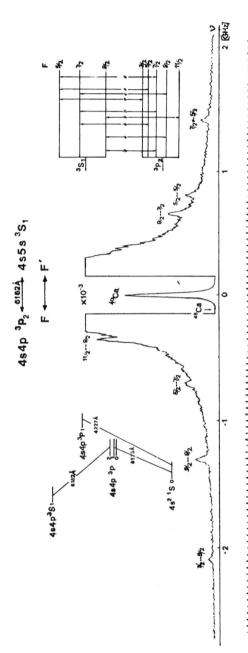

Fig. 1. High-resolution laser scan of the 6162 Å Ca line, displaying the hyperfine structure for ^{43}Ca.

References

1. K. Frederiksson, H. Lundberg and S. Svanberg, Phys. Rev. A. in press.
2. L. Nilsson and S. Svanberg, Z. Physik A, in press
3. T.F. Gallagher and W.E. Cooke, Phys. Rev. A18, 2510 (1978).
4. P. Grundevik, M. Gustavsson, I. Lindgren, G. Olsson, L. Robertsson, A. Rosén and S. Svanberg, Phys. Rev. Lett., 42, 1528 (1979).
5. P. Grundevik, H. Lundberg, L. Nilsson and G. Olsson, to be published.

Observation of Transient and Stationary Zeeman Coherence by Polarization Spectroscopy

J. Mlynek, K.H. Drake, and W. Lange

Institut für Angewandte Physik, Universität Hannover
D-3000 Hannover, Fed. Rep. of Germany

H. Brand

Institut A für Experimentalphysik, Universität Hannover
D-3000 Hannover, Fed. Rep. of Germany

The method of Doppler-free laser polarization spectroscopy [1] relies on the sharply resonant behaviour of the complex susceptibility of a sample under conditions of velocity-selective holeburning.It is the aim of this paper to demonstrate that making use of light induced birefringence and dichroism connected with the existence of coherence between substates of atomic or molecular states is another useful approachto high-resolution laser spectroscopy. In the simplest type of experiment following this scheme, a first laser beam creates coherence in a sample and the resulting optical anisotropy is detected by transmitting a second beam through a system of crossed polarizers with the coherently excited sample placed in between. Geometrical arrangements can be found, where the transmission is governed by the coherence only.

Using this polarization technique we have recently shown that the free propagation of coherence between nondegenerate substates of an excited atomic state, induced by a short resonant broadband 'pump'-pulse, can be monitored sensitively by use of a weak delayed light pulse [2]. Alternatively a cw probe beam might be used. A simple density matrix calculation reveals that excactly the same experimental set-up can be used in order to create and detect coherence in atomic or molecular ground states. The coherent effects in the upper ('excited') or lower ('ground') state show up in the same order of perturbation theory. In both cases the detector signal exhibits beats, whose frequency is determined by the splitting of the levels engaged. Obviously this technique is closely related as well to the 'quantum beat method' as to the technique of 'coherent Raman beats'. As a demonstration Zeeman beats on the $4f^66s^2\ ^7F_1 - 4f^66s^2\ ^7F_0$ transition of Samarium were observed. In sodium vapor transient ground state coherence was detected in a single shot measurement (Fig. 1).

Resonant interaction of a sample with a stationary light field can efficiently induce coherence between degenerate levels only. Thus the experimental set-up described above can be used to observe 'level-crossings' either in the upper or the lower state, by just switching from a pulsed to a cw 'pump'--beam and applying a suitable external electric or magnetic field. The widths of the signals are determined by the lifetimes of the levels and are affected neither by Doppler-broade-

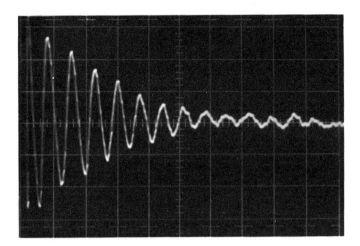

Fig. 1: Single-shot measurement of Zeeman quantum beats in the $^2S_{1/2}$ ground state of Na (time scale 1µs/div., magnetic field 130 A/m, sodium density 10^{13} cm^{-3}).

ning nor by the bandwidth of the laser. As a demonstration the method was applied to zero field level-crossings in the SmI spectrum.

This work was supported in part by the Deutsche Forschungsgemeinschaft.

References
/1/ C. Wieman and T.W. Hänsch, Phys. Rev. Lett. 36, 1170 (1976)
/2/ W. Lange and J. Mlynek, Phys. Rev. Lett. 40, 1373 (1978)

Experimental Evidence for Light-Shift Induced Zero-Field Level Crossing („Optical Hanle Effect")

C. Delsart, J.-C. Keller

Laboratoire Aimé Cotton, C.N.R.S. II, Bâtiment 505
F-91405 Orsay Cedex, France

and

V.P. Kaftandjian

Université de Provence, Centre de St. Jérôme
F-13000 Marseille, France

Two types of zero-field level crossing (Hanle effect) have been observed in the past using either a static magnetic field or a static electric
field [1]. The physical basis of the phenomena is found in the lifting of
the m-state degeneracy by the external static field (Zeeman effect or Stark
effect) which breaks down the zero-field quantum interference. A lot of lifetimes and relaxation rates have been measured for atomic and molecular species by this method with broad-band coherent excitation of the fluorescent
levels and magnetic field scanning. An optical analogue of the Hanle effect
has been recently proposed where the light-shift induced by a high power
laser beam replaces the level shift produced by the static fields [2] ; the
dynamic Stark shift plays the same role as the Zeeman shift in the magnetic
Hanle effect or as the Stark shift in the electric Hanle effect. The theory
of the so-called "optical Hanle effect" has been previously worked out for a
$(J=0 \rightarrow 1 \rightarrow 0)$ three-level system [2]. In this letter, we present the first experimental observation of the light-shift induced zero-field level crossing
in a $(J=0 \rightarrow 1)$ two-level system of Ba I, as well as the main results of the
corresponding calculations.

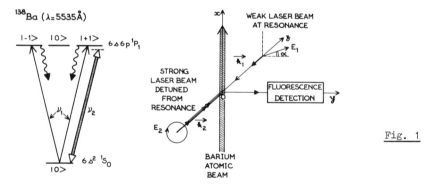

Fig. 1

The level scheme and the light polarization directions are indicated on
Fig. 1. The Ba atomic beam is crossed at right angle by two counter-propagating beams from CW single mode dye lasers. The frequency of the weak,
linearly polarized, beam is locked to the ^{138}Ba resonance line $(\lambda = 5535 \text{ Å})$.
The frequency of the powerful, circularly polarized, beam is detuned of $\Delta\omega_2$
from exact resonance thus producing a light-shift δ of the level $|+1\rangle$.
Provided that the conditions $|\Delta\omega_2| \gg \beta \gg \gamma$ are valid (2β is the Rabi

nutation frequency and γ is the natural width of the transition), the
light-shift is equal to $\delta = -\beta^2/\Delta\omega_2$. The fluorescence emitted perpendicularly both to the atomic beam and to the laser beams is detected by a photomultiplier. To eliminate the incoherent part of the fluorescence light,
the polarization direction is rotated at a frequency ν and a lock-in amplifier operating at a frequency 2ν is used. The in-phase and in-quadrature signals are then proportional to the real part ("absorption-shaped"
signal) and to the imaginary part ("dispersion-shaped" signal) of the
Zeeman coherence ρ_{+1-1} [3].

The corresponding calculations for the degenerate two-level system have
been achieved using a non-linear response function method. Within the above
approximations the optical Hanle effect signals can be written :

$$\mathcal{R}e(\rho_{+1-1}) \propto (1+8y^2) / (1+4y^2)(1+16y^2)$$
$$\mathcal{I}m(\rho_{+1-1}) \propto 2y / (1+4y^2)(1+16y^2) \quad \text{where } y = \delta/\gamma .$$

The above situation differs from that of the magnetic Hanle effect with
monochromatic excitation because the light-shift affects both the ground
state and the excited state of the $0 \rightarrow 1$ transition. This is of no importance in the case of a broad-band excitation for which classical lorentzian
Hanle curves are found as expected.

The absorption-shaped and dispersion-shaped curves have been recorded
experimentally as a function of the non-resonant beam power (proportional to
β^2) and for different values of the detuning $\Delta\omega_2$. Typical experimental
curves are shown on Fig. 2 ; as expected from the above expressions, the
relative amplitude of the dispersion curve appears to be rather small. The
extremum on the dispersive curve is obtained for $\delta \simeq 5$ MHz as expected (the
natural width γ is about 20 MHz). Similar curves have been obtained for
$|\Delta\omega_2|$ going up to 6000 MHz ; they show a typical $1/\Delta\omega_2$ change for the X-
axis unit. Nearly lorentzian curves have also been recorded using broad-band
excitation (obtained by frequency modulation of the weak beam) instead of
narrow-band excitation.

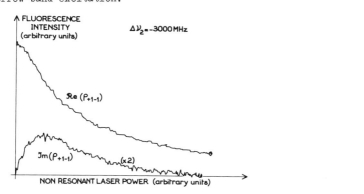

Fig. 2

[1] Laboratoire associé à l'Université Paris-Sud.

[1] W. Hanle, Z. Phys. 30 (1924) 93 ; 35 (1926) 346.
[2] V. P. Kaftandjian, L. Klein, Phys. Letters 62A (1977) 317 ;
    V. P. Kaftandjian, L. Klein, W. Hanle, Phys. Letters 65A (1978) 188.
[3] D. Lecler, R. Osterman, W. Lange, J. Luther, J.de Physique 36 (1975)647.

Diamagnetic and Collisional Perturbations of Caesium Rydberg States

L.R. Pendrill[1], D. Delande, and J.C. Gay

Laboratoire de Spectroscopie Hertzienne de l'ENS
Université Pierre et Marie Curie - Tour 12 - E01, 4,place Jussieu
F-75230 Paris 5e, France

The goal of the present experiments is to study the fine structure of the "quasi Landau" spectrum of atoms in strong magnetic fields [1] with high resolution techniques and the character and modifications of the spectrum in the additional presence of a crossed electric field [2][3][4].

For such purposes the highly excited n^2F levels of atomic Cs present an interesting case because of the large atomic mass and particularly because of the small quantum defect of such states. The situation is then almost that of the hydrogen atom but for the fact that S, P and D levels are perturbed by the atomic core out of the hydrogenic manifold. Moreover, Cs n^2F states can be easily excited in a vapor even with the relatively low light power available from a c.w. single mode dye laser.

The n^2F Rydberg states are produced by a two step process :
i) the production of Cs 5^2D states via the photodissociation of Cs_2 molecules
ii) a fundamental series transition from the 5^2D states to the F levels [5]. The Rydberg atoms are detected with a thermoionic detector installed in the vapor cell which itself is placed in the core of a superconducting solenoïd producing fields up to 8 Teslas.

By this means we have been able to produce easily in a vapor Rydberg atoms with n $\sim$ 100 (Fig. 1), thus permitting the extension of the range of measured term values for this series. We find that the results for the

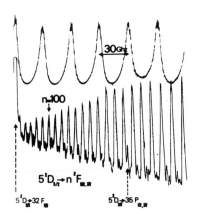

[1]J.J. Thomson Physical Laboratory - University of Reading (U.K.)

Fig. 1 F Rydberg series of Cs atoms

620

quantum defect of the nF are in good agreement with previous measurements
[5] for n values in the range n = 35 to 60. Near the value n = 100 however,
systematic variations in the quantum defect are found to occur. The corresponding shift of the lines could be the result of Cs* - Cs collisions :
there is rough agreement with the predictions of a collision model using
a Fermi contact potential and a single scattering length approximation [6].

Diamagnetic shifts of the n^2F levels (n ∿ 60) of several cm^{-1} have been
observed in magnetic fields of just a few kG (Figu 2). They are in good
agreement with the predictions of a simple hydrogenic model, as expected.
But, at the moment, we have found no experimental evidence for the existence of quasi Landau resonances in strong fields of up to 80 KG.

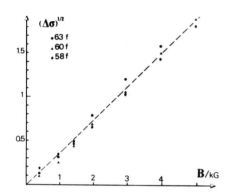

Fig. 2 Diamagnetic shift of
Rydberg n^2F lines against magnetic field (Δσ in cm^{-1})

1   R.J. FONCK, D.H. TRACY, D.C. WRIGHT and F.S. TOMKINS - Phys. Rev. Lett.
    40, 21 (1978) 1366
2   A.R.P. RAU - J. Phys. B, Atom. Molec. Phys. 12, 6 (1979) L 193
3   H. CROSSWHITE, U. FANO, K.T. LU and A.R.P. RAU - Phys. Rev. Lett. 42,
    15 (1979) 963
4   J.C. GAY, L.R. PENDRILL and B. CAGNAC - Physics Letters - to be pu-
    blished
5   C.B. COLLINS, B.W. JOHNSON, P.Y. MIEZA, D. POPESCU, I. POPESCU -
    Phys. Rev. A 10, 3 (1974) 813
6   A. OMONT - J. Phys. (Paris) 38, 1343 (1977)

Collisional Effects in the Saturation Spectroscopy of Three-Level Systems: Theory and Experiment

P.R. Berman[1]

Physics Department, New York University, New York, NY 10003, USA
and
P.F. Liao and J.E. Bjorkholm

Bell Telephone Laboratories, Holmdel, NY 07733, USA

We report on a theoretical and experimental study of the influence of collisions on the saturation spectroscopy line shapes associated with three-level gas vapor systems. The study is carried out with the goal of gaining new information concerning (a) the collisional processes that occur in atomic vapors, (b) the nature of the interatomic potential between a ground state and an excited-state atom and (c) the possibility of collision-induced enhancement of the absorption of radiation by an atomic system. In each of these areas, new results are obtained.

Theory Using a model in which collisions are assumed to be phase-interrupting in their effect on level coherences and velocity-changing in their effect on population densities, we calculate the absorption profiles associated with three-level atoms that are subjected to two incident radiation fields while undergoing collisions with structureless perturber atoms. One of the fields (pump) is of arbitrary strength and acts on a given transition while the other field (probe) is weak and acts on a transition sharing a common level with the first. In the absence of collisions, the probe absorption profile can exhibit many well-known features [1], including narrow Doppler-free resonances and strong-pump field induced ac Stark splittings. Collisions distort these profiles and can actually lead to enhanced probe absorption in cases of either large pump detunings or strong pump fields. Moreover the collisional modifications of the profiles may be used to extract information on both total and differential scattering cross sections. Theoretical profiles illustrating these features have been derived.

Experiment The theory is applied to explain the $3S_{1/2} \to 3P_{1/2} \to 4D_{3/2}$ excitation spectra that we have obtained for Na atoms undergoing collisions with foreign gas perturbers. A pump laser is detuned either 4.0 GHz or 1.6 GHz below the $3S_{1/2} \to 3P_{1/2}$ transition frequency and a probe laser beam, counter-propagating with the first, completes transitions to the $4D_{3/2}$ state. The population of the $4D_{3/2}$ state is monitored (via fluorescence) as a function of probe frequency for pump detunings of -4.0 GHz and -1.6 GHz, using various pressures of He, Ne, and Kr perturbers. With a pump detuning of -4.0 GHz, which is greater than the Doppler width ≈ 1.6 GHz, we are able to systematically study collisional redistribution [2] (resulting from

[1] Supported by the U.S. Office of Naval Research.

collisionally-induced excitation of the $3P_{1/2}$ state) and to obtain a fit to theory containing essentially no free parameters. For a detuning of -1.6 GHz, the pump laser excites a given longitudinal velocity class of atoms. Velocity-changing collisions cause this velocity group to relax back towards equilibrium, and the probe absorption monitors the progress of this relaxation. Attempts to fit the data were made using both the Keilson-Storer and classical hard sphere collision kernels to describe the velocity-changing collisions. The theory includes the effects of $3P_{1/2} \rightarrow 3P_{3/2}$ state-changing collisions, which significantly modify the excitation line shapes.

References

1 See, for example, I.M. Beterov and V.P. Chebotaev, Prog. Quantum Elec. <u>3</u>, 1 (1974) and references therein.
2 D.L. Huber, Phys. Rev. <u>178</u>, 93 (1969); A. Omont, E.W. Smith and J. Cooper, Astrophys. J. <u>175</u>, 185 (1972).

Double-Quantum Saturation Spectroscopy with Lasers and rf

E.W. Weber

Physikalisches Institut, Universität Heidelberg
D-6900 Heidelberg, Fed. Rep. of Germany

Saturation spectroscopy of resonant transitions is normally limited in resolution by the natural lifetime of the broad upper level. Double-quantum saturation (DQS) transitions [1,2] between longer-lived equal parity states can be much narrower, for instance a factor of 30 for the (2S-3S) transition in atomic hydrogen compared to the H_α D_1 ($2S_{1/2}-3P_{1/2}$) line. The DQS transitions are induced by means of the simultaneous, correlated absorption of a laser and a rf photon (or two laser photons) both of which are resonant with an intermediate state of opposite parity (e.g. H_α,2S—laser $\rightarrow$ $3P_{1/2}$—rf $\rightarrow$ 3S). The saturated absorption of a transmitted probe laser beam (Fig. 1a) can be used as a convenient detection method.

For low laser and rf powers, the contribution from the width of the intermediate state is insignificant, and the resolution of the DQS transitions is limited only by the natural width of the two final states. Thus DQS spectroscopy can combine the high transition probabilities of saturated absorption, i.e. low necessary laser (and rf) intensities, with the narrow linewidth of conventional two (equal frequency)-photon spectroscopy.

The laser-rf double-quantum saturation method has been applied to investigate the (2S-3S,3D) transitions in atomic hydrogen, yielding a precise value for the ($3P_{3/2}-3D_{3/2}$) Lamb shift [2] and the pressure shift of DQS transitions in He buffer gas [3]. In the case of the (2S-3S) resonance (Fig. 1b) the small hyperfine structure (17.6 MHz) of the intermediate $3P_{1/2}$ state is resolved because of the selection rule that $3P_{1/2}$, F=0 to $3S_{1/2}$, F=0 transitions are forbidden. In the He-H gas discharge 5 times narrower (2S-3S) DQS signals have been observed so far compared to the $H_\alpha D_1$ line, with similar signal to noise ratio when using the sensitive polarization method[4]. In a metastable H^* (2s) atomic beam DQS transitions (2S-nS) can yield linewidth to frequency ratios of $2 \cdot 10^{-9}$(n=3) to $6 \cdot 10^{-10}$(n=6) with the laser transitions still in the visible. The double-quantum resonances are detectable via (nS-n'P, n-n'=1,2,3,...) fluorescence downstream from the excitation region. Possible applications include an order of magnitude improvement in the precision of the Rydberg constant [5] and in the electron-proton mass ratio (from a determination of the H-D isotope shift), and investigations of nuclear size and recoil effects.

A H_α laser on the transition $3S_{1/2}-2P_{3/2}$ induced by the (2S-3S) DQ transition may be feasible. Test measurements in a He-H_2(1%) gas discharge indicate a possible population inversion of the 3S level over the short-lived $2P_{3/2}$ state at pressures below 10 mTorr. A beat experiment of the two lasers involved can yield a precise value of the $2S_{1/2}-2P_{3/2}$($\approx$ 10 GHz) fine structure splitting. Further examples of lasers induced by DQ transitions for H and other atoms are conceivable.

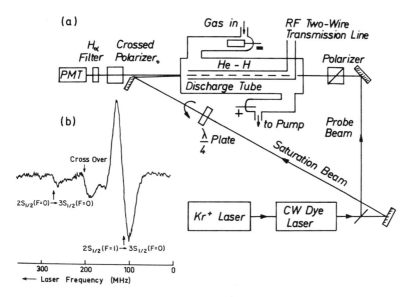

Fig. 1(a) Experimental setup for double-quantum saturation (DQS) spectroscopy of atomic hydrogen with polarized laser beams in a He-H dc discharge.
(b) DQS transitions ($2S_{1/2}$,F=1,O-$3P_{1/2}$,F=1-$3S_{1/2}$,F=O) obtained with polarization detection and amplitude modulation of the rf (271 MHz)

The DQS spectroscopy is not restricted to atomic hydrogen but can be extended very profitably to the study of high Rydberg states of other atoms and
to molecules. For molecules a variety of high precision experiments can be
carried out which supplement the infrared-microwave double-resonance method
[6] : laser-microwave (rf) DQS spectroscopy in electronically excited states,
assignment of vibrational-rotational transitions, study of collision processes, etc.. Furthermore, DQS resonances composed of an optical electric dipole and an rf magnetic dipole transition should be possible with sufficiently high rf power. Then extremely narrow transitions including and coupling
two adjacent equal parity states, e.g. (^1S_O-$^3P_1$-^3P_O), can become feasible,
with possible applications for frequency standards.

I wish to thank J.E.M. Goldsmith for his fruitful collaboration in the hydrogen experiment, A.L. Schawlow and T.W. Hänsch for stimulating discussion.

References

1. D.E. Roberts and E.N. Fortson, Phys. Rev. Lett. <u>31</u>, 1539 (1973)
2. E.W. Weber and J.E.M. Goldsmith, Phys. Rev. Lett. <u>41</u>, 940 (1978)
3. E.W. Weber and J.E.M. Goldsmith, Phys. Lett. <u>70A</u>, 95 (1979) and
   E.W. Weber, submitted to Phys. Rev.
4. C. Wieman and T.W. Hänsch, Phys. Rev. Lett. <u>36</u>, 1170 (1976)
5. J.E.M. Goldsmith, E.W. Weber, T.W. Hänsch, Phys. Rev. Lett. <u>41</u>, 1525 (1978)
6. K. Shimoda in Laser Spectroscopy of Atoms and Molecules (H. Walther, ed.)
   Springer, New York (1976)

High Resolution Spectroscopy in Helium

E. Giacobino, E. De Clercq, F. Biraben, G. Grynberg[1] and B. Cagnac

Laboratoire de Spectroscopie Hertzienne de l'ENS, 4, place Jussieu
F-75230 Paris Cedex 05, France
[1]Laboratoire d'Optique Quantique, Ecole Polytechnique
F-91128 Palaiseau, France

Although helium is of fundamental theoretical interest, only a limited
number of experiments have been performed up to now on the even parity states, especially on the S states. In particular, the isotope shifts and hyperfine structure constants are rather poorly known (except in the metastable 2^3S state).

Using Doppler-free two-photon absorption starting from the metastable
2^3S state, we are able to excite these levels and obtain directly the values of the hyperfine structures and isotopic shift with an accuracy of a
few MHz.

The helium atoms are excited in the 2^3S state using a pulsed discharge.
Then they are excited to other S or D states by absorbing two photons from
a cw single mode laser. Indeed the cell containing the atoms is placed in
the standing wave produced inside a Fabry-Perot cavity. This cavity is
piezo-electrically tuned to the laser wavelength. The fluorescence light
emitted from the level under investigation is detected sidewards during the
afterglow and recorded as the laser wavelength is scanned.

The adjacent figure shows a recording of the two-photon spectrum of ^{3}He
for the transition $2^3S - 4^3D$. All the hyperfine components allowed by the
selection rules of two-photon excitation are present. As we know the hyperfine structure of the 2^3S state, such a recording permits to deduce accurate values of the hyperfine intervals in this level.

We obtain the following values of the hyperfine intervals of the 4^3D
level :

3,5/2 - 2,3/2   274.4±4. MHz
2,3/2 - 1,1/2   289.1±5. MHz
2,3/2 - 2,5/2   5903.8±3. MHz
2,5/2 - 1,3/2   422.5±3. MHz

The distance between the components
$2^3S_1 \rightarrow 4^3D_3$ in ^{9}He and 2^3S_1 (F=3/2)
$\rightarrow 4^3D$ (F=7/2) in ^{3}He is 44728±6 MHz.

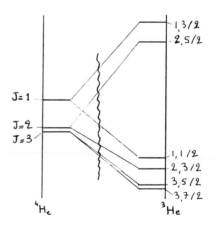

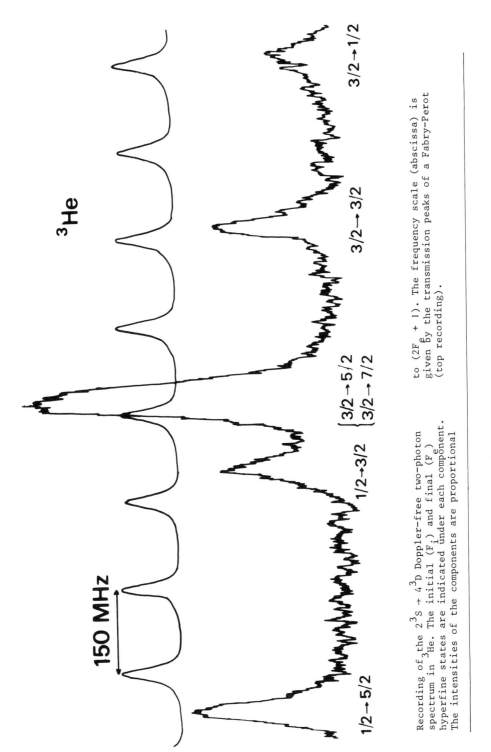

³He

150 MHz

1/2→5/2

1/2→3/2

3/2→5/2
3/2→7/2

3/2→3/2

3/2→1/2

Recording of the $2\,^3S \to 4\,^3D$ Doppler-free two-photon spectrum in ³He. The initial (F_i) and final (F_e) hyperfine states are indicated under each component. The intensities of the components are proportional to ($2F + 1$). The frequency scale (abscissa) is given by the transmission peaks of a Fabry-Perot (top recording).

Doppler-Limited Spectroscopy of the 3 ν_3 Band of SF$_6$[1]

A.S. Pine

Lincoln Laboratory, Massachusetts Institute of Technology
Lexington, MA 02173, USA

Abstract

The strongest portion of the 3 ν_3 band of SF$_6$ has been recorded at T = 160 and 295 K with Doppler-limited resolution using a tunable laser difference-frequency spectrometer. The structure in this band has been identified with the P, Q, and R branches of one F_{1u} sublevel (with essentially ℓ = 1 character) within the 3 ν_3 vibrational manifold. Preliminary effective rotational constants have been obtained for this band. The role of hot bands and of the other anharmonic sublevels is discussed in relation to prior interpretations of low resolution spectra and isotope selective CO$_2$ laser photodissociation of SF$_6$.

---

[1] This work was sponsored by the Department of the Air Force.

Stimulated Level Shifting and Velocity Inversion in UV-Laser-Excited Photofragments [1]

D.J. Ehrlich and R.M. Osgood

Lincoln Laboratory, Massachusetts Institute of Technology
Lexington, MA 02173, USA

We have recently begun to study radiative processes that occur in excited atoms which have been created by molecular photodissociation with UV excimer lasers. [1,2,3] This is a particularly useful technique for producing high densities of excited atoms in specific atomic states in low density gases. Because bound-free continua are relatively broadband, intense excimer lasers can be used for excitation. Further, since the pumping process does not involve a direct optical transition in the resultant atom, many of the limitations on direct optical pumping can be avoided, e.g., competing multiphoton processes, necessity for dipole-allowed one or two photon transition, etc.

An example of a stimulated optical effect which is observed in a photodissociated gas is provided by dissociation of NaBr molecules with 193 nm radiation. In this case 0.5 eV of translational energy is imparted to the Na atom that is produced in one of the 3P states, and as a result the Doppler width for the 589 nm transition is 10.6 GHz (FWHM), ~ 6 times the thermal value. However, the UV pump pulse also creates cold ground-state atoms, probably as a result of dissociation of $(NaBr)_2$ dimers. Thus the net effect of the dissociation pulse is to produce Na(3P) atoms which are inverted only for atoms with velocities in the wings of the Doppler profile. As a result the laser at 589 nm consists of two frequencies separated by ~ 9 GHz (Fig. 1). The two frequencies are asymmetric in intensity as a result of focusing effects due to anomalous dispersion. The anomalous dispersion is observed to saturate with stronger excitation, and the two-frequency components then become equal in intensity.

Zare and Herschbach (4) and Wilson and coworkers (5) have shown that in the photodissociation process the projection of the electric field vector on the dissociating dipole of the parent molecule results in an angularly anisotropic distribution of the recoil velocities of the molecular fragments, and thus in a strikingly non-Gaussian Doppler-broadened lineshape in the fragment emission spectra. In a "cell" experiment this effect is only observable if the fragment velocity is much greater than its thermal value. The dissociation of NaBr with 193 nm radiation gives a sufficiently translationally energetic Na atom that a non-Gaussian lineshape should be observable. We are currently attempting to observe this effect.

---

*This work was sponsored by the Department of the Air Force.

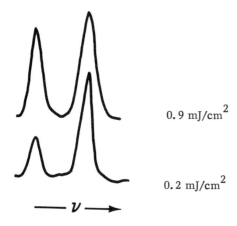

0.9 mJ/cm^2

0.2 mJ/cm^2

$\longrightarrow \nu \longrightarrow$

Fig. 1. Fabry-Perot scan of Na 589 nm
laser output for two values of
pump-laser energy. The spacing
of the peaks is 9.1 GHz.

A second effect which can be seen by photodissociative pumping is obtained by irradiating T 1 I with 193 nm light from an ArF laser. This results in the production of Tl($7S_{1/2}$) atoms, with approximately 100% quantum yield, and hence stimulated emission at 535 nm on the Tl($7S_{1/2} \rightarrow 7P_{3/2}$) transition. At low inversion densities the green laser consists of a strong and a weak hyperfine component, each of which is composed of two peaks separated by 1.7 GHz, the isotope splitting for the two isotopes of thallium. As the laser intensity is raised by increasing the inversion density, each isotope generates a field which is near resonant with the corresponding transition in the other isotope.

This causes a level shift (6) of each isotopic level away from the other, and as a result the two components in each hyperfine transition exhibit a field-dependent splitting which is ~ 4.8 GHz at the highest pump power observed (~ 60 kW/cm^2).
This phenomenon is considerably different from previous observations of level shifting since the shift is observed in the same medium which produces the optical field via stimulated emission. At the highest fields which we attained, $\mu E/\hbar$ becomes comparable to the zero-field separation of the hyperfine-level splitting. In this case, the hyperfine spacing changes with optical field, and, in addition, a further alteration of the lineshapes occurs.

We would like to acknowledge the very considerable contribution of A. Sanchez in understanding of the level-shifting phenomenon. In addition, the comments of T. F. Deutsch and P. L. Kelley and the technical assistance of D. J. Sullivan are greatly appreciated.

REFERENCES

(1) D. J. Ehrlich and R. M. Osgood, Appl. Phys. Lett. <u>34</u>, 655 (1979).

(2) D. J. Ehrlich and R. M. Osgood, IEEE J. Quantum Electron. (to be published).

(3) D. J. Ehrlich and R. M. Osgood, unpublished.

(4) R. N. Zare and D. Herschbach, Proceedings IEEE, <u>51</u>, 173 (1963); R. Zare, Mol. Photochem., <u>4</u>, 1 (1972).

(5) K. R. Wilson, "Photofragment Spectroscopy of Dissociative Excited States", in Chemistry of the Excited State, J. K. Pitts, Jr., ed., (Gordon Breach, New York, 1973).

(6) See for example, P. F. Liao and J. E. Bjorkholm, Phys. Rev. Lett. <u>34</u>, 1 (1975).

Multiphoton Ionization Mass Spectrometry

J.P. Reilly, K.L. Kompa

Projektgruppe für Laserforschung der Max-Planck-Gesellschaft zur
Förderung der Wissenschaften e.V.
D-8046 Garching, Fed. Rep. of Germany

Multiphoton ionization, which began with experiments probing the interaction of intense laser radiation with gas phase atoms and was later exploited by molecular spectroscopists, has most recently been applied to mass spectrometry [1-3]. Results thus far indicate that intense lasers can induce abundant ion yields of selected species. A considerable degree of ionic fragmentation has also been observed. Using a time of flight mass spectrometer in order to obtain an entire ion spectrum in a single laser shot, we have been investigating the UV excimer laser induced multiphoton ionization of benzene. A description of the apparatus has been previously published [2], although it should be mentioned that the strong hydrocarbon background evident in this earlier work has been eliminated by thorough cleaning and improved vacuum conditions.

Typical ion spectra obtained using KrF and ArF lasers are shown in Fig.1. Note the strong intensity dependence which the fragmentation patterns exhibit. It is apparent that the ArF generated fragments are similar to those induced by KrF, but relative peak heights and the overall laser intensity dependence differ. The exact mechanism yielding the numerous observed ion fragments is clearly complex. As a start in interpreting the data we divide the problem into two parts: the initial ionization and the subsequent fragmentation. Only the former will now be considered. The basic assumption which we make is that only the parent molecule, benzene, undergoes ionization; the neutral fragments generated after the benzene ion absorbs additional photons and breaks up are not also ionized. If this is correct then all of the observed ionic fragments derive from $C_6H_6^+$ ions and the total, mass integrated ion yield should be calculable based on an ionization model of benzene alone. Simple rate equations can be legitimately employed [4]. Benzene is depicted as a four level system with a ground state, a resonant intermediate state, an ionization continuum and an additional population trapping state to which the intermediate state relaxes and from which it cannot be ionized. Relaxation from the intermediate state thus reduces the total ion yield. The equations and specific values of parameters employed will be discussed in a later publication. For now it is only pertinent to note that the calculated ion yield is a function of the lifetime of the intermediate state. At the KrF wavelength, benzene's $^1B_{2u}$ state has a known lifetime of 50 ns [5]. Comparison of our experimentally measured integrated ion yield vs. laser pulse energy with the predicted results was used to test the validity of the model. Agreement in curve shape was found to be excellent. At the ArF wavelength, the $^1B_{1u}$ intermediate state lifetime has not previously been measured. Comparison of the experimental ion yield with the model calculation can lead to an estimate of the lifetime of the $^1B_{1u}$ second excited singlet state. Far less ionization is experimentally observed at the ArF wavelength than at KrF.

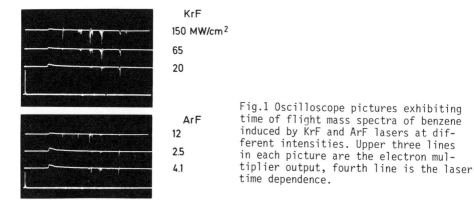

KrF

150 MW/cm²

65

20

ArF

12

2.5

4.1

Fig.1 Oscilloscope pictures exhibiting time of flight mass spectra of benzene induced by KrF and ArF lasers at different intensities. Upper three lines in each picture are the electron multiplier output, fourth line is the laser time dependence.

Considering that the cross sections for stimulated transitions are much larger in the vacuum ultraviolet region, it appears that relaxation of the $^1B_{1u}$ state must be much faster than that of the $^1B_{2u}$ state. This is of course no surprise since fluorescence is not observed from the $^1B_{1u}$ state. Comparison of the ArF experimental results with the model predictions for various possible relaxation lifetimes are presented in Fig.2. Best agreement obtains with a ^1B1u lifetime of 20 ps. This should be construed as only an estimate because of the assumptions that we make concerning the laser's temporal and spatial pulse shape and the lack of fragment ionization processes. Neverthe-

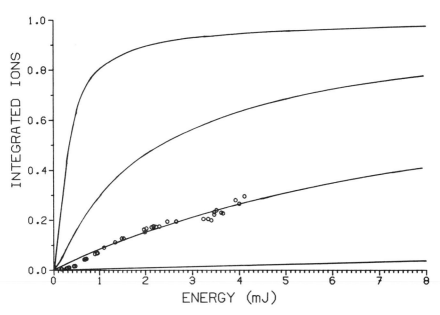

Fig.2. Mass integrated ion yield as a function of ArF laser energy. Circles are experimental points, solid lines are rate equations model predictions for $^1B_{1u}$ lifetimes of 1000, 100, 20 and 1 ps (from top to bottom).

less the basic idea here of measuring the rate of fast kinetic processes by competing against them with stimulated transitions seems to have potential for studying systems which cannot be accessed by more conventional fluorescence lifetime or absorption linewidth techniques.

1. V.S. Antonov et al., Optics Lett. 3, 37 (1978).
2. S. Rockwood, J.P. Reilly, K. Hohla, K.L. Kompa, Opt. Comm. 28, 175 (1979).
3. L. Zandee, R.B. Bernstein, J. Chem. Phys. 70, 2574 (1979).
4. J.R. Ackerholt, J.H. Eberly, Phys. Rev. A. 14, 1705 (1979).
5. K.G. Spears, S.A. Rice, J. Chem. Phys. 55, 5561 (1971).

Standing Wave Saturation Effects in Coupled Three-Level Systems: IR-FIR Transferred Lamb-Dip in Optically Pumped Molecular Lasers

M. Inguscio, A. Moretti, and F. Strumia

Istituto di Fisica dell'Università di Pisa
and
Gruppo Nazionale Struttura della Materia del CNR

From the quantum-mechanical point of view optically pumped molecular lasers can be considered as a coupled three-level system interacting with applied radiation fields. Many phenomena typical of saturation and nonlinear spectroscopy were predicted and experimentally observed (1). In this paper we report a new effect which is the consequence of the standing wave saturation of the pump transition in optically pumped FIR molecular laser. The theoretical model considers a three-level system composed of a IR pump transition, 2-0, center frequency ω_2 coupled to a FIR transition 0-1, center frequency ω_1, via the common level, 0 (fig. 1 and ref.2). The IR pump transition is always inhomogeneously Doppler broadened, while the lasing FIR transition can be either Doppler or homogeneously pressure broadened, depending on the wavelength. The active medium in the resonator interacts simultaneously with forward and backward propagating pumping laser beams (standing wave). When a detuning Δ_2 is introduced between the pump frequency and the absorbing transition line center, in case of fully Doppler-broadened systems, the FIR gain curve was predicted to split into the sum of two Lorentzian curves centered at frequencies:

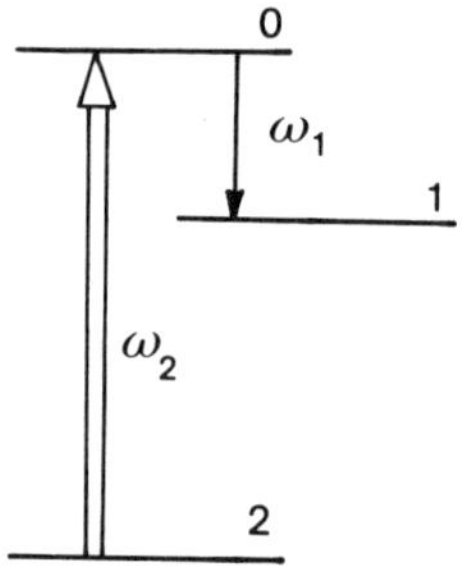

Fig. 1

$$\Omega\left(\pm\right) = \omega_1 \left(1 \pm \frac{\Delta_2}{\omega_2}\right). \qquad (1)$$

Also an asymmetry in the forward (+) backward (-) gain was predicted. For the short wavelength CH_3OH laser lines the effect was actually observed (2,3) by scanning the FIR frequency in the presence of fixed pump detunings. A more general effect was expected when the pump detuning Δ_2 was scanned at fixed FIR frequency. When $\Delta_2=0$, a Lamb-dip is generated in the Doppler broadened IR molecular transition. Since the FIR intensity depends on the number of excited molecules (4), a dip was expected also in the laser emission, as explicitly predicted in ref. (5) using a rate-equation model. Optoacoustic measurements were carried out on different gases in order to directly investigate the IR absorption. The Lamb-dip in the pump transition was demonstrated at pressures typical for FIR operation (6). Also the

emission Lamb-dip was observed on the CH$_3$OH 70.5, 119 μm and CH$_3$F 496 μm
laser lines. The emission Lamb-dip was demonstrated to be the consequence
of the Lamb-dip in the IR pump transition itself and could be explained as
a IR-FIR transferred Lamb-dip (TLD). It is worth noting that this effect
was observed both for Doppler and Pressure Broadened FIR transitions. We
report here further results on the CH$_3$OH laser lines at 251 (Pressure broadened) and 37.5 μm (Doppler broadened). The main experimental apparatus was
similar to that described elsewhere (6,7,8). In fig. 2 is shown a typical recording of the 251 μm line as a function of the pump detuning.

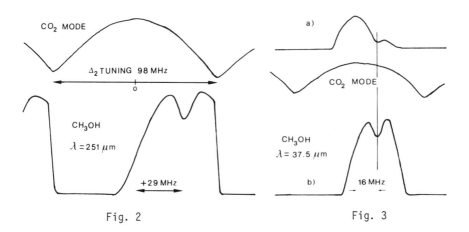

Fig. 2   Fig. 3

The 9P(38) CO$_2$ laser line mode is also monitored. The TLD allows the determination of the center of the absorbing transition which is located at
+ 29 MHz from the CO$_2$ laser line maximum. The gas pressure was 75 mTorr and
the measured value for the dip FWHM ($\sim$ 8 MHz) was compatible with the pressure and saturation broadening parameters in methyl alcohol. Due to the
detection through the FIR resonator, the TLD contrast strongly depended on
the threshold conditions. Contrast so high as nearly 100% could be monitored at low CH$_3$OH pressures ($\sim$ 30 mTorr). The TLD practically disappeared
for pressure values higher than 300 mTorr. Typical results for the 37.5 μm
line are shown in fig. 3. The FIR emission was recorded together with the
9-P(32) line pump mode. The TLD was detected at +16 MHz from the CO$_2$ laser
line maximum. Neither the TLD position nor the width were affected by the
FIR cavity length: in b) the FIR cavity was tuned at the center frequency
while in a) a small cavity detuning was introduced. These results are in
agreement with the interpretation as a TLD. For a comparison the theoretical
gain profiles in the case of the 37.5 μm line are plotted in fig. 4.
For the dashed lines the equations of ref. 2 were used. No dip is forecast
for exact tuning of the cavity. A dip seems to be observable for significa-
tive cavity detuning but also in that case the widths are much larger than
those experimentally observed. When the standing wave saturation is intro-

duced in the equations, the dip in the emission is observable also for exact FIR cavity tuning (continuous lines). The position and the widths of the TLD are not affected by the cavity tuning, as experimentally observed. In conclusion the TLD has been observed, explained and it has been shown to be a powerful tool for spectroscopic investigations (6). For instance the Stark effect on the vibrational molecular transitions can be investigated and information complementary to those inferred by the intracavity IR laser Stark spectroscopy can be obtained (6,8). Also transitions too weak to be investigated by the IR laser Stark spectroscopy can be carefully studied through the coupled FIR emission and consequent TLD (9).

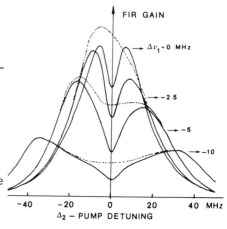

Fig. 4

(1) V.S. Letokhov and V.P. Chebotayev:"Nonlinear Laser Spectroscopy", Optical Science (Springer-Verlag, Berlin, Heidelberg, New York 1977), chapter 5.

(2) D. Seligson, M. Ducloy, J.R. Leite, A. Sanchez, M.S. Feld: IEEE J.Quantum Electron. QE-13, 468 (77)

(3) J. Heppner, C.O. Weiss and P. Plainchamp: Optics Commun. 23, 381 (77)

(4) B.J. Feldman and M.S. Feld: Phys. Rev. 12A, 1013 (75)

(5) G.A. Koepf and K. Smith: IEEE J. Quantum Electron. 13, 425 (78)

(6) M. Inguscio, A. Moretti, F. Strumia: Optics Commun., received March 7th, 1879

(7) M. Inguscio, P. Minguzzi, A. Moretti, F. Strumia and M. Tonelli: Appl. Phys. 18, 261 (1979)

(8) M. Inguscio: Ph.D.Thesis, Scuola Normale Superiore, Pisa, 1979

(9) M. Inguscio, A. Moretti, F. Strumia: to be published.

Laser Spectroscopy in Intense Magnetic Fields

M. Rosenbluh, R. Panock[1], and B. Lax[1]

Francis Bitter National Magnet Laboratory[2], MIT Cambridge, MA 02139, USA
and
T.A. Miller[3]

Bell Laboratories, Murray Hill, NJ 07974, USA

Although the behaviour of atoms in intense magnetic fields has been the subject of investigators since the pioneering work of Jenkins et al. (Jen 39, Sch 39), significant new advances have been made only in recent years. Using conventional spectroscopic techniques (Fon 78) and laser-atomic beam techniques (Eco 78) the Landau-level regime has been examined in a number of atoms. Laser-atomic beam techniques have also been used to probe the "n-mixing" region (Zim 78). Our work has been concerned with understanding the behaviour of the thermal distribution of atoms (as opposed to an atomic beam) in an intense magnetic field in the "diamagnetic-perturbation" and "L-mixing" region (Ros 77, Ros 78, Pan 79). Our experimentation and theoretical understanding of our findings is advanced enough to allow us to obtain very precise spectroscopic information from our data. Furthermore, the implications of our results to certain astrophysical spectra and plasma diagnostics seem significant.

Our experiments use step-tunable fixed frequency infra-red and far infra-red lasers, to drive transitions between various excited states of He. A high field (140 kG) homogeneous Bitter magnet is used to tune the levels into resonance with the laser frequency. Our data has shown the presence of a number of effects related to the high magnetic fields used and to the atomic motion in the presence of this field.

The first observation we can make concerns the Zeeman tuning of the levels, which for the high fields and moderately high quantum numbers of our experiments, is extremely nonlinear. Since the resolution of our experimental technique is high ($\sim 10^{-4}$ cm^{-1}) our data provides the most stringent test for Zeeman tuning theory to date, in the "L-mixing" and "diamagnetic perturbation" regimes.

The second observation relates to a new lineshape for these resonances, which is drastically different from the normally observed Gaussian or Voigt profiles. The change in the lineshape is caused by a second order Stark effect--the "Motional Stark Effect" (MSE). The effect is velocity dependent, and thus alters the lineshape, since the electric field causing it is just the electric field that the atom senses in its rest frame due to its motion perpendicular to the magnetic field. A complete and detailed understanding of this lineshape, which possesses a Doppler-free edge, has enabled us to measure

---

[1] Also Physics Dept., MIT
[2] Supported by the National Science Foundation
[3] Visiting scientist at FBNML

energy separations between various Rydberg states of He with a precision of ~ 2 parts in 10^7.

Finally the MSE couples states of differing parity and magnetic quantum number ($\Delta L = \pm 1$, $\Delta M_L = \pm 1$). In addition, at high magnetic fields the diamagnetic perturbation mixes states of the same parity and same M_L ($\Delta L = 0, +2, \Delta M_L = 0$). Utilizing this breakdown of the usual dipole selection rules, we have observed these normally forbidden transitions with ~ 1 watt/cm^2 CO_2 laser intensities.

In order to obtain spectroscopic information from these transitions, a new lineshape which now includes a velocity dependent transition moment and a velocity dependent saturation intensity had to be developed. Excellent quantitative agreement between this theory and experiment was obtained, and these measurements have yielded the zero field energies of all the L states (including some previous unobserved states) in n=9 manifold of He.

Eco 78: N. Economou, R. Freeman, P. Liao, Phys. Rev. A 18, 2506 (1978)
Fon 78: R. Fonck, D. Tracy, D. Wright, F. Tomkins, Phys. Rev. Lett. 40, 1366 (1978)
Jen 39: F. Jenkins, E. Segré, Phys. Rev. 55, 52 (1939)
Pan 79: R. Panock, M. Rosenbluh, B. Lax, T. Miller, Phys. Rev. Lett. 42, 172 (1979)
Ros 77: M. Rosenbluh, T. Miller, D. Larsen, B. Lax, Phys. Rev. Lett. 42, 175 (1977)
Ros 78: M. Rosenbluh, R. Panock, B. Lax, T. Miller, Phys. Rev. A 18, 1103 (1978)
Sch 39: L. Schiff, H. Snyder, Phys. Rev. 55, 59 (1939)
Zim 78: M. Zimmerman, J. Castro, D. Kleppner, Phys. Rev. Lett. 40, 1083 (1978)

Real-Time Picosecond Measurements of Electronic Energy Transfer from DODCI to Malachite Green and DQOCI

M.C. Adams, D.J. Bradley, W. Sibbett, and J.R. Taylor

Optics Section, Blackett Laboratory, Imperial College
London SW7 2BZ, England

Introduction

The addition of malachite green to the saturable absorber DODCI has been reported (1) to increase the stability of operation of a mode-locked CW Rhodamine 6G dye laser. It has been suggested (2) that this result could be due to the resonance transfer of excited electronic state energy(3) between DODCI and malachite green. DQOCI, which gives subpicosecond pulses when used by itself to mode-lock the CW Rhodamine 6G dye laser (4), also produces an improvement in performance when added to the DODCI solution. The extension (5) of the 'Synchroscan' streak-camera (6) for use with pulses from CW mode-locked lasers, greatly facilitates the carrying out of electronic excitation energy transfer measurements. Using low peak power ($\sim300W$) picosecond laser pulses for excitation, nonlinear effects in the experimental solutions are avoided while precisely superimposed successive fluorescence streak-records on the image-tube phosphor permit integration of $>10^8$ pulses (in $\sim0.5s$) with a vidicon optical multichannel analyser (OMA) (7). Thus direct linear detection in real time of complete fluorescence decay curves, obtainable with single-shot streak cameras (4) and, effectively, photon-counting with digitization and storage can be combined in one convenient system. With the low values of instantaneous photo-electron currents involved no further image intensification is needed to avoid photocathode saturation or image-tube space charge effects (8). We report detailed measurements, of the effects on the lifetime of DODCI dissolved in the low viscosity solvent ethanol, when DQOCI or malachite green is added to the solution.

Experimental Arrangement

The arrangement used is shown in Figure 1. The synchronously pumped (9) CW rhodamine 6G laser was tunable from 565 to 630 nm, and produced $\sim2ps$ pulses ($\sim0.6nJ$) at a repetition rate of 140MHz. The train of excitation pulses was focused into the sample cell to give peak power densities of $\sim5MW\ cm^{-2}$. A polarizer, P, compensated for any fluorescence depolarization (10). Successive identical streak records were superimposed on the image tube phosphor at the 140 MHz repetition rate while maintaining an integrated time-resolution of down to 3ps. The variation from linearity of the streak was measured to be better than 3%. A dynamic range of 4×10^3 was demonstrated for the OMA system used to record and store the streaked images, before display on a VDU or printing on a chart recorder. At its normal streaking speed in repetitive mode the S20 Photochron II image tube has a resolution of <3ps for light at 600nm. To record the complete fluorescence decay curves of the samples the tube was operated at a writing speed of $5 \times 10^8 cm\ sec^{-1}$ giving a resolution limit of $\sim20ps$.

639

Results

Various concentrations of the additive were used (see Table 1) for a constant DODCl concentration of 10^{-4}M. The excitation laser wave-length was tuned to 575nm, and the centre of the observed fluorescence bandwidth (1.5nm) was set to 615 nm. The average fluorescence lifetime of a 10^{-4}M ethanolic solution of DODCl at room temperature, taken from several traces similar to Figure 2(a), gave a value of 1150 $\pm$ 40 ps, in good agreement with values previously reported (11). Increasing the malachite green concentration led to significant decreases in the measured lifetime. The departure from exponential decay became more dominant as the malachite green concentration approached and became greater than that of the DODCl. A similar behaviour was also observed when DQOCl was added. For concentrations of 10^{-4}M DQOCl and above, a contribution due to DQOCl fluorescence became evident with a lifetime of 15ps. DODCl lifetimes were determined after allowing for the effect of the short lifetime DQOCl fluorescence on the recorded traces.

TABLE I

Acceptor Concentration	Malachite Green		DQOCI	
	τ (psec)	R_o (Å)	τ (psec)	R_o (Å)
-	1148			
1 x 10^{-5}M	1060	152	1036	150
5 x 10^{-5}M	982	120	965	96
1 x 10^{-4}M	961	91	814	68
2.5 x 10^{-4}M	921	84	751	65
5 x 10^{-4}M	862	69	697	60
7.5 x 10^{-4}M	811	47	592	60
1 x 10^{-3}M	751	48	555	56
2.5 x 10^{-3}M	510	46	-	-
5 x 10^{-3}M	434	45	-	-
7.5 x 10^{-3}M	260	42	-	-

Discussion

For resonance energy transfer under conditions for which Forster kinetics apply, the rate of energy transfer from excited donor to acceptor molecule is given by

$$K_{D*A} = (1/\tau) \ (R_o/R)^6 \qquad (1)$$

τ is the fluorescence lifetime of the donor in the absence of excitation energy transfer, R is the intermolecular distance and R_o is the critical transfer distance (3). The time evolution of the fluorescence intensity, $I_F(t)$, is modified to a non-exponential decay of the form

$$I_D(t) = I_D (o) \ \exp \ (-t/\tau) \ \exp \ (-2\gamma \ (t/\tau)^{\frac{1}{2}}) \qquad (2)$$

A graph of $\log_{10} (I_D (t) / I_F(t))$ against $t^{\frac{1}{2}}$ should give a straight line dependence of gradient $(-2 \gamma/ 2.303\tau^{\frac{1}{2}})$. This linear dependence can be seen from figure 3 which relates to 5 x 10^{-4}M malachite green in 10^{-4}M DODCl. Similar graphs were plotted for the other concentrations. Assuming that Forster kinetics were valid over the complete range of acceptor concentrations used, the critical transfer distance R_o was calculated from the linear plots above. Table 1 shows that the change in R_o with concentration is much smaller at the higher concentrations and tends to level out. Computer plots of the

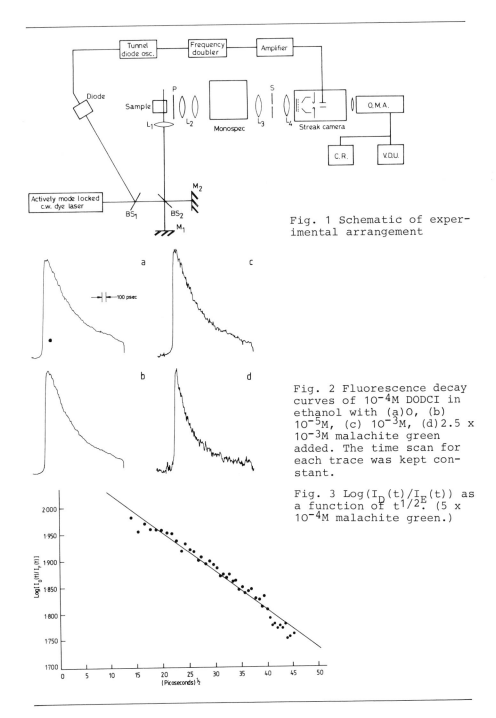

Fig. 1 Schematic of experimental arrangement

Fig. 2 Fluorescence decay curves of 10^{-4}M DODCI in ethanol with (a)0, (b) 10^{-5}M, (c) 10^{-3}M, (d)2.5 x 10^{-3}M malachite green added. The time scan for each trace was kept constant.

Fig. 3 Log($I_D(t)/I_E(t)$) as a function of $t^{1/2}$. (5 x 10^{-4}M malachite green.)

function (2) for $I_D(t)$ were drawn and fitted against the experimental data. For a $2.5 \times 10^{-3}M$ solution of malachite green the computed form of $I_D(t)$ fits the experimentally recorded fluorescence trace very well. However, at low concentrations of the acceptor, deviations from the theoretical Forster expression become apparent, showing that the correction term introduced by the Förster theory is not valid in this region. A similar effect was also found for DQOCI as the acceptor. Taking the values of R_o obtainable for the five highest concentrations used for both acceptors gives average R_o values of 45.6 A^o for malachite green and 61.8 A^o for DQOCI respectively.

The ability of the synchroscan picosecond streak-camera to resolve small changes in fluorescence lifetimes of dye under low power excitation greatly facilitates accurate direct measurements of resonance energy transfer from donors with nanosecond and subnanosecond fluorescence lifetimes, since the excited state decay kinetics can be studied with a time resolution of $\sim$3ps. Further studies with the synchroscan system, of excitation transfer on a picosecond time scale are planned, in particular between DODCI and its photoisomer (4).

References

1. E.P.Ippen and C.V.Shank: Appl. Phys. Letters 27 (1975) 488
2. G.Porter and C.J.Tredwell: Chem. Phys. Letters 56 (1978) 278
3. Th. Förster, Discussions Faraday Soc. 27 (1959) 7
4. D.J.Bradley: Topics in Appl. Phys. Vol. 18 "Ultra-short Light Pulses" Ed: S.L.Shapiro (Springer Verlag, Berlin 1977) 57
5. D.J.Bradley: J.Phys. Chem. 82 (1978) 2259 and references therein
6. D.J.Bradley: UK Patent Application 34544 (1972)
7. M.C.Adams, W.Sibbett and D.J.Bradley:Opt. Commun.26 (1978) 273
8. D.J.Bradley, S.F.Bryant, J.R.Taylor and W.Sibbett: Rev. Sci.Instrum. 49 (1978) 215
9. J.P.Ryan, L.S.Goldberg and D.J.Bradley: Optics Commun. 27 (1978) 127
10. H.E.Lessing and A.vonJena: Chem. Phys. Letters 42 (1976) 213
11. J.C.Mialocq. A.W.Boyd, J.Jaraudias and J.Sutton: Chem. Phys. Letters 37 (1976) 236

Quantum Beats in Triplet States of Phosphorescent Molecules Following Laser Flash Excitation

J. Schmidt

Huygens Laboratory, University of Leiden, Leiden, The Netherlands

We have recently performed an experiment which proves that the radiation-less intersystem crossing from the first excited singlet state S_1 to the lowest triplet state T_0 of a polyatomic molecule may carry the ensemble over into a coherent superposition of two zero-field spin components of T_0 [1]. In the system tetramethyl pyrazine (TMP) dissolved in a single crystal of durene we have observed the free induction signal of the $T_z - T_y$ zero-field transition at 859.5 MHz following flash excitation at 266 nm with a Nd glass laser system in the singlet manifold. Since the coherence between the systems in the ensemble is preserved, relaxation to the vibrationless triplet state of TMP after the laser flash must occur in a time short relative to the inverse of the beat frequency, i.e. within 1 ns. From the quantum mechanical point of view our experiment shows a close similarity to an experiment by Dodd et al., in which they observed quantum beats in the resonance fluorescence emitted by mercury following pulsed excitation by linearly polarized light to the $m = \pm 1$ Zeeman components of the $6\,^3P_1$-state [2].

[1] C.J. Nonhof, F.L. Plantenga, J. Schmidt, C.A.G.O. Varma and J.H. van der Waals, Chem. Phys. Lett. 60 (1979) 353.
[2] J.N. Dodd, R.D. Kaul and D.M. Warrington, Proc. Roy. Soc. (London) 84 (1964) 176.

Doppler-Free Optoacoustic Spectroscopy

E.E. Marinero and M. Stuke

Max-Planck-Institut für biophysikalische Chemie, Abteilung Laserphysik
D-3400 Göttingen, Fed. Rep. of Germany

The first observation of Doppler-free optoacoustic spectroscopy is reported. As a first example the P(93) line of the 11-0 band of the B$\leftarrow$X transition of $^{127}I_2$ is used. The output of a cw single mode dye laser is split into two equal intensity beams chopped at frequencies ω_1 and ω_2 (Fig. 1).

The nonlinear component of the optoacoustic signal at the frequency ($\omega_1 + \omega_2$) is detected and Doppler-free resolution is obtained (Fig. 2). Comparing the Doppler-free optoacoustic and fluorescence spectra of iodine measured under similar conditions, good agreement is found. Since optoacoustic and fluorescence methods complement each other, this opens up new possibilities for weakly or nonfluorescing molecules.

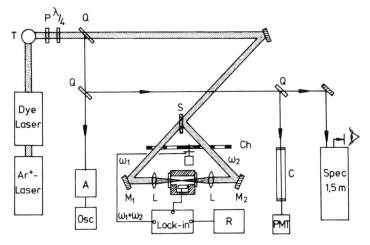

Fig. 1

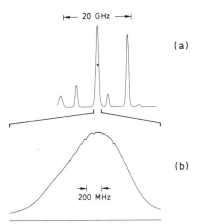

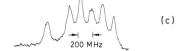

Fig. 2

Pulsed Opto-Acoustic Spectroscopy of Liquids

C.K.N. Patel and A.C. Tam

Bell Laboratories, Murray Hill, NJ 07974, USA

It is well known that opto-acoustic (OA) spectroscopy is the most sensitive technique to detect small absorptions in the gas phase. In the condensed phases, sensitivity has been severely limited by the use of gas phase microphones[1] for acoustic detection. In the past few years, various authors[2-7] have realized that piezoelectric transducers, in direct contact with a liquid, give much higher OA detection sensitivities than conventional microphones, because of the large acoustic impedance of liquids. We have recently demonstrated,[8,9] for the first time, that the combination of a pulsed dye laser, submersed piezoelectric transducer, and gated OA detection can be advantageously used to measure weak absorption spectra (linear or nonlinear absorptions) in liquids. Our pulsed OA spectroscopy method can presently provide a detection limit of 10^{-9} J of absorbed optical energy per cm path length, corresponding to an absorption coefficient of 10^{-6} cm^{-1} for our typical laser pulse energy of 1 mJ. Improvement by at least a factor ~ 10 in the detection limit seems possible. We have applied this technique to a variety of novel liquid state spectroscopy measurements including (1) determination of the absorption spectra of high harmonics (up to 8th) of the C-H stretch in organic liquids,[10] (2) a first accurate determination of the visible spectra of water and heavy water,[11] (3) OA Raman gain spectroscopy,[12] and (4) two-photon OA absorption spectroscopy.[13]

Our method is not only very sensitive, but the experimental arrangement is also very simple and inexpensive. There are no delicate alignments involved (which are needed in various other spectroscopic techniques like intracavity measurements,[14] thermal blooming technique using two laser beams,[15] or direct gain measurements in stimulated Raman scattering[16]). Also no elaborated instrumentation is necessary. An experimental OA cell can be constructed as shown in Fig. 1. The pulsed dye laser beam that is passed through the quartz windows of the cell should be normally unfocused for linear absorption measurements, and be focused for nonlinear absorption measurements. An experimental arrangement suitable for linear absorption measurements is shown in Fig. 2.

A simplified theoretical analysis[17] of the pulsed OA effect for our experimental configuration shows that our normalized OA signal $S(\nu)$ (i.e., the ratio of the OA signal amplitude and the laser pulse energy at a laser frequency ν) for the case of a weak linear absorption is given by

$$S(\nu) = K\alpha(\nu)$$

where

$$K = K'c\beta/C_p .$$

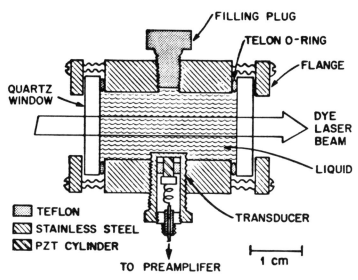

Fig.1: An OA spectroscopy cell for liquids

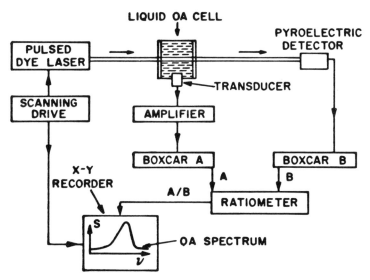

Fig.2: An arrangement for linear absorption spectroscopy

Here $\alpha(\nu)$ is the absorption coefficient at frequency ν, K' is a constant dependent on geometry, c is the ultrasonic velocity, β is the coefficient of thermal expansion and C_p is the specific heat of the liquid. We have demonstrated[10,11] that absolute calibration of the OA spectrum (i.e., the evaluation of the constant K for a fixed geometry and for a fixed liquid) can be readily done by a comparison technique involving the use of a dilute nonfluorescing dye solution.

The pulsed OA spectroscopy method appears quite promising for many new applications in the linear or nonlinear spectroscopy of liquids, solids or their surfaces. Besides the applications mentioned above, further possible applications include studies of collision-induced absorptions, quenching collisions, lifetimes and fluorescence quantum yields of electronically excited states, and absorption spectra of thin films and monolayers. These novel applications to the spectroscopy of condensed matter at various temperatures (including cryogenic temperatures) are currently being investigated.

REFERENCES

1. See for example W. R. Harshbarger and M. B. Robin, *Acc. Chem. Res.* **6**, 64 (1973); A. Rosencwaig, *Optics Commun.* **7**, 305 (1973).

2. Y. Kohanzadeh, J. R. Whinnery, and M. M. Carroll, *J. Acoust. Soc. Am.* **57**, 67 (1975).

3. A. M. Bonch-Bruevich, T. K. Razumova, and J. O. Starobogatov, *Opt. Spectrosc.* **42**, 45 (1977).

4. P. Sladky, R. Danielius, V. Strutkaitis, and M. Boudys, *Czech. J. Phys. B,* **27**, 1075 (1977).

5. W. Lahmann and H. J. Ludewig, *Chem. Phys. Lett.* **45**, 177 (1977).

6. S. Oda, T. Sawada and H. Kamada, *Anal. Chem.* **50**, 865 (1978).

7. M. M. Farrow, R. K. Burnham, M. Auzanneau, S..L. Olsen, N. Purdie, and E. M. Eyring, *Appl. Optics,* **17**, 1093 (1978).

8. C. K. N. Patel and A. C. Tam, *Appl. Phys. Lett.* **34**, 467 (1979).

9. A. C. Tam, C. K. N. Patel and R. J. Kerl, *Optics Lett.* **4**, 81 (1979).

10. C. K. N. Patel, A. C. Tam and R. J. Kerl, *J. Chem. Phys.* (in press).

11. C. K. N. Patel and A. C. Tam, *Nature* (submitted).

12. C. K. N. Patel and A. C. Tam, *Appl. Phys. Lett.* **34**, 760 (1979).

13. A. C. Tam and C. K. N. Patel, *Nature* (submitted).

14. W. Wernick, J. Klein, A. Lau, K. Lenz, and G. Haunsalz, *Optics Commun.* **11**, 159 (1974).

15. R. L. Swofford, M. E. Long and A. C. Albrecht, *J. Chem. Phys.* **65**, 179 (1976).

16. A. Owyoung, *IEEE J. Quant. Electronics,* **QE-14**, 192 (1978).

17. A. C. Tam and C. K. N. Patel, *Appl. Optics* (submitted).

Design and Performance of a Broadband "Optical Diode" to Enforce One-Direction Traveling Wave Operation of a Ring Laser

T.F. Johnston, Jr. and W.P. Proffitt

Coherent, Inc., 3210 Porter Drive, Palo Alto, CA 94304, USA

A traveling-wave, CW ring dye laser system has been developed which produces a peak single-frequency output of 1.75W, at 6.0W of 514 nm optical pump power, with Rhodamine 6G dye dissolved in ethylene glycol solvent. The system servo-locks the frequency of this output to an external interferometer to produce 0.2 MHz rms frequency jitter with a 30GHz locked scan range and less than 100 MHz per hour total-frequency drift. Also used was thickened water[1,2] as the R6G dye solvent, which improves the heat-loading capabilities of the optical jet stream but makes a microscopically less uniform dye mixture. With this, the system produced 3.5W of single-frequency output at 2MHz rms frequency jitter with 24W of all-lines input power from a Coherent CR-12 argon laser.

Ring dye lasers are capable of watt-level single-frequency outputs[3] because they may be pumped in this manner with the full power of available ion lasers. In contrast, an input power limit exists in a standing-wave dye laser due to the regions of unsaturated gain in the pumped volume of the dye jet at the nodes of the standing wave. Marowsky and Kaufman[4] have shown that the fraction of the total volume the unused portion represents, decreases as the dye beam intensity increases. The drop in volume, however, is less rapid than the linear rise in pump power. Thus a mode at a second frequency, which has antinodes where the first mode has nodes[5], must eventually reach threshold and oscillate as the pump level is increased in the standing wave case. This limit does not exist in a ring laser, and typically a ring will be pumped four times harder than a standing wave laser, with ten times the single-frequency output due to the higher conversion efficiency in a traveling-wave system.

The crucial, new optical element in a ring laser is the "optical diode", the device which forces the laser to operate stably in the preferred, or forward-traveling-wave direction. This device consists of a Faraday cell (with samarium-cobalt permanent magnets) and a section of quartz-crystal cut for optical activity [6]. The rotation of polarization of the dye beam through a small angle in the transit of the Faraday cell is undone for the forward wave in traversing the crystal. The backward wave undergoes a polarization rotation of approximately double the Faraday angle in transit of this pair of elements, and suffers a subsequent reflection loss at the Brewster surfaces in the cavity. Both the Faraday rotation angle for a fixed magnetic field, and the rotation angle from the optically-active plate, vary roughly as the inverse-square of the wavelength. The problem in designing a wide-band optical diode is one of keeping a close enough match of the Faraday and back-rotation angles to have an acceptably low insertion loss for the forward wave on the blue end of the spectrum, and at the same time to have enough Faraday rotation remaining on the red end to give the minimum required differential loss between forward and backward waves. These dispersion properties are a prime consideration in the choice of Faraday cell material. Two different types of optical diode configurations were built and studied, and their characteristics will be discussed in showing how the final design was reached. This is based on SF-2 glass as the Faraday rotator material, and has given with a single device stable, one-direction single-frequency operation from 425 nm to 780 nm, with satisfactory operation in the 3.5 W laser.

References

1. S. Leutwyler, E. Schumacher, and L. Wöste, Optics Comm., vol. 19, pp. 197-200, Nov. 1976.
2. P. Anliker, H.R. Luthi, W. Sedig, J. Steinger, H.P. Weber, S. Leutwyler, E. Schumacher, and L. Wöste, IEEE J. Quantum Elec., vol. QE-13, pp. 547-548, July 1977.
3. H. W. Schröder, L. Stein, D. Frölich, B. Fugger, and H. Welling, Appl. Phys., vol. 14, pp. 377-380, 1977.
4. G. Marowsky and K. Kaufman, IEEE J. Quantum Elec., vol. QE-12, pp. 207-209, 1976.
5. C.T. Pike, Optics Comm., vol. 10, pp. 14-17, Jan. 1974.
6. S.M. Jarrett and J.F. Young, Optics Letters 4, (June' 79) 176-178.

Index of Contributors

Tunable Lasers and Applications

Proceedings of the Loen Conference, Norway, 1976

Editors: A.Mooradian, T.Jaeger, P.Stokseth

1976. 238 figures. VIII, 404 pages
(Springer Series in Optical Sciences,
Volume 3)
ISBN 3-540-07968-8

Contents:
Tunable and High Energy UV-Visible
Lasers. – Tunable IR Laser Systems. – Isotope
Separation and Laser Driven Chemical Reactions. – Nonlinear Excitation of Molecules. –
Laser Photokinetics. – Atmospheric Photochemistry and Diagnostics. – Photobiology. –
Spectroscopic Applications of Tunable
Lasers.

Advances in Laser Chemistry

Proceedings of the Conference on Advances
in Laser Chemistry, California Institute of
Technology, Pasadena, USA, March 20–22,
1978

Editor: A.H.Zewail

1978. 242 figures, 2 tables. X, 463 pages
(Springer Series in Chemical Physics,
Volume 3)
ISBN 3-540-08997-7

Contents:
Laser-Induced Chemistry. – Picosecond Processes and Techniques. – Non-Linear Optical
Spectroscopy and Dephasing Processes. –
Multiphoton Excitation in Molecules. –
Molecular Dynamics by Molecular Beams.

Laser Spectroscopy

Proceedings of the 2nd International
Conference, Mégève, France, June 23–27,
1975

Editors: S. Haroche, J. C. Pebay-Peyroula,
T. W. Hänsch, S. E. Harris

1975. 230 figures, 30 tables. X. 468 pages
(5 pages in French)
(Lecture Notes in Physics, Volume 43)
ISBN 3-540-07411-2

Contents:
Spectroscopy 1. – Tunable Lasers 1. – Spectroscopy 2. – Spectroscopy 3. – Tunable
Lasers 2. – Laser Isotope Separation. – Spectroscopy 4. – Spectroscopy 5. – Titles and
Abstracts of Post-Deadline Papers.

Laser Spectroscopy III

Proceedings of the Third International
Conference, Jackson Lake Lodge, Wyoming,
USA, July 4–8, 1977

Editors: J. L. Hall, H. L. Carlsten

1977. 296 figures. XII, 468 pages
(Springer Series in Optical Sciences,
Volume 7)
ISBN 3-540-08543-2

Contents:
Fundamental Physical Applications of Laser
Spectroscopy. – Multiple Photon Dissociation. – New Sub-Doppler Interaction
Techniques. – Highly Excited States, Ionization, and High Intensity Interactions. – Optical Transients. – High Resolution and Double
Resonance. – Laser Spectroscopic Applications. – Laser Sources. – Laser Wavelength
Measurements. – Postdeadline Papers.

Springer-Verlag Berlin Heidelberg New York

Laser Spectroscopy

of Atoms and Molecules

Editor: H. Walther

1976. 137 figures, 22 tables. XVI, 383 pages
(Topics in Applied Physics, Volume 2)
ISBN 3-540-07324-8

Contents:
H. Walther: Atomic and Molecular Spectroscopy with Lasers. – *E. D. Hinkley, K. W. Nill, F. A. Blum:* Infrared Spectroscopy with Tunable Lasers. – *K. Shimoda:* Double-Resonance Spectroscopy of Molecules by Means of Lasers. – *J.M. Cherlow, S.P.S. Porto:* Laser Raman Spectroscopy of Gases. – *B. Decomps, M. Dumont, M. Ducloy:* Linear and Nonlinear Phenomena in Laser Optical Pumping. – *K. M. Evenson, F. R. Petersen:* Laser Frequency Measurements, the Speed of Light, and the Meter.

High-Resolution Laser Spectroscopy

Editor: K. Shimoda

1976. 132 figures, XIII, 378 pages
(Topics in Applied Physics, Volume 13)
ISBN 3-540-07719-7

Contents:
K. Shimoda: Introduction. – *K. Shimoda:* Line Broadening and Narrowing Effects. – *P. Jacquinot:* Atomic Beam Spectroscopy. – *V. S. Letokhov:* Saturation Spectroscopy. – *J. L. Hall, J. A. Magyar:* High Resolution Saturated Absorption Studies of Methane and Some Methyl-Halides. – *V. P. Chebotayev:* Three-Level Laser Spectroscopy. – *S. Haroche:* Quantum Beats and Time-Resolved Fluorescence Spectroscopy. – *N. Bloembergen, M. D. Levenson:* Doppler-Free Two-Photon Absorption Spectroscopy.

Excimer Lasers

Editor: C. K. Rhodes

1979. 59 figures, 29 tables. XI, 194 pages
(Topics in Applied Physics, Volume 30)
ISBN 3-540-09017-7

Contents:
P. W. Hoff, C. K. Rhodes: Introduction. – *M. Krauss, F. H. Mies:* Electronic Structure and Radiative Transitions of Excimer Systems. – *M. V. McCusker:* The Rare Gas Excimers. – *C. A. Brau:* Rare Gas Halogen Excimers. – *A. Gallagher:* Metal Vapor Excimers. – *C. K. Rhodes, P. W. Hoff:* Applications of Excimer Systems.

Laser-Induced Processes in Molecules

Physics and Chemistry

Proceedings of the European Physical Society, Divisional Conference at Heriot-Watt University, Edinburgh, Scotland, September 20–22, 1978

Editors: K. L. Kompa, S. D. Smith

1979. 196 figures, 31 tables. XIV, 367 pages
(Springer Series in Chemical Physics, Volume 6)
ISBN 3-540-09299-4

Contents:
Study of Lasers and Related Techniques Suitable for Applications in Chemistry and Spectroscopy. – Spectroscopic Studies With and Related to Lasers. – Multiphoton Excitation, Dissociation and Ionization. – Laser Control of Chemical Reactions. – Molecular Relaxation.

Springer-Verlag
Berlin
Heidelberg
New York